Ten Thousand Birds

TEN THOUSAND BIRDS

BIRDS

Ornithology since Darwin

TIM BIRKHEAD   JO WIMPENNY   BOB MONTGOMERIE

PRINCETON UNIVERSITY PRESS   PRINCETON AND OXFORD

Published by Princeton University Press,
41 William Street, Princeton, New Jersey 08540
In the United Kingdom: Princeton University Press,
6 Oxford Street, Woodstock, Oxfordshire OX20 1TW

press.princeton.edu

Jacket art: *Magnificent Bird of Paradise*, linocut print,
2013 © Robert Gillmor

ISBN 978-0-691-15197-7

Library of Congress Control Number 2013939390

British Library Cataloging-in-Publication Data is
available

This book has been composed in PF Monumenta and
Verdigris MVB Pro Text

Printed on acid-free paper.

Printed in the United States of America

10 9 8 7 6 5 4 3 2 1

CONTENTS

PREFACE

The body of a bird is not just a prodigiously complicated machine, with its trillions
of cells—each one in itself a marvel of miniaturized complexity—all conspiring
together to make muscle or bone, kidney or brain. Its interlocking parts also conspire
to make it good for something—in the case of most birds, good for flying. An aero-
engineer is struck dumb with admiration for the bird as flying machine: its feathered
flight-surfaces and ailerons sensitively adjusted in real time by the on-board computer
which is the brain; the breast muscles, which are the engines, the ligaments, tendons
and lightweight bony struts all exactly suited to the task. And the whole machine is
immensely improbable in the sense that, if you randomly shook up the parts over and
over again, never in a million years would they fall into the right shape to fly like a
swallow, soar like a vulture, or ride the oceanic up-draughts like a wandering albatross.

—RICHARD DAWKINS, IN *THE WASHINGTON POST* ON 23
AUGUST 2011, IN RESPONSE TO TEXAS GOVERNOR PERRY'S
CLAIM THAT "EVOLUTION IS JUST A THEORY"

THERE ARE CURRENTLY VERY CLOSE TO TEN thousand species of birds in the world, both beautiful and improbable, and they have contributed more to the study of zoology than almost any other group of animals (Konishi et al. 1989). The reasons are obvious: birds are diurnal, they are often easily observed and studied, and we like them. As a result, the study of birds goes back at least as far as ancient Greece, although it is generally recognized that scientific ornithology began in the mid-1600s with the publication of John Ray's *Ornithology of Francis Willughby* (Ray 1676). Since then, the study of birds has continued apace, with by far the greatest increase in ornithological knowledge occurring since the middle of the twentieth century. We estimate that there have been no fewer than 380,000 ornithological publications since Darwin published *The Origin of Species* in 1859.[1] The temporal pattern reflects the change in numbers of ornithologists: increasing slowly between 1860 and 1960, but then more rapidly as more academic positions for zoologists became available in the 1960s. In 2011 there were as many papers on birds published as there had been during the entire period between Darwin's *Origin* and 1955.

Several "histories of ornithology" have been written (appendix 1)—especially in the last few years, suggesting that the subject has come of age. Few of these, however, have included the twentieth century, possibly because of the sheer volume of information. Yet residing within this enormous mass of literature is a small number of wonderful, groundbreaking discoveries, and it is these that form the basis for this book. This isn't to say that most of what has been done is of little value but rather that, as in most areas of

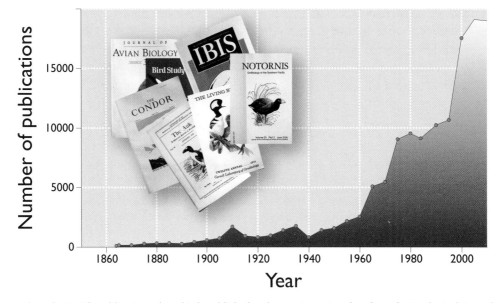

The number of scientific publications about birds published each year since 1850; data from the Zoological Record and Google Scholar. Inset shows some covers of ornithological journals.

science, the few individuals that make major breakthroughs have relied consciously or unconsciously on the substantial foundations provided by generations of ornithological foot soldiers.

Science in its broadest sense has a long history, but modern science began only in the seventeenth century, with the scientific revolution, as logic and experimentation gradually swept away the folklore, alchemy, and old wives' tales that had persisted since the time of Aristotle. As Jürgen Haffer (2007a) points out, the renaissance in science in the mid-1600s—and the work of Francis Willughby and John Ray in particular—provided not only a firm scientific foundation for ornithology but initiated what were to become the two major strands in the study of birds: systematics and field ornithology.

The first of these strands, beginning with the naming and description of all known bird species—which at the time was thought to number about five hundred—formed the basis for Ray's *Ornithology of Francis Willughby* (1676, 1678), so named because Willughby, Ray's protégé and patron, died at just thirty-six years of age, before their book was completed. Ray's second, field-based approach was presented later in his book *The Wisdom of God*, published in 1691, long after Willughby's death. Here Ray introduced the concept of physicotheology (later known as "natural theology"), which used the exquisite fit between an animal's design and its lifestyle as evidence of God's wisdom. In modern terms, *The Wisdom of God* is about adaptation, which for Ray was mediated through God. The book caused a revolution both in religious thinking and in natural history. With extraordinary prescience Ray asked, for example, why some birds produce a clutch of one egg, while others produce clutches of ten or more; why some birds breed early in the year, while others breed later. Not only did Ray pose important biological questions, he anticipated their answers with uncanny insight and common sense (Birkhead 2008).

Ray's ingenious ideas were appropriated by others, most notably William Paley, whose *Natural Theology* (1802) became essential reading for nineteenth-century Cambridge undergraduates intending to enter the church—as was Darwin before he went off on his *Beagle* voyage in December 1831. Paley's rich examples captivated Darwin, who went on to call them adaptations. Paley is best known now—thanks to Richard Dawkins's *Blind Watchmaker* (1986)—for his parable of the watch. Imagine finding a watch, he said: its intricate design tells you that it must have a designer. Now look at nature: the exquisite fit between an organism and its environment tells you that it too must have had its designer, and that designer could only have been God. Paley's writings shaped Darwin's thinking, not about God but about adaptation, and as he later said, "The old argument from design in Nature [natural theology], as given by Paley, which formerly seemed to me so conclusive, fails, now that the law of natural selection has been discovered."[2]

Despite the genius of Ray's double-barreled approach, the next two hundred years of ornithology were dominated by systematics: the naming and describing of species, as well as determining their position in God's grand scheme of things. Only after Darwin seeded the idea that the behavior and ecology of animals might have evolved through natural selection did Ray's second idea begin to take hold. But it was a slow change. Until the 1920s, ornithology, like the rest of zoology, consisted almost exclusively of museum work—the study of skins, skeletons, and eggs—and the museum ornithologist's idea of "fieldwork" was the killing and collecting of specimens for study. In the late nineteenth century, Elliott Coues (1896) identified the shotgun as the ornithologist's most important piece of field equipment. His contemporaries—like Edmund Selous, who opposed museum-based ornithology and attempted to promote the study of the living bird—were castigated. As we'll see, genuine field ornithology was not reunited with museum ornithology until the period from 1920 to 1940—a union that pulled ornithology from the sidelines into mainstream biology (Birkhead 2008). This revolution, which forms an important part of the current book, transformed zoology and fueled the extraordinary explosion in ornithological knowledge.

We take Darwin as our starting point because "nothing in biology makes sense, except in the light of evolution,"[3] and because Darwin made so many perceptive observations and comments on birds that inspired a number of pioneers to test his ideas. In constructing our overview of ornithology since Darwin, how did we decide what to include—and what to omit? It is quite clearly impossible to summarize every relevant person or idea in a book of this (already large) size. Instead, we decided to focus on a selection of the major contributions of ornithology to general science—that is, on areas where the endeavors of ornithologists have influenced the course of scientific progress. In doing so, we had to identify what we considered the most exciting and interesting findings in ornithology and how those subjects and the people that worked on them helped to transform biology. Deciding how best to do this occupied us for several months.

To help us decide on the book's scope we did two things. First, we made a database of 325 ornithologists who were prominent since the 1960s, and for each of these we created a citation report from the Web of Science (citation reports were not available for earlier ornithologists). We ranked these reports according to the total number of times that each ornithologist's work had been cited in scientific publications (excluding self-citations). Second, we conducted a survey of thirty-one senior ornithologists—from a variety of countries and with diverse research interests—asking them to name both the most influential ornithologists and books written by ornithologists since Darwin.[4] Why books, you might ask? Obviously, there are many fewer books than scientific papers, but books provide authors with the intellectual freedom to express their ideas in a way that is usually impossible in scientific, peer-reviewed papers. Furthermore, while today's researchers concentrate on publishing research papers, this was less often the case in the middle of the twentieth century; thus, relying solely on citation reports based on scientific papers biased our survey in favor of "modern" researchers. Books also have the potential to make authors an "authority" because they provide a synthesis of old and new ideas and, deliberately or inadvertently, often point the way forward.

Of the ornithologists considered to be the most influential, David Lack was the clear leader (30 votes), followed by Ernst Mayr (23), Niko Tinbergen (21), Robert MacArthur (11), Peter Grant (11), Nick Davies (11), Erwin Stresemann (11), Charles Sibley (11), Konrad Lorenz (9), and Donald Farner (8).

Of the books considered to be most influential, David Lack's again came out on top, taking the first three places: *Ecological Adaptations for Breeding in Birds* (1968), *The Natural Regulation of Animal Numbers* (1954), and *Population Studies of Birds* (1966). In chronological order, the others in the top-ten list of books were *Systematics and the Origin of Species* (Mayr 1942), *The Study of Instinct* (Tinbergen 1951), *The Herring Gull's World* (Tinbergen 1953b), *Animal Species and Evolution* (Mayr 1963), *The Theory of Island Biogeography* (MacArthur and Wilson 1967), *Ecology and Evolution of Darwin's Finches* (Grant 1986), and *Sperm Competition in Birds* (Birkhead and Møller 1992).

We initially considered the straightforward option of writing a chapter on each of our top ten ornithologists, or of adopting a chronological approach, recounting the major ornithological discoveries by each of those individuals decade by decade. Both of these alternatives seemed a bit tedious, so we decided instead that a topic-based series of chapters was more interesting for both us and our readers, and more meaningful in a broader biological sense. Using the achievements of our top ten ornithologists and books as a guide—but also consulting colleagues and relying upon our own experiences as professional ornithologists—we identified eleven topics that encompass much of ornithology since Darwin.

We had several criteria for deciding what kinds of discoveries to include. Discoveries had either to have broad biological relevance, to change the course of ornithology, to make an important point, or simply to appeal to our interests. Our account comprises what we consider to be the major advances in scientific ornithology over the past 150

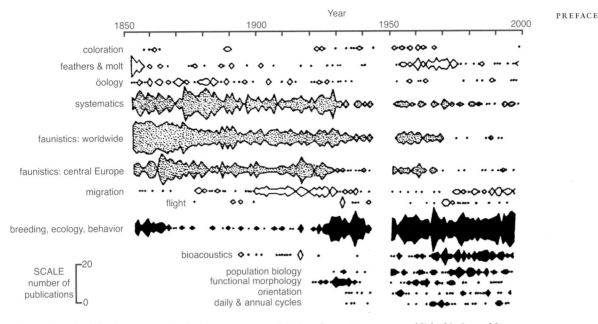

The number of publications per year in different areas of ornithology from 1850 to 2000 published in *Journal für Ornithologie.*

years: a broad introduction that includes an overview of the main discoveries and those who made them, as well as myriad other research programs that extended, refined, and built upon those major advances. We expect that most knowledgeable readers will agree with the major discoveries that we cover, but the others that we describe here are personal choices that we thought were both interesting and informative about the voyage of discovery, the people involved, or the scientific findings themselves. In a way, then, what we have written is a set of essays on key ornithological topics whose development we explore from Darwin to today. Darwin was more than an ornithologist—he was too broad for that—but he had good credentials as an ornithologist because he raised birds and he wrote extensively about their biology. Many of today's ideas have their genesis in his writings.

It is inevitable that some readers will consider our account biased—and it is, for bias is unavoidable. An obvious bias is that much of the ornithological knowledge acquired since Darwin has come from Europe and North America, far less from other parts of the world, although wherever possible we have identified important people and advances from Australasia, Africa, Asia, and South America. Our account is also biased by our choice of topics, of stories, and of the people involved in those stories, all of which reflect our own interests and our interpretation of the available information. Others with different backgrounds and expertise would have undoubtedly written a different account—indeed, we hope they will.

For us, ornithology is the *scientific* study of birds, and an ornithologist is someone who studies birds and writes up their findings for publication in scientific journals (Haffer

2001). Although bird watching was a precursor of scientific ornithology and many ornithologists began their careers as bird watchers (Fisher 1940; Weidensaul 2007), this book is not a history of bird watching.

Histories can be dull. But our experiences teaching undergraduates show us that histories are brought to life by stories about the people that populate them. The history of ornithology is overflowing with extraordinary individuals and intriguing stories. Science—ornithological or otherwise—is conducted by real people with real human attributes, including ambition, integrity, jealousy, obsession, and deception. In telling their stories we encounter the full gamut of human frailties from fraud to murder. Some individuals make a name for themselves from a single moment of insight, whereas for others fame emerges only after decades of labor. Our emphasis here is on people— the ornithologists who created the wonderful and extraordinary body of knowledge that we so often take for granted. Michael Brooks, author of *Free Radicals: The Secret Anarchy of Science* (2011), perfectly captures our view: "Scientists have a habit of airbrushing science's great moments to smooth out the human wrinkles and flaws in the process of discovery. Ultimately, though, scientists did themselves a disservice when they dehumanised their field. No wonder we have had such trouble keeping schoolchildren interested in science."[5] Knowing about history is important too. As the nineteenth-century French philosopher Auguste Comte said, "It is true that a science cannot be completely understood without a knowledge of how it arose."[6] Many great biologists have said the same. Here's the evolutionary biologist, R. A. Fisher, writing in 1959:

More attention to the History of Science is needed, as much by scientists as by historians, and especially by biologists, and this should mean a deliberate attempt to understand the thoughts of the great masters of the past, to see in what circumstances or intellectual *milieu* their ideas were formed, where they took the wrong turning or stopped short on the right track. A sense of the *continuity* and the progressive and cumulative character of an advancing science is the best prophylactic I can suggest against the manic-depressive alternations of the cult of *vogue* and *boost*, which threatens to smother the scientific efforts, gigantic as they are, of at least one great nation.[7]

The value of an historical perspective on a scientific discipline like ornithology is not always immediately obvious. Many young researchers feel they do not have sufficient time to plough through the original texts and so do not bother. We believe very strongly that an understanding of the history of a topic has several advantages. First, it allows researchers to see their own work in context: scientists search for "the truth," but what seems to be the truth can change in the light of new evidence—evinced by the ongoing refinements in avian taxonomy and systematics that we describe in chapter 3— and hence the more appropriate description of science as "truth for now." In other words, on the basis of what we know, this is what we currently believe to be true, but as scientists we are prepared to change our minds if convincing alternative evidence comes to light. Looking back on his career, this was one of the traits that Darwin felt had contributed to his success (Darwin 1887; Barlow 1958).[8]

Second, knowing "the literature"—what one's predecessors have done—is an essential part of scholarship and at the very least helps to avoid reinventing the wheel. The problem, of course, is how much of the previous literature can a researcher be expected to know. As we've already indicated, the volume of ornithological articles from the twentieth century is overwhelming. For this reason most young biologists assume that going back as far as the year 2000 is far enough. It isn't, but such a strategy is perhaps the only way to survive in the current academic environment where the acquisition of research grants is so essential for a scientist's career. It is precisely because most scientists cannot afford the luxury of learning history that we hope our synthesis of ornithology will be useful and at the very least provide a guide to the literature of a recently passed era.

But there's a third reason why history is of value: it can be a crucible of creation, triggering new ideas and new ways of looking at old problems. Thus it can be immensely stimulating to see how our ornithological predecessors grappled with particular topics; how they behaved or misbehaved; how they organized their lives; how they failed to recognize the significance of certain facts or data because they had no useful frame of reference at the time. Scientists get their inspiration from a variety of sources, but looking at the history of one's own discipline can be the most rewarding of all.

The topics we have chosen to cover in this book each have their own chronology, their own characters, ideas, and stories. There's no particular logic to the order of chapters other than what we thought would make an interesting read, which means that, in a way, each chapter can be read in isolation. Having said that, the influence of some ornithologists—David Lack and Ernst Mayr, for example—is so far reaching they appear in several chapters.

For consistency, and because the common names of birds can vary from country to country, we have used the International Ornithological Congress (IOC) list of world bird names (Gill and Donsker 2012) for the vernacular names of birds, capitalized for full species names—such as American Robin and Common Cuckoo—but lowercased when referring to robins or cuckoos in general. For scientific names refer to the latest online version[9] of that IOC world list. We have made every effort to be scientifically correct and to document all our sources, citing references in the text in the standard scientific manner. To avoid cluttering the text we tried to use no more than two key references at a time, and these should provide the reader with a portal into the relevant literature. This means that we included references in the text—such as "Lack (1954)"—in part because this is the convention in the scientific literature. We recognize that this style can sometimes seem intrusive to the reader, but its advantage is that it allows one to instantly attribute information to a particular person or a particular publication; giving credit where credit is due is an important feature of scientific practice. A list of all the references is provided at the back of the book. We have also included additional notes at the back of the book, identified by superscripts in the main text. We have kept these notes brief to save space, but we provide more scholarly detail on the book's website at http://myriadbirds.com. At the end of each chapter we have also written a "coda" to present a summary of the historical significance of the topic, as well as our own speculations and opinions.

The graphical timelines in each chapter are an important feature of the book. At a glance, these provide a chronological summary of events, key concepts, discoveries, publications, and so on. The late Jürgen Haffer, a superb historian of ornithology—a geologist by profession—urged us to use the geologist's bottom-up timeline, with the most recent events at the top.

A list of most of the ornithologists that we refer to explicitly in the text is presented in our gallery of five hundred ornithologists at the back of the book. Our aim here is to provide some more details, such as birth and death dates, of those people whose work we describe; it is not intended as a list of the most influential ornithologists since Darwin.

Jürgen Haffer. An oil geologist by profession, Haffer was an amateur ornithologist who undertook pioneering studies of speciation in Amazonian birds (photo in 2008 or 2009 at age 75 or 76).

Each chapter opens with a painting or illustration of a bird or particular group of birds relevant to that chapter. In each case we have chosen an artist whose work we find inspiring, and to highlight the fact that artists have made an enormous contribution to our appreciation of birds.

Our primary goal has been to present the history of modern ornithology in a readable fashion. In doing so we have avoided historical fiction, in that we do not pretend to know what people might have said or thought. Instead we have used direct quotations, either from published sources or from our own interviews, experiences, and interactions with people who study birds. Indeed, one of the most enjoyable aspects of this project was meeting and talking to a wide range of eminent ornithologists. The quotes are important because they constitute empirical information: this is what was said. To maintain the flow of the text, all quotes are referenced with a superscript in the notes at the back of the book.

The audio recordings of those ornithologists that we interviewed are available at http://myriadbirds.com. These interviews provided us with a wonderful web of connections between ornithologists of different eras, and we used them both to inform the text and as a source of quotes. We hope that they will be useful to future historians of ornithology. In each chapter we also present some brief autobiographies, featuring key researchers who were involved in the development of each topic. Some of these were constructed from our interviews, but in most cases individuals wrote a brief account for us, detailing what and who influenced their ornithological career as well as a summary of their main achievements. Our instructions

Chestnut-mandibled Toucan, one of several species whose systematic relations Jürgen Haffer explored (e.g., Haffer 1974). Mayr (1983) referred to Haffer's work on this group as the finest research on bird speciation.

were deliberately vague, since we wanted the authors themselves to determine what they wrote—the results speak for themselves.

Between us we have lived through at least half of the twentieth century, and the busiest half at that. Our origins (two in Britain, one in Canada) have helped reduce any geographic bias; our ages (two around sixty, one in her late twenties) have helped minimize any ageism, and our genders (two male, one female) have helped, we hope, to avoid any sexism. We have been practicing ornithologists ourselves for more than a hundred years in total (we started young!), and we know or have known many of the ornithologists mentioned in the book—an enormous privilege that has provided us with an intimate and inspirational view of ornithology.

ACKNOWLEDGMENTS

In 1959, as part of the centenary of the British Ornithologists' Union (BOU), Max Nicholson wrote this: "The recent successes of British ornithology have largely been based on new ideas with new organizations to foster and serve them, but these would not have been enough without the spread of a sense of common purpose and the growth of innumerable friendships which it has brought about."[10] His comment about friendship is as true today as it was then, and in completing this volume we have been overwhelmed by the generosity of our fellow ornithologists across the world in helping us achieve our goals.

We started this project by conducting a survey of the most influential ornithologists and the most influential ornithological books of the twentieth century. The following kindly provided nominations: Malte Andersson, Peter Berthold, Jacques Blondel, Jerry Brown, Andrew Cockburn, Fred Cooke, John Coulson, John Crook, John Croxall, Nick Davies, André Dhondt, Peter and Rosemary Grant, Jürgen Haffer, Richard Holmes, Ellen Ketterson, Walt Koenig, John Krebs, Kate Lessells, Anders Møller, Pat Monaghan, Ian Newton, Gordon Orians, Chris Perrins, Theunis Piersma, Morné du Plessis, Robert Ricklefs, Uli Reyer, Karl Schulze-Hagen, Claire Spottiswoode, John Wingfield, Roswitha and Wolfgang Wiltschko, and Robert Zink.

We also interviewed and/or obtained autobiographies from Thomas Alerstam, Pat Bateson, Peter Berthold, Walter Bock, Terry Burke, Nicky Clayton, Andrew Cockburn, Nigel Collar, Joel Cracraft, Nick Davies, Steve Emlen, John Fitzpatrick, Brian Follett, Rosemary and Peter Grant, Jack Hailman, Mike Harris, Ben Hatchwell, Geoff Hill, Robert Hinde, Peter Hudson, Alex Kacelnik, Ellen Ketterson, Walt Koenig, Kate Lessells, Ian Newton, Fernando Nottebohm, Peter O'Donald, Colin Pennycuick, Chris Perrins, Richard Prum, Robert Ricklefs, Wolfgang Schleidt, Peter Stettenheim, Bridget Stutchbury, Arie van Noordwijk, Sarah Wanless, Adam Watson, Roswitha and Wolfgang Wiltschko, John Wingfield, and Amotz Zahavi.

The following patiently provided answers, comments, and other personal communications: Ted Anderson (David Lack's biographer), George Barrowclough, Carla Cicero, Fred Cooke, Nick Davies, Jack Dumbacher, John Fanshawe, John Fitzpatrick, Donald Forsdyke, Robert Gillmor, Martyn Gorman, Peter Grant, Chris Guglielmo, John Harshman, Volker Heine, Robert Hinde, Wes Hochachka, Rudy Jonker, Euan Kennedy, Alan Knox, John Krebs, Andrew Lack, Peter Lack, Mary LeCroy, Bernd Leisler, Jere Lipps, Steve Lougheed, Irby Lovette, Jim Lowe, Bruce Lyon, Melanie Massaro, Craig Moritz, Ian Newton, Ian Nisbet, Chris Perrins, Rick Prum, Steve Redpath, Karl Schulze-Hagen, Susan Smith, Mary Sunderland, and Kathy Wynne-Edwards. Our apologies to anyone we have overlooked.

Other colleagues read and commented on either entire chapters or parts of chapters. We are most grateful to Thomas Alerstam, Malte Andersson, Allan Baker, Jerry Brown, Alan Brush, Joel Cracraft, Nick Davies, Scott Edwards, John Fitzpatrick, Jim Flegg, Brian Follett, Frank Gill, Rhys Green, John Harshman, Geoff Hill, Robert Hinde, Pat Monaghan, Ian Newton, Trevor Price, Jens Rolff, Wolfgang Schleidt, Ben Sheldon, Peter Stettenheim, Bill Sutherland, Brian Switek, Charles Wellman, and Tony Williams. We are especially grateful to Frank Gill, Jeremy Mynott, and Ian Newton, who read and commented on the entire manuscript.

We obtained archive and library assistance, translations, photographs, and other information through the help of a number of people, including Rupert Baker, Emma Bedoukian, Alex Best, Karen Bidgood, Patricia Brekke, Deirdre Bryder, Clair Castle, Danielle Castronovo, Isabelle Charmantier, Stamati Crook, Linda DaVolls, Elaine Engst, Jens Rolff, George Franchois, Peter Gallivan, Paul Heavens, Nicola Hemmings, Andrew Lack, Peter Lack, Mary LeCroy, Eleanor MacLean, Cara McQuaid, Margaret Schuelein, Andrew Selous, Ann Sylph, Jamie Thompson, Francis Willmoth, and Mike

Wilson. We are especially grateful to those individuals and organizations that provided us with images with a minimum of hassle—they know who they are.

Special thanks to Al Bertrand, our editor, and his team at Princeton University Press (including Hannah Paul, Dimitri Karetnikov, and Ali Parrington) for their efficient and enthusiastic support.

We thank all of those listed above for their help: we couldn't have completed this project without them. We are also very grateful to the Leverhulme Trust, which awarded TRB a research grant that covered the salary of postdoc JW, as well allowing us to travel to libraries and to interview people. BM received funding from the Natural Sciences and Engineering Research Council of Canada, Queen's University, Université Paul Sabatier (Toulouse, France), and Station d'Ecologie Expérimentale du CNRS (Moulis, France).

Ten Thousand Birds

CHAPTER I

Yesterday's Birds

The road from Reptiles to Birds is by way of Dinosauria to the Ratitae.
—THOMAS HENRY HUXLEY, IN A LETTER TO ERNST HAECKEL ON 21 JANUARY 1868[1]

THE TERRIBLE CLAW

LATE ONE HOT AUGUST EVENING IN 1964, near Bridger, Montana, the paleontologist John Ostrom and his assistant, Greg Meyer, made a discovery that revolutionized the study of ancient birds. Toward the end of a hard day in the field, they spotted, in the slanted light, some claws and bones protruding from the reddish-brown soil. Scrambling to the spot, they began digging with the only tools they had at hand—a jackknife, a small paintbrush, and a whisk broom. Rapidly running out of natural light, they marked the location so they could resume work the next morning. Given the fossil's sickle-like claws, Ostrom was convinced this was a carnivorous dinosaur: "I was almost certain, although still wary, that we had discovered something totally new."[2] And they had, as the subsequent week of excavation revealed—a specimen considered by some[3] to be the most important dinosaur discovery of the mid-twentieth century, an animal Ostrom called *Deinonychus*, "the terrible claw." This was a seventy-kilogram bipedal runner with sharp claws on all four feet and an especially outsized retractable claw on the second toe of each hindlimb. *Deinonychus* was a killing machine, and its study revolutionized our understanding of how dinosaurs lived and breathed and how birds evolved. *Deinonychus* was a member of the *Dromaeosauridae*, a family of theropod dinosaurs—including *Velociraptor*, made famous by the movie *Jurassic Park*—that proliferated in the Cretaceous.

Like so many others who influenced ornithology in the early twentieth century, Ostrom started out studying medicine. Growing up in Schenectady, New York, he began his premed studies there at Union

A pair of *Archaeopteryx lithographica*. Painting by Rudolf Freund for an article in *LIFE* magazine on evolution (Barnett 1959). In 1959 nothing was known about the colors of plumages and bare parts of fossil birds, so Freund was guessing (probably incorrectly, as it turns out).

College in the late 1940s. Prophetically, one of his course requirements was to study evolution, so—keen student that he was—he started to read the course text, Simpson's (1949) *The Meaning of Evolution*, the night before the first lecture. Enthralled, he spent the night reading, then wrote to the author, the eminent paleontologist George Gaylord Simpson, to say how excited he had been by what he had read. Much to Ostrom's surprise and delight, Simpson answered right away, inviting Ostrom to come and study the paleontology of mammals with him at Columbia University in New York City. To the chagrin of his parents, Ostrom abandoned his medical studies and moved to the big city, in 1951, to begin a PhD on the paleontology of reptilian dinosaurs, in the end working with a leading dinosaur specialist, Edwin H. Colbert, rather than Simpson. Six years after obtaining his PhD, Yale hired Ostrom as their curator of vertebrate paleontology at the Peabody Museum, a post held a century earlier by one of the great American paleontologists, Othniel Charles Marsh.

Marsh held the first chair of paleontology at Yale, a post created especially for him in 1866. Ever the entrepreneur, he persuaded his wealthy uncle, George Peabody, to donate funds[4] to establish a museum at Yale so Marsh would have a place to store and display his fossil discoveries. And discover he did—in twenty years of exploration he and his crew found more than a thousand new species of fossil animals, including eighty new dinosaurs,[5] the first pterosaurs from North America,[6] and a new group of fossil birds, with teeth, which he called the "Odontornithes." Marsh's "Odontornithes," as presented in his 1880 monograph, included *Hesperornis regalis* ("the royal bird of the

Deinonychus antirrhopus was discovered by John Ostrom in 1965. This species was originally depicted as naked (top), but recent evidence suggests that it was covered with "dino-fuzz" as shown here (bottom). A fossil of the sharp hind "killing" claw is also shown (bottom left).

west") and *Ichthyornis* ("fish bird"), both of which he had described for science.[7] These new birds were related to *Archaeopteryx*—one of the most famous fossils ever found—and all three of these early birds had teeth, suggesting to Marsh that birds had descended from the toothed reptiles, especially the dinosaurs. Charles Darwin was thrilled: "Your work on these old birds, and on the many fossil animals of N. America, has afforded the best support to the theory of evolution, which has appeared within the last 20 years."[8]

It was not until Marsh and Edward Drinker Cope (from the Academy of Natural Sciences in Philadelphia) began exploring the western United States in the 1870s that the American badlands began to relinquish their biological secrets.[9] Cope and Marsh were both brilliant scientists who laid the foundations of modern paleontology. They are probably most famous, though, for their lifelong feud—aptly called the "Bone Wars"—involving intrigue, chicanery, and insanely intense competition to be first, best, and most famous at everything they attempted, and to have the biggest and most significant collections of discovered-in-America fossils at their home institutions. As we shall see, controversy is a hallmark of paleontology, even today, but the scope, intensity, and nature of the Bone Wars belongs among the great tales of the Wild West, albeit in the name of science. John Ostrom also generated considerable controversy, which continues to this day (2013).

Unlike most of his paleontological predecessors and contemporaries, Ostrom thought about dinosaurs, like *Deinonychus*, as living, breathing animals, not just as a jumble of bonelike rock embedded in a geological stratum. Even his own PhD supervisor, Colbert, considered them to be "sad, slow, stupid creatures that deserved to be extinct."[10] By focusing on how these animals once lived and evolved—including consideration of their behavior, physiology, development, and ecology—Ostrom's approach revolutionized paleontology. Ostrom reasoned that *Deinonychus* must have walked on its hind legs—as its forelimbs were built for killing, not walking—and its posture (based on bone and joint reconstruction) was likely upright, bipedal. As a predator, *Deinonychus* would have pounced on its victims, ripping them open with its razor-sharp claws, possibly using the extra-large claws on its back feet to hold its prey down (Fowler et al. 2011), much in the manner of raptorial birds today. Contrary to the standard (albeit Victorian) image of dinosaurs as enormous, plodding, dimwitted beasts, *Deinonychus* was a relatively small—3.4 meters (11 feet) long—nimble

predator with an active lifestyle, and was almost certainly warm blooded.

Just about everything that Ostrom suggested about *Deinonychus* was unorthodox: here was a dinosaur more like a small ostrich than the lumbering giants usually depicted in books. Could it be that birds and dinosaurs were more closely related than had previously been thought? To explore this possibility, Ostrom needed to reexamine both the oldest known fossil bird, *Archaeopteryx*, to learn about the origins of birds, and the pterosaurs, to learn about the origins of flight in the vertebrate animals.

ARCHAEOPTERYX

When Ostrom began his study of *Archaeopteryx* in 1970, only four specimens were known—a lone feather and three partial skeletons—arguably the most important, valuable, famous, and beautiful fossil animal ever found. Ostrom traveled to Europe to study the original specimens kept in London, Berlin, and Maxburg (Germany), and to visit the vast Solnhofen quarries, where the only *Archaeopteryx* specimens ever have been found. To put *Archaeopteryx* into context of the evolution of both reptiles and flight, Ostrom also went to the Teylers Museum in Haarlem, Netherlands, where some of the world's most complete pterosaur fossils were housed. Pterosaurs—the group that includes the pterodactyls—were flying reptiles, contemporaries of *Archaeopteryx* but not closely related to birds. However, they had several anatomical adaptations for flight that Ostrom wanted to study in detail. With its neoclassic architecture, the Teylers Museum was (and is) a lovely place to work. At the time Ostrom was working, there was no artificial lighting in the galleries, so the museum had to close earlier in winter when the sun set. It was here in the setting sun that Ostrom made one of his greatest discoveries.

Ostrom was examining the type specimen of a pterodactyl called *Pterodactylus crassipes*; yet as he looked over the rock he knew that something was not right. As an expert, he could see that this was no pterosaur. He took it to the window for a clearer view, and the slanting, natural light picked out very faint—but very clear—impressions of feathers. It was an *Archaeopteryx*, mislabeled ever since its discovery in 1855,[11] even earlier than the "first" specimen known, and hidden in plain view for more than a century. Ostrom was beside himself, torn between keeping his discovery a secret, lest the museum curator stop his examination, and announcing that the museum actually had this most valuable of specimens. Integrity triumphed, but as Ostrom feared might happen, the curator whisked the specimen away. Ostrom's immediate reaction—"You blew it, John, you blew it"—was short lived, as the curator soon returned with the specimen in a battered shoebox, saying: "Here, here, Professor Ostrom, you have made the Teylers Museum famous."[12] Even better, he was allowed to borrow the fossil for detailed examination in his own lab. Ostrom was thrilled, but nervous to be carrying such a valuable specimen. He insured the fossil for one million dollars as a precaution, and flew back to Yale with the box on his lap the whole way.

To put Ostrom's work in perspective, we need to go back more than a century to the first reported discovery of an *Archaeopteryx* fossil. The year was 1861, only two years after the publication of Darwin's *Origin of Species*—an amazing coincidence, really, as Darwin had been plagued by the absence of transitional forms, writing: "Why then is not every geological formation and every stratum full of such intermediate links? Geology assuredly does not reveal any such finely graduated organic chain; and this, perhaps, is the most obvious and gravest objection which can be argued against my theory. The explanation lies, as I believe, in the extreme imperfection of the geological record."[13] Darwin noted that many key animal groups were missing from the rather limited fossil record that had been documented by the middle of the nineteenth century. By 1859 a few dinosaur fossils had been found, named, and debated, and there were thousands of fossil invertebrates from around the world, but no obvious intermediates between some of the major, and clearly related, present-day animals, like birds and reptiles. There will always be gaps in the fossil record, but the big ones fuel scientific hypotheses and are grist for the creationist, antievolution mills.

How fortunate, then, that one of the most interesting and useful "transitional forms" ever found should be discovered so soon after Darwin had highlighted the issue of gaps in the fossil record. Here was a fossil with a combination of traits, both reptilian (a long bony tail) and avian (feathers), clearly indicating that it was an intermediate form between birds and reptiles. Here was the best candidate so far for the title of "first bird."

The British Museum of Natural History (BMNH) in London bought the specimen[14] for the then princely sum of £450 in 1862, equivalent to about $55,000 in today's dollars. The BMNH purchase was especially significant because the leading dinosaur expert of the day—Richard Owen, the man who coined the term "dinosaur"—was curator of

A cast of the Berlin specimen of *Archaeopteryx lithographica*, discovered in 1874/75. This is the most complete specimen found so far, and the first with a complete head.

paleontology there, and he was keen to make a detailed study[15] of what he immediately recognized to be an important specimen. Owen was a brilliant man, but he was also nasty, incredibly ambitious, politically connected, very influential—and very much opposed to Darwin's new ideas. Here was his chance to show Darwin wrong. His analysis of *Archaeopteryx*, published in 1863, proclaimed the fossil "unequivocally to be a Bird,"[16] and not a transitional form at all. Owen even renamed the species, unnecessarily, as *Archaeopteryx macrura*, on the (shaky) grounds that it was likely a different species from the fossil feather that had been found in those same beds just a few months earlier, and that *lithographica* was a poor species name anyway. Or was he trying to snatch some glory as the naming authority[17] of this outstanding species?

Owen was at odds with many people, one of whom was Darwin's great friend, Thomas Henry Huxley. In contrast with Owen, whose attempts at public discourse were often both awkward and malicious, Huxley was an articulate, charming raconteur (Desmond and Moore 1991). Huxley had read Owen's account of *Archaeopteryx*, and noticing Owen's errors, may have seen this as an opportunity to embarrass the man who so opposed Darwin's views. To put the record straight, Huxley embarked on his own careful study of the specimen, completing and publishing his analysis in 1868, "which in part intended to rectify certain errors which appear to me to be contained in the description of the fossil" by Owen.[18] Among other things, Owen had mistaken both the left leg for the right and the dorsal for the ventral side, misidentified the right scapula, and had misoriented the furcula (wishbone) and the vertebral column. Huxley's account

made Owen look sloppy and foolish. Even though this first specimen had no head, Huxley (1868a) speculated, correctly as it turns out, that *Archaeopteryx* would have teeth. Owen, on the other hand, was sure that *Archaeopteryx* would have a beak so it could preen its feathers. In a separate paper published that same year, Huxley concluded that the specimen was a wonderful example of a creature "intermediate between reptiles and birds"[19]—a transitional form of just the sort that Darwin had predicted. Darwin was ecstatic: "The fossil Bird with the long tail & fingers to its wings (I hear from Falconer that Owen has not done the work well) is by far the greatest prodigy of recent times. It is a grand case for me; as no group was so isolated as Birds; & it shows how little we know what lived during former times."[20] Despite his enthusiasm for this fossil, Darwin actually never really made much of a fuss about *Archaeopteryx* being a transitional form, at least not in print (Kritsky 1992), nor did Huxley (1968b), who focused his attention on birdlike reptiles and flightless birds.

Archaeopteryx was immediately significant and controversial (e.g., Wagner 1862)—as it continues to be—and straightaway suspicions were raised about its validity (Chambers 2002). How convenient that a fossil so perfectly supporting Darwin's theory should appear just as his ideas were being so hotly debated! In the 1860s, though, Huxley was widely considered to be the ultimate scientific authority, so claims that *Archaeopteryx* might be a fake were not taken very seriously. But 120 years later, in the 1980s, a more serious claim of fakery emerged—serious not because it was valid but rather because it was made by a highly respected scientist at Cambridge University, the cosmologist and mathematician Sir Fred Hoyle, fellow of the Royal Society and former president of the Royal Astronomical Society. This sad incident is more a tale of hubris, and the influence of religion on the dark closets of the human mind, than a legitimate claim, but the drama did play out on the pages of scientific journals and the popular (particularly religious) press.

The accusations began in 1980 at a conference of orthodox Jewish scientists, where Lee Spetner suggested—based on photographs—that the feathers on the *Archaeopteryx* specimen had been applied to a reptile fossil by some modern-day forger. Spetner was a creationist, well known for both his attacks on macroevolution and his belief that there were only 365 bird species on the day of creation. His claims that *Archaeopteryx* was a fake would probably have gone unnoticed had they not been embraced by Hoyle and his former student, Chandra Wickramasinghe.[21] This was not the first—nor will it be the last—time that famous scientists have ventured outside their area of expertise to make pronouncements about subjects they know nothing about.[22] Hoyle and Wickramasinghe joined forces with Spetner and others to publish a series of four papers[23] in the *British Journal of Photography* in the mid-1980s, claiming to show—ironically, with surprisingly poor quality photographs—that the feathers of the BMNH specimen were fake: "Our contention is that the feather impressions were forged onto a fossil of a flying reptile."[24] The fossil reptile, they claimed, had been "enhanced" with chicken feathers[25] pressed onto a paste made from ground-up Solnhofen limestone. They even suggested that Owen had likely perpetrated this fraud to get back at, and disprove, Darwin and Huxley. Sometimes, as we shall see, scientific controversies can lead to productive debate and discovery, but in this case the claims of Hoyle and company were so ill founded that proving them wrong without a shadow of doubt was merely tedious work. In the end, the definitive study, published in *Science*, took researchers at the BMNH in London almost eighteen months to complete (Charig et al. 1986)—time the scientists could better have spent on discovery. Their analysis confirmed what virtually all scientists—ornithologists, paleontologists, geologists, and evolutionary biologists—knew all along, that *Archaeopteryx* was an important real fossil, represented today by eleven excellent specimens excavated over a century and a half from the same extensive limestone beds in Germany.

As of early 2013, all eleven known *Archaeopteryx* specimens (plus one feather) are thought to be the same species, or at least very closely related species, and all have come from Solnhofen. Ostrom's discovery was not the only instance of mislabeling: two specimens were similarly rediscovered, having been originally labeled *Compsognathus*, a small bipedal, theropod dinosaur that had interested Huxley (1968b). Until the 1990s the *Archaeopteryx* specimens were the most studied fossils of birds, providing perhaps the most useful insights into the origins of feathers, birds, and flight, but this was not the only fossil bird known at the time. Huxley, Marsh, Cope, Elliott Coues, Karl Gegenbaur, and others in the nineteenth century had all tried to make sense of the limited fossil record—including *Compsognathus* and the early birds *Archaeopteryx*, *Hesperornis*, and *Ichthyornis*—without much consensus. Then, in the early 1900s, Gerhard Heilmann, an artist with no formal scientific training, began a brilliant attempt at a comprehensive analysis of bird origins, one that dominated the thinking about fossil birds until the 1960s.

THE ORIGIN OF BIRDS

In the century following the 1861 discovery of *Archaeopteryx* by German quarrymen, fossil birds continued to be uncovered at a slow but steady pace. There were some clear ancestors of extant birds—*Hesperornis, Ichthyornis, Aepyornis, Donornis*—particularly in the London Clay Formation in Britain and the Green River Formation in Wyoming, USA. The majority of these fossils were fragments of bones and bits of feathers from which paleontologists used all their ingenuity to imagine the entire bird, with surprisingly good success based on later finds. With one notable exception, there were few attempts to use the fossil record to reconstruct the evolutionary origin of birds. That exception was Heilmann's five comprehensive papers, published from 1913 to 1916 and then gathered together in 1916 and published as a book, *Fuglenes Afstamning*, in Danish. Ten years later this was updated, translated into English, and published to worldwide acclaim as *The Origin of Birds* (Heilmann 1926).

Heilmann was middle aged before he took up the serious study of birds,[26] and we have no clear idea how that interest developed. Certainly, in the early years of the twentieth century, paleontologists were actively debating the origins of birds and flight, and Heilmann was well read and curious:

> During my entire childhood I suffered from the fact that I did not think what others thought, but I had to suppress it. In those days [1880s] casting doubts on religious affairs was not tolerated. But then, a book arrived. It was Lüken's [Lütken 1893] portrayal of animal life in past and present, the first book in our house to deal with paleontology. One of the questions he treated was whether birds descended from the crawling animals, and his conclusion was that he did not believe so. That came as a bitter disappointment to me, and it drove me to addressing the issue myself.[27]

Heilmann was born in 1859 in Skelskør, Denmark, but spent much of his youth boarding unhappily at a strict Catholic school near Copenhagen. In 1877 he enrolled at Copenhagen University to study medicine but quit after six years to pursue his real love, art. For years he worked mainly at the Royal Porcelain Works in Copenhagen, where he became one of their best decorative designers.

By 1902, at the age of forty-two, Heilmann had become tired of painting ceramics and decided instead to spend the rest of his life as a freelance painter and illustrator[28]: "A union of art and science I should consider the highest attainable ideal, and paleontology, in particular, would furnish but a meagre and deficient image of the past without the aid of art."[29] By this time he had also become interested in the origins of birds and flight, devouring every bit of information about these subjects he could get his hands on, using his artistic skills to draw what he examined and to re-create what he thought the whole animal must have looked like in real life. The first product of this labor was a massive paper, *Vor nuværende Viden om Fuglenes Afstamning* [Our Present Knowledge about the Origin of Birds], published in 1913 in Danish. In those days, professional scientists in Denmark showed little interest in ornithology, but in 1906 the wealthier class, keen on collecting, studying, and writing about

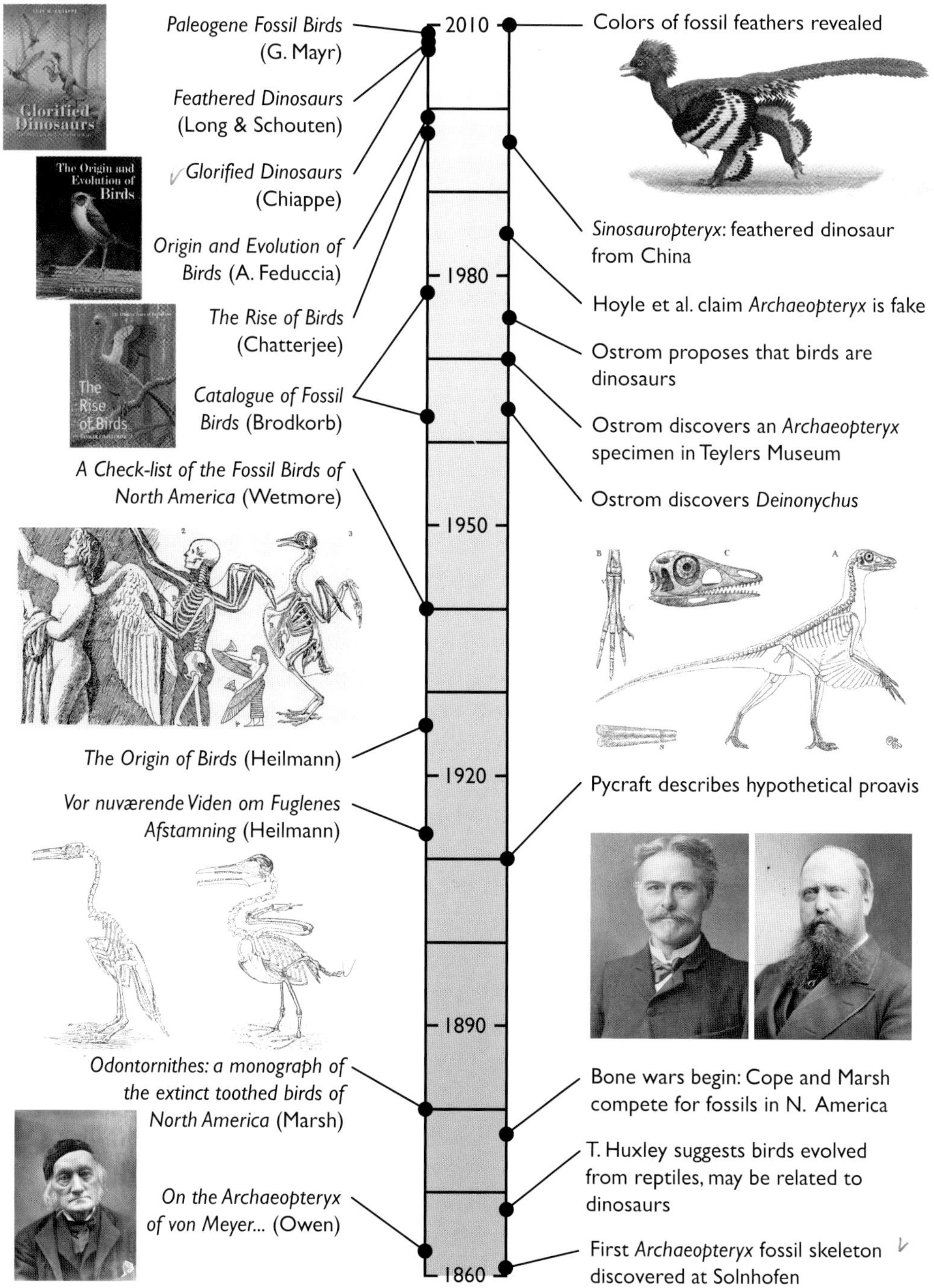
2010
1980
1950
1920
1890
1860
Paleogene Fossil Birds (G. Mayr)
Feathered Dinosaurs (Long & Schouten)
Glorified Dinosaurs (Chiappe)
Origin and Evolution of Birds (A. Feduccia)
The Rise of Birds (Chatterjee)
Catalogue of Fossil Birds (Brodkorb)
A Check-list of the Fossil Birds of North America (Wetmore)
The Origin of Birds (Heilmann)
Vor nuværende Viden om Fuglenes Afstamning (Heilmann)
Odontornithes: a monograph of the extinct toothed birds of North America (Marsh)
On the Archaeopteryx of von Meyer... (Owen)
Colors of fossil feathers revealed
Sinosauropteryx: feathered dinosaur from China
Hoyle et al. claim Archaeopteryx is fake
Ostrom proposes that birds are dinosaurs
Ostrom discovers an Archaeopteryx specimen in Teylers Museum
Ostrom discovers Deinonychus
Pycraft describes hypothetical proavis
Bone wars begin: Cope and Marsh compete for fossils in N. America
T. Huxley suggests birds evolved from reptiles, may be related to dinosaurs
First Archaeopteryx fossil skeleton discovered at Solnhofen
Glorified Dinosaurs
The Origin and Evolution of Birds
ALAN FEDUCCIA
The Rise of Birds

TIMELINE for BIRD ORIGINS. Left: Covers of books by Long and Schouten (2008), Feduccia (1980), and Chatterjee (1997); Heilmann's (1940) diagram to show that angels could not fly, based on his knowledge of bird anatomy; Marsh's skeletons of *Hesperornis* and *Ichthyornis*; Sir Richard Owen. Right: *Anchiornis huxleyi* (a small, feathered, theropod dinosaur) was the first fossil to have its feather colors determined (Vinther et al 2010); Pycraft's (1910) conception of the proavis; Edward Drinker Cope (left) and Othniel Charles Marsh (right).

Gerhard Heilmann (self-portrait in 1906 at age ca. 47).

birds, started an ornithological society, the Dansk Ornitologisk Forenings Tiddskrift (DOFT), with its own journal. It was in this journal—a journal for which he had already provided both the cover (in 1907) that graced every issue for the next ninety years and articles about Darwin and bird conservation—that Heilmann published his papers.

Heilmann's five papers on bird origins in DOFT ran to more than four hundred journal pages, but despite this work being revolutionary, comprehensive, and published in a scientific journal, the Danish ornithological establishment either ignored or disparaged it. A lesser man would have been discouraged. But Heilmann was particularly fortunate to be corresponding with Robert Shufeldt, a prolific and well-known American expert on avian osteology and paleontology. Shufeldt (1914) wrote a brief note in the journal *The Auk* about Heilmann's first four "origin of birds" papers and asked readers for help in locating and lending hard-to-get specimens, papers, and books that could be sent to Heilmann. Shufeldt wrote a further note in 1916, alerting his English-speaking audience that all five of Heilmann's DOFT papers had now been collected together

into a book and that Shufeldt and his wife had started work on an English translation. For some reason—possibly the onset of the First World War—the English version was not published until 1926, with no apparent assistance from (or acknowledgment to) the Shufeldts. In preparing this new English edition, Heilmann was able to study the original *Archaeopteryx* specimens, adding additional details and drawings. At the time, George Gaylord Simpson said the book was "a work at once readable, stimulating, useful and beautiful—a rare achievement. It is surely destined to be one of the classics for paleontologists and ornithologists . . . as for the scientific public."[30] Simpson was then just twenty-four years old and a beginning graduate student at Yale, but he was destined to become probably the most influential paleontologist of the twentieth century, an architect of the Modern Synthesis of evolutionary biology (chapter 2) and, as we have already seen, a mentor to John Ostrom.

Heilmann had done it. A brilliant amateur working with little help from professionals, and certainly none from anyone in his own country, he had produced what was quite probably the most comprehensive, original, and influential treatise on the origin of birds ever written. The book was illustrated with 140 photographs and his own drawings and paintings showing details of reconstructed skeletons and whole birds, often based on very incomplete and two-dimensional fossils. As if that was not enough, Heilmann wrote major sections on embryology, anatomy, and flight to construct his case about avian evolution. Like T. H. Huxley and others, Heilmann demonstrated the close relations between birds and reptiles by detailed comparisons of their anatomical structures; this relationship had really never been in much doubt, but Heilmann brought so much information to bear on the problem that his conclusions were definitive. The final section of the book was devoted to the search for the "proavian," the ancient—but unknown—group of reptiles from which the birds must have evolved. He concluded that the *Pseudosuchia* (now sometimes known as the Crurotarsi) must be that group, one of the two major branches of the tree of "ruling reptiles" (*Archosauria*) that appeared 250 million years ago in the fossil record. This branch lead to modern-day crocodiles and birds, he concluded, whereas the other major branch (*Ornithosuchia*) of the reptile tree evolved into the dinosaurs and pterosaurs. Heilmann reasoned that some lineages in the *Pseudosuchia* branch developed into terrestrial, bipedal runners and some of those eventually into arboreal climbers, then to the common ancestor of *Archaeopteryx* and modern flying birds. This hypothesis was first proposed by Huxley based on much less evidence. Heilmann had seemingly identified the origin of birds, and for the next forty years his book was the bible for avian paleontology—a remarkable achievement and one that nobody, apparently, had the knowledge or enthusiasm to challenge.

Though he worked alone, Heilmann appears to have relied quite heavily on the ideas of contemporary paleontologists in Europe. For example, Robert Broom, an eccentric Scottish paleontologist, had, in 1913, also identified the *Pseudosuchia* as the ancestors to birds, and Heilmann essentially championed Broom's hypothesis: "It is evident that all our requirements of a bird ancestor are met by the Pseudosuchians, and nothing in their structure militates against the view that one

of them might have been the ancestor of the birds."[31] In other words, Heilmann settled on a bird origin in the *Pseudosuchia*, not because there was an abundance of positive evidence for this but because there was nothing to suggest that that theory was wrong. Broom, on the other hand, seemed to be promoting the *Pseudosuchia* to draw attention to his own discoveries in South Africa; he was well known for grandstanding by presenting controversial ideas, often with little supporting evidence. Similarly, Heilmann's search for a "proavis" may have been motivated by a hypothetical bird ancestor of that name illustrated a little earlier by Baron Franz Nopcsa (Nopcsa 1907). Nopcsa, whom we will meet later, was interested in the origins of flight and proposed this ancestor as a model for the sort of bird that must have preceded *Archaeopteryx*.

Heilmann railed against Danish ornithologists, rebuking them roundly in his book: "It is in Denmark a difficult and thankless task to study Paleontology. . . . It is no wonder that, even amongst highly cultivated men, dense ignorance as to the importance of these subjects prevails. . . . Anything like the very great benevolence and good-will which I have met with from several foreign scientists, I am sorry not to be able to record of my own countrymen."[32] He was nothing if not blunt, and many considered him to be an uncultured bully who lacked respect for authority (Reis 2010). But working independently outside the bounds of academia, he really had nothing to lose. In fairness, his treatment by the Danish establishment would have made anyone grumpy. Here is part of Robert Stamm's letter to the journal editor after Heilmann had published the first of his four papers on bird origins: "May

I offer my condolences to the latest volume? It must have been hard for you—who must know birds well . . . to include in the journal the dilettantish mess which occupies most of the issue."[33] Most important, Heilmann's independence gave him the freedom to take an original approach to the study of bird origins, based on his own meticulous study and drawings. Evolutionary principles like Louis Dollo's Law of Irreversible Evolution and Ernst Haeckel's Law of Biogenesis guided his reasoning. Dollo's Law, in particular, led Heilmann to the conclusion that birds had evolved from what were then called "thecodonts"—early archosaurs related to the dinosaurs—and thus had not evolved from dinosaurs themselves, despite overwhelming evidence to the contrary. Dollo's Law stated that, once lost, a trait could not reevolve, and dinosaurs apparently lacked a furculum, one of the key bird-defining traits: "When strictly adhering to this law, we shall find only a single reptile-group can lay claim to being the bird-ancestor."[34] Heilmann was led astray by Dollo's Law, an early evolutionary principle that we now know is incorrect.

Despite the triumph of his *Origin of Birds*, Heilmann never again wrote about birds or paleontology. He was, after all, sixty-seven years old when the English edition of his book was published, a time when less ambitious men might be devoting their remaining years to golf and grandchildren. Instead, Heilmann was busy applying his energies to painting, including an excellent series of raptors for a three-volume work on the birds of Denmark (Heilmann and Manniche 1928–30). Toward the end of his life, he published a book on the relation between science and religion, *Universet og Traditionen* [The Universe and the Tradition], in which he offers some personal insights into his own history, demons, and philosophy. The first part is a popular account of life on Earth, but the final section is an angry attack on Christianity, born perhaps from his unhappy childhood experience at the Christian boarding school. In a final flourish, drawing on his extensive knowledge of avian anatomy, he presents a critique of the supposed morphology of divine beings: "If angels had any reality, they would be very clumsy and awkward fliers with a slow heavy flight, lacking as they are in aerodynamic shape."[35]

BIRDS ARE DINOSAURS

Once Ostrom had examined his newly acquired specimens of *Deinonychus* and *Archaeopteryx*, he knew Heilmann was wrong. He was well versed in Heilmann's ideas, as—even in 1964, when he discovered *Deinonychus*—*The Origin of Birds* was still the bible for anyone interested in avian paleontology. Despite excellent work on fossil birds by such luminaries as Alexander Wetmore (1940), George Gaylord Simpson (1946), William Swinton (1958), and Pierce Brodkorb (1963) in the intervening half century, no new fossils shed any obvious light on the origins of birds, and everyone seemed to accept Heilmann's arguments. However, Ostrom's careful study of his fossils led him to the conclusion that birds are not just related to dinosaurs, they *are* dinosaurs, descendants of the group called "theropods" (*Theropoda*). Theropods are a lineage of dinosaurs that first appeared in the fossil record about 230 million years ago (MYA), in the Late Triassic, and were really the only large terrestrial carnivores on Earth for 135 million years, from the Early Jurassic (200 MYA) until the

Late Cretaceous (65 MYA). Thus Ostrom concluded that birds were not descended from the thecodonts, as had previously been thought by Heilmann and others. Instead, he reasoned that birds were descendants of theropods, one of the great dinosaur lineages (including *Tyrannosaurus rex*) that was thought to have gone extinct. According to Ostrom, *Archaeopteryx* and *Deinonychus* were both theropod dinosaurs—*Archaeopteryx* near the base of the lineage that evolved into modern birds and *Deinonychus* in a different lineage, the dromaeosaurs, that went extinct with most of the other dinosaurs at the end of the Cretaceous 65 million years ago.

Others had earlier suggested that birds had evolved from dinosaurs, but Heilmann appears to have put an end to that kind of talk. For example, Huxley (1868b, 1870) certainly hinted at this possibility, though his treatment of this subject is complex and never particularly straightforward (Switek 2010). Instead, Huxley devoted most of his attention to the evolution of birds from the reptiles, without explicitly claiming in print that birds are dinosaurs. He focused instead on *Compsognathus* because, to him, it "affords a still nearer approximation to the 'missing link' between reptiles and birds. This is the singular reptile . . . *Compsognathus longipes* . . . some of the more recondite ornithic affinities of which have been since pointed out by Gegenbaur. Notwithstanding its small size (it was not much more than 2 feet in length), this reptile must, I think, be placed among, or close to, the Dinosauria; but it is still more bird-like than any of the animals which are ordinarily included in that group."[36] In 1879 two professors, Samuel Williston and Benjamin Mudge, had a celebrated debate on the possible evolution of birds from dinosaurs without really coming to a resolution (Williston 1879, Mudge 1879), but indicating that the idea was certainly worth considering.

By the time Ostrom made his discoveries in the 1960s, the evolutionary principles that had guided much of Heilmann's thinking had been shown to be wrong, or at least overly simplistic. Ostrom took an implicitly cladistic approach (chapter 3), in which derived traits that are shared between lineages are used to indicate close evolutionary relationships. He published his ideas in a series of papers, culminating in a comprehensive review in 1975. Immediately recognizing the potential of Ostrom's work and excited by its potential to revolutionize the study of bird evolution, Joel Cracraft, then at the University of Chicago, wrote a "Special Review" of it for the Wilson Bulletin in 1977. Cracraft was enthusiastic but cautious: "I am not trying to create a bandwagon over Ostrom's papers, but they are exciting. Some of his findings may eventually be refuted, but there is no doubt that much of his meticulous work will last and that our ideas on avian evolution will be significantly influenced by his results. Ornithologists owe this nonornithologist a great deal for this contribution."[37] There was now no excuse for Ostrom's ideas being ignored by the ornithological community, a community that was, at the time, mostly out of touch with developments in paleontology.

Ostrom's conclusions were threefold, and threefold controversial. First, as we have seen, he argued that dinosaurs had not all gone extinct at the end of the Cretaceous after all, because one type of dinosaur had evolved into modern-day birds. As if that was not revolutionary enough, he also suggested that the dinosaur ancestors of birds, and possibly other dinosaurs as well, might have been warm blooded. Dinosaurs were not ponderous giants; many, he argued, were nimble creatures with fast metabolisms. Heilmann had illustrated them this way as well, even though he considered them cold blooded, but Ostrom's warm-blooded dinosaurs had more behavioral possibilities and more niches to fill. Finally, Ostrom suggested that feathers may have evolved before flight, for insulation, and that wings therefore may have preceded flying, possibly as adaptations for catching prey or balancing while running. As Cracraft said, "I find most of his arguments persuasive not in the sense that they are necessarily true, but that they "explain" far more than previous hypotheses."[38] Ostrom's ideas were new and controversial, but they had the added virtue of being based upon sound principles of evolutionary and systematic biology, careful study, and new specimens. Most important, Ostrom had formulated testable hypotheses and was able to show that there was really not much evidence to support anything but birds being descended from dinosaurs. In 1986 Jacques Gauthier, also at Yale, provided a formal cladistic analysis that clearly placed birds among the theropod dinosaurs, just as Ostrom had argued a decade earlier.

The ensuing controversy divided ornithologists and paleontologists into two camps—the birds-as-dinosaurs (BAD) group, who followed Ostrom's lead, and the smaller birds-are-not-dinosaurs (BAND) faction, some of whom[39] argued that birds are descended from a branch of the *Pseudosuchia*, as Heilmann had originally proposed. Although the data and analyses supporting the BAD hypothesis today seem overwhelming, a small but vocal minority still disagrees and have unfortunately become the darlings of the creationist–intelligent design–antievolution camp, who latch onto anything that smells of controversy among scientists.

Paralleling, and indeed a descendant of, the "systematics wars" of the 1960s and 1970s (chapter 3), the debate about bird origins has been fractious, public, and at times downright nasty. Where the systematics wars had little traction for nonscientists, the debate about dinosaur origins has played out in newspapers and magazines around the world. Here is Storrs Olson, never one to mince words, reviewing a symposium volume honoring John Ostrom:

> One of the rituals of the Birds-Are-Dinosaurs-Movement (BADM) is to hold periodic symposia to reaffirm the belief that birds really are dinosaurs, much as Southern Baptists hold revival meetings.... Just as a revival tent is not the haunt of free-thinkers, there are few authors in this book who depart from the true path and numerous papers consist of the cladogram-thumping dogma we have come to expect from the more insistent proponents of the BADM. Kevin Padian, the Elmer Gantry of the theropod crusade, is an author on no fewer than four contributions, which does nothing to diminish the impression of the whole volume as a dreary, sectarian tract from the Kingdom Hall of Hennig's Witnesses.[40]

Like much of the debate about cladistics in general (chapter 3), the issues about bird origins revolve around evidence, logic, and analytical methods, with the BAD advocates complaining that their opponents are basing their conclusions on weak evidence and advocacy rather than sound scientific principles addressing testable hypotheses. As with the proponents of an approach to classification called "cladistics," BAD scientists never

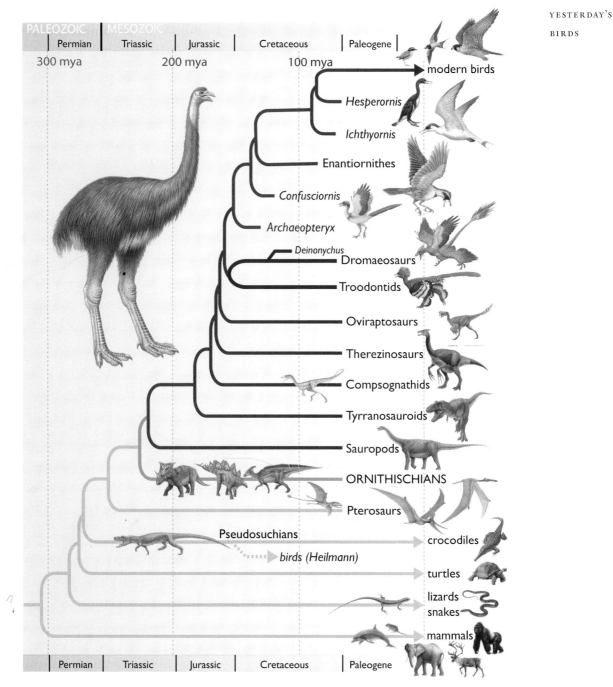

Evolutionary history of the birds (red), a selection of other dinosaurs (purple and blue), and some other groups of vertebrate animals (gray). The dinosaurs comprise two major groups: Ornithischians (blue) and Saurischians (purple and red). Representatives of each named taxon, except *Confusciornis* and *Deinonychus* (see page 2), are shown (not drawn to scale). The large bird in the upper left is a moa (*Dinornis maximus*) from New Zealand, one of an entire genus of birds that was extinct by 1800 due to human activities. Heilmann and others originally thought that birds evolved from Pseudosuchian reptiles (dashed gray line).

argued that their answers are always correct and immutable, but just that their methods generate logical, testable hypotheses, the hallmarks of good science.

Alan Feduccia, in particular, has been a strong, prolific, and vocal opponent of the BAD hypothesis. With serious credentials as both ornithologist and paleontologist, Feduccia was, initially at least, taken seriously as presenting a reasonable alternative hypothesis. While he was still an undergraduate in the 1960s at Louisiana State University, expeditions to Honduras, El Salvador, and Peru catalyzed his interests in birds. When he graduated, Michigan was a leading school for ornithology, so Feduccia went there to do a PhD with Bud Tordoff. On obtaining his PhD in 1969, he took up a faculty position at the University of North Carolina, where he remained for his entire career. An accomplished paleontologist, and one of the few working on birds, he became interested in bird origins when writing his 1980 book, *The Age of Birds*, a wide-ranging treatise on the evolutionary history of birds. During that same period he published two papers, one with Tordoff and one with Olson, criticizing the BAD hypothesis. Feduccia has consistently argued that birds are not theropod dinosaurs, instead echoing Heilmann that birds are evolved from other ancient reptiles and that feathers must have evolved in an aerodynamic context, with flight beginning in arboreal, not terrestrial, birds (Feduccia 2002).

Opponents of the BAD hypothesis suggested that it fails on three main fronts. First, they said, there was a temporal paradox—many theropod dinosaur fossils were too recent (80 MYA) to be the ancestor of *Archaeopteryx* (148 MYA)—"You can't

be your own grandmother,"[41] Feduccia is often quoted as saying. But this ignores the fact that birds are just one of several lineages of theropod dinosaurs, many persisting long after *Archaeopteryx* appears in the fossil record. Second, there were the standard arguments about the methods of constructing phylogenies, as we will see in chapter 3, particularly with respect to the use of cladistics. As Richard Prum has noted, however, "Phylogenetics [cladistics] is universally recognized in systematics and evolutionary biology, and even by U.S. courts. . . . But, critics of the theropod origin of birds have been a singular exception to that nearly universal intellectual trend."[42] Third, there seems to be an important difference between the skeletal structure of the "hands" of theropods and birds, with birds having digits 2, 3, and 4 but theropods 1, 2, and 3. It was long thought that such differences were so fundamental that the bird "hand" could never have evolved from the theropod "hand." For decades the debate was about the actual identity of these digits, but the answer is just now (2013) emerging from the new field of evo-devo—the marriage of evolutionary and developmental biology. Using the latest techniques of molecular genetics, it has now been shown that the digits in the domestic fowl are the result of a mutation in the genes controlling development, such that the ancestral genome that resulted in the development of the 1-2-3 digit pattern is now producing what looks like a 2-3-4 pattern (Young et al. 2011). This sort of analysis could barely have been dreamed of in Ostrom's day, and it will undoubtedly emerge as one of the great triumphs of evolutionary developmental biology. Thus, a simple change in the genes that control the development of the "hand" is all that is needed to make what looks like a very different structure.

By the end of the twentieth century it would be fair to say that Feduccia's views had largely fallen out of favor and that both the evidence and current consensus support a theropod origin for birds: birds are dinosaurs. Nonetheless, to evaluate that consensus using all of the morphological evidence currently available, a recent detailed cladistic analysis by Fran James and John Pourtless (2009) did not find unequivocal support for the BAD hypothesis that birds are theropod dinosaurs. They concluded that there are some uncertainties in both the data and the relevant hypotheses that appear to have been glossed over by previous researchers, suggesting that more work needs to be done to resolve the issues. Despite the cautious tone of the James and Pourtless paper, the BAD hypothesis received a spectacular boost in the 1990s from a flood of fossils of early birds discovered in China and elsewhere, owing as much to a change in the geopolitical winds as to any scientific progress.

CHINA OPENS HER DOORS

Late in the summer of 1996, Ji Qiang, director of the National Geological Museum in Beijing, opened a green silk box presented to him by Li Yinfang, a farmer from Liaoning Province who supplemented his income by selling fossils to collectors and museums. In the box was a fossil that would usher in a new era in paleontology: "When I got the specimen . . . I had never seen a creature like this . . . the first real evidence that dinosaurs gave rise to birds."[43] A canny operator, Li had recognized right away that the specimen was unique, so he separated the two slabs (called

"part" and "counterpart" by paleontologists) and sold them separately to Ji and his rivals at the Nanjing Institute of Geology and Palaeontology. By chance, Phil Currie, head of dinosaur research at Alberta's Royal Tyrrell Museum of Paleontology, was visiting the Beijing Museum in early October 1996 and immediately recognized the significance of this new specimen, later named *Sinosauropteryx prima* (which translates loosely as "the first feathered dinosaur from China"). Currie took a photograph of the specimen to the Society of Vertebrate Paleontology conference in New York that same month, where it caused a hubbub of speculation; John Ostrom was said to be "in a state of shock."[44]

Ji Qiang first published on the discovery of *Sinosauropteryx* in Chinese in 1996 (Ji and Ji 1996), then made a public announcement of their findings to the West at a press conference at the Academy of Natural Sciences of Philadelphia[45] on 24 April 1997, then a year later in *National Geographic* magazine (Ackerman 1998) and *Nature* (Chen et al. 1998). *National Geographic* was involved because it had been instrumental in bringing these amazing specimens to the West. Here is Ji Qiang: "For me it was very difficult. . . . Not even my museum has displayed them! . . . National Geographic is famous in China, and Chinese people love this magazine."[46]

China had been closed to Western exploration for almost half a century. In the 1920s, while Roy Chapman Andrews, from the American Museum of Natural History (AMNH), was discovering and exploring the incredible fossil dinosaur fields in Mongolia, political unrest was fomenting, and further exploration in that country soon became too dangerous. When Mao Tse-tung came to power in the 1940s, China closed her doors

Chinese paleontologists Xu Xing (left; photo in 2010 at age 41) and Zhou Zhonghe (right; photo in 2007 at age 42).

to Westerners seeking fossils. The progress of science within the country was also suppressed, especially during the cultural revolution of the 1960s. Following the death of Mao in 1976, the rise of Deng Xiaoping to the leadership of China, and the increasing move to globalization, both science and international relations began to flourish there in the 1980s.

When Dong Zhiming, sometimes called "China's Mr. Dinosaur," graduated from university in 1962, he immediately joined the Institute of Vertebrate Paleontology and Paleoanthropology (IVPP) in Beijing to pursue his dinosaur-hunting dreams. Like many intellectuals, however, Dong was sent to a farm to study agriculture for the state in 1965, then for the next decade to southwestern China on geological surveys to help develop irrigation systems. Undaunted, Dong continued to collect fossils, and by 1980 he was back at the IVPP full time, contacting dinosaur researchers in the West to tell them about his many discoveries and hoping to develop cooperative projects that would provide the much-needed funding for exploration, and particularly excavation, in China. In 1985 his diligence paid off when he and his Canadian colleagues, Phil Currie and Dale Russell, obtained largely private funding[47] to embark on a massive expedition that would run for four years and employ fifty field workers. This "Sino-Canadian Dinosaur Project" uncovered eleven new species and, most important, alerted the world to the fabulous Chinese fossil discoveries.

At a press conference in June 1998 at the National Geographic Society in Washington, DC, Ji and Currie, with Mark Norell of the AMNH, presented three important discoveries relevant to the origins of birds and feathers: (1) *Protarchaeopteryx robusta*, an animal 90 centimeters (3 feet) long that they thought might actually be a precursor to *Archaeopteryx*; (2) *Caudipteryx zoui*, a turkey-size, fast-running dinosaur with simple feathers, and (3) *Sinosauropteryx prima*, mentioned above. Ji Qiang said:

Currie and Norell believe that *Caudipteryx* is a dinosaur; I believe that anything with wings and flight feathers is a bird. I think *Caudipteryx* was running and jumping and trying to fly, but he couldn't, because his feathers were symmetrical [without the narrower leading edge needed for true flight] and too short. I think there is generally an evolutionary tendency from *Sinosauropteryx* to *Caudipteryx* to *Protarchaeopteryx* to *Archaeopteryx* to modern birds.[48]

Sinosauropteryx is particularly noteworthy because it was clearly a theropod dinosaur, closely related to *Compsognathus*, being in the same family (*Compsognathidae*). But now here was a nonbird dinosaur with a trait—feathers—that had previously been that most bird-defining trait of all. *Sinosauropteryx* was a tiny—60–100 centimeters (2.0–3.3 feet) long—Late Cretaceous (124 MYA) dinosaur with short arms, a large first digit, and a long tail—a feathered dinosaur that preyed on fast-moving lizards and small birds, prey whose remains are clearly visible in the guts of some fossils.

Sinosauropteryx, more than any other discovery, drew attention to the fossil treasures of China. Where there is treasure, though, there are often pirates, and the world of fossil collecting has had its share. In the most celebrated case of "piracy," *National Geographic* was again at the forefront, in an article titled "Feathers for T. Rex?" (Sloan 1999), announcing that *Archaeoraptor lianoningensis* was "a true missing link in the complex chain that connects dinosaurs to birds."[49] This fossil, also from Liaoning Province, appeared to have the body of an early bird but the teeth and tail of a small terrestrial dromaeosaur—another apparent gap in the fossil record filled by a transitional form. Currie was thrilled: "We're looking at the first dinosaur that was capable of flying."[50] His enthusiasm was short lived, though, as it was soon discovered that *Archaeoraptor* was a fake, a chimera: "After observing a new, feathered dromaeosaur specimen in a private collection and comparing it with the fossil known as *Archaeoraptor*, I have concluded that *Archaeoraptor* is a composite."[51] Further detailed study confirmed this conclusion, and the eventual *Nature* paper was a report of a fraud rather than a missing link: "Sadly, parts of at least two significant new specimens were combined in favor of the higher commercial value of the forgery. . . . Knowing the history of human handling can be critical to proper evaluation and scientific interpretation of specimens."[52] The good news is that the controversial nature of fossil birds has heightened the level of vigilance normally applied to discovery in science; the bad news is that creationists and others who would denigrate science see such incidents as proof that scientists will go to any length to argue their case, not realizing, apparently, that it was scientists who uncovered this fraud, and that the fraud itself was perpetrated by greedy dealers. How could the experts have been so readily deceived? On the one hand, the fake *Archaeoraptor* was extremely well done, but it is also true that scientists are sometimes so excited by new discoveries that they initially just believe what they see.

Since 1998, hardly a month has gone by without another amazing discovery of fossil birds and feathered dinosaurs from China, and also from Spain, Argentina, and Canada. As the paleontologist Dale Russell said, "John [Ostrom] has to be very pleased. He showed the right way. He laid the foundation."[53] One upshot of all this activity and interest has been a spate of wonderful books about bird origins and the relations between birds and dinosaurs—Chatterjee's *The Rise of Birds* in 1997, Shipman's *Taking Wing* in 1998, Feduccia's *The Origin and Evolution of Birds* in 1999, Paul's *Dinosaurs*

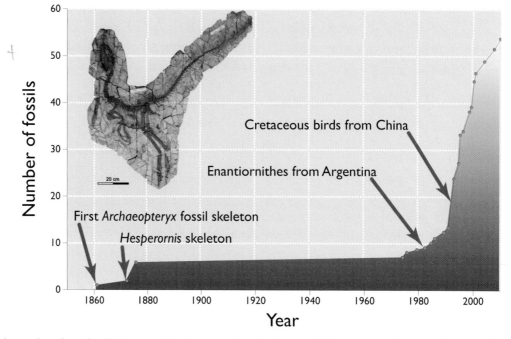

The number of new fossil bird species found per year since the first fossil *Archaeopteryx* skeleton was discovered in 1861. The inset photo is the recently discovered fossil of *Sinocalliopteryx gigas*, a large compsognathid dinosaur. This specimen had fossils of the bird *Confusciornis sanctu* in its stomach.

of the Air in 2002, Currie's *Feathered Dragons* in 2004, Chiappe's *Glorified Dinosaurs* in 2007, Kaiser's *The Inner Bird* in 2007, Long and Schouten's *Feathered Dinosaurs* in 2008, and Feduccia's *Riddle of the Feathered Dragons* in 2012—and new ideas on the evolution of feathers and flight. Echoing Cracraft's appeal to take Ostrom seriously, Rick Prum provided an insightful review of these new developments in 2002, detailing why a theropod dinosaur ancestor for birds—as well as new findings about feathers and flight—heralds a new direction for modern ornithology: "Because the theropod origin of birds is relevant to almost all aspects of avian biology and should influence the way we think about, study, and teach avian anatomy, behavior, physiology, ecology, and evolution."[54] These recent developments

also indicate that we still have a lot to learn but that, with careful analysis and fruitful debate, we can now realize the dream that Huxley had 150 years ago on examining *Archaeopteryx*, when he said that "it can hardly be doubted that a lithographic slate of Triassic age would yield birds so much more reptilian than *Archaeopteryx*, and reptiles so much more ornithic than *Compsognathus*, as to obliterate completely the gap which they still leave between reptiles and birds."[55]

THE ORIGIN OF FEATHERS

Since birds and reptiles share a common ancestor, it seemed, for a long time, eminently logical that feathers evolved from scales. So logical in fact—like wings evolving from forelegs—that it is difficult to say who first suggested it. From Darwin through the entire twentieth century the feathers-from-scales idea was the basis for virtually all theories about the origin of feathers. Even Ostrom's revolutionary ideas, decoupling the origins of birds from the origins of feathers and flight, had no immediate influence on theory that feathers evolved from scales. Darwin and Huxley both seemed to accept this as if it was common wisdom, and they made no explicit mention of the origin of feathers in their work. The belief that feathers evolved from scales was also implicit in the writings of Victorian ornithologists (Alfred Newton, Hans Gadow, Max Fürbringer), but it was not until the early twentieth century that explicit models for this evolutionary transition were developed, mainly in connection with propositions about the origins of flight (next section). Indeed, the idea that feathers probably did *not* evolve from scales emerged only toward the end of the twentieth century, the brainchild of a three-way marriage of Ostrom's insights, sound evolutionary principles (as we will see in chapter 3), and plain logic. Before we deal with this truly revolutionary idea, though, we need some context. What was the feathers-from-scales idea, and how did it hold sway for more than a century in the scientific literature?

The fabulously named Baron Franz Nopcsa von Felsö-Szilvás presented one of the first feathers-from-scales scenarios in 1907. Nopcsa was a brilliant amateur paleontologist who pioneered many techniques for fossil analysis and was one of the first to depict dinosaurs as living organisms. As a teenager he became interested in fossils in 1895, when his sister discovered some dinosaur bones on one of the family estates in Transylvania (in those days part of Austria-Hungary). Only eighteen at the time, Nopcsa worked hard to learn anatomy and within a year wrote a paper describing this fossil find as *Telmatosaurus*, a new species of dinosaur in the *Ornithischia*.[56] Nopcsa was a small man, openly gay and often nattily dressed, and perhaps better known for his flamboyant personal life and tragic history than for his paleontological work. He was wealthy enough to travel freely around Europe searching for fossils, visiting museums and personal collections to look at their fossils. While in the field in Albania, he often dressed in local shepherd's garb so he would not attract attention, working for months at a time with only his secretary/lover for companionship. Nopcsa accumulated an incredible range of geologic, meterologic, paleontologic, and ethnographic data, eventually writing more than a hundred papers on fossils. During the First World War, he operated as a spy and also made a vain attempt to become king of Albania. Then, when Transylvania was ceded to Romania after the war, he lost all of his estates and wealth. To make ends meet, he moved to Budapest to take up a paid position as head of the Hungarian Geological Institute. Soon tiring of that sedentary lifestyle, Nopcsa quit his job after a few years and headed off to Italy on a motorcycle with his lover, in search of fossils and to map geological formations. To fund this work he sold his fossil collections to the BMNH, but by 1933, at the age of fifty-six, he was both destitute and depressed. Despondent that they would suffer in sickness and poverty[57] Nopcsa killed his lover and took his own life.

In a paper on the origins of flight, Nopcsa (1907) included his own drawing to show what he thought the hypothetical "proavis" might have looked like. Nopcsa suggested that proavis might have achieved some lift from extended scales at the rear margin of the forelimbs: "By gradually increasing in size, the enlarged but perhaps still horny hypothetical scales of the antibrachial margin would . . . ultimately develop to actual feathers . . . [and] could attain quite a considerable size without essentially altering the underlying bones of the arm, a fusion of the carpal phalanges being only then necessary, when in flight rigidity of this region became requisite."[58] This theme was picked up by J. Versluys (1910), William Pycraft (1910), Friedrich von Huene (1914), and, of course, Heilmann (1916, 1926). Though Pycraft and Heilmann proposed no plausible mechanism for such an evolutionary change, both thought that the margins of these elongated scales would have become frayed to form "protofeathers" that gradually evolved into the feather structures we see in fossils like *Archaeopteryx* and in living birds today. Here is Pycraft on proavis:

> The body clothing at this time was probably scaly, but with scales of relatively large size. Those covering the hinder border of the incipient wing, growing longer, would still retain their original overlapping arrangement, and along its hinder border would, in their arrangement, appearance, and function, simulate the quills of modern birds; as their length increased they became also fimbriated and more and more efficient in the work of carrying the body through space. There is less of imagination than might be supposed in this attempt at reconstructing the primitive feather, inasmuch as there is a stage in the development of the highly complex feather of to-day which may well represent the first stage in this process of evolution.[59]

Baron Nopcsa (left) attached to an inflated goat bladder so that he can safely cross the River Drin (Albania); seen here with his secretary/lover (photo from ca. 1907 when Nopcsa was ca. 30 years old).

For the next eighty years, many suggestions were made about feather origins, based on the theme of scales-to-feathers first described by Nopcsa and Pycraft. The great German paleontologist Friedrich von Huene,[60] for example, postulated that longish, loosely attached scales on the forelimbs would have bent upward and spread out when early, flightless birds jumped from tree to tree, with the pressure of the air on the scales stimulating more growth and elongation. He suggested that such protofeathers would have first formed on the limbs and tail then spread to the body at some later time. William Beebe (1915) proposed that the lizard ancestors of birds would have benefited from their scales becoming longer, allowing them to parachute as they leaped from trees and eventually evolving into feathers. Heilmann (1916, 1926) weighed in to suggest that as those scales became longer, friction would fray their outer edges and these frayed edges would gradually have evolved into longer horny processes that became increasingly like flight feathers. Hans Steiner developed the most comprehensive early model in his doctoral work, published in 1917, where he outlined the details of transition from scales to feathers, including the evolution of afterfeathers and interlocking barbules. While Steiner's hypothesis was the most detailed, it fell short of explaining how the microscopic details of barbules and hooklets might have reached their final form in birds. Much later, Kenneth Parkes (1966) and Phil Regal (1975) added some details to this scales-to-feathers idea, but the general feeling was that most of these hypotheses "lack detail on how the scale-to-feather transformation might have occurred and how the intermediate stages might have been adaptive."[61]

As Rick Prum and Alan Brush (2002) recently recognized, the twentieth century saw four conceptual approaches to the problem of feather origins: (1) functional theories; (2) the analysis of so-called primitive feathers of modern birds; (3) scale-to-feather transformations; and (4) theories based on the details of the structural protein "keratin" and the process by which it is produced in the body. As we shall see in the next section, functional theories arose from considerations about whether feathers evolved for flight, thermoregulation, water repellency, communication, or tactile sensation, but such an approach was, in retrospect, unnecessarily confining. The feathers-from-scales idea also led researchers down the wrong path, as H. R. Davies (1889) initially suggested it might, but it was not until Brush (1993) revealed that feathers and scales were made from different forms of keratin that it was obvious to many that something was wrong with this idea, despite it being the focus of a century of investigation. A major symposium on the origin of feathers in 2001 presented such a diversity of opinions and approaches that it became clear that the question of feather origins was far from answered.

In what may be the finest eureka moment in twentieth-century ornithology, Rick Prum came upon a plausible solution while teaching his ornithology class as a new professor at the University of Kansas in the late 1990s. He had put feather development and evolution into the syllabus with some trepidation, knowing full well that he had little knowledge or interest in either subject. He struggled with that first lecture on development: "What was I thinking? . . . It was a Sunday night. We had a new baby at home, nobody's sleeping, and I'm far behind—typical disorganized professor—and I cracked open Lucas and Stettenheim [1972], perhaps one of the most boring books in the history of ornithology, and I tried to understand what was going on."[62] Then, two days later, he gave the lecture on feather origins: "I basically repeated what was a very cloudy twentieth-century view. And I criticized it, saying that it doesn't make sense for a bunch of reasons. But it's not like I had an alternative."[63] Standing at the chalkboard, he resolved then and there to sort it out. Within two years he had the answer, eventually able to show his class how feathers must be an evolutionary novelty, the follicle evolving from the epidermis at a deeper layer than the scales, and in five stages from a hollow tube through downy plumes to the vaned feather. At the end of that lecture a visiting graduate student from China suggested that Prum should publish the idea. The result was a comprehensive theory of feather origins that was radically different from previous ideas and made good biological sense (Prum 1999). That Chinese student was Zhou Zhonghe, who later became a prominent paleontologist, appointed in 2008 to direct the IVPP at the Chinese Academy of Sciences. Zhou may have realized during Prum's lecture that newly found fossils from China would support Prum's ideas.

The Lucas and Stettenheim (1972) volume that Prum referred to above is just the kind of "boring" book that scientists love—rich in detail, wide ranging, and accurately illustrated: "Nothing in the recent literature on avian anatomy can compare with this work in scope, thoroughness, and attention to detail. The authors and illustrators have worked together to provide a masterpiece of anatomical research and presentation."[64] This instant classic was the product of many years' work from the US Department of Agriculture's Avian Anatomy Project at the University of Michigan, under the direction of Alfred M. Lucas, with Peter Stettenheim as the lead researcher. The goal of that project was to learn as much as possible about the anatomy of the domestic fowl in the hope that this information would increase the production of meat and eggs. The result was the most comprehensive treatise—both literature survey and original research—on the avian integument ever done. Unfortunately, the project ran out of funds in 1972, with only the integument volumes completed. Nevertheless, Lucas and Stettenheim provided a nice summary of the scales-to-feathers literature on feather origins, and most important for the genesis of Prum's ideas, exquisite detail on feather development. Prum's theory had a sound basis in avian anatomy, and he was able to draw upon his extensive knowledge of birds, phylogenetics (cladistics), and evolutionary

biology. Like any theory, however, the proof was in the testing.

Just as the discovery of *Archaeopteryx* was so well timed for Darwin's musings about transitional forms, fossils of birds from China soon provided evidence to support Prum's ideas. While still a graduate student, Xu Xing described two new Chinese fossils in 1999 and 2000: *Beipiaosaurus inexpectus*, which had unbranched simple quills (Stage I feathers) along its back (Xu et al. 1999), and *Microraptor zhaoianus*, which had both the fully modern feather type (Stage V) and those with interlocking barbules (Stage IV) (Xu et al. 2000). In 1998 Ji Qiang and his colleagues described *Caudipteryx zoui*, a theropod dinosaur with feathers having a distinct rachis and a symmetrical vane on the "hands" and the tip of the tail (Stage III feathers), and so-called dino-fuzz on the body (Stage II). As of 2013, at least twenty species of feathered dinosaurs have been discovered, all of which have Stage II filaments on their body. Dissenters in the BAND camp argued initially that Stage I filaments were actually collagen, and thus unrelated to feathers, but the discovery of pigments in these fibers put an end to that idea, as collagen is never pigmented.

This most recent development in the BAD story is in many ways the most remarkable. In 2008 Prum's graduate student Jakob Vinther, with Prum and two of their Yale colleagues, identified melanosomes (tiny organelles that contain melanin) in fossil feathers from the Lower Cretaceous (100–65 MYA) of Brazil and the Early Eocene (56–49 MYA) of Denmark. They were thus able to show that those feathers were colored with black and white stripes. Indeed, they concluded that most fossil feathers are actually preserved in such a way that it might be possible to determine the colors of extinct birds and feathered dinosaurs. Then, in 2009, Vinther and colleagues applied their technique to fossil feathers from the Middle Eocene (49–37 MYA) of Germany, discovering this time organelles in ordered arrays inside the feather barbs that would have given the feathers an iridescent blue sheen. All well and good, but these were single feathers—what did the whole feathered bird look like? It did not take long to find out, but the first answers came not from a bird but from a feathered theropod dinosaur. Taking twenty-nine samples from different regions of the body of a 155-million-year-old *Anchiornis* fossil, Vinther and colleagues were able to paint a picture of the entire animal, comparing fossil melanosome structures with those of living birds whose colors they could observe and measure directly (Li et al. 2010). It won't be long before we have a full-color field guide to at least a few fossil birds.

THE ORIGIN OF FLIGHT IN BIRDS

Baron Nopcsa was the first person to write a paper focusing explicitly on the origins of flight in birds. While others had mentioned in passing some ideas about how flight may have evolved, Nopcsa's papers, "Ideas on the Origin of Flight" (1907) and "On the Origin of Flight in Birds" (1923), were among the first to deal with the issue head on, and in a characteristic comprehensive, thoughtful, and highly original way. At the end of the nineteenth century, an arboreal origin of flying seemed logical and was the only idea that anyone really considered (e.g., Hurst 1895, Pycraft 1894), on the assumption that the ancestors of birds must have lived in trees.

O. C. Marsh was the first to clearly articulate this idea in 1880: "In the early arboreal birds, which jumped from branch to branch, even rudimentary feathers on the fore limbs would be an advantage, as they would tend to lengthen a downward leap, or break the force of a fall."[65] Later, Pycraft (1894) and Heilmann (1916, 1926) picked up on this theme, arguing that the long, clawed fingers on the forelimbs of *Archaeopteryx* were clearly adapted for tree climbing. Their books and papers thus established a scientific basis for arboreal theories for the origins of flight that dominated thinking on this subject until the 1970s.[66]

Nopcsa—and Williston (1879) before him—took a completely different approach, arguing that birds originated from a bipedal dinosaur whose forelimbs became modified into wings that provided an advantage in running.[67] In Nopcsa's view, protobirds became bipedal first and fliers second, and his ideas gained support from *Archaeopteryx*, which was clearly bipedal. The apparent flaw in Nopcsa's argument was that wings seemed to provide no assistance to a running animal. William Pycraft (1910) certainly supported the arboreal theory, as did William Beebe (1915), who proposed his own four-winged model of what an early bird might have looked like, an idea that turned out to be quite prophetic ninety years later.[68] By the 1920s this cursorial model had fallen by the wayside, and Heilmann (1926) barely mentioned it in his book. Almost sixty years after Nopcsa had presented his cursorial model, John Ostrom (1974) resurrected it with a different functional argument for the first feathered wings.

Ostrom, you will recall, revolutionized the study of bird origins with his evidence that birds are descendants of theropod dinosaurs. Then, in 1974, he used his knowledge of theropod dinosaur anatomy to devise a new cursorial theory wherein protobirds used their forelimbs to capture prey. He argued that the lengthening of feathers in the wings would form two large fans that could be clapped together to catch insects or pin prey to the ground. Ostrom's intriguing "predation hypothesis" did not survive the scrutiny of its many critics, but it opened the door to a cursorial origin of flight, by showing that feathers and bipedalism undoubtedly preceded the first birds and by suggesting that *Archaeopteryx* was as likely to be a cursorial runner as a tree climber.

Possibly the most interesting and provocative idea about early birds being cursorial is the wing-assisted incline running (WAIR) hypothesis formulated by Ken Dial (2003). Dial is a professor of biology at the University of Montana–Missoula, where he has spent his academic career studying the flight performance of birds. His new idea is

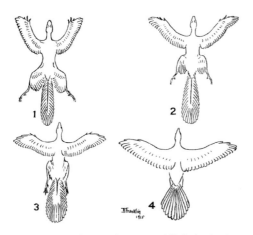

William Beebe's ideas on the origin of flight beginning with an unknown tetrapteryx, or four-winged stage (1), progressing through *Archaeopteryx* (2) and some unknown intermediate with a long bony tail (3), to the modern bird (4).

that feathers could have assisted protobirds in escaping from predators even if they could not fly. The WAIR hypothesis thus provides a distinct advantage for even rudimentary feathering—half a wing—long before birds could actually fly or even glide. It is also unique among ideas about the origins of avian flight in that it can be readily tested on living birds, and test it Dial has, in his lab, with birds as diverse as partridges and pigeons, confirming that feathers can help a bird escape even if that bird cannot fly.

Dial and his son Terry first noticed this flap-running behavior of birds when filming flightless Chukar Partridge chicks navigating obstacles. As a result, Dial decided in the late 1990s to turn his attention to the origins of flight in an experimental fashion rather than just theorizing, as most others had done. He chose chukars because, like theropod dinosaurs, they are mainly bipedal runners, starting to run just after hatching and not flying properly until they are two months or more old.[69] Thus, with chukars, he could examine in the lab the entire process of anatomical development involved in getting proficiently airborne. Instead of running around objects in their escape path, his chicks ran up and over them, flapping their stubby wings. Even when they could fly quite well, the birds still usually ran up the vertical sides of bales of hay to get back to their nests or to rejoin their siblings. Dial was astonished and set up high-speed cameras to record exactly how the birds used their wings to achieve this previously undocumented form of locomotion. The birds gained traction when running up inclines by changing their wingstroke from the typical dorsoventral stroke used by flying birds to one oriented in a more head-to-tail (anteroposterior) plane. Dial likens the effect to that of a spoiler on a racing car pushing the rear wheels to the road to increase traction.[70]

Like Prum's model of feather evolution, the WAIR model has a developmental parallel: chicks of many species use WAIR before they can fly, their rudimentary feathers giving them stability and increasing their traction on inclined surfaces. Even the most rudimentary feathering gives energetic and traction advantages during WAIR, and as in early birds, the growing flight feathers of chukars are initially symmetrical. In 1871 St. George Jackson Mivart challenged Darwin by asking how part of a wing or an eye could be useful,[71] questioning the role of intermediate forms in the evolution of traits that have obvious current utility. The WAIR model provides a plausible answer about the utility of a wing on a bird that cannot fly, amply supported by the recent Chinese fossils of the dinosaurs *Sinosauropteryx*, *Caudipteryx*, and *Protarchaeopteryx*, which have feathered forelimbs but could not possibly have accomplished powered flight. The WAIR hypothesis also counters early objections to Nopcsa's ideas about the advantages of wings for running. Like virtually all hypotheses about the origins of birds and flight, the WAIR model has been heavily criticized—but WAIR combines elements of both cursorial and arboreal models. If correct, the WAIR model suggests that the ground-up versus trees-down debate was based on a false dichotomy that, like so many other debates in biology, might potentially be resolved by a modal theme (Stebbins 1982) that combines the useful elements of both hypotheses.

Since Ostrom, there have been several variations on the cursorial model for avian flight origins. Luis Chiappe, now at the Natural History Museum of Los Angeles,

suggested a twist on the idea proposed by Nopcsa whereby feathered forelimbs aid in running, not so much to increase running speed as Nopcsa proposed, but to provide balance and the ability to make rapid turns without falling (see Burgers and Chiappe 1999). With the evolution of longer forelimb feathers, gliding and even takeoff would have been possible. "Even *Archaeopteryx*, which is often cast as a poor flier, could have taken off from the ground,"[72] says Chiappe. In a similar vein, John Videler (2006) of Leiden University suggested the "Jesus Christ dinosaur" model of flight origins, whereby protobirds may have gained advantages for both escape and foraging by running over the surface of water rather than land. Videler supported his ideas with a detailed analysis of the flight mechanics of *Archaeopteryx* and the observation that this early bird, at least, seems to have been associated with wetlands.

Our focus on cursorial models here may give the impression that the arboreal model was shown to be implausible as a result of Ostrom's insights, but the debate continues to this day. Using evidence from other gliding animals and aerodynamic models, several scientists have argued that powered avian flight is most likely to have evolved from adaptations for leaping, then parachuting, gliding, and flapping from tree to tree (Geist and Feduccia 2000). Interestingly, however, no living bird flies solely by gliding, suggesting either that this is not a logical transitional phase in the evolution of flapping flight, or that it is an unstable intermediate stage. As some have argued, "No modern gliding animal can even approach the conditions necessary for flapping flight. It is highly unlikely that an ancestral glider could do likewise."[73]

Our best hope for a resolution to the debates about the origin of flight in birds is to adopt a multidisciplinary approach incorporating paleontological evidence, phylogenetic analysis, developmental biology and genetics, aerodynamic theory, and what we know about flight energetics, combined with the sorts of experiments with living birds that Colin Pennycuick, Vance Tucker, and Ken Dial have pioneered. Multidisciplinarity is a "buzzword" of twenty-first-century biology, often offered as a vague panacea to solve complex problems. But few problems in ornithology are as complex as bird flight itself, and we are still struggling to fully understand how the avian flying machine actually works.

FLYING MACHINES

When Otto Lilienthal was a boy, in the 1850s, he wanted to fly, so he turned to birds for inspiration, design, and technique. He was determined to fly, or to die trying, and he did both. Growing up in Anklam in what is now northeastern Germany,[74] Otto and his brother, Gustav, made a variety of wings that could be strapped onto their backs but never got airborne. Needing to know more about mechanics, Otto trained to be a professional design engineer and spent his working life designing and building boilers and steam engines. His passion, though, was birds and flight, inspired—obsessed, even—by the White Storks that nested each spring in his home village. Convinced that these birds nested near humans because God wanted the storks to show humans how to fly, Lilienthal studied their flight mechanisms in detail and designed some novel instruments to measure lift forces on wings of various shapes and sizes. He was also an unrepentant

experimenter on live birds, tying the wing feathers of pigeons in various configurations to see if the pigeons' flight performance would be affected. Surprisingly, he concluded that this drastic reduction in wing area caused very little impairment.

Lilienthal's (1889) book *Der Vogelflug als Grundlage der Fliegekunst* [Birdflight as the Basis of Aviation] is a classic, revising ideas about avian flight mechanics originally outlined by Giovanni Alphonso Borelli[75] in 1680—ideas that had remained largely unchallenged for two hundred years. Using what he learned about aerodynamics from studying birds, Lilienthal began in 1880 to build a variety of what we would today call "hang gliders"—wide, flat wings attached to and controlled by his shoulders while his body hung below. With these early flying machines, he achieved more than two thousand flights from both a conical hill he had constructed for this purpose[76] and some natural hills nearby, getting more than 20 meters (65.5 feet) off the ground, gliding for up to 250 meters (820 feet), and becoming a worldwide sensation in the process. On 9 August 1886, however, at an altitude of about 17 meters (56 feet), his glider stalled: the fall broke his spine and killed him.

Possibly inspired by the promise of manned flight, as well as the interest in bird and flight origins, the late 1800s saw a flurry of interest in avian flight mechanics by ornithologists as well as engineers. Étienne-Jules Marey, for example, was a polymath French scientist and inventor who turned his attention to the study of flight—first in insects, then in birds—in the 1860s,[77] focusing particularly on the interactions between wings and air to determine how animals could move forward aloft. Living and working in

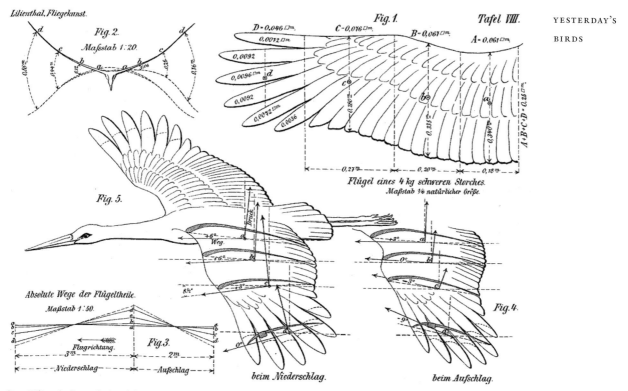

Otto Lilienthal's analysis of the aerodynamically important dimensions of storks.

Paris, Marey built a series of physiological laboratories with outdoor tracks where he could study birds and flying machines and photograph them with his mobile cameras. There, in the 1870s, he helped his friend Victor Tatin design and build one of the first powered model aeroplanes—called "ornithopters" or *oiseaux mechaniques*—based on his studies of animal flight.

In his 1873 book, *La Machine Animale* [Animal Mechanism], Marey describes an apparatus he designed to trace the wingtips of a flying bird (or insect) onto smoked paper. The results corrected some errors in Borelli's work and provided surprising new details about the movement of bird wings and bodies during flight. The book reviewer in *Nature* was clearly impressed, if understated:

"The way in which the author invents means for reproducing and originating any quality of movement he may want to develop, must be a source of admiration and almost astonishment to all readers of his work."[78] Then, in 1882, Marey invented his famous *chronotographe*, or chronophotographic gun, able to record twelve images per second on the same photographic plate.[79] His first photographs with the chronotographe were of a "sea-gull" and were published in an article he wrote for *La Nature* in 1882. Though today Marey's ingenuity and early photographs seem remarkable, this time the reviewer in *Nature* was less impressed with the results: "The photographic process . . . did not seem to add much to our knowledge of the mechanism of flying."[80] Marey, however, was just getting

29

going and went on to film[81] the flights of pigeons, herons, pelicans, ducks, and more gulls. He had some of these photographic images translated into a series of plaster and bronze models so he could examine—and show off—the results of his photography in three-dimensional detail. The results were published in *Le Vol des Oiseaux* [The Flight of Birds] (Marey 1890), a landmark in the annals of both bird flight and photography.

In 1899 Marey began a new venture, building wind tunnels and smoke machines to figure out "how the air behaves as it provides support to the wing."[82] His first smoke machine had twenty smoke trails that would interact with objects in their path, but the second, built in 1901, had fifty-eight and produced the most amazing images, clearly showing the eddies and vortices produced by wings. It would be another seventy-five years before these vortices were studied by those interested in avian flight, using what they learned to build sophisticated models of avian aerodynamics (Rayner 1979).

All of this interest in bird flight machinery came abruptly to a halt with the Wright brothers' triumph of powered flight in 1903. In fact, their first airplane looked nothing like a bird, and its design owed relatively little to the study of avian flight mechanics; it was a biplane with four wings, not two, and two tails; its wings were flat sheets of muslin, not a cambered airfoil, and were fixed, not flapping; and the whole thing was driven by propellers. As Marey had presciently noted, "The most perfect examples of locomotion which man has achieved are in general obtained by methods quite different from those of nature."[83] While the Wrights were inspired by—and acknowledged—the work and success of Lilienthal, their approach was one of

trial and error, working with existing materials to refine their designs, an approach that would be used by aircraft designers for some time to come rather than turning to the study of bird flight. Instead, aircraft design informed the study of bird flight mechanics —rather than the other way around—for the next eighty years.

From the Wright brothers until the end of the Second World War, talented engineers and scientists interested in aerodynamics focused on aircraft design, and relatively little attention was given to understanding how birds fly. In Lucien Warner's (1931) review, "Facts and Theories of Bird Flight," he had little to say, mainly quoting work done in the nineteenth century. Even in 1960, August Raspet, in a paper on the biophysics of bird flight, concluded that "we probably can sum up the state of our present knowledge by saying that we know very little."[84] As an aerophysicist and head of the Aeronautics Department at Mississippi State College, Raspet knew his stuff. He also had a lifelong interest in birds and observed wild birds to get ideas for his aeronautics research. In the late 1940s he flew his glider 5 to 10 meters (16 to 33 feet) behind soaring Turkey Vultures to photograph the birds and record airspeed and altitude so that he could estimate the power required to remain aloft. Raspet knew that measurements in wind tunnels were poor predictors of flight performance in the wild and thought his method of following birds in the air, which he called "comparison flight studies," was the way forward. He further recognized that mathematical aerodynamic models, based on both clay birds and airplanes, must be inaccurate for birds because the porosity of feathers would offer some sort of boundary layer control.

Just three months before that paper was published, Raspet died when his Piper Cub crashed while he was trying to demonstrate the influence of such boundary layer modifications on airplane flight.

Not long after Raspet wrote his final paper, the confluence of innovation, ideas, technologies, and analytical tools that characterized the 1960s began to completely revolutionize the long-dormant topic of bird flight mechanics and energetics. By 1960 aircraft design was a mature subject—the Douglas DC3 and the Boeing 707, two of the most successful commercial planes ever built, were already in service, and the advantages and aerodynamics of delta-winged craft had already been discovered and described. Thus, the aerodynamics of fixed wings had been thoroughly explored by engineers and was readily accessible to interested biologists. Taking the lead, Vance Tucker and Colin Pennycuick, two young academics on either side of the Atlantic,[85] turned their attention to bird flight, their backgrounds defining their different approaches. Tucker was a physiologist who had previously studied pocket mice, frogs, and lizards, measuring oxygen consumption, torpidity, and osmotic regulation. Pennycuick, on the other hand, was a bird enthusiast and a keen aviator interested in engineering and computers. Their approaches defined the two methods scientists would use to study and argue about bird flight for the next forty years.

Based on an idea proposed, but never executed, by Raspet (1950), Pennycuick used a wind tunnel, which he designed and built out of plywood, to investigate the mechanics of both steady-state gliding and horizontal flapping flight of pigeons in the 1960s. The tunnel was large enough to provide airflow past a flying pigeon with its wings fully extended but small enough to fit into a stairway, the only available space at the university. Hung from an overhead beam, the wind tunnel could be tilted on its long axis to provide airflow at different angles of attack to the horizontally flying bird. The bird being tested actually flew outside the tunnel itself, at the end opposite the motor and fan that provided the airflow. Pennycuick trained the birds to fly in the airstream by providing food on a teaspoon. Initially the birds could sit on a perch beside the spoon, but as he increased the airspeed, the perch was gradually moved away from the spoon until the bird had to fly to obtain food: "Pigeons seemed to find feeding on the wing somewhat unnatural, and at first "pedalled" vigorously with their feet whenever they took food from the spoon."[86]

So that he could calculate lift and drag at different airspeeds flown by the birds, Pennycuick also photographed the birds while measuring various features of their flight and the windspeed. Then, applying well-established aerodynamic principles based on "helicopter theory," he calculated the power required for flight.[87] His calculations revealed a U-shaped power curve with two local optima, the speeds at which the bird minimizes the cost of traveling per unit distance and per unit time. His model thus allowed him to predict just how fast a bird might fly in different situations, as well as the rate of fuel consumption and the distance that could be traveled by migratory birds if their fuel load was known. Using this model and a scaling law, he determined that about 12 kilograms (26 pounds) was probably the upper limit to body size for a vertebrate that could fly under its own power. The results of these measurements and calculations were published back

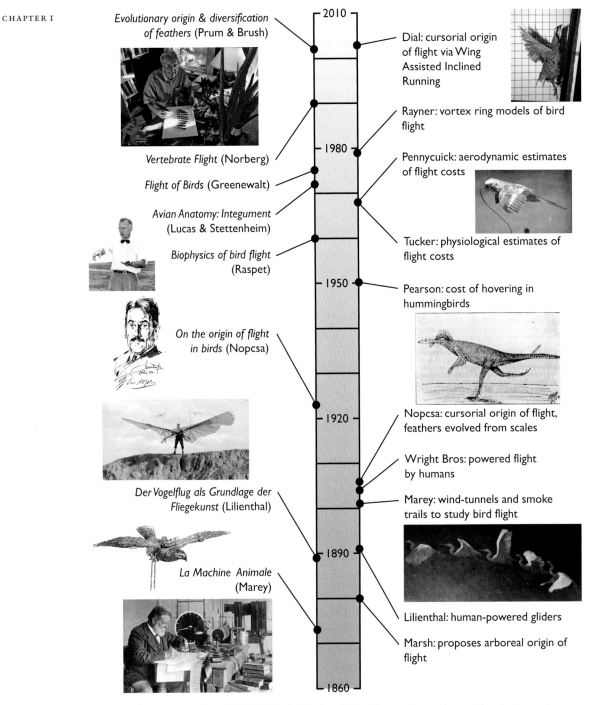

Evolutionary origin & diversification of feathers (Prum & Brush)

Dial: cursorial origin of flight via Wing Assisted Inclined Running

Rayner: vortex ring models of bird flight

Vertebrate Flight (Norberg)

Flight of Birds (Greenewalt)

Pennycuick: aerodynamic estimates of flight costs

Avian Anatomy: Integument (Lucas & Stettenheim)

Tucker: physiological estimates of flight costs

Biophysics of bird flight (Raspet)

Pearson: cost of hovering in hummingbirds

On the origin of flight in birds (Nopcsa)

Nopcsa: cursorial origin of flight, feathers evolved from scales

Wright Bros: powered flight by humans

Der Vogelflug als Grundlage der Fliegekunst (Lilienthal)

Marey: wind-tunnels and smoke trails to study bird flight

La Machine Animale (Marey)

Lilienthal: human-powered gliders

Marsh: proposes arboreal origin of flight

2010 1980 1950 1920 1890 1860

TIMELINE for FEATHERS and FLIGHT. Left: Alan Brush; Gus Raspet; Baron Nopcsa (drawing in 1926 at age ca. 49); Otto Lilienthal and one of his hang gliders; a pigeon in an apparatus made by Étienne-Jules Marey to measure the action of its wings; Marey with some of the apparati he invented. Right: Chukar running up an incline in Ken Dial's lab; Budgerigar fitted with a mask by Vance Tucker so that oxygen consumption could be measured; Nopcsa's conception of a proavis; Marey's stop action photograph of a heron taking off (photo ca. 1883).

to back in two landmark papers in the *Journal of Experimental Biology* (Pennycuick 1968a, b).

Vance Tucker had published the results of his own wind tunnel studies of avian flight energetics earlier that year in the same journal, using a completely different approach but coming to essentially the same conclusions. As a physiologist, Tucker sought to measure a bird's energy expenditure more directly by determining how much oxygen a flying bird—this time a Budgerigar—consumed, rather than estimating the costs of flight from aerodynamic principles and measurements, as Pennycuick had done. Two years earlier Tucker (1966) had published, in *Science*, the results of his first measurements from these birds, but by 1968 he had refined his technique to reduce measurement error and to measure birds in horizontal, ascending, and descending flight. For the 1966 *Science* paper he extracted air from the wind tunnel itself to determine the rate that the flying bird used oxygen; for the 1968 study he fixed a lightweight—less than 1.5 grams (0.05 ounces)—mask to the bird's bill so that he could measure the oxygen consumed and the carbon dioxide exhaled, more or less in real time. In both sets of experiments he used two Budgerigars obtained from a local pet shop, training them to fly inside a small wind tunnel[88] lined with an electrified grid. The birds' feet were painted with a calcium chloride solution to improve electrical conduction so that the birds received a mild shock whenever they tried to land on the grid—a technique that seems startlingly inhumane today. The birds soon learned to stay aloft for twenty minutes or more at airspeeds from 19 to 48 kilometers (12 to 30 miles) per hour. The results were clear and striking, qualitatively

matching Pennycuick's independently measured and calculated U-shaped curves from pigeons.

Tucker went on to measure the oxygen consumption of Laughing Gulls, hawks, and parakeets in his wind tunnels, and he spent the rest of his career studying and writing about bird flight. Though his results from other species showed that his conclusions were general with respect to rates of energy expenditure during normal flight, only those Budgerigars exhibited the U-shaped curve, leading researchers to question its generality for the remainder of the century. Pennycuick himself eventually expressed confidence in being able to estimate the costs of flight only over a narrow range of speeds around the average flight speed of a bird. Then, in 2003, Ken Dial's lab measured muscle lengths, electromyographic activity, and bone strain directly from Cockatiels and ringed turtle doves (African Collared Doves) flying in wind tunnels and found strong support for the Tucker-Pennycuick U-shaped curves (Tobalske et al. 2003).

Before Tucker, direct measures of oxygen consumption during flight had been accomplished only for hovering hummingbirds, so Tucker's methods, applied to forward flight, were groundbreaking. The previous work had been done by Oliver Pearson (1950), who trained Anna's and Allen's Hummingbirds to hover in a sealed bell jar so he could measure their oxygen consumption. The results revealed that hovering was energetically costly for the birds—about six times the cost of just sitting at rest—but provided no insight into the costs of forward flight.

Also studying hummingbirds, Crawford Greenewalt photographed and measured a wide variety of species and used the comparative method and some aerodynamic equations to estimate both hovering and flight costs from a simple set of morphological measurements. Greenewalt was an amateur ornithologist who had an interest in aerodynamics and engineering and—as CEO of chemical company DuPont—great resources at his beck and call. By 1950 he had become a wealthy man with lots of energy and a passion for birds. Photographing nuthatches and chickadees taking off from a feeder at the window of his home in Wilmington, Delaware, he became intrigued by the possibility of using photography to determine their rates of acceleration. He had his engineers at DuPont design and build[89] an apparatus that, triggered by the bird, set off a high-speed flash that took pictures at intervals of thirty milliseconds. The results, published in *The Auk* (Greenewalt 1955), demonstrated how photography could help understand bird flight mechanics, much as Marey had done sixty-five years earlier but with greater precision and shorter intervals between photographs.

Continuing to develop his equipment and photographic prowess, Greenewalt next turned his attention to photographing hummingbird species "frozen" in flight, using high-speed flash equipment that was better than anything commercially available. In 1960 he produced both a coffee table book, *The Hummingbirds*, and some amazing photographs for *National Geographic* magazine.

Then, in 1962, using published measurements and analyses from a wide variety of flying animals, Greenewalt showed that there were different kinds of aerodynamic structures that animals used for flying, almost certainly related to their different lifestyles. He developed that theme more

comprehensively in *The Flight of Birds*, focusing just on birds and showing that there were at least three different allometric models for flight energetics, which he called the "duck, shorebird, and passeriform models" (Greenewalt 1975). Taking a different approach from Tucker and Pennycuick, he also proposed that flight costs could be estimated by measuring body mass, wing span, and wing area. Tucker later demonstrated that Greenewalt's estimates were probably too low, compared to the similar predictions made by Pennycuick and himself, but his approach was thoughtful, and his 1962 monograph pioneered the application of comparative methods to the study of avian behavior, ecology, and energetics.

Melding the approaches of Pennycuick, Tucker, and to some extent, Greenewalt, behavioral ecologists began to incorporate estimates of flight costs into models and tests of optimal foraging theory, which began to be developed in the 1970s (chapter 8). Larry Wolf (an ecologist) and Reed Hainsworth (a

Green Violetear.

physiologist), both at Syracuse University, and Peter Feinsinger (an ecologist) exemplified this approach in their many excellent studies of tropical hummingbirds, analyzing foraging strategies from an energetics perspective. Feinsinger and colleagues (1979), for example, noticed in Costa Rica that the Green Violetear used different foraging tactics at different altitudes. At relatively low altitudes it was largely sedentary and defended territories, but at higher elevations it foraged on dispersed flowers and rarely defended territories and therefore required less energy for hovering and maneuvering to chase off intruders. In a flash of insight, the researchers applied the Pennycuick-Tucker equations, incorporating air densities at different altitudes, and showed that flying to defend territories would just be too costly at the higher altitudes. Moreover, they showed that the costs of flight correctly predicted an altitudinal change in wing morphology across thirty-eight Peruvian hummingbird species, with those at higher altitudes having lower wing disc loading. In addition to their influence on studies of foraging behavior, Pennycuick's ideas about flight range and the models of flight costs led to the development by Thomas Alerstam of optimal migration theory (chapter 4), wherein clear predictions can be made about the altitudes, distances, and timing of migratory flights.

While it would have been very tempting to continue to test and refine the Tucker-Pennycuick approach to modeling the flight of birds, scientists, to their credit, are rarely satisfied with the status quo, especially when there is a disconnect between theory and empirical evidence, as there appeared to be regarding bird flight in the 1970s. Taking a completely different approach, and a leaf from Marey's book, Jeremy Rayner (1979), then at Bristol University, began to look seriously at how the airflow vortices over wings might be modeled and used to predict avian flight mechanics and costs. Others had already shown that the flapping flight of relatively simple wings generated vortices that provided lift, so why not birds, whose wings and wing beats were decidedly more complex? In an elegant series of studies of birds in wind tunnels, and a careful analysis of these trailing vortices, Rayner was able to construct models that captured some of that complexity. Harkening back to what Lilienthal had pointed out almost a century before, birds are very different from man-made flying machines—more complex, more energetically efficient, and more maneuverable. Despite Rayner's success, ornithologists have generally found the vortex-ring models of bird flight too complex for application in their studies and have largely, to this day, relied on Pennycuick's models, equations, and computer programs to calculate the costs of flight, realizing that they are sometimes only approximately correct.

CODA

Early in the twentieth century ornithologists felt that the origins of birds and feathers, as well as the mechanics and evolution of flight, were well understood. Considering what we know today about how wrong they were, it seems remarkable that almost nobody paid attention to these interesting problems for more than sixty years. Instead, ornithologists focused mainly on systematics, adaptive radiation, behavior, and migration, but even in the past decade the study of bird origins, feathers, and flight has seen a stunning renaissance of interest and discovery, fueled by new technologies, new fossils, and new insights. Since the year 2000, for example, one-quarter of the papers published in *Science* and *Nature* have been on the subjects covered in this chapter.[90]

Indeed, we have just recently entered a new era in the study of bird origins, with the incredible explosion of new fossil finds. In one of the most recent of these, published in *Science* on 16 September 2011, a team led by Phil Currie reported on the discovery of eleven feathers in amber from the Late Cretaceous, 70–80 MYA (McKellar et al. 2011). These specimens, gleaned from the detritus of a coal-mining operation in southern Alberta, Canada, include examples of simple unbranched filaments (Prum's Stage I feathers) and modern-looking barbed feathers (Stage V), with pigments clearly visible in both. Since these feathers were not associated with skeletons, we cannot know whether they belonged to one of the fifteen or more genera of nonbird dinosaurs that we now know had feathers or to an ancient bird. Most important, though, they add yet another piece of evidence in support of Prum's hypothesis about feather origins.

Even *Archaeopteryx* has once again proven controversial. In July 2011 Xu Xing of the IVPP in Beijing reported on a new feathered fossil dinosaur, *Xiaotingia zhengi*, that, when included in a phylogenetic analysis, appeared to challenge the long-held notion that *Archaeopteryx* was an ancient bird (Xu et al. 2011). Instead, Xu and his colleagues suggested that both *Archaeopteryx* and *Xiaotingia* are closely related on a sister clade to birds that includes *Velociraptor*. It did not take long for the Australian paleontologists Michael Lee and Trevor Worthy to reanalyze these new

Feathers in amber found at Grassy Lake, Alberta, dated to the Late Cretaceous (70 to 85 million years ago). These pennaceous barbules have all of the structures needed to form vaned feathers with mottled pigmentation. They were probably borne by a bird or small theropod dinosaur capable of flight (McKellar et al. 2011). Black scale bar is 0.2 mm long.

data and publish a contrary view in 2012, suggesting that *Archaeopteryx* is an ancient bird after all. As we have seen in this chapter, controversy makes the study of bird, flight, and feather origins both fascinating and frustrating, particularly because new fossil discoveries make new kinds of information available to study.

The other topic of this chapter—the mechanics of flight—has also seen major developments during the twentieth century, mainly due to advances in technology. Thus developments in technologies from photography, aeronautics, and physiology to computers, wind tunnels, and electromyographs have dramatically influenced how we study bird flight. The consensus now is that bird flight is much more complex than we once thought, but the new tools hold some promise for unraveling that complexity.

BOX 1.1 Colin Pennycuick

When I first took an interest in flight, I wanted my toy airplanes to fly and let me see how they worked. I never did see the point of a nonflying model. I was a keen bird watcher at school and began to see how birds' bodies work when I was a zoology undergraduate at Oxford. I acquired the rudiments of flight theory when I learned to fly Chipmunks with the University Air Squadron, and I realized that there must be a basis of theory that applies to anything that flies. I have been following that idea up ever since, starting with pilot training on Provosts and Vampires in the RAF, when I did my two years' National Service. My PhD at Cambridge was about the properties of muscles (which I regarded as engines) under John Pringle, who was also president of the gliding club. By the time I finished I had learned to design and build equipment from military surplus scrap, and I was also a gliding instructor, proficient at planning and carrying out cross-country flights, either with an engine or without. As a postdoc in the animal behavior lab at Madingley, I trained a team of pigeons to home to a mobile loft, in an unsuccessful attempt to find out how they navigate.

As a zoology lecturer at Bristol from 1964 to 1968, I learned how to program the university's first computer, and I used it to design and build a small tilting wind tunnel, calculating the shapes of the plywood panels before I marked them out, cut them, and glued them together. I installed a (static) pigeon loft on the roof and trained pigeons to fly in the tunnel, before releasing them to find their own way home. The flight theory that I had learned in pilot training clearly also applied to birds, and I used the computer to adapt helicopter theory into a basic theory of level flapping flight that described what my pigeons did in the wind tunnel. I have been following this idea up ever since, integrating theory with ground and satellite tracking and observations of birds from aircraft. I got two winter trips to the Serengeti, where Hugh Lamprey, the first director of the research institute there, had acquired a primitive glider. He allowed me to fly it, and thus I got my first good look at storks and vultures by flying alongside them and doing what they did.

I got myself seconded from Bristol to Nairobi University in 1968 and spent three years there, and two more in the Serengeti. I dismantled the wind tunnel and installed it in Nairobi, where I trained a fruit bat to fly in it and made 3D contour maps of its wings. I bought my first airplane, a 1946 Piper Cruiser, in 1968 for £400 and later replaced it with another Cruiser with a bigger engine. It was a terrific way to get around East Africa, besides being in constant demand for projects involving counting and radio-tracking creatures from flamingos to elephants, through which I was able to cover my costs. Two TV companies, Anglia and Okapia, provided me with a Schleicher ASK-14 motor glider, which I based first at Magadi in the Rift Valley, and later at the Serengeti Research Institute, and used for studying vultures, storks, pelicans, and other soaring birds. At the end of my time in Africa, Hugh Lamprey bought the motor glider, while I installed a long-range tank on the back seat of the Cruiser, and flew it back to Bristol in nine stages, via Addis Ababa, Cairo, and Crete.

For the next ten years, 1973 to 1983, the Cruiser did sterling work towing gliders at the Bristol Gliding Club, and I used it for trips to Shetland and France, as well as for studying cranes with Thomas Alerstam as they migrated into Sweden in the spring of 1978. In the southern summer of 1979–80 I got a trip with John Croxall and the British Antarctic Survey to Bird Island, South Georgia, where I developed an optical device, the ornithodolite, for measuring 3D tracks and speeds of albatrosses and other southern seabirds. This was controlled by a Nascom 1 computer, which I built from a kit and installed in an attaché case, programming its 1K of memory directly in machine language, one byte at a time.

In 1983 I sold the Cruiser (and the folding bike that I carried in the back) and visited the frigatebird colony at Barbuda on my way to take up the Maytag Chair of Ornithology at Miami. Before long I acquired a Cessna 182 jointly with Kathleen Sullivan, whose interests in the Bahamas and Puerto Rico were in the same locations as mine, but below the surface (coral reefs) while mine were above (seabirds). I also had a DG-400 motor glider for studying soaring birds over the Everglades and at other locations in Tennessee, Pennsylvania, and Idaho. I upgraded the ornithodolite with a more capable computer and used it to measure bird speeds from a boat in Florida Bay and in various locations in Florida and the Northeast. Many of these projects were in collaboration with Mark Fuller, who encouraged me to develop software that could apply aeronautical theory to ornithological problems.

When I left Miami in 1992 I got in the Cessna, headed northeast to West Greenland, and spent a couple of months there radio-tracking breeding peregrines with Bill Seegar. Then I flew up and over the ice cap to Iceland, following a busy migration route for geese. Back in Bristol, I kept the Cessna for another year, during which I used it to observe migrating swans in Sweden. I had a hand in the development of the Lund wind tunnel, which was the first low-turbulence wind tunnel designed specifically for bird flight experiments, and I was there when the king of Sweden inaugurated it in 1994. In the late 1990s I joined forces with the Wildlife and Wetlands Trust in a three-year project satellite-tracking migrating Whooper Swans, whose large size makes them ideal for checking the upper limits of flight by muscle power.

An ornithodolite project at Falsterbo in Sweden in 2000 demonstrated that both the speeds and wing-beat frequencies of birds flying steadily along were well predicted by my software, which was available by then on the internet as a QBasic program. Beginning in 2002, the biennial migration course at Lund has provided a continuing stimulus for developing the *Flight* program for Windows, which now does performance calculations for both flapping and gliding flight and numerical simulations of long-distance migration. Julian Hector of the BBC Natural History Unit brought me into his *World on the Move* radio series in 2008, to track the fuel state of migrating geese with *Flight*'s migration simulations. It remains to be seen whether the published account of this project will encourage ornithologists to venture beyond plotting lines on maps and doing theory-free statistics. The program is published for anyone to use, and I am currently (2012) converting it into a web application.

BOX 1.2 Richard Prum

I started bird watching as a ten-year-old boy living in rural southern Vermont. In retrospect, I was an amorphously nerdy, science-oriented, and myopic kid without any clear direction. Before birds, I was passionately engaged in memorizing unusual facts from the *Guinness Book of World Records*. I was particularly interested in the extremes of human form and performance, such as the heaviest, tallest, and shortest people, and the greatest number of whelks eaten in fifteen minutes.

Then, in the middle of fourth grade, I got my first pair of glasses. Within six months of the world coming into focus, I was a bird watcher. I never looked back or ever considered any other possible option in my life.

The first moment of ornithological inspiration that I remember was seeing a copy of Peterson's *Field Guide to the Birds of Eastern North America*. I was captivated by the birds in the plates and especially by the Atlantic Puffin on the cover. I realized immediately that the purpose of the book was to get you to *see* these birds for yourself, in the wild. I expressed interest in the book to my mother. My curiosity was duly noted but not immediately satisfied; she

said, "Hmmm, you have a birthday coming up. We'll see." My mother did her own comparative study of available guides and decided that the Robbins Golden Guide, with its then revolutionary organization of text and maps opposite the plates, was a better fit for a ten-year-old.

Starting with my family's pair of old binoculars, I soon discovered the challenges and thrills of field identification, the insatiable desire to see new birds, and the great joy finally seeing a bird you had long hoped to find. My elementary years became an elaborate obsession with the Robbins guide and all the possibilities it proposed to my imagination. Travel, adventure, and a very long life list became the central organizing principles of my life. I regretted that I hadn't started birding when I was eight and my family lived in Massachusetts near Newburyport and Plum Island. I regretted, as well, that I had not been a birder at the age of six, when my family had visited El Yunque rain forest when my Dad was stationed in Puerto Rico with the US Navy.

For my next birthday, I received my first bird song record. It was a dramatized synopsis of "one day in an eastern deciduous woodland" from the Cornell Laboratory of Ornithology. It had one side with narration and the other side without, so it was perfectly designed for quizzing yourself to learn the songs. I soon graduated to the exhaustive two-LP set of the Eastern Peterson. I practically wore the grooves off those records on my family's compact hi-fi. Before I was out of elementary school, I had learned the bird songs of northern New England cold.

Of course, I had lots of adult bird-watching friends who encouraged and helped me as a kid. Some of them were garden club ladies—like "Bird" McCormack, Doris Dolt, Mimi

Crakoff, and Ruth Stewart. They had cars and I did not, so I got to see much of the state through them. An important early mentor was Tom Will, then a natural historian birder from Peru, Vermont, and now a regional wild-life biologist for the US Fish and Wildlife Service in Minnesota. Fifteen years older than I, Tom always treated me like a colleague. We did some pioneering southern Vermont hawk watching on Mt. Equinox, recording an astounding (to me!) number of sixty-six Broad-winged Hawks in a single day. I skipped school with Tom one May day to chase and find a Lawrence's warbler (hybrid between Golden- and Blue-winged Warblers) nearby. Tom was the first to take me birding by the ocean at Plum Island Wildlife Refuge, in Newburyport, Massachusetts, in November 1973, where I had dozens of lifers in a single day including Snowy Owl, Glaucous Gull, and all three scoters, among others. Years later, Tom and I would end up as colleagues in the PhD program at the University of Michigan.

Through Tom Will, I met other internationally renowned birders, including Frank Oatman from Craftsbury, Vermont. Their influence led me to discover the birds of the rest of the world, beyond the pages of the Golden and Peterson guides. I soon began to plan entirely imaginary birding trips to far-off places. The combination of sea eagles, skuas, Harlequin Ducks, and geothermal energy (it was the 1970s!) made Iceland an early travel obsession. But soon the Neotropics would take over my imagination.

By the time I arrived at college at Harvard in fall 1979, I was pretty sure I was going to be an ornithologist. What is interesting to me now, though, is how little understanding I had of what that was. I thought of myself as

becoming a park ranger, or running a wildlife refuge. I really had no real notion of what the real opportunities were for research or the breadth of scientific possibility.

Within a week, I was enrolled in a freshman seminar with Dr. Raymond Paynter on the biogeography of South American birds. Paynter introduced me to the world-class ornithological collections at Harvard's Museum of Comparative Zoology. The beautiful specimens provided a new, tangible connection to avian diversity, and the MCZ ornithology collection would become my second home as an undergraduate. I set up a little office in room 507 on a table beside and under a thirteen-foot-high moa skeleton (close to both the toucans and my other favorites, the suboscine passerines). In the more than thirty years since, I have been consistently associated with world-class collections of bird specimens, first at Harvard and then at the University of Michigan, the American Museum of Natural History, the University of Kansas, and Yale University, as a student, postdoc, or professor and curator.

In the MCZ I also discovered that evolutionary biology was the field of science that organized around the questions that I had always found inherently most fascinating. Birding was an elaborate exploration of biodiversity, and evolution was the process that had given rise to all these diverse forms, their behaviors, and their ecologies. I became dedicated to evolutionary biology.

At the same time I became interested in the exciting, controversial, and revolutionary new field of phylogenetics, which aimed to reconstruct the explicit history of evolutionary diversification and shared ancestry of organisms. For complex intellectual and sociological reasons, this commonsense, empirical, intellectual program was actually the subject of a great debate. Excitingly, Harvard's MCZ was home to some of the most recalcitrant members of the intellectual Old Guard, like Ernst Mayr (who coined the pejorative term "cladism"—like communism and fascism—only to have it embraced with pride by the opposing side), *and* a host of rebellious and brilliant graduate students who were leading the charge to change the way systematics and classification were studied and conducted. These students included many who would go on to make major contributions to phylogenetics, including botanists Michael Donoghue and Brent Mischler and arachnologists David Maddison and Jonathan Coddington.

I began to attend the weekly meetings of the Systematics and Biogeography Discussion Group in the Romer Library of the MCZ. An added incentive was that the drinking age had gone up to twenty-one only a few months before I turned eighteen, and the discussion group was always supplied with ample free beer and chips. It was also an exciting, freewheeling intellectual discussion. It was a lot more inspiring than organic chemistry! This early training in the logic of phylogenetics and its applications would remain useful throughout my career.

Following the suggestion of ichthyology professor Bill Fink, I wrote my senior honors project on the phylogenetics systematics and vicariance biogeography of the Neotropical toucans (*Ramphastidae*). It was a critical test of Jürgen Haffer's refugia theory, which I had consumed without reservation in the freshman seminar. Although it took some years, this undergraduate research would eventually be published in a series of papers on lowland Neotropical biogeography and the phylogenetic relationships of toucans and barbets.

After graduating from Harvard, I was inspired by the work of then Harvard graduate students Jonathan Coddington, on the evolution of spider webs, and Kurt Fristrup, on Guianan Cock-of-the-Rock ecology. I used a small nest egg from my grandma Agnes to fund a six-month field trip to Suriname to observe courtship displays of manakins. Using the classic manakin papers of David Snow as my guide, I spent the next six months observing and describing the lek courtship display behavior of the White-fronted Manakin and the White-throated Manakin. This fieldwork would lead to my first scientific publications in 1985 and later to my dissertation research on manakin phylogeny and behavioral evolution at the University of Michigan in Ann Arbor.

I remember in 1976 finding a copy of Robert Ridgely's brand-new field guide to the birds of Panama in the Dartmouth College bookstore in Hanover, New Hampshire. It was one of the first of a new generation of real field guides that would revolutionize birding in the Neotropics. Just holding it was exciting. When I asked my father for help to buy it, he asked, "But Ricky, when are you going to go to Panama?" This was actually a pretty reasonable question to ask a high school sophomore whose life's earnings consisted of income from a paper route and $300 from working in the kitchen of a summer camp the year before. I said, "Don't you see, Dad? *First* you buy the book, and *then* you go!" We bought the book. I was proud when years later some of my own systematic and behavioral research was cited in the Ridgely's 1989 second edition.

H. Douglas Pratt
© 1991

The Origin and Diversification of Species

I sat up late that night to read it: I shall never forget the impression it
made on me. Herein was contained a perfectly simple solution of all
the difficulties which had been troubling me for months past.

—ALFRED NEWTON (1859), ON DARWIN'S AND WALLACE'S ACCOUNTS OF
NATURAL SELECTION IN THE *JOURNAL OF THE LINNEAN SOCIETY* IN 1858

THE CHARM OF FINCHES

DAVID LACK'S *DARWIN'S FINCHES*, PUBLISHED
in 1947, was, like many of his other books
(chapter 5), utterly transformative. The mere
act of collecting and analyzing data for the
book metamorphosed Lack from a virtually
unknown Devonshire schoolmaster to director of the Edward Grey Institute of Field
Ornithology at Oxford, arguably the most
prestigious post in the world of ornithology
in those days. The book itself was a powerful
confirmation of the strength of the emerging
Modern Synthesis of evolutionary biology
described later in this chapter. Lack popular-
ized the common name[1] of these finches and
promoted the legend of Darwin's discovery
in both the public imagination and scientific
discourse (Sulloway 1982). In so doing he

initiated a vigorous enterprise of research
on Darwin's finches that would dominate,
influence, and excite evolutionary biologists
for the next seventy years. Most important,
Lack argued forcibly that competition for
food was an important factor driving the
evolutionary diversification of species—a
process that W. L. Brown and E. O. Wilson
(1956) would later call "character displacement," highlighting Lack's work on the
finches as a canonical example. Yet despite
the sophistication of Lack's case for competition, a young graduate student, Bob Bowman, disagreed.

Robert Irvin Bowman grew up in Saskatoon, Canada, and as the Second World War
was drawing to a close, he went to Queen's

Hawaiian Honeycreepers: 1 = adult (a) and juvenile (b) 'Ākohekohe; 2 = adult (a) and first basic plumage (b) 'I'iwi; 3 = Hawai'i Mamo; 4 = Black Mamo (extinct in 1898). Painting by H. Douglas Pratt.

University in Ontario to pursue his interests in biology. Queen's was one of the best places to study biology in Canada, in part because it had just established a field station in a wilderness of lakes and forests a short drive north of the campus in the city of Kingston. As an undergraduate Bowman conducted the first biological survey of the new field station, and that became his first "publication" (Bardach et al. 1947). In 1948 he moved to California and two years later began graduate work with Alden Miller, one of America's leading ornithologists, at Berkeley, one of the best schools. We do not know for sure why Bowman decided to do his graduate work on Darwin's finches, but it seems likely that he was inspired by Lack's book, which had just been published to great acclaim, and possibly also by the extensive collection of those finches[2] at the California Academy of Sciences, where he worked as a research associate in 1948–49.

By 1951 Bowman had decided that Lack's interpretation of the role of competition in the evolution of the finches was far from convincing, and that he wanted to settle the question by studying the birds in the field. His supervisor, Miller, may also have been an influence, as his writings about speciation—including an influential paper in the journal *Evolution* (Miller 1956)—credited no role whatsoever to competition between species. So, in 1952 Bowman and his wife, Margret, set out for the Galápagos Islands on their honeymoon, spending five months studying finches on that first trip, then making thirteen more field trips there over the next four decades. For his PhD research, Bowman mainly studied the finches on Indefatigable Island (now called Santa Cruz), where he documented their feeding habits and collected birds in order to obtain stomach

Bob Bowman in the Galápagos (photo in 1986 at age 60).

contents and to perform detailed analyses of cranial skeletons and jaw musculature. On that first trip he collected more than fifty finches to bring home for dissection, depositing them alongside the extensive collection already at the Cal Academy.

Bowman completed his thesis in 1957, and it was later published by the University of California as a 302-page monograph (Bowman 1961), richly illustrated with his wife's exquisite drawings of the bird's muscles and skeletons. Bowman meticulously measured skulls, beaks, and cranial musculature so that he could quantify the birds' functional feeding morphologies and calculate an index of the biting force of their beaks. He also designed an instrument to measure the hardness of seeds, and he found that bill sizes of the different species were related to the hardness of the seeds in their diets. Lack (1947a) had previously discovered that finch species with bigger bills ate bigger seeds, but Bowman was sure that seed hardness—and not size—was the key evolutionary force on bill size. In fact, seed size and hardness are closely aligned, so separating the two effects is tricky.

From the outset, Bowman had decided that Lack was wrong about many of his conclusions. Thus, shortly after returning from his first field season, he wrote to Lack to tell him of his concerns:

> It has become apparent to me, as a result of my 5 months stay on Indefatigable Island, that it is absolutely necessary to consider carefully all available data on weather conditions pertaining to any one collection of birds, when considering problems of moutling [*sic*] and distribution within any one island.... Suffice it to say here that I do not agree with your interpretation of the function of the bill differences within the genera ... Indeed, you were not residing on the islands at a time when this problem could be most satisfactorily studied. It was my luck to be present on the islands during an exceptional year.[3]

Unfazed, Lack replied, "One point I should perhaps make clear because, although it should not be necessary to do so in scientific circles, things do not always work out that way:—Namely, that if you do disprove my interpretation of the beak differences of the Galapagos finches, no one will be better pleased than myself."[4] As any good scientist should, Lack was more interested in the truth than in being right.

In his 1961 monograph, Bowman stated explicitly that he was embarking on a "new attempt to explain some of the structural variations in the Galápagos finches as adaptations for food getting."[5] He went on to say:

> In view of the magnitude of the differences in relative size and position of the adductor muscles between closely related species of *Geospiza* and *Camarhynchus*, it would seem that the suggestion made by Lack (1947:63–64) attributing the differences in the bills of these species primarily to their taking foods of different size, is not substantiated by the myological evidence presented here. Rather, these differences in musculature reflect differences in adducting potentiality, which may be better correlated with differences in feeding habits and availability of food, as well as in morphology of bill and skull.[6]

The claim here is that he is rejecting Lack's conclusions, but Bowman's work from today's perspective seems a rather slight refinement, showing that the muscles were as important as the shapes and sizes of bills.

Where Bowman and Lack really differed was in their ideas about the role of competition in structuring animal communities and influencing the course of evolution by natural selection. Bowman agreed with the

David Lack in the Galápagos (photo in 1939 at age 29).

entomologists Herbert Andrewartha and Charles Birch,[7] whose influential book *The Distribution and Abundance of Animal Numbers* was published in 1954 (chapter 10). Andrewartha and Birch believed strongly that animals of different species did not compete for resources and thus that any ecological differences between even closely related sympatric species must have evolved as adaptations to their environments during the speciation process. In contrast, Lack felt that competition could at least sometimes result in selection favoring differences between closely related species when they occupied the same habitat.[8] Bowman was certain that his study showed otherwise—"The anatomical differences between closely related species of *Geospiza* living in the same locality may be thought of as biological adjustments (adaptations) that prevent these species from competing with each other. The mechanisms by which these adjustments have evolved is unknown"[9]—but he presented precious little evidence in support of that claim. Instead, he argued simply that the match between bill sizes, jaw musculatures, and diets on the different islands was sufficient to account for the differences between species where they occupied the same habitats. For him, the possibility that there might be competition between species over scarce food resources offered no insights into the adaptive radiation of Darwin's finches; the available foods and the adaptations for eating them explained it all.

Bowman was enthralled with the Galápagos Islands from the start, unlike Lack and Darwin before him: "I can truthfully say that our trip to Galapagos was most pleasant and comfortable. We enjoyed fresh fruits and vegetables all the while and an abundance of fresh meat. Galapagos has impressed me more than any other area I have ever visited. There still remain unlimited biological problems of first class nature to be studied, including the finches."[10] Darwin was much less enthusiastic: "The black rocks heated by the rays of the Vertical sun like a stove, give to the air a close & sultry feeling. The plants also smell unpleasantly. The country was compared to what we might imagine the cultivated parts of the Infernal regions to be."[11] Lack agreed: "The Galapagos are interesting but scarcely a residential paradise. The biological peculiarities are offset by an enervating climate, monotonous scenery, dense thorn scrub, cactus spines, loose sharp lava, food deficiencies, water shortage, black rats, fleas, jiggers, ants, mosquitoes, scorpions, Ecuadorian Indians of doubtful honesty, and dejected, disillusioned European settlers."[12]

Seeing the advance of agriculture on the islands and some local destruction of habitats during his PhD work there, Bowman dedicated considerable time and effort to conservation in the Galápagos Islands and was instrumental in the making and promoting of popular films about the wildlife. He was influential in the creation of both a large national park and the Charles Darwin Foundation in 1959 and the Charles Darwin Research Station at Puerto Ayora on Isla Santa Cruz in 1964. For his efforts on behalf of the Galápagos, the government of Ecuador awarded him its Medal of Honor.

Inspired by the experiences of his PhD research, Bowman continued to study the finches but turned his attention to tool use and song. To study the tool-using behavior of the Woodpecker Finch at first hand in an aviary, he brought seventeen of these birds back to the Cal Academy. Bowman was fascinated by the foraging behavior of the birds—at the time the only known example of tool use in birds—and decided it would be worth studying experimentally. His aviary study established the ability of this finch to solve problems using tools (Millikan and Bowman 1967) and is a landmark study of avian cognition. The Austrian ethologist Irenäus Eibl-Eibesfeldt, inspired by Bowman's studies and encouraged by Bowman, went on to study these behaviors in the field (chapter 8).

During his PhD work on the functional anatomy of the finches, Bowman began recording their songs in the field. He noticed early on that, contrary again to Lack's conclusions, he could recognize differences in the songs of coexisting species on different islands, so he set out to carefully document those differences as a basis for studying the adaptive radiation of songs. Using the newly available sound spectrograph (chapter 7), he conducted characteristically meticulous studies of song structure, showing clearly that different species living in the same habitat had distinctly different songs. While doing this work, he made two further discoveries that were important to the development of birdsong research. First, he revealed that songs, like cranial morphologies, were adapted to the habitats occupied by the birds, with the details of song structure (the pitch, duration, and cadence of song syllables) correlated with vegetation structure such that the degradation of sounds over distance was minimized. To do this study he played tape-recorded songs in the birds' habitats and then recorded those songs at different distances from the playback recorder so that he could analyze the effects. His approach was inspired and encouraged by the young Gene Morton, then a graduate student with Charles Sibley (chapter 3) at Yale, with whom

he corresponded; Morton was at the time just beginning to formulate his own original ideas about sound transmission and ranging (see Morton 1975). Bowman's second discovery, with his captive finches at the Cal Academy, was to show that song characteristics are learned by young birds during a relatively short period after they leave the nest. Bowman reared finches in acoustic isolation so they could not hear the other birds' songs, then played them the songs of their species at different intervals thereafter. While this work was not published until the 1970s, the research was actually conducted at the same time as some of the more famous work on this phenomenon by Peter Marler and Bill Thorpe at Cambridge (e.g., Thorpe 1958a, b).

By the time Bowman's finch monograph was published, he was already well established in a faculty position at San Francisco State University, just down the road from the fabulous finch collections at the Cal Academy. Except for a small symposium paper in 1963, though, Bowman never published again on the adaptive radiation of beak and muscle morphologies. His monograph was not particularly well received by the scientific community, as summarized by Walter Bock: "General reactions I have . . . are certainly mixed. On the one hand, Bowman has presented a wealth of new data about the Geospizinae, but on the other he fails to convince me that his interpretations . . . represent an advance over the ideas presented by Lack. . . ."[13] As Bock makes clear, Lack and Bowman had taken strikingly different approaches to the same problem:

Lack was more concerned with general principles and frequently included examples from other groups of birds, while Bowman emphasizes detailed problems within the Geospizinae. Although Bowman's investigation is the more ambitious and includes far more detail, Lack presents a more complete picture of adaptive radiation in the Geospizinae. . . .The gathering and quantifying of the great mass of detailed data as done by Bowman for the Geospizinae is most desirable, but unless backed by a critical understanding of the subject matter, no amount of detail will lead to sound generalizations.[14]

Perhaps surprisingly, the controversy between Lack's and Bowman's interpretations of the evolution of these interesting finches lay dormant for another decade—until 1973—when Ian Abbott and Peter Grant decided to take up the challenge.

Inspired by Ernst Mayr's work on bird evolution and systematics in the South Pacific (see chapter 3), Peter Grant realized in the early 1960s that islands were ideal for studying evolutionary processes, especially where comparisons could be made among islands in a group, or between island forms and those on the nearest mainland. For his PhD work at the University of British Columbia, Grant studied the entire avifauna of the Islas Tres Marías, four small islands off the west coast of Mexico that had endemic subspecies of many birds that lived in the nearby mainland state of Nayarit. Following Mayr's approach to the birds of the South Pacific, Grant (1965a) conducted a detailed taxonomic analysis to determine if island-mainland differences were sufficient to warrant subspecific status; they were. Most important, though, he applied the emerging principles of island biogeography, ecology, and evolutionary biology to a number of interesting species and problems, including a

possible case of interspecific territoriality in *Thryothorus* wrens (Grant 1966); the tendency for island birds to be bigger but have duller plumage colors (1965b) and less fat stores (Grant 1965c) than their mainland counterparts; and the structure of island bird communities compared with those on nearby mainland sites (Grant 1968).

Upon obtaining his PhD in 1965, Grant took up a faculty appointment at McGill University, a position held previously by V. C. Wynne-Edwards in the early 1950s, before he had published his flawed but influential ideas on group selection (chapter 10). Unlike Wynne-Edwards, Grant had pitched his tent in the individual selection camp of George Williams and David Lack, and like Lack and Robert MacArthur (chapter 10), he felt that competition had an important role in the structuring of communities and in the evolution of at least some differences between species that lived in the same areas. Because he had a young family and wanted to work close to home on a project where he could do experiments, Grant focused his empirical studies on competition among three coexisting rodent species (*Microtus, Clethrionomys,* and *Peromyscus*) that lived in a bit of wild land not far from the university. The results of those experiments (Grant 1972) convinced him that competition was a potent ecological force, despite growing criticism of the general failure to consider alternative explanations that emerged in response to Lack and MacArthur in the 1960s and continued throughout the 1970s (e.g., Wiens 1977).

Still keenly interested in studying birds, however, Grant devoted part of his first sabbatical (in 1971–72) to investigating an interesting case of character displacement in the rock nuthatches of Eurasia, where

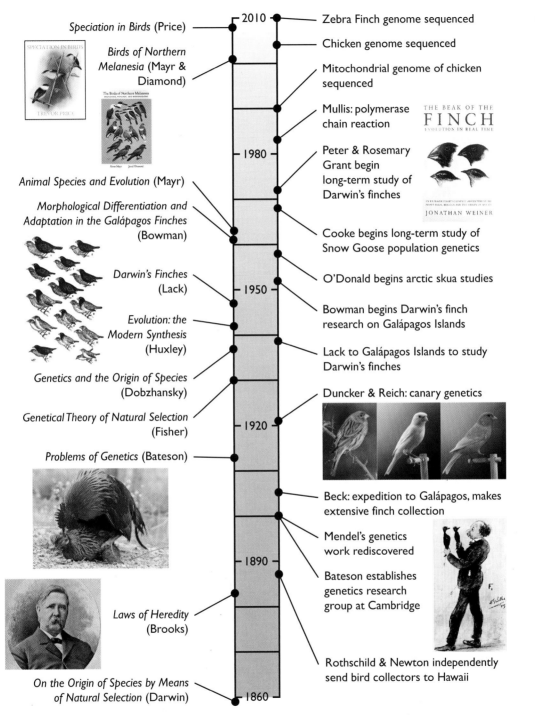

Speciation in Birds (Price) ——— ● — 2010 — Zebra Finch genome sequenced

——— Chicken genome sequenced

*Birds of Northern
Melanesia* (Mayr &
Diamond) ——— Mitochondrial genome of chicken
sequenced

Mullis: polymerase
chain reaction

THE BEAK OF THE

FINCH
EVOLUTION IN REAL TIME

— 1980 — Peter & Rosemary
Grant begin
long-term study of
Darwin's finches

Animal Species and Evolution (Mayr)

*Morphological Differentiation and
Adaptation in the Galápagos Finches*
(Bowman)

JONATHAN WEINER

Cooke begins long-term study of
Snow Goose population genetics

Darwin's Finches
(Lack) ——— O'Donald begins arctic skua studies

— 1950 —

*Evolution: the
Modern Synthesis*
(Huxley) ——— Bowman begins Darwin's finch
research on Galápagos Islands

Lack to Galápagos Islands to study
Darwin's finches

Genetics and the Origin of Species
(Dobzhansky)

Genetical Theory of Natural Selection
(Fisher) ——— Duncker & Reich: canary genetics

— 1920 —

Problems of Genetics (Bateson) ———

Beck: expedition to Galápagos, makes
extensive finch collection

Mendel's genetics
work rediscovered

Bateson establishes
genetics research
group at Cambridge

— 1890 —

Laws of Heredity
(Brooks)

Rothschild & Newton independently
send bird collectors to Hawaii

*On the Origin of Species by Means
of Natural Selection* (Darwin) ——— ● — 1860 —

TIMELINE for EVOLUTIONARY BIOLOGY. Left: Covers of Price (2008) and Mayr and Diamond (2001); drawings of Darwin's finches from Lack's book; Red Jungle Fowl copulating; William K. Brooks. Right: Cover of Weiner
(1994); Canaries—wild type (left), yellow, domesticated (center), and red, artificially selected (right); William Bateson
holding two woodpeckers (crayon drawing from 1906).

two species, the Eastern and Western Rock Nuthatches, have different bill sizes where their ranges overlap. This pattern suggested that competition for food between the two species in the region of overlap might have caused an evolutionary divergence in bill size that reduced the costs of competing (Brown and Wilson 1956). As a follow-up to his studies on the Tres Marías avifauna, Grant had reviewed the evidence for and against character displacement, concluding that the nuthatches represented one of the clearest and most interesting examples (Grant 1972). Yet he also felt that there were no examples that were completely convincing, because the evidence for competition was slim. On measuring a large sample of nuthatches both inside and outside the region where their ranges overlapped, he discovered to his amazement that the differences between the species in sympatry were the result of clinal variation in bill size—bigger in the east, smaller in the west—causing them to be more different where their ranges overlapped and perhaps not the result of competition after all. His paper on these birds is one of the classics of modern evolutionary biology (Grant 1975).

When Australian biologist Ian Abbott wrote to Grant in 1972 asking about the possibility of postdoctoral research on competition and community structure in Darwin's finches, Grant saw a golden opportunity to return to bird research, especially now that his two daughters were old enough to possibly make the fieldwork a family affair. Peter and Rosemary Grant devoted most of their adult lives to the continuing study of these finches, often spending months in the field on isolated islands with their girls, and recruiting and training an outstanding coterie of graduate students and colleagues to study the birds

with them. As of 2013 the Grants themselves have been to the Galápagos Islands on forty-three separate trips, spending more than 120 months in the field.[15] In the course of their research they moved from McGill to the University of Michigan to Princeton University, winning more than thirty awards, medals, and accolades for their research. Their finch research group—Grupo Grant—by 2013 has spent more than thirty person-years in the Galápagos Islands collecting data for one of the most in-depth long-term field studies ever conducted on birds.

In May 1972 Peter Grant made his first trip to the Galápagos Islands with Ian and Lynette Abbott—a six-week-long sojourn to find a suitable study site, to band and measure some birds, to collect some preliminary data on foraging behavior, and to check out the working conditions and the potential for testing their ideas about interspecific interactions. Their initial focus was on the birds of Isla Daphne Major, where the Medium Ground Finch was abundant and the Small Ground Finch—its most obvious competitor on many of the other islands—was rare. The idea was to examine the foods, bill structures, and foraging behaviors of the two species on islands where they co-occurred as well as on islands like Daphne Major, where only one of the two lived in any numbers—a natural experiment not unlike the controlled experiments that Grant had previously conducted with rodents. A further trip to the islands at the end of 1973 convinced Grant that this was a tractable study; little did he know at the time that it would be his life's work and that it was more than just tractable, though not only in the directions initially conceived. The initial plan was simple enough: study marked individuals; measure birds, bills, and seeds; and determine breeding success and recruitment into the population. With a relatively small island[16] to work on, marking and following the fates of almost every single finch was perfectly possible. As both Lack and Bowman had noted, the birds were very tame, often coming into camp and eating the researchers' food.

As the study progressed, it occurred to Grant that he could not really understand the process of evolution without some genetic analyses, so in 1975 he recruited Peter Boag—who already had some training in population genetics—to do PhD research under his supervision. That turned out to be serendipitous, as a drought befell the islands in 1977, with just short of 24 milliliters (1 inch) of rain falling on Daphne Major during the "wet" season, about one-fifth of the typical level. The finches did not breed at all that year, and before the next breeding season 85 percent of the marked birds had disappeared, presumed dead. The researchers were initially despondent, as it looked like their long-term study was ruined, but they soon realized that the drought had created a golden opportunity, a natural experiment in natural selection. As Herman Bumpus had done in 1898, when 64 of 136 House Sparrows died after experiencing a severe winter storm near his home in Providence, Rhode Island (Bumpus 1899), Boag and Grant (1981) compared the beak sizes (among other things) of finches that lived and died. Unlike Bumpus, Boag and Grant had already marked and studied many of the birds that disappeared—only one of the 388 Medium Ground Finch nestlings banded in 1976 lived to see the 1978 breeding season—and they could follow the breeding success of survivors. During the drought on Daphne Major, small seeds disappeared

faster than large seeds from the available food supply because the birds were eating them, and the finches that survived on the island were larger than those that disappeared, showing clearly that this episode of selection favored birds that could most efficiently eat the available seeds. Moreover, beak size was heritable; thus they had the data necessary to demonstrate Darwin's three requirements for evolution by natural selection: variation, heritability, and differential survival. This became what Peter Grant might himself have called "a classical case of natural selection in the wild." The advantages of a long-term study of marked birds was now abundantly clear, and the Grants went from strength to strength, eventually showing how selection in this one population of the Medium Ground Finch varied from year to year over the following decades.

Peter Grant's wife, Rosemary, herself a biologist, joined Peter and the Abbotts on that second field trip to the Galápagos, and she soon became a full partner with Peter in the finch research program. Rosemary shared the fieldwork, data analysis, and writing and contributed her own fair share of ideas, insights, and energy to studying the birds. For many years they took their young daughters to the Galápagos Islands for months at a time, making the Galápagos their home, isolated from the outside world. It would be fair to say that the Grants are unique in the annals of ornithology for their dedication to a lifelong study of a single group of birds in the field, especially in such a remote locale. Husband-wife teams are more common in the long-term field study of primates and large mammals, for some reason, and even then they rarely tackle questions of such breadth and depth. Continuing to recruit talented PhD students, the Grants went on to study in more detail competition and character displacement (with Dolph Schluter), sexual selection (with Trevor Price), song (with Laurene Ratcliffe), and genetics (with Lisle Gibbs and Kenneth Petren), to mention just a few of the many aspects of finch evolution, behavior, and ecology that they have studied. The story of this amazing long-term study of Darwin's finches—an important milestone in evolutionary biology—was beautifully told by Jonathan Weiner in 1994 in his Pulitzer Prize–winning book *The Beak of the Finch.*

Despite the success of the Grants' project, it has not always been smooth sailing. When Dolph Schluter began his PhD work in 1978, for example, the intellectual climate in population ecology was decidedly lukewarm about the evidence for competition, a subject that Schluter wanted to focus on for his thesis. After the initial influences of Lack, MacArthur, Martin Cody, and other ornithologists extolling the importance of competition, biologists from other fields weighed in on both the theory and the evidence, suggesting at the very least that documenting ongoing competition in the field would be extremely difficult. Sticking to the core values of the project—understanding natural selection, marking individuals, measuring carefully, and collecting data annually—Schluter succeeded admirably, comparing populations among islands in relation to the resources available and showing a clear pattern of competitive effects. Moreover, with Grant and Trevor Price, he was able to document strong evidence for character displacement between the Medium and Small Ground Finches—an initial focus of the research in 1972—with evidence that in the absence of the Small Ground Finch, the Medium Ground Finch has a smaller bill that enables it to exploit food resources otherwise used by Small Ground Finches. The team was cautious in interpreting their results, underlining once again how difficult it is to study character displacement—and competition—in the field (Schluter et al. 1985).

In 1936 Percy Lowe, who was curator of birds at the Natural History Museum in London, said that the adaptive radiation of Darwin's finches was "... only likely to be solved by experiments and observations on the spot, and that ought to be comparatively easy, for these Finches are so tame that they can almost be picked off the bushes."[17] He was right that "on the spot" research was needed, though nobody would now claim that the task has been easy. Others have argued that the Hawaiian honeycreepers might be a better—and certainly more colorful—example of adaptive radiation, but the wheels that Percy Lowe and David Lack set in motion have made Darwin's finches the exemplars of evolutionary biology. As the Grants observed in 2006:

> The evolutionary changes that we observed are more complicated than those envisioned by Lack. Nevertheless, they provide direct support for his emphasis on the ecological adjustments that competitor species make to each other, specifically in the final stages of speciation and more generally in adaptive radiations.... They also support models of ecological community assembly that incorporate evolutionary effects of interspecific competition, in contrast to null or neutral models.[18]

Thirty-four years after Grupo Grant began their research, it seems that Lack was right about the role of competition.

Lack had initially argued that the diversity of finches and the sizes and shapes of their bills was simply the result of genetic

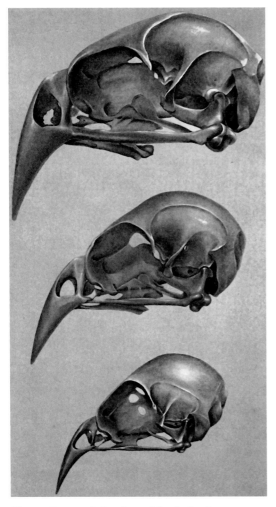

Margret Bowman's drawings of the skulls of some Darwin's finches for her husband's monograph.

some quite peculiar evolutionary agency. As on other oceanic islands, the almost complete absence of food competitors and predators has decreased the intensity of selection, so that peculiar types or habits have a greater chance of persisting. The existence of a number of islands [in the Galapagos archipelago] has promoted non-adaptive differentiation of island subspecies.... The genera show a minor adaptive radiation.[19]

How different from his conclusions in *Darwin's Finches* in 1947 and from the subsequent studies by the Grants! What caused him to change his mind?

Lack was in the Galápagos Islands between December 1938 and April 1939 to study the birds' natural behavior and ecology; he then went directly to the California Academy of Sciences to measure specimens and write up his findings for publication. He had attempted to get different finch species to interbreed in captivity on the islands but was unsuccessful, so he brought some live birds back to the Cal Academy to continue those experiments. Before Lack, Rollo Beck (see chapter 3) and others had collected so many finches for the Cal Academy that it was the obvious place to go to measure specimens; Beck alone brought back more than 2,400 finch specimens from eighteen of the islands in his expedition of 1905–1906. In preparing his 1940 paper for the journal *Nature*, Lack was apparently heavily influenced by three publications: Percy Lowe's (1936) study of speciation in the finches, based on collections made in the late 1800s by Walter Rothschild and Ernst Hartert; Rensch's (1933) claim that he could not see how the finches could have evolved by geographic speciation, and Sewall Wright's (1931) theories about the role of

drift, and that the biological significance of the differences among coexisting species might be useful in preventing inbreeding (Lack 1940a). Like Bernhard Rensch (1933) and Percy Lowe (1936), who had written previously about these birds, Lack had initially claimed that the islands appeared to be ecologically quite similar:

While the Geospizinae present certain unusual features, there is no need to postulate

53

genetic drift in evolution. Lowe, for example, had made a strong case for the finches being actually a "hybrid swarm," based on the similarities in plumage and the graded differences between bill sizes and shapes. With the weight of these previous ideas and what appears to have been a rather superficial assessment of the specimens at hand, Lack dashed his paper off to *Nature* in 1940, claiming that the adaptive part of the finch radiation was minor, the differences arising largely from genetic drift in small isolated populations. In his 1945 monograph he provided more detailed support for that view, concluding that "the most important factors in the evolution of the Geospizinae have probably been the almost complete absence of food competitors and predators, and the existence of several islands which provided partial, but not complete, isolation for island forms. Differences between island forms of the same species are considered nonadaptive, and due primarily to the 'Sewall Wright effect,' while colonization by an atypical sample may be a subsidiary factor."[20] In 1947 he expressed a very different opinion in *Darwin's Finches* about the role of competition—the result of more careful study, and possibly his own rise in stature. He had also discussed his findings with Ernst Mayr, but Mayr gives full credit to Lack for this focus on the importance of species interactions: "Since the earlier account was greatly delayed in publication (owing to the war), the two accounts were published close together, and this rather confused many readers. The new interpretation was the manifestation of a major revolution in David Lack's thinking."[21]

With this attention on Darwin's finches it would be tempting to conclude that the entire focus on bird evolution in the twentieth century was on these interesting birds. This

is not true. Studies on a variety of birds contributed to the development of genetics, as well as to the Modern Synthesis of evolution that brought genetics and systematics together in the 1940s, and an increasing focus on the study of speciation, the biological species problem, and the use of the comparative method, as we shall see in the remainder of this chapter.

SILKIE FOWL, RED CANARY

The silkie fowl is a very strange bird, at once both unbirdlike and adorable. The breed is more than a millennium old, artificially selected by the ancient Chinese from the domesticated version of the Red Junglefowl. First reported by Marco Polo in China in the thirteenth century, the silkie fowl soon after appeared in Europe and remains a very popular ornamental breed today, probably numbering in the tens of millions worldwide, kept mainly as pets, show birds, and incubators for the eggs of other chicken breeds and other birds. These are relatively small chickens,[22] usually with entirely black or white plumage, and they have a calm, friendly temperament. One is struck immediately, though, by the bird's fluffy plumage and apparently furry topknot. The plumage actually feels like fur—silky even—as the feathers lack the barbicels that hold barbs together on typical bird feathers. All of the silkie's feathers are like down, and without vaned feathers the birds cannot fly. If that was not strange enough, the silkie's skin, flesh, bones and connective tissues are all black from heavy deposits of melanin—in Chinese they are called *wu gu ji* (烏骨雞), the "dark-boned chicken"—a rare attribute in chickens and unheard of in other birds.

They also have five toes on each foot, in contrast to the normal four in birds, blue "ear-lobes,"[23] dark wattles, and a small dark comb that resembles the flesh of a walnut. No wonder the early geneticists studied this bird—here was an animal that was easy to keep, would freely interbreed with wild-type breeds, and had, through artificial selection, been endowed with an array of bizarre traits whose genetic provenance could be studied by crossbreeding.

William Bateson, one of the founders of genetics, was already using the silkie to study the nature of variation when Mendel's work was rediscovered in 1900.[24] Bateson grew up at Cambridge—his father was master of St. John's College for twenty-four years— eventually getting a first-class degree there, specializing in zoology. After graduation he went to America to work with William K. Brooks[25] to study the embryological development of marine worms in Chesapeake Bay. Brooks had just published an influential book (Brooks 1883) on heredity that undoubtedly influenced the young Bateson, and from then on his work on morphology was all about the study of variation and inheritance. Bateson was initially interested in both development and the origins of vertebrates, but he soon turned to the study of evolutionary processes. On returning to Cambridge in 1884, he worked with the great ornithologist Alfred Newton, employed on what would today be called "soft money." Though his research was focused on the study of variation, his official duties—the ones that paid the bills—were often menial. Newton eventually hired Bateson to manage the kitchens at St. John's College, ordering food and supplies so the students and masters could be fed in residence.

Until 1900 Bateson's work focused on studies of structural variation, using the quantitative methods being developed by Darwin's cousin, Francis Galton. In Bateson's (1894) book *Materials for the Study of Variation*, he showed that some traits varied continuously whereas others were discontinuous, coining the terms "meristic" and "substantive," respectively, for those kinds of traits. He was convinced that "substantive" traits were the main targets of natural selection, and that meristic (quantitative) traits were not easily "perfected" by natural selection because of the "swamping effects of intercrossing." Even then, though, he knew that "the only way in which we may hope to get at the truth [about the underlying genetic mechanisms] is by the organization of systematic experiments in breeding, a class of research that calls perhaps for more patience and more resources than any other form of biological enquiry. Sooner or later such an investigation will be undertaken and then we shall begin to know."[26]

Then, in 1900, Newton secured Bateson an academic appointment at the college, still on soft money, where he could devote his time to research, establishing his own informal research group to study genetics. That group was composed largely of women associated with Newnham College, Cambridge (Richmond 2001). In the 1870s women students at Cambridge had been granted access to science lectures, but even though they did well in exams, they were denied the right to obtain a degree in science.[27] Thus when Bateson began to build his research group in 1900, the intellectual climate for women at Cambridge could best be described as hostile, making the opportunity to work in Bateson's group all the more remarkable. Together Bateson and his students studied plants, birds, and rodents, laying many of the foundations of modern genetics, both supporting and extending Mendel's newly rediscovered laws of heredity. Bateson also coined the word "genetics" to describe the subject in 1905, but still the university refused to recognize it as worthy of study or to grant Bateson a permanent appointment. Frustrated by the lack of recognition from the institution where he had spent his life, Bateson moved to the John Innes Horticultural Institute at Merton Park, Surrey, in 1910 to be its first director, where he continued his genetics work for the rest of his career.

Bateson is perhaps most famous for his opposition to the "biometrics school" of Karl Pearson and Francis Galton, who saw continuous variation as the key to evolutionary change (Provine 1971). Bateson could not see how such continuous variation had any role to play in the origin of species, famously stating in 1913 that "the transformation of masses of population by imperceptible steps guided by selection is, as most of us now see, so inapplicable to the facts . . . that we can only marvel both at the want of penetration displayed by the advocates of such a proposition, and at the forensic skill by which it was made to appear acceptable even for a time."[28] As Richard Dawkins said in 2012, "It is rare to find a distinguished scientist so catastrophically and comprehensively and clearly wrong."[29]

The work of Bateson's group on the silkie fowl yielded tremendous insights into Mendelian genetics in general and the genetics of birds in particular, though the parallel studies of *Drosophila* begun by Thomas Hunt Morgan in 1908 soon overshadowed the fowl research. With Reginald Punnett,[30] for example, Bateson crossed white silkies with white Dorkings and discovered that their offspring were colored, not white, the first example of an epistatic interaction. In 1908, they demonstrated that the hookless condition of the silkie's feather barbules was caused by recessive alleles. Their work on the silkie fowl is often cited as the first demonstration of Mendelian genetics in the animal kingdom (e.g., Siegel et al. 2006). Though there was some obvious potential for the work on fowl to have financial benefits to society, Punnett was not interested in the agricultural aspects of their work, instead realizing that the fowl was an excellent model organism for the study of genetics. For the next century, many of the most famous geneticists would devote some of their research to the domestic fowl—Charles Davenport, Thomas Hunt Morgan, Alfred Sturtevant, J. B. S. Haldane, Ronald Fisher, Morley Jull, Arend Hagedoorn, L. C. Dunn, F. B. Hutt, Alexander Sergeevich Serebrovsky, Boris Zavadovsky, and I. I. Schmalhausen. Silkie fowl were not the only birds that Bateson's group studied, though, and it was not long before the duck and the canary took their place in the development of genetics.

Canaries were an obvious choice for the early geneticists—readily available as cage birds and, like the domestic fowl, modified into different forms by artificial selection, especially with respect to plumage color and the presence of a crest (e.g., Davenport 1908). In Bateson's group, Florence Durham[31] and Dorothea Maryatt pioneered the canary research, showing that some traits were sex-linked and also that offspring sex ratios were sometimes female biased.

Originating from the Canary Islands, the Atlantic Canary was esteemed for its song and has been a popular cage bird for the past five hundred years—the first captive breeding and the beginnings of domestication occurred around 1500. The wild bird is greenish gray, but by the 1650s breeders had created—probably by careful backcrossing—the familiar pure yellow variety. By the mid-1800s domesticated canaries had become immensely popular as pets and also for amateur breeders hell bent on creating new mutations—birds that differed in size, posture, plumage, and song.

A particular fantasy was a red canary, and in the 1920s the unlikely team of Karl Reich—a shopkeeper and canary enthusiastic—and Hans Duncker—a biology teacher with a talent for genetics—began a project to create such a bird. Their starting point was the knowledge gained from breeders that in captivity canaries readily bred with other finches, often producing startlingly colored hybrids. Duncker had previously worked out the genetic bases for the color variations—blue, violet, green, and yellow—in the Budgerigar, which proved to be as straightforward as the discrete and discontinuous traits in Mendel's peas. He imagined the same would be true if he crossed yellow canaries with Red Siskins: that some of the offspring would bear the siskin's red plumage. It was an ambitious project because it was well known among breeders that the offspring from crosses between different species were almost always sterile. Reich and Duncker produced numerous canary-siskin hybrids, but instead of being either yellow or red like their parents as Duncker anticipated, most were an ugly reddish-gray color.

By extraordinary luck and much perseverance they found several male hybrids that were fertile, and then, over several generations, they backcrossed these to canaries, actively selecting the offspring with the reddest plumage each generation. They made little progress toward a pure red canary, however, and the project came to abrupt end with the onset of the Second World War. In the 1940s, when canary breeders in the United States picked up the red canary project, they discovered that red plumage had both a genetic and an environmental component—via the intake and expression of carotenoids. Red canaries had existed for several years, but the absence of sufficient carotenoids in their diet meant their color remained hidden in their genes rather than expressed in their feathers. The red canary thus provided a powerful example of the role of both genes and environment in creating particular phenotypes. By the 1960s the red canary—Reich and Duncker's fantasy—had been established and remains a popular cage bird to this day (Birkhead 2003).

THE MODERN SYNTHESIS

From today's perspective, it is clear that the rediscovery of Mendel in 1900, and the subsequent development of genetics, provided the essential unifying framework for Darwin's theories of evolution by natural selection. But during the first decades of the twentieth century evolutionary biology was in disarray. Here is German biologist Wolfgang Freiherr von Buddenbrock-Hettersdorf in 1930: "The controversy [about heredity] . . . is as undecided today as it was 70 years ago . . . neither party had been able to refute the arguments of their opponents and one must assume that this situation is not going to change very soon."[32] The main problem was that, at least to some, the results of genetic experiments and analysis seemed to say that evolution by natural selection was impossible, and the gradual evolution of traits unlikely. In the decades following the publication of Darwin's *Origin of Species*, biologists and most learned people accepted the fact of evolution, but the mechanism proposed by Darwin and Wallace—natural selection—was hotly debated. The new research on genetics following the rediscovery of Mendel merely heightened the controversy, formalized in the debate between the Mendelians, like Bateson, and the Biometricians, like Karl Pearson (Provine 1971). The resolution began to emerge in the 1930s, from a remarkable group of naturalists—three of them stationed in New York City (Ernst Mayr, George Gaylord Simpson, and Theodosius Dobzhansky), three of them ornithologists (Mayr, Julian Huxley, and Bernhard Rensch), and two of them students of Erwin Streseman (Mayr and Rensch)—as well as the theoretical geneticists Sir Ronald Fisher and J. B. S. Haldane in Britain and Sewall Wright in Chicago.

Beginning with the publication of Dobzhansky's (1937) *Genetics and the Origin of Species*, these men created a truly pioneering marriage of genetics, population biology, paleontology, systematics, and evolution—a mix that Julian Huxley called the "modern synthesis" in the title of his 1942 book. Variously also called the "new synthesis," the "evolutionary synthesis," and "neo-Darwinism," this approach to the theory of evolution once and for all did away with the old controversies and set a new course for evolutionary biology, culminating in 1947 at a meeting in Princeton and the founding of the journal *Evolution*. The main architects of this synthesis are usually considered to be the small group of men mentioned above, but many other scientists also contributed substantially to this new

understanding of evolution, each bringing their own perspective and expertise to the table. In the preface to Dobzhansky's book, L. C. Dunn said, quite perceptively, that "the work symbolized something which can only be called the Back-to-Nature Movement."[33] Interestingly enough, there was relatively little communication among these "architects" until that 1947 meeting, after the synthesis itself had really been forged (Mayr 1982). The Modern Synthesis was really the first attempt to bring the findings of both theoretical and empirical geneticists to bear on the problems of evolution in nature, problems that Darwin had begun to revolutionize almost a century earlier by marshaling evidence from free-living organisms.

Mayr's contributions to the Modern Synthesis are legendary (Haffer 2004b, 2006b, 2008; Haffer and Bairlein 2004), interesting for both the insights and the biases that he brought to the subject as a leading ornithologist, systematist, and student of Stresemann. Mayr's early work on avian systematics (chapter 3) naturally led him to evolutionary questions, especially with regard to what

Ernst Mayr (age 62) and Stresemann (age 76) at the 14th International Ornithological Congress in Oxford in July 1966.

constitutes a species and the mechanisms by which species arise. In 1940, partly in response to Dobzhansky's book, the geneticist Richard Goldschmidt published his own book, *The Material Basis of Evolution*, in which he proposed that species evolved mainly by large mutations—hopeful monsters— rejecting Darwin's (and Dobzhansky's) idea that subspecies are actually incipient species evolving slowly by small mutational steps, and incidentally supporting Bateson's concerns about gradual evolutionary change. Mayr was irate and decided to provide a public rebuttal of Goldschmidt when invited to give a series of talks[34] on speciation at Columbia University in 1941.

Mayr's lectures were published in expanded form in 1942 as a book, *Systematics and the Origin of Species*, that became the most cited and arguably the most influential volume of the Modern Synthesis.[35] Ironically, although the title of his book was based on Darwin's, Mayr (1980) later admitted that he did not read *Origin* until the 1960s. In his own book, Mayr focused mainly on modes of speciation, introducing terms like "allopatry," "ring species," and "sibling species"—all terms that are part of the enduring lexicon of evolutionary biology. Mayr also canonized his definition of a species—a definition that guided work for the rest of the century and beyond: "Species are groups of actually or potentially interbreeding natural populations, which are reproductively isolated from other such groups."[36] This was remarkably similar to a definition proposed two decades earlier by his mentor Stresemann (Haffer 2006b).

While the main focus of the Modern Synthesis was, quite obviously, synthesizing— bringing together the emerging principles of genetics, biogeography, paleontology, and systematics to form a unified evolutionary biology—the architects of this approach also paid particular attention to the subject of speciation. Surprisingly, speciation had been more or less ignored since Darwin raised the issue in the title of his 1859 book, and even Darwin devoted relatively little to the subject in any of his writing (Wilkins 2009). Now, however, Dobzhansky, Mayr, Huxley, and the others were defining what a species is and speculating on how the process of speciation might proceed, based on the study of birds and other animals in the field. Thus not only did the Modern Synthesis point the way forward for the study of natural selection, it also heralded a new era in the study of speciation and adaptive radiation that became a major focus of evolutionary biologists in the latter part of the twentieth century. We have already seen how David Lack rekindled an interest in adaptive radiations with *Darwin's Finches*, and we will return to this subject in the final section of this chapter.

Like many revolutionary scientists, Mayr was not shy about arguing his case and promoting his own ideas. By the 1940s he was sufficiently well established, productive, and confident to take on the major scientists of the day when he disagreed with them. While Mayr certainly appreciated the role of genetics in the study of evolution, he initially had little use for Fisher's mathematical models, or Haldane's "bean-bag genetics," and he famously argued with Haldane about the latter in the 1960s (Dronamraju 2011). Though he later appreciated the value of both approaches, it is interesting to see how his own biases, based on his ornithological work (chapter 3), and his focus on systematics, clouded his judgment. Though he had been trained by Stresemann, Mayr was now beginning to distance himself from Stresemann's views: "As progressive as Stresemann was in practicing population systematics and in his concepts of species and speciation, he was rather backward in his understanding of the mechanisms of evolution. He probably would have called himself an orthodox Darwinian, but he felt quite strongly that there were severe limits to the power of natural selection."[37] Mayr's focus remained on birds, though, and for good reason: "In the 1930s–40s no other group of organisms was better known taxonomically than the birds. Therefore the understanding of geogr[aphical] speciation was documented better than for any other higher taxon."[38]

Birds are hardly a representative organism. They comprise less than 1 percent of the known species of animals, and they have many characteristics that make them, in a way, rather poor models for informing us about the processes of evolution. They are, for example, large, sexually reproducing, diploid, warm-blooded vertebrate animals, with separate sexes, color vision, parental care, internal fertilization, and relatively large brains. These traits make them more like us than like the majority of other living things, which may in part account for our attraction to birds for pleasure and study. The evolutionary biology of birds is rather straightforward because speciation appears to be largely due to the geographic isolation of populations and hybridization between species is relatively rare. Combined with Mayr's insights and personality, the study of birds thus had a seminal influence on the Modern Synthesis, and it continues to this day to have a major impact on evolutionary biology in general and the study of speciation in particular, due to the obvious geographic variation within and between species. Some of our colleagues have said that this overemphasis on bird biology has been in many ways misleading, but it could also be argued that evolutionary theory would have moved forward much more slowly if Mayr and birds had not been at the forefront. The sheer complexity of evolutionary processes unmasked as we have learned more about other kinds of organisms should make us grateful that the study of birds offered a simple beginning. Birds provided both examples of and support for many of the ideas emerging from the understanding of evolutionary processes that genetics brought to natural history during the Modern Synthesis.

As early as 1918, for example, Fisher had worked out, in theory at least, how polymorphisms could be maintained in a population. There was some clear support for his theory of balanced polymorphisms from lab studies of insects and rodents, but it was to be fifty years before genetic polymorphisms would be studied well in the field, especially in snails and the peppered moth. Mayr was well aware of the existence of plumage polymorphisms in birds, having documented several interesting cases in 1963 in his *Animal Species and Evolution*. Later in that decade, studies of plumage polymorphisms in birds finally provided some support for Fisher's theory.

POLYMORPHIC BIRDS

Before Peter Grant began making important discoveries about natural selection in finches near the equator, two other recent Cambridge graduates—Peter O'Donald and Fred Cooke—were already taking a different approach to selection in the Arctic, an approach in which the focus was genetics right at the outset and the genetic variation was manifested in discrete color polymorphisms. O'Donald was the last PhD student to work with Ronald Fisher, one of the leading geneticists and statisticians of the twentieth century. Starting his PhD at Cambridge in 1958, O'Donald was intrigued by sexual selection and dismayed to see how the subject had been neglected in general, and especially since Huxley's (1938) critique twenty years earlier (see chapter 9). Fisher had proposed an elegant verbal model of the genetic basis for sexual selection in just four paragraphs of his classic book, *The Genetical Theory of Natural Selection*, in 1930, but this too had been largely ignored by the time O'Donald began his studies.

A chance visit to the bird observatory on Fair Isle in 1957 set O'Donald's course. Lying midway between Orkney and Shetland off the northeastern tip of Scotland, Fair Isle was famous for its bird migration observatory, established in 1948, with Kenneth Williamson as its first resident warden. Williamson was keen to learn about the birds on the island—both migrants and breeders—and set about studying molt, ectoparasites, the relation between migration and weather, and the breeding biologies of Northern Fulmars and Parasitic Jaegers (long called "arctic skuas" in Europe).

On that first trip, O'Donald saw the potential to study the genetics of mate choice in the Parasitic Jaeger, as these birds came in three color morphs that clearly interbred—"dark," with dark brown plumage throughout and a black cap; "pale," with dark brown plumage on its back, white underparts and neck, and a black cap; and a relatively rare "intermediate" that looks like the pale morph except that the brown feathers have white bases. These color morphs were a

potential genetic marker that could readily be studied, and in a fully banded population where the causes and genetic consequences of mate choices could be assessed quantitatively. In 1948, when Williamson began studying them, there were just fifteen pairs of Parasitic Jaegers on Fair Isle, but by 1958 the population had grown to sixty-one breeding pairs, and—best of all— Williamson had banded the birds and kept track of pairings over the years. O'Donald recognized this as a unique opportunity to test in the field some of Fisher's and Darwin's ideas about sexual selection—something that had not been done before. He spent the 1958–61 breeding seasons on Fair Isle, documenting—with the new warden, Peter Davis—the jaegers' pair combinations, the color morphs of their offspring, and details about their ecology and behavior that might influence their mate choices and reproductive success. The result was a series of papers in the journals *Heredity* and *Nature* on imprinting, inbreeding, and assortative mating in these birds (O'Donald 1959, 1960, 1973; Davis and O'Donald 1976), providing the first field evidence for patterns that had been suggested by genetic work on mate choice in *Drosophila* conducted in the lab in the 1940s and 1950s (Milam 2010).

On completing his PhD, O'Donald stopped working on the jaegers for a few years but returned in the 1970s, when he took up a faculty position at Cambridge. He then devoted the period from 1973 to 1979 to intensive fieldwork on Fair Isle, working again with John Davis. The result was a remarkable series of papers on the genetics of sexual selection in these birds, culminating in two influential books, one on the theory (*Genetic Models of Sexual Selection*), in 1980, and one on the birds (*The Arctic Skua:*

A Study in the Ecology and Evolution of a Seabird) in 1983. Both books received mixed reviews at the time[39]—the subject of sexual selection was still contentious with respect to both the mechanisms and the evidence—but O'Donald helped bring the subject back to the forefront of scientific interest, particularly by ornithologists (chapter 9), and laid some of the foundation for a modern, genetics-based development of sexual selection theory. One reviewer of *The Arctic Skua* liked the data and the quantitative testing of genetic models but lamented the lack of information on the underlying processes, particularly behaviors involved in mate choice (Harvey 1983). That review ended by saying, "Field studies on sexual selection in the arctic skua have finished and this is likely to be (almost) the last word for quite some time."[40] Fortunately, the reviewer was wrong, as field studies on Parasitic Jaegers and Great Skuas were already being conducted by Robert Furness on the nearby island of Foula. Furness's long series of papers and his 1987 book, *The Skuas*, provide information on some of the behavioral processes that Harvey found lacking in O'Donald's work. But Furness's analyses found only weak support for many of O'Donald's conclusions, and in the end Furness could only conclude that "perhaps colour phase is but a secondary consequence of the genes' actions, and of rather little ecological importance in itself. Doubtless further research will eventually shed light on this confusing and complex issue."[41]

While O'Donald was working on jaeger genetics, Fred Cooke was establishing his own genetics research program in Canada, at Queen's University. Cooke's research at that time was focused on fungal genetics, but he was a keen birder and soon joined the local naturalists' club. One evening in 1967 a government biologist, Graham "Gus" Cooch, spoke to the club about his work on what were then considered to be two species of goose—snow and blue—nesting in the Canadian Arctic. Cooch had observed them interbreeding on the Boas River, Southampton Island, and noticed that the "hybrid" offspring were either "blue" or "white."[42] After the meeting, Cooch asked Cooke about the genetic implications of his observations and Cooke was hooked: here was a chance to combine his vocation with his avocation, and he immediately restructured his research program to start working on the geese—which he did for the next twenty-five years.

Cooch convinced Cooke to conduct his studies in a relatively young, small mixed colony of blue and white geese at La Pérouse Bay just east of Churchill, Manitoba, one of the most accessible localities in arctic Canada. Establishing a field camp right at the edge of the colony, Cooke and his students and collaborators monitored the mating patterns and reproductive success of the geese each summer, quantifying the frequencies of mating combinations and the colors of the

Ian Newton (left, age 32) and Fred Cooke (right, age 36) banding Snow Geese at the La Pérouse Bay colony in 1972.

resulting offspring. When Ian Newton visited the camp in 1972, he suggested to Cooke that it would be most useful to band as many birds as possible so that they could be followed through their lifetimes, thereby establishing, as it turned out, one of the most intensive long-term studies of a bird species ever attempted (Cooke et al. 1995). With marked individuals, Cooke and crew were able to show that not only were the blue and white geese freely interbreeding, and thus are a single species (Snow Goose) with two color morphs, but also that there was no assortative mating of the morphs, that there were no lifetime fitness differences between the morphs, that the morphs were determined by a fairly simple two-locus genetic system, and that females were more philopatric—more likely to return to their natal colony to breed—than males. Natal philopatry in females, and dispersal by males, had important influences on gene flow in this species. With some clever experiments on captive birds, Cooke's group also demonstrated that parental imprinting influenced the color morph that an individual bird preferred when it came to choosing a mate (Cooke and McNally 1975). For the first two decades this research program employed mainly the classical tools of population genetics, relying on the analysis of morph frequencies to deduce the underlying genetics. Early attempts by the researchers to study protein variants met with limited success, but later they pioneered the use of DNA markers to get a better handle on the underlying genetic structure of the goose population (Quinn et al. 1987). Eventually this Snow Goose colony became so "successful" that it grew several fold, eventually eating itself out of house and home by destroying its own nesting habitat.

By 2013 the La Pérouse Bay colony was virtually gone, with most of the geese dispersing to other sites along the Hudson Bay coast.

SPECIATION

As we saw at the start of this chapter, Darwin's finches played a central role in the development of evolutionary biology. But as is now well documented, Darwin himself did not actually notice the diversity of finch species in the Galápagos, nor did he properly document where his specimens of those birds were collected (Sulloway 1982). It is unlikely that he would have made those mistakes had he collected birds on the Hawaiian Islands instead. For here—in the Hawaiian honeycreepers—is an adaptive radiation of birds so striking and so beautiful that many have extolled its value for the study of evolution and adaptive radiation. It was only in the 1890s that ornithologists started to pay attention to this fascinating group, though by then some of the known species had already gone extinct.

Alfred Newton was one of the first to take a real interest in the Hawaiian avifauna, publishing a short paper on those birds in *Nature* in 1892. Newton arranged for the collectors, Scott B. Wilson and R. C. L. Perkins to spend years in the 1890s on the islands, where they discovered several new species (e.g., Wilson and Evans 1890–99). Not to be outdone, Walter Rothschild (chapter 3) also sent his own collector, Henry C. Palmer, to Hawaii to amass his own collection over a period of years, again discovering new species, some on the verge of extinction. By the time this thinly veiled competition between Newton and Rothschild came to a close at the turn of the century, the honeycreepers had been well

63

described and illustrated in color, and much was known about the behavior of most species. Only one more species of honeycreeper (family *Drepaniidae*) was ever discovered, the Nihoa Finch, in 1917. Dean Amadon, who was to become the head curator of ornithology at the American Museum of Natural History, studied the honeycreepers for his PhD thesis at Cornell in 1947, the same year that Lack's *Darwin's Finches* was published, and came independently to many of the same conclusions as Lack about processes of adaptive radiation in birds (Amadon 1950).

We now know there were at least fifty-three species of Hawaiian honeycreepers before man came on the scene (Pratt 2005; Lerner et al. 2011), all descended from a single finch species that colonized the archipelago four to five million years ago (Tarr and Fleischer 1995). We know this because new methods of genetic analysis and computation allow us to reconstruct both the structure and timing of evolutionary changes within a lineage to an ever increasing level of accuracy and sophistication, a level far beyond what those architects of the Modern Synthesis could have dreamed possible in the 1940s. Using the mitochondrial cytochrome-*b* gene, for example, Rob Fleischer of the US National Museum and colleagues determined that in the honeycreepers the rate of change in DNA is about 2 percent every million years, that is, two base pairs out of every hundred have changed during that time period (Fleischer et al. 1998). The researchers accomplished this determination by comparing the genes of species across the volcanic islands in the Hawaii group that were known to have risen out of the sea at different times.

Striking young adaptive radiations are relatively rare in birds, with the Darwin's finches and Hawaiian honeycreepers being the best-studied examples (Schluter 2000). As Schluter has so well documented, the study of adaptive radiations has much to tell us about the processes of evolution in general, but they are a specific example of the more universal phenomenon of speciation. As Stresemann predicted in the 1930s, it is the study of speciation that eventually fulfilled Darwin's promise of a unifying evolutionary theory, as outlined in Mayr's (1963) second seminal book, *Animal Species and Evolution*. Although Mayr had laid the groundwork in that volume, it was the advent of modern genetic analysis in the 1980s that allowed the study of speciation to flourish. Molecular genetic analysis gave us a clearer picture of both the phylogenetic relations between taxa and a reasonable estimate of the timing of speciation events. The 1960s also saw an improved understanding of the timing of geological processes like continental drift, glaciation, and changes in sea levels—events that split species ranges into independent populations.

Robert Mengel (1964) was one of the first to attempt an analysis of bird speciation on a continental level, looking at speciation in the North American wood warblers. Using only morphology (and not genetics) to build phylogenies, Mengel argued that successive continental glaciations split the ranges of many species, creating barriers to gene flow and allowing the independent evolution of each population on either side of the divide, resulting in the creation of many new species over a period of a hundred thousand years. We now know, based on genetic information and a better understanding of the timing of glaciations, that some of the details of Mengel's conclusions were wrong, but his paper exemplifies the beginning of a new integrated approach to the study of avian speciation.

In 1974 Joel Cracraft showed how the marriage of cladistics and continental biogeography could tell us about the splitting of early bird lineages during the breakup of Gondwanaland. Analyzing the morphologies of ratites (emus, ostriches, rheas, cassowaries, and kiwis) alive today, Cracraft demonstrated both that cladistic analysis (chapter 3) could be used to formulate a testable hypothesis about bird phylogeny and how that phylogeny mapped onto the newly confirmed phenomenon of continental drift. He thus provided both a pattern of bird evolution and a plausible mechanism to explain that pattern by allopatric speciation. Here, in a stroke, Cracraft (1974) exhibited the value of cladistics—then still shunned by most ornithologists (chapter 3)—for the study of evolutionary patterns in birds, and the importance of biogeography for understanding evolutionary processes. Subsequent work on the DNA of these birds by Charles Sibley and Jon Ahlquist (1990; chapter 3), as well as by Oliver Haddrath and Allan Baker (2001) of the Royal Ontario Museum, has confirmed the patterns that Cracraft uncovered and revealed that the timing of population differentiation into new species closely matches the time course of continental separation.

While most examples of speciation and its effects on morphology and behavior are inferred about events that happened long ago, occasionally we get to see speciation in action, resulting in barriers to gene flow that have an effect in just a few generations. In less than thirty generations, for example, populations of the Eurasian Blackcap that

had different, newly established migratory routes (see chapter 4) have diverged both morphologically and genetically (Rolshausen et al. 2009). Populations that migrate northwest from Germany to England have narrower beaks, rounder wings, and browner back plumage than those that migrate southwesterly to Spain. Because populations from the two wintering areas arrive back on the breeding grounds at different times, there is assortative mating by individuals taking different migration routes, resulting in the rapid evolution of ecotypes of the sort that result in speciation. Though these two migratory groups are not yet reproductively isolated, as is required for recognition as separate species, this is a clear example of the speciation process in action. Remarkably, it is also an example of the process of speciation in sympatry, where the two migratory groups co-occur during the breeding season, a process once thought to be nonexistent or at least extremely rare in animals.

On the Galápagos Islands the Grants witnessed a different kind of speciation event in progress in the finches (Grant and Grant 2010). During their research on Medium Ground Finches and Common Cactus Finches on the tiny island of Daphne Major, they witnessed an instance of incipient speciation that could have been documented only in such a long-term study. After eight years' work on the island, during which they had banded and measured more than 90 percent of the finches, in 1981 they caught an unusually large finch with an unusual song. Subsequent genetic analysis suggested that this bird, a male, was likely to be an immigrant from the nearby island of Santa Cruz and was also probably a hybrid between a Medium Ground Finch and a Common

Cactus Finch. The researchers followed the survival and reproduction of this bird and all of its known descendants for the next twenty-eight years, through seven generations. The original immigrant hybrid male mated first with a hybrid female, and all subsequent matings of all family members after 2002 were to members of this lineage. The unusual song—learned, culturally transmitted from father to son—seems to have isolated this lineage reproductively from the resident populations of finches on the island, maintaining the birds' morphological and genetic distinctiveness. Here was an example of incipient speciation resulting from both hybridization and a culturally transmitted premating isolating mechanism, providing at least preliminary support for an ecological process of speciation different from the allopatric model that Mayr had initially proposed and that formed the foundation of the Modern Synthesis. It is as likely as not that this small, distinct lineage of finches on Daphne Major will go extinct, but its existence provides an intriguing window on a role for song and behavior in the speciation process.

The study of adaptive radiation and speciation in birds is now quite mature and has been the subject of two comprehensive books since the year 2000—by Ian Newton (2003) and Trevor Price (2008)—summarizing both the principles and many examples. Reading these books, we get a sense that ornithologists have come a long way from the early days of the Modern Synthesis, in part due to the revolutions in systematics, molecular biology, and computer analysis. There is also a sense that we still have much to learn, aided by new techniques in genome analysis, fossil discovery, and the analysis of continent-wide patterns using comparative methods.

Ornithology played a leading role in both the development and the application of evolutionary principles before and after the Modern Synthesis was hammered out. In this chapter we have barely scratched the surface of interesting evolutionary studies of birds, in part because there are so many, but also because many are discussed in other chapters. Some subjects like the origin of birds (chapter 1), systematics (chapter 3), the ecological adaptations for breeding (chapter 5), behavioral ecology (chapter 8), and sexual selection (chapter 9) are by their very nature evolutionary, and others are increasingly incorporating an evolutionary approach.

Since 2000 there have been so many technological advances in molecular genetics that recent studies might give us a glimpse into the future of evolutionary studies of birds. We highlight below a few of these that have been published very recently in high-profile journals, with examples directly related to the subjects of each subsection in this chapter.

Darwin's finches: Using a microarray analysis of gene transcription in different species of ground, cactus, and warbler finches, Arkhat Abzhanov and colleagues (2006) discovered that expression of the protein calmodulin is related to the development of beak shape. This means that they were able to identify the genes that influence beak elongation in a Darwin's finch embryo as it grows in the egg. The result is that there is evidence not only for how competition and the available food supply affect the evolution of bill shape in these birds but also for the underlying genetic mechanisms behind such an evolutionary change. It turns out that a simple regulatory pathway might hold the key via its effect on development. This is a nice example of the emerging field of evo-devo, looking at how evolution and development are intricately related.

Silkie fowl: William Bateson and Reginald Punnett (1911) studied the genetics of black skin pigmentation (called "hyperpigmentation") in the silkie fowl and discovered that it was controlled by a dominant allele and a sex-linked modifier gene. Exactly one century later, Benjamin Dorshorst and colleagues (2011) used newly developed genomics technology to find exactly where the mutation that causes this hyperpigmentation occurs in the silkie's DNA. Interestingly, they found that (1) the mutation is really a duplication and inversion of two regions of the genome that are relatively close together on the same chromosome, and (2) the mutation is a key factor in the regulation of melanin production. Thus where the earliest genetic work on this species looked for patterns, the latest work is focused on processes and the underlying biochemical mechanisms.

Modern Synthesis: By the time that the Modern Synthesis was put together, in the 1930s and 1940s, a great deal was known about Mendelian genetics, and that became the foundation for a truly integrative theory of evolution. While the basic tenets of the Modern Synthesis still appear to be rock solid, new discoveries about genes and the interactions between genes and environment are broadening our knowledge of evolutionary processes (Danchin et al. 2011). Two subjects that were virtually unknown during the 1940s—epigenetics and plasticity—have been the focus of recent

attention. Epigenetics is the study of environmental influences on DNA that result in changes to an organism's phenotype but, importantly, those changes are passed on to their offspring. There is evidence from mammals (Rosenfeld 2010)—but not yet birds—that a mother's condition can influence the genes in her growing embryo (i.e., maternal effects), changes that the embryo will pass on to its own offspring via its genes. Mother birds vary the levels of testosterone that they put into their eggs (Groothuis and Schwabl 2008), and it will be interesting to see if that has an epigenetic effect. Plasticity, on the other hand, refers to the ability of an organism to change phenotypically—not genetically—in response to its environment. But how does such plasticity evolve as an adaptive response to environmental variation? Anne Charmantier and colleagues (2008), for example, using data from a forty-seven-year-long study, found that Eurasian Blue Tits in Britain responded rapidly to a change in spring temperatures, nesting earlier as the average monthly temperature warmed up during that period. Such plasticity has important implications for the responses of animals to global warming and can tell us which species are most likely to survive rapid environmental change.

Polymorphisms: It turns out that both of the genetic polymorphisms that we described—in Snow Geese and Parasitic Jaegers—are the result of different mutations in the same gene (the melanocortin-1 receptor, MC1R) that influences the production of melanin in feathers (Mundy 2005). The Bananaquit that occurs on islands in the Caribbean comes in yellow and black morphs due to mutations in the MC1R locus as well, as does the black morph of the White-winged

Fairywren that occurs on two small islands off the coast of Western Australia (Doucet et al. 2004). In all four species, the dark morphs evolved from the light or colored morphs, with the mutations increasing the production and deposition of melanin. These four species are only distantly related, with different mutations in the MC1R gene causing the polymorphisms, thus providing a clear example of independent convergence in the general mechanism producing more melanin, with subtle differences in the genetic mutations. By understanding the underlying genetic and biochemical mechanisms for such polymorphisms, we begin to really understand the details of the evolutionary processes.

Speciation: The ratites have generated a lot of interest among ornithologists who study both systematics and evolution, in part because those birds have an ancient history but also because they pose some interesting problems about the early diversification of birds. Using an exhaustive analysis of DNA from the nuclear genes of these birds, John Harshman and colleagues (2008) have now answered two outstanding questions. First, they showed that the tinamous belong among the ratites, being most closely related to the emus, cassowaries, rheas, and kiwis and more distantly related to the ostriches. Second, this new phylogeny of the ratites (including now the tinamous) suggests that flight has been lost at least three times in this lineage—in the ostriches, the rheas, and the Australasian ratites (cassowaries and emus). As Harshman et al. (2008) show, this pattern of the loss of flight fits nicely with what we know about the breakup of Gondwanaland and subsequent continental drift.

Mengel's (1964) fascinating hypothesis about speciation in the North American wood warblers was based on earlier ideas about how glacial events might have promoted speciation and assumptions about relatedness among species based on plumage patterns and present-day geographic ranges. By the 1990s there were enough molecular tools available to test Mengel's ideas more rigorously by constructing phylogenies where the timing of branch nodes (speciation events) could be estimated reliably. Using mitochondrial DNA (mtDNA) analysis of the five species in the Black-throated Green Warbler complex, Bermingham et al. (1992) found some support for Mengel's model but, importantly, discovered that not all species had diverged directly from their eastern ancestor as Mengel had proposed. Then, using both mtDNA and a restriction fragment length polymorphism (RFLP), John Klicka and Robert Zink (1997) discovered that most speciation events in North American passerines apparently occurred before the Pleistocene glaciations, contrary to Mengel's proposal for the warblers. Many other studies engaged in this debate, but it seems to have been resolved by Jason Weir and Dolph Schluter (2004, 2007), who looked at speciation and extinction rates of North American birds in relation to latitude by studying the cytochrome-b gene of mtDNA. They found that speciation rates were higher at higher latitudes, with all of the speciation events that they examined occurring in the Pleistocene, as Mengel's model required. This sort of marriage of ideas and technologies to solve interesting questions is a pattern that occurred time and again in the development of modern ornithology.

BOX 2.1 Peter and Rosemary Grant

ORIGIN AND
DIVERSIFICATION

Rosemary grew up in the Lake District of England. Her choice of career as a biologist is not surprising, as she was strongly influenced by her early experience in the family garden and on the fossil-studded limestone hills. Her mother encouraged an interest in nature; birds were only a part of it. Her father encouraged an interest in medicine, and she helped him in his practice as a country doctor for a year. Her career began in earnest when she went to Edinburgh as an undergraduate and entered Conrad Waddington's genetics department. She learned quantitative genetics from Douglas Falconer as he tried out in his undergraduate lectures the chapters of his forthcoming and, as it turned out, highly influential book. The Edinburgh atmosphere was exciting, stimulating, and encouraging to the students lucky enough to be among the chosen few for the diploma course: six from Britain and six from overseas. For a PhD degree Rosemary decided to study genetic differentiation and speciation in land-locked char in Iceland. Before that happened, she took a year off to teach embryology, cytology, and genetics at

the University of British Columbia. It took a long time for her to visit Iceland, and then only as a tourist first and as a lecturer later. Her husband-to-be had complicated her life.

Peter's bird watching began when he was about four years old, but for quite some time it took third place to butterfly catching and bouncing a ball. I was a general naturalist, and I was encouraged in this branch of biology at school and at university. At Cambridge I was influenced by two ornithological ethologists, Bill Thorpe and Robert Hinde, and a naturalist, Hugh Cott, but also, not far offstage at Oxford, by David Lack and Niko Tinbergen and, more distantly, by E. B. Ford. Their combined work in animal behavior, ecology, and genetics became the heterogeneous framework in which Rosemary and I now seek an understanding of evolution in the natural world. David Lack's *Life of the Robin* stimulated me to catch one in a makeshift trap fashioned out of a fire-guard, and I still remember the thrill of holding that first bird. While an undergraduate I joined the Cambridge Bird Club. I learned from contemporaries that there were two grades of ornithologists, A and B. A grades caught migrating birds in mist nets on the frozen fens every weekend of winter and knew everything there was to be known about age, sex, and molts of each of the species that flew in from Scandinavia. I was grade B: keen to learn but without a clue as to how many species I had ever seen in my town, county, or Britain, in that year or lifetime. I was allowed to take birds out of a mist net once.

Rosemary and I met at the University of British Columbia, where I was a graduate student in ecology and evolution specializing in birds. Our joint incursions into ornithology took us on excursions into the forests and alpine meadows of British Columbia, as well as far away on the Tres Marías islands and adjacent mainland of Mexico for my PhD research. Part of the research involved getting specimens of birds for the University Museum. Once, having just gone to bed in a motel in Tepic, we were awakened by loud knocking on the door. Disturber of the peace was a tobacco executive from Rhodesia (as it then was) on vacation. We had done him a favor in giving him advice on where to find quail to hunt and then forgotten about it. There he was in the doorway at nine p.m., holding two Elegant Quail as a present. So we skinned them in record time, Rosemary in a state of fury close to outright rebellion (very dangerous with a scalpel in her hand), and went back to sleep. Months later the UBC curator (Lazlo Witt) told us apologetically, with a long face, that he had been forced to throw away one very badly prepared quail specimen, whereas the other one was in fine condition; the survivor was the specimen prepared by she-who-had-cursed-it! Perhaps cursing is a preservative. It was certainly caustic.

After PhD research we spent a year at Yale on a postdoctoral fellowship. Evelyn Hutchinson, my mentor, was a kindly genius. He taught by example to think imaginatively and creatively, for the sheer intellectual fun of it, not speculating without foundation but drawing upon (in his case) an unequaled breadth of knowledge and deep insights into ecological interactions. He influenced me more than any other single person. Conrad Waddington and Douglas Falconer were the equivalent for Rosemary. At Yale Rosemary, conspicuously pregnant, taught introductory biology to the all-male undergraduates and continued research in embryology and genetics.

After Yale we left ornithology for a decade-long experimental study of interspecific competition in the use of habitat by mice and voles. After that, a sabbatical leave gave us the opportunity of traveling to Europe and thinking about what would follow. With eyes turned back to birds, and minds directed toward ecological niches, character displacement, and the coexistence of competitors, we visited many museums to measure specimens and the Azores and Canary Islands to observe the living birds. These studies did not yield much. We also carried out a combined field and museum study of rock nuthatches in Iran, Greece, and former Yugoslavia. They had been described as the classical example of character displacement, which was crying out for a field and partly experimental study. The research was a fascinating exploratory exercise, but in the end we decided there were too many obstacles in the way of doing further meaningful fieldwork. For example, we could not enter Turkey or Afghanistan without substantial personal risk. As it was, I was once detained in Iran for half an hour while I explained to the military police why I was using a tape recorder on an artillery range while the country was at unofficial war with Iran. As if that was not risky enough, on the last day of fieldwork at Kotor in southern Yugoslavia I managed to simultaneously tape-record rock nuthatches and machine-gun fire on a firing range, with bullets ricocheting from the karst limestone above my head. We had a few anxious moments while crossing the border into Italy with the tapes.

Galápagos has been very different. Apart from a single theft from fishermen, apparently specializing in female underwear, we have been left in peace in the National Park to conduct research. The initial stimulus for the research was a letter of inquiry from Ian Abbott, a prospective postdoctoral fellow from Australia, and his wife, Lynette. Funding for it was in the balance until rescue came in the form of $4,000 from McGill University for two people to spend four months in the field (plus travel)! There were a few faltering financial steps after about twenty years, but those apart we have been fortunate to have financial support for an ultralong field study. David Lack's book on Darwin's finches provided the intellectual launching pad with provocative ideas about the role of competition in evolution. We had met him more than once and talked about island birds, but sadly he died in the month our fieldwork began. It would have been valuable to discuss finch evolution in depth with him. We look back on forty years of field research and still marvel at his acute insight based on a thorough understanding of birds. He was a grade A ornithologist. And so, I should mention, was another highly influential person, Ernst Mayr, who from our graduate student days onward became the authority on general matters of speciation. He seemed to approve of what we were doing when we last

saw him at the age of ninety-nine, lively as ever and full of strong convictions.

A personal highlight for us was the opportunity to take our two daughters, Nicola and Thalia, into the field, beginning in 1973 when they were eight and six years old respectively. It meant that we could camp and live as a family in our outdoor "laboratory," engaged in both professional and domestic activities in just one location. Those professional activities led to Rosemary's PhD degree. Nicola went off to college in 1983, thence to a medical degree, and has not returned to Galápagos, whereas Thalia continues to revisit Galápagos every year. There cannot be many families so fortunate to have such an experience as ours. We can say this because we have had no serious accidents. In the old days we had our fair share of delays in fishing boats arriving to collect us, and boats with us on board going astray, although not far. Other scientists have been in boats that got completely lost or ran aground. We have had no appendicitis or broken limbs on remote and uninhabited islands, nor have we fallen off an island or into a volcanic crater (although I once tried on Genovesa). And we will end the research in almost the same state of excitement as we began. At the beginning it was the discovery that simple, arid, island environments offer great scope for ecological research into the kinds of evolutionary questions about biological diversity that grip us. Now the excitement is the fate of hybrids we have been following for years, in particular a hybrid lineage of finches that, by not breeding with the other species, is functioning as an independent species! More than fifty years ago as British undergraduates we could never have foretold our career paths.

BOX 2.2 Arie J. van Noordwijk

I grew up in Amsterdam, and one of my early memories is that the teacher in nursery school had obtained a Montessori set of beads that allowed me to play with numbers up to 1,000. Playing with numbers was fun!

My mother is a biologist who later specialized in environmental education for primary schools, so I suspect my sister, my brother, and I must have had an atypical upbringing; we all ended up as biologists. I think there are three more elements that played an important role when I was a teenager. First, my mother was in politics, so pollution as well as conservation issues were often discussed at the dinner table. Second, around 1965 my father came home one evening and very enthusiastically told about a demonstration of a (mechanical) calculating machine that allowed you to calculate a square root at the press of one button (followed by two minutes of rattling). He was a pharmacologist and taught me that statistics can be great fun. Dealing with variation has

always remained tricky but rewarding. The third element was being a member of a youth organization of naturalists that made excursions into the surrounding areas almost every weekend. I never became very specialized—plants, insects, birds, or aquatic life were all interesting to me. Setting up little research projects or organizing bird counts was what really interested me, and I remember many heated arguments over total numbers seen.

When I entered university in 1967, the dean of biology gave a speech telling us all to go away, because there would never be enough jobs for so many biologists. I was convinced that he was wrong, because environmental and conservation problems were getting more and more urgent. In my undergraduate years I was mainly inspired by two newly appointed professors in two new disciplines: theoretical biology, including both philosophy and mathematical biology, and population genetics. Both of these combined biology and playing with numbers.

The possibility that I could follow large numbers of individuals with known family trees in natural populations brought me to work mainly with birds. It was, in fact, the population geneticist Wim Scharloo who first thought we could use the data from long-term population studies to test whether heritable variation in life history traits could be demonstrated. In the decade before, enzyme polymorphisms had been demonstrated in all species investigated, but it was an open question whether there was any selection on these enzymes. I hoped to be able to demonstrate that ecologically important traits also showed significant genetic variation. If I could not find this in one of the best possible data sets, from the long-term study of Great Tits initiated by Huijbert Kluijver and continued under

Hans van Balen, there would be a contrast between enzyme variation and life history traits. In that period, the prevailing idea was still that one could separate ecological and evolutionary timescales. My major challenge was to separate a family resemblance due to common environment and a resemblance due to shared genes. At that time there was a fierce debate about the heritability of human IQ scores, a debate that was quite helpful to me, because all the methodological problems arising from interpreting family resemblance in heterogeneous and structured environments apply equally to data from natural populations.

The documentation of heterogeneity, the consequences for maintaining genetic variation, and methodological problems due to heterogeneity have remained major elements in all of my subsequent work. I have always been amused and intrigued by the very different conclusions that people have drawn from this same simple theory. For example, a minority of scientists draw the conclusion that trade-offs are not so important in shaping the variation that we can observe, claiming that variation in the amount of resources available to individuals is nearly always more important. Far more people draw the conclusion that in order to demonstrate trade-offs, one has to control for variation in resource acquisition. Both conclusions are correct, but they illustrate the different approaches.

Kluijver started to study Great Tits by ringing all nestlings and identifying all breeding birds. His stated aim was to bridge the gap between processes at the level of individuals and those at the population level. Thus he thought that dissecting population level processes into the actions of all individual members of the population would draw out the importance of interactions and make the variation among individuals visible. Eighty years later, hole-breeding passerines in general, and Great Tits in particular, are still an excellent model system to study. And there are still many important questions to be answered about the connection of processes at the individual level and their population level consequences.

After our initial demonstration of heritable variation in life history traits, a number of PhD students worked with me to determine how much of the environmental variation in body size, laying date, and clutch size could be explained by environmental factors measured at the level of the individual territory. Another main line of our research was based on the discovery that a substantial part of the individual variation in behavior was not random but, rather, followed distinct patterns that allowed one to predict how a particular individual would behave in a new situation. The great similarity with the description of human personalities by psychologists led to the study of animal personality.

Over the past four decades, electronic miniaturization and advances in molecular biology have created enormous new opportunities for our research. However, long-term studies where the basic data are collected in a uniform way are still essential in providing insights into evolution at work. Human-made changes in land use and climate also give us the opportunity to study evolution at work at high speeds and allow us to make rapid progress in answering the question as to what should be measured in order to predict the limits of evolutionary adaptation. Birds have and will contribute disproportionally to our insights in basic ecological, behavioral, and evolutionary processes.

CHAPTER 3

Birds on the Tree of Life

... so many distinguished investigators have labored in this field in
vain, that little hope is left for spectacular break-throughs.

—ERWIN STRESEMANN (1959), ON THE DIFFICULTIES INVOLVED IN FIGURING
OUT THE RELATIONSHIPS AMONG THE VARIOUS ORDERS OF BIRDS

LORD ROTHSCHILD'S FOLLY

WALTER ROTHSCHILD WAS AT ONE TIME A very rich man. Heir to the Rothschild banking empire, he was born to wealth and privilege. His father, Natty, was one of the great Jewish bankers of nineteenth-century Britain, head of the family's merchant bank,[1] who built a vast fortune from venture financing, diamonds, the Suez Canal, and loans to needy governments. By the time Walter was born, the family was arguably the wealthiest in the world—but Walter was passionate about birds and butterflies, and it soon became obvious that he would not follow in his father's footsteps. Fortunately for Walter, his father was quite indulgent, giving him the family estate at Tring in 1908 and the almost unimaginable sum of £1 million[2] for his museum and his growing collections

(Rothschild 1983). By 1930 those collections were among the biggest and best in the world. But suddenly a looming bankruptcy—due in part to blackmail—forced Walter in 1931 to sell his bird collections to the American Museum of Natural History (AMNH). In this incredible stroke of bad luck for Walter, the power base in bird systematics—together with Ernst Mayr, one of the key figures in our story—moved from Europe to America.

Walter was an imposing figure at six feet three inches tall and weighing more than three hundred pounds, but a debilitating stutter kept him relatively shy, shunning social interactions and instead devoting himself to his collections (Rothschild 1983). He was not without confidence and opinion, though, and he used his considerable wealth and power

Cassowaries: clockwise from upper left, *Casuarius casuarius, loriae, uniappendicularis*, and *philipi*. These were all named as separate species by Rothschild (1900), but *philipi* is now considered to be a subspecies of *uniappendicularis*, and *loriae* a subspecies of *bennetti*. Paintings by John Gerrard Keulemans.

to influence both politics and zoology. He worked hard and had a prodigious—probably photographic—memory that served him well in his pursuit of ornithological prestige.

As a boy Walter had spent much of his time in the countryside around Tring, collecting things for a small museum in his bedroom. Having decided upon a life of building collections, Walter began to establish his private museum in 1889 at Tring, where his parents had built him a separate museum building[3] as a twenty-first birthday present. Tring Park, the family estate, 50 kilometers (31 miles) northwest of London, was vast, with the main building set in 121 hectares (about 300 acres) of forest and farmland. Walter housed his collections in drawered cabinets made of fine mahogany and glass cabinets for display. No expense was spared to build, house, and curate those collections, and Walter clearly wanted them to be the best in the world, particularly with respect to the diversity of species and the abundance of rare and exotic material. While the vast majority of his collections were in the form of standard museum specimens, quite a few were mounted in lifelike poses and from 1892 displayed to the general public, with three new wings added between 1906 and 1912 as the collection grew.

Walter Rothschild at his home in Tring with his zebra-drawn carriage (photo in 1895 at age ca. 27).

To make these collections, Walter hired men to explore the far corners of the world, under instruction specifically to search out new and rare species. He also sponsored a few full-scale expeditions to remote locations, including the Galápagos Islands, Australia, Timor, and New Guinea. In those days[4] some men made their living collecting zoological and botanical specimens for museums and wealthy private collectors, commanding high prices for rare and exotic species—between 1890 and 1908 Rothschild employed more than four hundred collectors worldwide. Among his favorites were Rollo Beck and Albert Stewart Meek, as both men were uncommonly successful at obtaining new species for his collections, collections that were gradually becoming the largest and most prestigious in the world. By 1930 he had 300,000 bird skins and 200,000 eggs in the drawers at Tring, and more than two and a half million butterflies (Gray 2006).

In 1892 Rothschild realized that he needed to hire a curator; specimens were arriving too quickly for him to process. Each specimen needed a label and had to be put into its proper place alongside closely related species in the specimen cabinets. Like other museum curators, Rothschild wanted to use his collections to learn more about evolution and evolutionary relationships. He was particularly interested in discovering and naming the birds of the South Pacific and in speculating upon the relationships among different species that were clearly closely related because they shared so many features. He wrote more than 150 scientific papers and monographs based on the birds in his collection, including a comprehensive treatise on the cassowaries (*Casuaridae*) in which the morphology and plumage of every known species and subspecies was described in exquisite (excruciating!) detail, their ranges delimited, and the possible evolutionary relationships among them hypothesized based on plumage and skeletal similarities (Rothschild 1900). The cassowary work alone must have taken Rothschild months (or years) of painstaking effort, even though this is a small group of only three extant species. What about the other two thousand bird species in his collections? Enter Ernst Hartert, Rothschild's curator from 1892 to 1930, to whom we shall come back in more detail a little later.

With Hartert in charge of the bird collections, Rothschild was able to devote his considerable energies to analysis and writing, but also to a wide variety of political, social, and philanthropic endeavors. He was, for example, one of the architects of the 1917 Balfour Declaration for the establishment of a Jewish homeland in Palestine (Mallison 1973). Based on concerned reports from collectors, he also worried about the fate of the Galápagos tortoises, as they seemed destined for extinction at the hands of those very collectors. To save the tortoises, he hatched a scheme to transport live individuals of each species to Tring for breeding and preservation; fortunately, this wildly ambitious plan never got out of the drawing room.

From the 1880s until the 1920s, Rothschild was on top of the bird world with his unparalleled collections and phenomenal productivity. By 1929, however, trouble was looming on two horizons. First, tax problems threatened his ability to keep paying his collectors and curators. If this was not worrisome enough, one of his mistresses[5] began to blackmail him. The details of this blackmail have never emerged, and Rothschild certainly never identified the blackmailer.

It seems clear, however, that he continued to pay the blackmailer for forty years, either from sheer embarrassment or possibly because he had business or political dealings with the mistress's husband (Rothschild 1983). We will probably never know. Thus Rothschild needed money, and in sheer desperation he resolved to sell off his beloved collection of birds. We don't know why he sold the birds instead of the butterflies, but it is certainly possible that the birds were more readily salable, and he may have thought that his future happiness and fame would be more assured by the butterfly collection.

The upshot was that he sold almost the entire bird collection[6] to the AMNH in New York, in October 1931, for $225,000 US,[7] about a dollar per bird.[8] In the following months his collection of birds was boxed up in 185 wooden crates and shipped across the Atlantic. Rothschild was understandably depressed by the loss and only ever published one more paper on birds. By then Hartert had already retired to Germany in declining health. On learning of Rothschild's selloff, Hartert was devastated, as related by his lifelong friend and colleague Erwin Stresemann: "I can never forget how, on a gray February morning, he [Hartert] came staggering in to me with an envelope in his fingertips, and sank into a chair. "My collection! My collection!" he stammered out, his chest heaving and his clear eyes swimming with tears."[9]

How was the AMNH able to swing this amazing deal, bringing the world's best private collection of birds to America? In the 1920s the AMNH was among the finest museums in the world, but it still vied for status with several others in America. Leonard C. Sanford was on the board of directors and was determined to make the AMNH the very best museum, especially in friendly competition with the Peabody Museum at Yale. When Rothschild told Sanford that his collection was for sale, Sanford approached one of the museum's benefactors, Gertrude Vanderbilt Whitney, to put up the funds, which she did without hesitation (Gray 2006). Inadvertently, Rothschild is both hero and villain of this chapter as, without his dalliance gone wrong and his pending bankruptcy, his collections would probably have stayed in Europe, and the development of both systematics and evolutionary biology in the twentieth century would no doubt have taken a very different course.

COLLECTING BIRDS

The great bird collections, like Rothschild's, depended upon legions of collectors brave—or mad—enough to explore the wildest corners of the world for new and rare species. As early as the seventeenth century, men like George Marcgrave traveled the tropics in particular, bringing home specimens and paintings of the new species they discovered. That sort of explorer saw its heyday in the 1800s, with Alfred Russel Wallace, William Henry Hudson, Henry Walter Bates, and others financing their own expeditions through the sale of specimens and books about their travels and adventures—and making important scientific discoveries[10] themselves. The difficulties faced by these explorers makes one simultaneously wonder about their sanity and appreciate their love for exploring nature (Conniff 2010).

What drove these men to build collections and to explore dangerous and unforgiving lands, in the pursuit of new and exotic species? Certainly ego, or at least its more polite cousin, fame, was a factor for some. Sanford at the AMNH was clearly driven by his competitive spirit to build that museum. Rothschild was determined to have the best collection, carelessly sparing no expense to get specimens: "Walter was relentlessly victimised, bamboozled and deceived by dealers and collectors, who felt they were poor fellows scraping around for a living, whereas he was a man backed by unlimited wealth and affluence."[11] But he also loved natural history: "His amazement at their richness and endless variety was renewed every morning, and endured all his life."[12] No doubt both collectors and museum men reveled in the discovery of new species and marveled at the beauty of the birds they killed. Paradoxically, until well into the twentieth century, the best way to see a bird close up, to really appreciate its beauty, was to shoot it.

As is often the case, new technology also fueled the collecting binge in the 1800s, with the development of taxidermic methods (Milgrom 2010). Prior to the 1800s, mice and the dermestid beetle were the scourge of dried specimens, devouring them within a few weeks of collection unless they were sealed under glass: "Those who had begun to make any [collections of birds] soon became weary of going on, having had the Mortification to see them every Day destroyed by ravenous insects."[13] In the 1740s the French pharmacist and ornithologist Jean-Baptiste Bécoeur discovered that wiping the inside of a skin with arsenic soap[14] would deter even the most determined pest, though his methods were kept secret until the end of that century (see Rookmaaker et al. 2006). Once the arsenic soap method became widely known, explorers no longer had to preserve their specimens in heavy, dangerous, and volatile

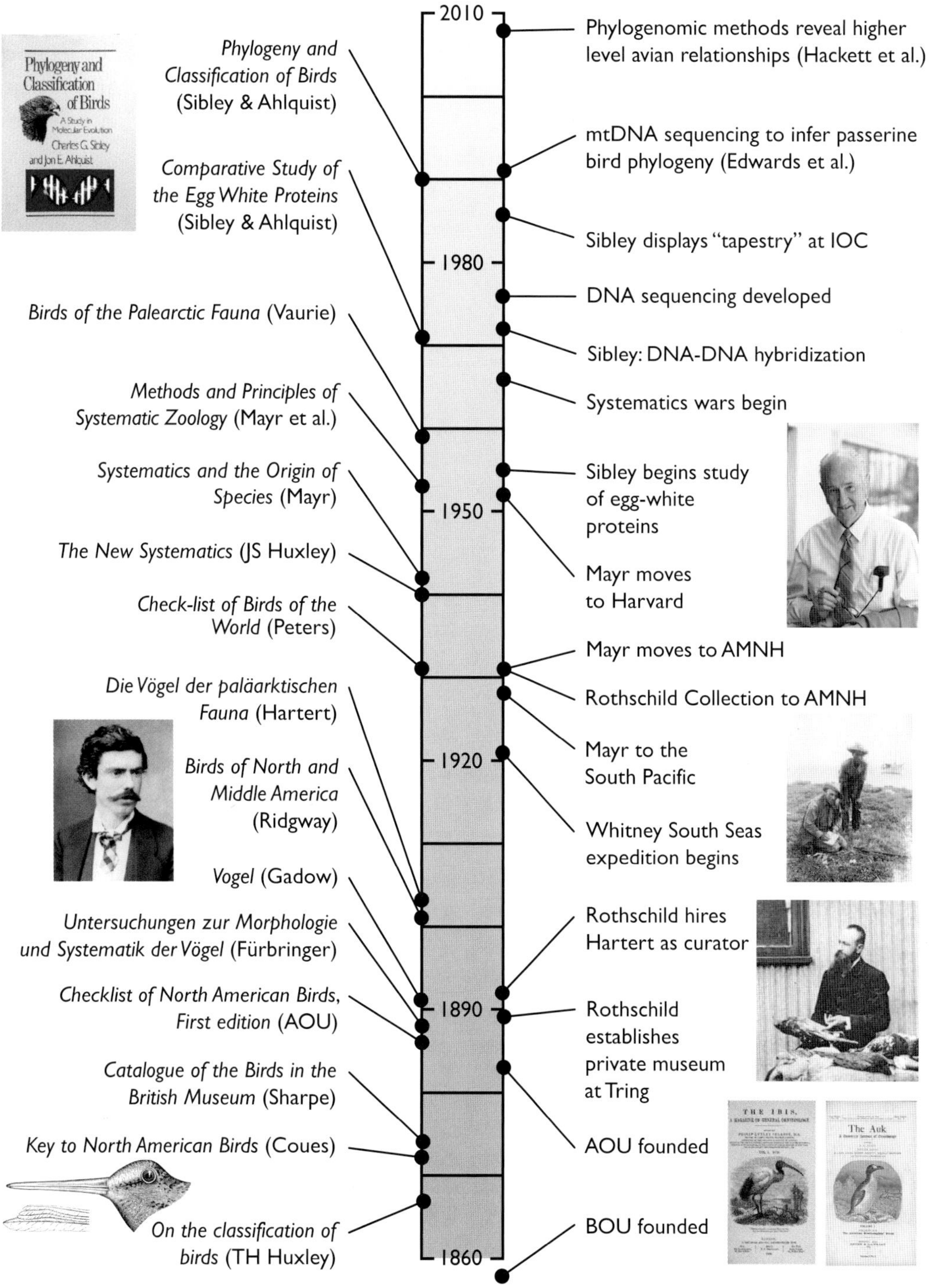
2010
1980
1950
1920
1890
1860
Phylogeny and Classification of Birds (Sibley & Ahlquist)
Comparative Study of the Egg White Proteins (Sibley & Ahlquist)
Birds of the Palearctic Fauna (Vaurie)
Methods and Principles of Systematic Zoology (Mayr et al.)
Systematics and the Origin of Species (Mayr)
The New Systematics (JS Huxley)
Check-list of Birds of the World (Peters)
Die Vögel der paläarktischen Fauna (Hartert)
Birds of North and Middle America (Ridgway)
Vogel (Gadow)
Untersuchungen zur Morphologie und Systematik der Vögel (Fürbringer)
Checklist of North American Birds, First edition (AOU)
Catalogue of the Birds in the British Museum (Sharpe)
Key to North American Birds (Coues)
On the classification of birds (TH Huxley)
Phylogenomic methods reveal higher level avian relationships (Hackett et al.)
mtDNA sequencing to infer passerine bird phylogeny (Edwards et al.)
Sibley displays "tapestry" at IOC
DNA sequencing developed
Sibley: DNA-DNA hybridization
Systematics wars begin
Sibley begins study of egg-white proteins
Mayr moves to Harvard
Mayr moves to AMNH
Rothschild Collection to AMNH
Mayr to the South Pacific
Whitney South Seas expedition begins
Rothschild hires Hartert as curator
Rothschild establishes private museum at Tring
AOU founded
BOU founded
Phylogeny and Classification of Birds
THE IBIS,
The Auk

TIMELINE for SYSTEMATICS. Left: Cover of Sibley & Ahlquist (1990); Robert Ridgway; American Woodcock and detail of outer three primaries from Coues's *Key to North American Birds* (1896). Right: Charles Sibley; Rollo Beck kneeling at a nest with eggs, on the Whitney South Seas expedition; Ernst Hartert holding a Spotted Eagle specimen (photo 1898 at age ca. 39); covers of the first edition of *Ibis* (1859) and *The Auk* (1884).

alcohol for the long trip home, and taxidermy blossomed (Milgrom 2010). Using arsenic soap, Darwin was able to preserve almost five hundred bird specimens during his five-year voyage around the world, most of which are still in reasonably good shape 180 years later. This taxidermic revolution was directly responsible for the popularity of the bird (and butterfly) display cabinets—often with dozens of birds perched in more or less life-like poses—that graced the drawing rooms of Victorian England, and to the creation of lifelike dioramas in the great museums. Now everyone could see even exotic birds, mammals, fish, and reptiles close up, resembling how they might appear in nature, in displays that would last more than a few weeks.

The great German ornithologist Ernst Hartert, Rothschild's bird curator, began as a collector. Born a month before the publication of Darwin's (1859) *Origin of Species*, he was fortunate to be the son of an army officer who moved around quite a bit, meaning that he was able to watch birds and collect their eggs in a wide variety of habitats and locations in present-day Germany, Lithuania, and Poland.[15] As a teenager he learned to prepare study skins, expanding his own collections and making some money working for others. At twenty-three he spent three or four months exploring the moors and marshes of Masurenland (now Poland), collecting birds and eggs for the prominent German ornithologists Eugen von Homeyer and Eugène Rey (Rothschild 1934).[16] At twenty-six he volunteered to go on a sixteen-month-long expedition, led by Eduard Flegel, to western Africa to collect specimens for the Berlin Museum. No sooner did the expedition arrive at the Benue River, in Nigeria, than everyone fell sick with fever. To add insult to illness, their steamer was too large to go up that tributary of the Niger, so Hartert and Flegel returned to Brass at the mouth of the Niger, while the rest of the expedition went on by canoe. Flegel and two expedition members were so ill they had to be sent home, but Hartert recovered and proceeded inland, collecting birds as he went (Rothschild 1934).

For the next five years (1887–92), Hartert was almost constantly on the move, commissioned to collect beetles and butterflies in the East Indies, as well as birds in Indonesia, India, Tibet, Aden, Venezuela, and the Dutch West Indies. The Venezuela-Caribbean trip was commissioned by Lord Rothschild, who wanted both birds and butterflies, "... in particular to find the marvellous Humming-Bird *Heliangelus mavors*, then known only from a single example...."[17] Upon Hartert's return from that trip in the fall of 1892, Rothschild hired him to be his curator of birds at Tring. Besides his expertise as a collector in particular, and as an ornithologist in general, Hartert had already distinguished himself as a brilliant systematist and a productive scientist. As we shall see in the next section, his eventual output of scientific publications made him one of the leading ornithologists of the early twentieth century. Even as full-time curator at Tring, he made many productive collecting trips to Morocco, the Channel Islands, the Engadine (a long valley in Switzerland), the Pyrenees along the France-Spain border, Algeria, the Sahara, and Cyrenaica (eastern Libya).

When they needed a bird collector, Rothschild and Hartert sometimes turned to Rollo Beck of California. Beck was a rare bird himself—an enthusiast, a marvelous naturalist, and a tireless field man. Like Hartert, he learned to make museum skins as a teenager, opening the door to earning a living as a collector. On his very first collecting trip, to the Channel Islands off the Southern California coast in 1897, he found the first nests of the endemic Island Scrub Jay and collected several clutches of their eggs. Later that same year he embarked on his first collecting trip for Rothschild, joining an expedition to the Galápagos to study both giant tortoises and land birds and to collect specimens for Rothschild's museum.

Back in California, he was hired by L. M. Loomis, director of the California Academy of Sciences, to collect seabirds in Monterey Bay and around the Channel Islands, as well as on the Revillagigedo Islands, Mexico (Pitelka 1986). Then, in 1905–1906, because of his prowess as a sailor, naturalist, and collector, the Cal Academy hired Beck to organize and command an ambitious expedition to Cocos Island and the Galápagos, with specialists in malacology, herpetology, entomology, botany, and ornithology on board. The result was the most important survey of these islands ever done, bringing to the Cal Academy the most comprehensive collections from Galápagos since Darwin's visit almost seventy years earlier. As they sailed home to San Francisco on the schooner *Academy*, the great earthquake of 18 April 1906 struck, devastating the city. The academy building, then on Market Street downtown, collapsed and burned, destroying most of the collections.[18] The specimens from Beck's Galápagos expedition thus formed the nucleus for rebuilding the museum's collections after the earthquake.[19]

In 1907–1908 Beck collected waterbirds for Joseph Grinnell's studies of California birds (e.g., Grinnell 1912), then went to Alaska to gather specimens and observations for Arthur Cleveland Bent's twenty-one-volume *The Life Histories of North American*

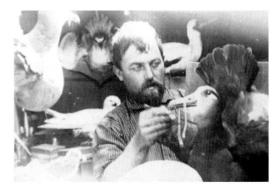

Rollo Beck (at age ca. 35) with albatross specimen on the Galápagos expedition of 1905–1906 for the California Academy of Sciences.

Birds (Bent 1919–68), for which he was paid handsomely by Sanford on behalf of the AMNH. Sanford also sent Beck and his wife, Ida, to South America for two years; there they collected specimens that provided a cornerstone of Robert Cushman Murphy's (1936) *Oceanic Birds of South America*. Impressed by Beck's prowess in the field, Sanford turned to him again in 1920 to lead the colossal Whitney South Seas Expedition for the AMNH (more on that below).

While collectors like Beck and Hartert stocked the world's ornithological collections, documenting avian biodiversity and biogeography and providing material for the systematists to ponder, their pursuit of the rare and vanishing was not without controversy. Museums wanted rare specimens, and the museum men urged their collectors to get everything they could, especially species that were already endangered. For example, here is Henry Fairfield Osborn, president of the AMNH, writing to Sanford in 1922:

> I want to express, too, my appreciation of the enthusiastic and persistent way in which you are filling the gaps in our Bird Collection, especially the rarities and antiques. This will make it, in time, one of the great collections

of the world, second only to the British Museum. The only way to get these rare things, like the king penguin, the auk, is to keep after it year after year until we get them; make out a list of what we want and never say die until they are actually in the Museum.[20]

For the museum men, Beck was the best collector because he could deliver everything they wanted, and they took little notice of the objections levied against him (and them) for his overzealous collecting. Beck had already been blamed for collecting eleven Guadalupe Caracara on Guadalupe Island off the Baja California coast (Abbott 1933). The small, island population of the Guadalupe Caracara was already being decimated by goat herders because they thought the bird was preying on their flocks and is now extinct. Beck also killed three of the four remaining individuals of the unique race of Galápagos tortoise (*Geochelone nigra abingdonii*) from Abingdon Island, also now, not surprisingly, extinct.

Even more eccentric and notorious was Colonel Richard Meinertzhagen, who occasionally collected for Rothschild (Garfield 2007). Meinertzhagen was an aristocrat and did not need the money, collecting instead for fame but eventually achieving infamy instead. He was a prolific author of scientific papers and popular books about birds and a valued friend, confidant, and colleague of many of the world's leading ornithologists, writers, and politicians. But he was also, as it turns out, a pathological liar and self-promoter, fabricating just about everything about his work, stealing specimens from major museums, changing labels on specimens, and quite likely murdering his wife so he could take up with a much younger woman. As one reviewer of Brian Garfield's

Richard Meinertzhagen holding a recently killed Kori Bustard in Kenya (photo in 1915 at age ca. 37).

(2007) book on Meinertzhagen pointed out, "Why, precisely, this man felt the need to embellish his life story so extravagantly is—despite Mr. Garfield's admirable labors—destined to remain a mystery."[21]

Through the first half of the twentieth century a few intrepid bird collectors[22] still sold specimens to museums, but the majority of discoveries and collecting was done by museum-sponsored expeditions. These were often incredible enterprises, involving tens, hundreds, or even thousands of people, extending for months or years at a time, funded by museum benefactors. One of the great museum expeditions was mounted in 1909 by the AMNH to explore the Congo River basin, with Herbert Lang as leader. Lang was a mammalogist and photographer who had worked at the museum as taxidermist and collector since 1903. He was only thirty years old at the time, and he chose the nineteen-year-old James Paul Chapin, a Columbia University student and museum volunteer, to be his assistant. Intended, and funded, to be only two years long, the expedition was terminated after six years because

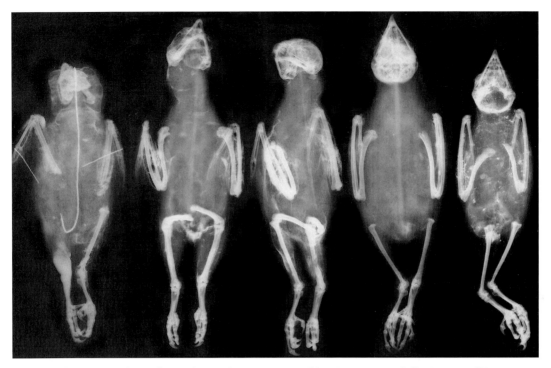

Until 1997 the Forest Owlet was known from only seven museum skins. Rasmussen & Collar (1999) used X-rays to establish the specimens' authenticity following Meinertzhagen's falsification of data on the labels.

of the start of the First World War. While their primary goal was to obtain specimens of the then mythical okapi (*Okapia johnstoni*) and square-lipped rhinoceros (*Ceratotherium simum*), Lang and Chapin worked tirelessly to survey the entire fauna of the region, eventually collecting 6,200 birds, 5,800 mammals, 4,800 reptiles and amphibians, 6,000 fishes, and more than 100,000 insects, some of them new to science. Lang was a superb photographer and Chapin an accomplished illustrator, so the expedition was well documented with 10,000 photos and 300 watercolor and ink drawings of the animals. Chapin's many paintings of fish are justly famous, but birds were his passion, and it was for his bird work that he made his name, eventually becoming a curator (and chairman) of the ornithology department

at the AMNH. His four-volume *Birds of the Belgian Congo*, published in 1932–39 was "a treatise on distribution, variation, and evolution which will be food for thought both for the field naturalist of the museum and the experimental zoologist of the university laboratory, happily linking, as it does, both aspects of the subject."[23] Like many great expeditions, the Lang-Chapin provided material that was studied for decades to come, resulting in more than 13,000 pages of research published in the scientific literature.

For more than a hundred years, starting in the late 1800s, the AMNH mounted dozens of expeditions to all corners of the globe to collect birds and other fauna.[24] By 1950 all this collecting had produced museum specimens—and sometimes lots of them—of males and females of virtually all of the world's 10,000 species of birds, a level of completeness that could be claimed for no other group of animals except the mammals. Several of the major museums worldwide[25] had more than 100,000 bird skins each—and a hundred smaller institutions and private collectors each held important collections of 10,000 or more skins. The result was an explosion of knowledge about bird distributions that accompanied the major strides in systematics that we describe in the rest of this chapter. From 1950 on, expeditions became much more targeted at specific taxa and habitats,[26] and collectors were encouraged to preserve whole birds—rather than just skins—for use in analyses of skeletal material, tissues, and DNA.[27] Starting in the 1980s, museum specimens began to be used for other purposes, to learn about diets by analyzing radioisotopes in feathers, to study historical incidence of pollutants, and to reveal the underlying mechanisms of color production. Museums also preserve birds' nests, clutches of eggs, and specimens in different stages of body and plumage development and molt. An old museum ornithologist even told one of our colleagues that "there's only so much you can learn from the living bird,"[28] suggesting that study specimens were the key to learning about birds.

By 1965 the era of independent bird collectors and private collections was just about over. Perhaps some thought the job was done—as most species had now been collected—and there was little fame and glory to be had in acquiring and naming new species. Three other factors probably had their influence as well. First, the increasing regulations in many tropical nations on both hunting and export has sometimes made collecting birds for museums a logistical

nightmare. Second, increasing awareness and concerns about conservation—some of which were poorly founded (Snyder 1958)—led many to shy away from collecting birds and also increased the regulatory burden. Finally, and maybe most significantly, the advent of high-quality field guides, binoculars, telescopes, and cameras meant that birds could be observed, identified, and "captured" at a distance, obviating the need to collect specimens just to be certain of species' identities for biogeographic and faunistic research.

As collecting became relatively less important, so did collections for the study of birds. As a result, bird study since the 1970s has been increasingly focused on behavior (chapters 7–9), ecology (chapters 5 and 10), physiology (chapter 6), and fossils (chapter 1). For complex reasons, the funding available for museums and museum work—work on the vast collections of bird specimens—declined precipitously toward the end of the twentieth century (McCarter et al. 2001; Thomson 2010). Nonetheless, recent work on bird coloration, evolution, biodiversity, and pollutants has shown the exceptional value of these collections, though wider recognition of this value is needed to ensure their expansion, and indeed in some cases even just their continued preservation. The days of massive bird-collecting museum expeditions that inspired the likes of Hartert in the 1880s, Beck in the early 1900s, and Ernst Mayr in the 1920s are probably gone forever.

GERMAN DYNAMOS

Hartert, along with Erwin Stresemann, laid many of the foundations of avian systematics during the first half of the twentieth century.

Though Hartert was thirty years older than Stresemann, and they worked in different countries, the two men were lifelong friends. Stresemann survived Hartert by forty years, but felt such affection for his old friend that he asked to be (and was) buried in the same grave,[29] with both names etched onto the tombstone (Haffer 2004a).

Systematics at the beginning of the twentieth century was a strange science. Through today's glasses most scientific endeavors from a century ago look primitive, especially with respect to technologies, but almost every other subject that we cover in this book was practiced in a way that still makes sense today, resulting in scientific publications that remain insightful and valuable for the data produced and the principles identified. Not so with systematics, as the subject was completely revolutionized in the 1970s with new ideas, then toward the end of the century with technological developments so amazing that they could not even be imagined in 1900.

Hartert and Stresemann were rather typical systematists of the early 1900s, but both were responsible for innovations that had some lasting influence on avian biology. Back in their day, the way to pursue a career in systematics was to become an expert on a particular group of birds, usually at the genus or family level, and publish your opinion on what distinguished one species (or subspecies) from another, maybe speculating how they were related—phylogenetically—both to one another and to other birds. To become such an expert, specimens were needed, and adult birds—particularly males in breeding plumage—were considered to be the most useful. A library was also needed, to see what others had written about the species of interest and their closest relatives; Hartert's

library at Tring had forty thousand volumes by the 1920s! The problem with this process is that it was highly personal and subjective, not the way science is supposed to be at all. Thus systematists became experts on this or that group of birds, and when they died much of their knowledge, and the methods by which they classified birds, died with them. No wonder there was so much controversy and so much reliance on authority rather than on hard data.

The result was that systematists in Hartert's day produced three kinds of publications: descriptions, catalogs, and faunistics. Hartert was an acknowledged master of all three, publishing more than 575 papers and books on birds during his thirty-eight-year-long career at Tring. Admittedly he, and many of his contemporaries, had a distinct advantage that today's scientists can only dream about in that they published largely without peer review, often in journals created specifically for the purpose by their research institution. Rothschild himself founded, published, coedited, and financed *Novitates Zoologicae* from 1894 to 1948, with most of the 42 volumes illustrated with exquisite drawings and color plates painted by the finest zoological illustrators of the day—no expense was spared. Hartert published 238 of his papers in *Novitates*, most with titles that sound desperately pedantic today (see Rothschild 1934).

Descriptions served two purposes. On the one hand, they provided extensive details about the type specimens of newly discovered species and subspecies. The type specimen is that single specimen of a species or subspecies that is first described in the scientific literature, and to which the scientific name is formally attached. Originally the type specimen was thought to be a typical representative by which all unclassified specimens could be compared, but that sort of typological thinking—wherein variation within species was not considered to be interesting—declined rapidly after Darwin (1859) presented his ideas on variation and natural selection (Wilkins 2009). The description of type specimens gave the author naming rights and thus considerable prestige in the business. Often those descriptive papers also included a key telling readers how to distinguish among closely related species and subspecies, thereby providing a foundation for classification.

Catalogs also served two purposes. Many simply summarized collections of birds, letting others know what specimens were held and where. But the best of them provided painstakingly detailed descriptions of species and subspecies and a historical record of all of the scientific names used previously for each species, a brief note on where the species could be found, and some attempt at classification, as described in the next section. One gets the distinct impression that all this was thought to be of the utmost importance, but these catalogs must, even then, have been useful to only a few specialists. The monumental *Catalogue of the Birds of the British Museum* (Sharpe 1874–98) ran to a staggering 27 volumes and included 374 superb color plates. That catalog was started by Richard Bowdler Sharpe in 1874 but was not completed until 1898, with the help of ten experts,[30] including Hartert, and four artists,[31] each of them devoting months or years to their tiny sphere of expertise. The ultimate goal was not only to provide a summary of the museum's holdings but also to list all of the known birds of the world and their taxonomy.

Faunistic works were designed to summarize all that was known about the bird fauna of a specific region, usually one defined by political rather than biotic boundaries. Like descriptions and catalogs, faunistic works presented detailed descriptions of plumages and soft parts but also summarized in much more detail where birds could be found, both geographically and ecologically, and made some attempt at organizing species taxonomically. Thus faunistic books and papers provided some of the earliest ideas on both the biogeography and ecology of birds, and they often put classifications in a biogeographic context. Though faunistic works were initially of interest mostly to scientists, by the mid-twentieth century they began to be popular among birders such that no region was too small to have its bird fauna summarized, and the market for these works remains unabated, as dozens more still appear every year. The best of these, like the recently completed (November 2011) sixteen-volume *Handbook of Birds of the World* (del Hoyo et al. 1992–2011), is a rich trove of information for both birders and professional scientists.

"Hartert knew more birds of the world probably than any other ornithologist, but, together with this general knowledge, he had specially complete knowledge of the avifaunas of several widely separated regions."[32] Just after returning to Germany from a collecting trip to India in 1890, he was hired to write a catalog of the birds in the museum of the Senckenberg Society in Frankfurt, then, at the relatively tender age of thirty,[33] he was commissioned to move to London to research and write the section on swifts and goatsuckers for Sharpe's *Catalogue of the Birds in the British Museum*. He completed that task in just eight months—an astounding accomplishment. Throughout his career at Tring he devoted much of his considerable energy to acquiring specimens of palearctic birds and thinking about their relationships to one another, resulting in his magnum opus, *Die Vögel der paläarktischen Fauna* (Hartert 1903–32), published in sections between 1903 and 1932 but never fully completed. In this faunistic work, in particular, he applied his revolutionary ideas on the recognition of subspecies and the use of trinomials (see below), eventually convincing everyone else to follow suit.

Erwin Stresemann was even more influential to ornithology but is almost as poorly known among twenty-first-century ornithologists. Why then has it been said that "Stresemann was clearly the most influential ornithologist of the twentieth century";[34] that his book *Aves* was "the most important single-volume compendium of ornithology ever published";[35] and that "none of his contemporaries had as great an impact as Stresemann on the study of birds and on the development of ornithology as an integral part of biological science."[36] And Stresemann was not short of famous contemporaries in the ornithological world—working professionally from 1921 until his death in 1972, he was a leading researcher in the heydays of Ernst Mayr, David Lack, Julian Huxley, Robert MacArthur, Charles Sibley, Konrad Lorenz, Niko Tinbergen, and Peter Marler, all of whom are much better known today. Why is Stresemann so poorly remembered? The answer is complex but includes the realities that most of his writings were in his native German; that after the Second World War Germany was unpopular in Britain and the United States; and that Stresemann was eclipsed by others, including Mayr and Lack.

Stresemann published his first paper—on a successful cross between a male Common Redpoll and a female European Goldfinch—at sixteen and appeared to be destined for a career in ornithology. He even worked for short periods between the ages of nineteen and twenty-three in Lord Rothschild's collection. At twenty-one he explored some islands in the South Pacific to look for birds, discovering (and collecting) several new species and subspecies on the islands of Bali, Seran, and Buru. But in those days, early in the twentieth century, the only way to make a living as an ornithologist was to work in a museum, and such positions were scarce indeed. So, as did many other eventually great ornithologists of the day, Stresemann started out studying medicine. Like the others he soon switched to a degree in zoology, graduating in 1920. Because of his expertise with birds—and, no doubt, his unbridled dynamism—he was asked in 1914, while still studying medicine, to write the *Aves* section of the *Handbuch der Zoologie* (Kükenthal 1923), and this may have been the stimulus that made him realize that he really wanted a profession in ornithology.

The *Handbuch der Zoologie* is a renowned German reference work, begun by Willy Kükenthal in Berlin, attempting to cover the complete animal kingdom in eight massive volumes.[37] Stresemann submitted his draft of *Aves* in 1920—having been delayed by the First World War—and it was well worth the wait: nine hundred pages long and the first major work to move away from simple descriptions to a more comprehensive approach, in which he stressed how important it is to study the entire biology of a species (Stresemann 1927–34). The editors were so impressed by Stresemann's contribution that they immediately offered him the headship of the bird department of the Berlin Museum für Naturkunde [Zoological Museum]. This is the largest natural history museum in Germany, today housing more than 30 million specimens of animals, minerals, and fossils, including the most complete *Archaeopteryx* specimen ever found (chapter 1). Stresemann's appointment to this prestigious post must have been an incredible snub to older ornithologists in the rigidly hierarchical German system; Stresemann was only thirty-one and had just been awarded his PhD. Nonetheless, he immediately took over editorship of the influential *Journal für Ornithologie*, inspired (and obtained the funding for) many of his students to embark on ornithological expeditions of their own, built his own intellectual "school" of scientific ornithology, and eventually transformed his department into one of the world's outstanding bird collections.

The young Erwin Stresemann in Malaysia, where he is getting a tattoo with the sign of a headhunter (photo in 1911 at age ca. 22).

During his fifty-year career Stresemann published six hundred papers and thirteen books and monographs on birds,[38] supervised twenty-two PhD students (including Ernst Mayr and Bernhard Rensch), and wrote, in German, a book on the history of ornithology, *Entwicklung der Ornithologie*, in 1951, translated posthumously into English (Stresemann 1975). It is emblematic of Stresemann's incredible knowledge that he wrote the first draft of that book almost entirely from memory while holed up in Berlin during and immediately after the Second World War.

As editor of the *Journal für Ornithologie* (*JfO*) for thirty-two years,[39] Stresemann changed the face of professional ornithology in Europe, deliberately giving less space in the journal to faunistic works and more to studies of bird behavior, morphology, and physiology, in addition to well-illustrated papers on avian life histories. Stresemann's editorial innovations were later emulated by ornithological journals in Britain and America. To promote greater international recognition of work that he thought was important, Stresemann also translated papers by American (Margaret Morse Nice) and British (Julian Huxley, Edmund Selous) ornithologists into German for republication in *JfO*. He had great personal charm but was a surprisingly poor lecturer, and that may, in part, be one "reason why he was not more widely appreciated among general biologists."[40] Also, because his major works—*Aves* and *Entwicklung der Ornithologie*—were not written in English, they had only a limited audience.

As a systematist, Stresemann focused on the species as a biological unit—anticipating, and undoubtedly influencing, the development of the Biological Species Concept in the 1940s. He was bent on incorporating knowledge of everything known about the biology of a species when assessing taxonomic relationships, rather than just what could be learned from a study skin. He thus espoused the study of anatomy, development, reproduction, molt, plumage coloration, life span, physiology, migration, population studies, systematics, and classification to really get a handle on how one species might be related to another. Here too Stresemann anticipated and influenced approaches to taxonomy that were later developed in the 1960s and 1970s. Using these principles, he revised the taxonomy of genera considered then to be "difficult,"[41] like *Accipiter, Buteo, Pericrocotus, Terpsiphone,*and *Zosterops* in a way that made some sense (e.g., Stresemann 1923), at least until the advent of modern molecular tools. These revisions are classic examples of the Stresemann approach. He also made important contributions to the study of intraspecific polymorphisms (chapter 2), using this knowledge to inform the taxonomies of polymorphic species as well. Toward the end of his life, he published a major study on the molt of birds, including a detailed exploration of its relation to ecology and its value as a taxonomic trait (Stresemann 1967).

Stresemann rejected the nineteenth-century traditions that had dominated avian systematics, and it was due to him that "systematic ornithology and field ornithology . . . were reunited in the New Avian Biology from the 1920s onward."[42] Stresemann also emphasized the importance of geographic isolation for the formation of new species and abandoned typological thinking that said there was a standard morphology for each species, caring instead about variation and stressing the importance of mutation in the speciation process. His students—Mayr and Rensch, especially—went on to incorporate these ideas into a revolutionary unification of disparate fields that we now call the Modern Synthesis of evolutionary biology (chapter 2). Stresemann never gained much recognition for his contributions to that synthesis because he was "rather backward in his understanding of the mechanisms of evolution . . . [and] felt quite strongly that there were severe limits to the power of natural selection."[43] Along with many of his contemporaries Stresemann instead supported Lamarckian inheritance and orthogenesis[44] as the engines of evolutionary change. That philosophical position and his interests in nothing but ornithology may have prevented his fame from spreading more widely. With his students and ideas, however, Stresemann laid the foundation for much ornithological work to come.

THE GREAT TRINOMIAL DEBATE

Hartert's main claim to fame—and the subject of a huge proportion of his scientific output—was the description of narrowly defined subspecies, recognizing that much of the morphological variation within a species had a geographical basis, even suggesting that incipient speciation was occurring where isolated populations were morphologically distinctive. To fully appreciate the significance of this approach, we need to go back, briefly, to both Carl Linnaeus and Carl Bruch, a rather obscure German ornithologist who worked in the early part of the nineteenth century. Birds have been classified since the beginning of recorded history—it's human nature to want to classify things—initially by their edibility, then by habitats, ecologies, and morphological features, especially in the seventeenth and eighteenth centuries (e.g., Ray 1676). In the mid-1700s Linnaeus's great contribution was to provide an internally consistent way to name groups of animals and plants with Latinized words, labeling species with a binomial—for example, *Lagopus lagopus* for the Willow Ptarmigan—the first word indicating the genus, and the second the species. The two names together are unique for every animal species in the world.

The extensive collecting of specimens around the world that began in the sixteenth and seventeenth centuries revealed all kinds of geographic variation—some scientists believing that each variant should be called a species, and others, including Darwin, calling each discernible variant a variety or race. The recognition of such variation at the population level led eventually to the overthrow of essentialism—the idea that species did not change through time—and the development of the Biological Species Concept. By the early 1800s it was clear that the Linnaean binomial system was inadequate to deal with the obvious within-species variation, so Carl Bruch proposed, in 1828, a trinomial system, adding a third word, the subspecies, to the scientific name of a species to indicate each different, geographically distinct, form. Thus the Scottish red grouse became *Lagopus lagopus scoticus*, one of the ten to twenty subspecies[45] of the Willow Ptarmigan recognized today. The idea caught on, possibly, in part, because by that time the vast majority of the world's bird species had already been discovered, and the trinomial idea breathed new life into the apparently universal human need to name things. There was lasting fame and prestige to be had, too, as naming subspecies allowed one to honor, and be honored by, one's colleagues, mentors, and benefactors.

Hermann Schlegel (1844) was the first to use trinomials routinely for birds, and the first AOU checklist (American Ornithologists' Union 1886) listed trinomials for each species where geographic variation was evident. Elliott Coues was the American champion of trinomialism, introducing it in his *Key to North American Birds* (Coues 1872).[46] In the words of the reviewer in *Science*, Coues's *Key* "unquestionably had a career of usefulness, and has helped on the advance that has so strongly characterized the last decade [1870s] of North-American ornithology; the object of the treatise being to enable any one … to identify his specimens without recourse to other information … [and is] to some extent, at least in its methods, an innovation in zoology."[47] Coues was a surgeon in the US Army when he did much of his bird work, retiring only in 1881 to devote himself full time to ornithology as professor of anatomy at Columbia University in New York City. He was a dominant force in American ornithology, being a founding member of the AOU, the editor of its journal (among others), and a gifted naturalist and systematist.

Largely due to tireless promotion by Coues in America, and his influence on ornithologists like Hartert and Henry Seebohm in Britain,[48] trinomialism was the order of the day in both scientific and popular works on birds—and as a result in most of zoology—by the early years of the twentieth century. By then the designation and naming of subspecies was both standard practice and the major industry of all working avian systematists. The effects of this revolution were both good and bad. On the good side, trinomialism codified the fact that species varied geographically by giving formal names to each distinctive geographic variant. No longer was each of these variants considered to be a separate species, and the apparent number of bird species worldwide was soon halved from 20,000 in 1920 to 10,000 by 1990. On the bad side, the gold rush to name subspecies consumed the lives of many professional ornithologists and filled the pages of faunistic and descriptive publications with information that seems to us to have had little lasting value. Even popular books on birds sometimes adopted the practice—Bent's *Life Histories of North American Birds* (Bent 1919–68) has a separate account for every American subspecies, with often boring or confusing redundancy. Similarly, Robert Ridgway's twelve-volume *Birds of North and Middle America* (Ridgway 1901–47) and James Peters's sixteen-volume *Check-list of Birds of the World* (Peters 1931–87) all focused on subspecies.

Controversy is endemic in the world of systematic biology, as we shall see, so it's not particularly surprising that trinomialism had its early critics and was initially even controversial among the converted. Coues and Ridgway felt that each population of a species that was morphologically distinguishable—but not too different—should be called a subspecies. Their operating principle was that the variation within each subspecies had to show some overlap with other designated subspecies—whereas those populations clearly related but not overlapping in appearance (or other morphological traits) should be designated as species. The 1886 report of the AOU's committee on classification and nomenclature (on which both Coues and Ridgway sat) stated explicitly that "intergradation is the touchstone of trinomialism."[49]

Hartert took exception. Based on a concept that had been proposed by the German ornithologist Otto Kleinschmidt,[50] Hartert felt that even populations that were morphologically distinctive—where every individual in a given population was different from every individual in every other population of the species—should be called subspecies if it was clear (to his subjective assessment) that all populations really did belong to the same species. Thus Hartert's classification system limited the number of new species being named, instead resulting in a greater number of subspecies. This was an early manifestation of what later became the Biological Species Concept. The Americans were furious about this heresy, but Hartert gained support from many of the prominent European ornithologists of the early 1900s[51] and finally Stresemann in the 1920s. Hartert and Stresemann eventually won out over the Americans, as their new approach became the standard operating principle for systematic ornithology. But even Stresemann and his acolytes sometimes went too far by today's standards, lumping together every allopatric population (even distinct species) into a single species. It was Stresemann's student, Bernhard Rensch, who called a halt to this incessant lumping, pointing out that closely related species that were recently evolved in reproductive isolation deserved species status. Rensch (1929) called these groups of closely related and virtually indistinguishable species *Artenkreise*, a term that Mayr (1931) translated as "superspecies" and Amadon (1966) renamed as "allospecies."

These seemingly arcane distinctions about what a species is were actually useful, as they recognized the importance of allopatry in the speciation process—at least for birds—and laid the foundations for the Biological Species Concept, which was to dominate evolutionary biology for the remainder of the twentieth century (Haffer 2006b). Fortunately for those early systematists, bird species limits actually have a simplicity not found in many other animals, and certainly not in plants, and this simplicity probably fueled much of the early development of both systematics and evolutionary biology. Today we realize how misleading that worldview, defined by ornithologists, can be when we look at plants, insects, and invertebrates, for example. We now know that even birds do not always fit that simple notion of what a species really is (if anything).[52]

Trinomialism reigned supreme in bird taxonomy for the first half of the twentieth century, but in 1953 two entomologists sowed the seeds of dissent. And not just any entomologists, but none other than the young E. O. Wilson,[53] destined to become arguably the best-known biologist of the twentieth century, and his mentor, W. L. Brown Jr. Their critique was devastating: "We are convinced that the subspecies concept is the most critical and disorderly area of modern systematic theory—more so than taxonomists have realized or theorists have admitted."[54] Their criticisms were taken seriously by ornithologists because many of Wilson and Brown's (1953) examples came from studies of birds, and because their new colleague at Harvard, Ernst Mayr, worked on birds and was the leading systematist at the time. Some of this criticism had been anticipated by David Lack: "It is simpler and more accurate to describe subspecific variation in terms of geographic trends, and to omit altogether the tyranny of subspecific names";[55] and by Mayr himself: "Instead of expending their energy on the describing and naming of trifling subspecies, bird taxonomists might well devote more attention to the evaluation

of trends in variation."[56] Wilson and Brown predicted that "we shall soon begin to observe the withering of the trinomial and its cumbersome appurtenances . . . the ponderous subspecies lists gravely entered in a thousand catalogues . . . that so unnecessarily consume the few effective working hours a modern taxonomist has."[57] Little did they know how stubborn and feisty the avian systematics community would be.

The AOU eventually stopped listing subspecies in 1957 in the fifth edition of their *Checklist of North American Birds*, and the incessant naming of new subspecies fell off sharply following Wilson and Brown's damning review. But good arguments for and against the subspecies concept in bird taxonomy have appeared in recently published debates (e.g., a 2010 issue of *Ornithological Monographs* was dedicated to the subject). Despite the potential for a revolution brought about by cladistic and molecular phylogenetics, the trinomial is far from forgotten today, as almost all major works on birds still list the known subspecies.

PASSERINES FIRST

Looking at ornithological publications at the turn of the twentieth century, you might well, and correctly, think that taxonomists in those days spent most of their time and considerable energy naming birds—describing new species and subspecies, specifying and chronicling the historical details of type specimens, and arguing about what constitutes a subspecies. This is the realm of what we now call "microtaxonomy," the study and definition of species limits. The other major task of the taxonomist concerns the higher categories—genus, family, order—that comprise a class, like birds (Aves); that is, determining how they relate to one another, how they can be grouped together into a natural classification, and how to make that classification informative and accessible in a list. This is macrotaxonomy, and the goal, post-Darwin, has been to discover what the relationships among the main branches of the avian tree of life might look like, based upon evolutionary principles. One reason for the early obsession with microtaxonomy is that macrotaxonomy is a different, much harder problem, a resolution of which eluded ornithologists for all of the twentieth century, mainly for lack of the appropriate technologies.

If the defining of species and subspecies is fraught with controversy, at least objective definitions are possible. Not so with the higher categories where whim, convenience, authority, and stubbornness are all at play. The genus, for example, is an aggregation of species that are most closely related, but where to draw the limits? Sometimes those limits are narrowly drawn by taxonomists, especially when there are many closely related species; otherwise the genus would be too large and unwieldy. But one person's unwieldy is another's life's work, and just like the oscillating trends of lumping and splitting with respect to species and subspecies, the numbers of recognized species, genera, families, and orders of birds have ebbed and flowed seemingly without rhyme or reason since Linnaeus first proposed these categories. The changing attitudes toward lumping and splitting persist today, long after the vast majority of bird species have been discovered. Thomas Huxley, for example, said there were 2 major groups of extant birds in 1867, Leonhard Stejneger said 4 in 1885, Hans Gadow said 2 in 1891–93, Richard Bowdler Sharpe said 3 in 1874–98, and

Robert Ridgway went back to 2 in 1901. With regard to families just in the order *Passeriformes*, Gadow said 3 in 1888 but Sharpe said 49 in 1891; more recently Charles Sibley and Jon Ahlquist (1990) said 29, though today the IOC World Bird List (2013) says 125.

Even if taxonomists could agree on what constitutes any of those higher taxonomic categories, how could the relationships among them be summarized in a list? This is no small problem. The issue, of course, is that the tree of life *is* a tree, and not a linear sequence. All things alive today are at the tips of that tree's branches. With the right tools, we can determine the entire branching sequences all the way to the trunk—which species is most closely related to which, and how long each branch is—but which tip goes first on the list and which last? Taxonomists have generally agreed that major branches that came lowest off the main trunk should be listed first and so on, but time and again as we work our way up the tree more than one branch appears to emerge at the same place. Moreover, a linear sequence of names loses so much useful information. It is difficult to reconstruct a tree from such a simple list of scientific names within higher categories if no other information is available. The potential modern solution is to provide digital lists with molecular data that can be accessed on the web, reconstructing trees as needed, using the latest computer programs for tree construction.

Ernst Haeckel (1866) was the first to attempt to put birds on a tree of life based on Darwinian ideas about evolution, in his book *Generelle Morphologie der Organismen*, depicting a few of the major groups and their phylogenies on a surprisingly accurate geological scale. Next, Thomas Henry Huxley (1867) focused just on birds in his *On the*

Classification of Birds, in which he compared most of the major groups but provided no explicit phylogeny. To compile natural taxa into a Linnaean hierarchy he used a single trait—the bony palate—largely because skeletons, and particularly skulls, of many species were readily available in collections close at hand. Huxley's classification was roundly criticized in the latter part of the nineteenth century for its reliance on a single character in the absence of any apparently logical justification (e.g., Streets 1870). Other avian taxonomists were inspired to try different suites of characters, often called the "beak-foot-feather approach" to bird taxonomy. For example, Gadow (1891–93) and Garrod (1876) focused on syringeal and muscle features that, though criticized vehemently at the time, are still considered valuable today for indicating relationships within orders.

Birders are often confused, yet intrigued, by the seemingly random order in which major groups of birds appear in field guides and checklists. Today, in northern hemisphere field guides, loons (divers) usually come first, followed by grebes, ducks, hawks, herons, shorebirds, owls, hummingbirds, and songbirds. The sequence is often called "evolutionary," with the widely held, but mistaken, notion that the more "primitive" species—species whose main branch came off the trunk earliest—are at the beginning of the list, and more recent "advanced" species at the end. Evolutionary biologists abhor this primitive/advanced characterization of the tree of life, but it has proven a convenient, if sometimes illogical, way to make a list. So who decides on the sequence in the list, and why not put the most recent species first? Huxley's (1867) arrangement had the ratites (ostriches, rheas, emus) first, and then Coues (1872) put the passerines (songbirds)

first in his *Key,* in direct conflict with European ornithologists, led by Richard Bowdler Sharpe at the British Museum. Sharpe had an unbelievably productive life, writing nearly four hundred papers, fathering ten daughters, and writing half of the *Catalogue of the Birds in the British Museum* while editing the rest. He was one of the world's most famous ornithologists as the twentieth century began, largely on the strength of that catalog, but also because he corresponded so widely. Following Huxley, Sharpe (1874–98) put the hawks and owls before the songbirds, and in the listing of songbirds, he put the crows last, presumably because their apparently high intelligence suggested that they were the pinnacle of avian evolution, the most "advanced" species. In direct contrast, Ridgway (1901–47) put the songbirds first in his *Birds of North and Middle America,* not because of evolutionary considerations but rather because museums housed more songbird specimens due to their smaller size, making them easier for him to study with some degree of completeness.

The problem with making ordered lists of species is nicely illustrated by the treatment of songbirds over the years, with three major sequences presented in world lists of birds. Sharpe, as we saw, put the crows last, and Hartert (1903–32) followed suit, establishing some authority that was obeyed by most European ornithologists for the next half century. In North America, Ridgway (1901–47) set the precedent by putting thrushes last, but by the mid-1920s this arrangement was challenged by two young rising stars, Alexander Wetmore and Waldron DeWitt Miller. Based partly on previous authorities, Wetmore and Miller (1926) started the passerine list with the crows, because that group seemed rather generalist in their habits, a

Alexander Wetmore shown with some of his first bird and egg specimens and holding a copy of *Bird-Lore* (photo in 1901 at age 15).

trait that they felt was more "primitive." At the end of the passerine list, they put the seed-eating finches because they felt that this group epitomized specialization and was most likely to see further evolutionary "progress." With twenty-first-century hindsight, this all seems eccentric in the extreme, but in those days tradition and authority, however attained, ruled avian taxonomy.

ERNST MAYR'S CENTURY

In the realms of ornithology, evolution, and systematics, the twentieth can certainly be called Ernst Mayr's century. Born in Germany in 1904 to a middle-class family, Mayr had privileges but he also capitalized brilliantly on some incredible coincidences

and plain good luck. He eventually came to dominate the study of avian systematics, was an architect of the Modern Synthesis of evolutionary biology in the 1940s (as we saw in chapter 2), and guided the course of evolutionary biology for the latter half of the century (Haffer 2007b). He published more than eight hundred works in his almost eighty years as a professional ornithologist, and even after formal retirement in 1975, he continued to produce major works on philosophy, ornithology, and biology right up until his death in 2005, just a few months shy of his 101st birthday.

Growing up in Dresden, Mayr was introduced to natural history by his parents, on frequent walks in the countryside. As a teenager he was already a keen and accomplished birder, and when deciding where to pursue medical studies, he chose the University of Greifswald because it was close to the Baltic coast, a good birding area. Just before departing for his first semester at med school, he sighted, in Moritzburg, a pair of Red-crested Pochard, a bird that had not been recorded in Germany for more than seventy-five years.[58] A member of the local natural history society suggested that he report this sighting to Erwin Stresemann at the Berlin Museum. Fortuitously, Mayr could pass through Berlin to get to Greifswald, so he stopped off to tell Stresemann about the pochards and to show him his detailed field notes on the sighting. Stresemann was immediately impressed with Mayr's exactitude and enthusiasm for bird study, inviting him to work as a volunteer in the museum during university holidays: "It was as if someone had given me the key to paradise" (quoted in Bock 2005). Mayr published his first paper on that sighting of the pochards (Mayr 1923a), and his second followed soon after his arrival in Greifswald, on his observations of Red-breasted Flycatchers (Mayr 1923b). He was already, at the age of nineteen, on a roll that would gather speed for the rest of the twentieth century.

Stresemann was soon urging Mayr to drop out of medical school and begin a career in ornithology. When his young helper quickly and accurately (and no doubt confidently) distinguished between specimens of two almost identical species of treecreeper, Stresemann told him, "You are a born systematist."[59] Those species, the Eurasian and Short-toed Treecreepers, are ones that even today birders find almost impossible to tell apart unless they hear them sing. Mayr was easy to convince about dropping out of med school, and on completing his preclinical courses,[60] moved to Berlin to begin PhD work with Stresemann. Stresemann sweetened the pot by promising to send Mayr on a tropical expedition once he completed his thesis. That thesis (published as Mayr 1926), on the historical biogeography of the European Serin, took Mayr only sixteen months to finish, rushed to completion in 1926 just before his twenty-second birthday so he could take up a newly vacated assistantship at the Berlin Museum. Though his first job was to catalog the museum's library of books, he worked diligently in the specimen collections as well, publishing his first taxonomic paper, on snowfinches, in 1927. Expedition, though, was on his mind, and he was eager to get to the tropics.

Stresemann had originally intended to get Mayr onto an expedition to Peru or Cameroons, but to no avail after months of trying to find either a sponsor or an expedition to join. Then, in 1927, Stresemann and Rothschild agreed to send a small expedition to New Guinea to collect for the Berlin Museum, the AMNH, and Rothschild's private collection at Tring (LeCroy 2005). Mayr was to be the ornithologist, departing Berlin in February 1928 on a steamer bound for the South Pacific via the Suez Canal—twenty-three years old, fresh out of grad school, and on his way to both a new world of wonders and hardships enough for a lifetime. Though exceptionally well prepared for the birds—"Before starting for the field, I went to several of the large European museums and worked through their New Guinea collections, with the result that when I arrived in New Guinea, I knew not only the name of every bird I might collect, but also whether it was rare, or desirable for my collection, and whether it showed any peculiarities of particular interest to science"[61] Mayr had no management experience for such a trip involving many local helpers. Nor did he have any idea how the tropics would enthrall him: "Soon darkness fell on my first night in the tropical forest. What this means only the man who has witnessed a tropical night himself, can appreciate. No words can describe the concerts produced by the cicadas, locusts, tree-frogs, and night birds, a symphony of peculiar and deeply impressive harmony. Listening and dreaming, I lay awake for a long time in spite of the fatigue caused by the march and all the exciting experiences of the day."[62]

From today's perspective—with GPS, Gore-Tex, cell phones, Google maps, iPads, aircraft, and modern medicine—the rigors of Mayr's fieldwork boggle the mind. He had to hire more than fifty native men and boys to carry his food and equipment, and they spoke neither German nor English. Medical treatment was also rudimentary and injuries abounded:

One of my mantris had a bad attack of malaria, and another suffered from arsenic poisoning, one of the boys developed pneumonia and the third mantri and myself sores, which forced me to stay in camp for quite a long time. . . . I had a fainting fit on account of fever. . . . The little bush-mites were very bad in our camp and we all got sores from too much scratching. After I had collected series of most of the birds I returned to Ifar, mainly because most of my hunters and Malays had left me before on account of sickness and the cold. It would take too long to describe all the troubles and difficulties one meets with in New Guinea.[63]

Surprisingly, to Mayr at least, the lowland natives couldn't take the cold (!) mountain air: "The night temperature went down to 18° C and my boys felt the cold and did not get much sleep, in spite of the big fires,[64] and the highland natives were sometimes hostile, sometimes just afraid: Women and children left the houses screaming when I appeared."[65] Despite all this, Mayr soldiered on, collecting, during his 200 days in the field, almost 3,000 birds and 260 mammals, many new to science and all useful to his later work:

> I value more than the discovery of many specimens and facts new to science, the education that it was for me. The daily fight with unknown difficulties, the need for initiative, the contact with the strange psychology of primitive people, and all the other odds and ends of such an expedition, accomplish a development of character that cannot be had in the routine of civilized life. And this combined with a treasury of memories, is ample pay for all the hardships, worries, and troubles that so often lead us to the verge of desperation in the scientific work that takes us into the field.[66]

Ernst Mayr (right) with a native assistant in New Guinea (photo in 1928 at age 24).

Mayr arrived back in New Guinea on 21 October 1928 to get his bearings, pack up his specimens, and head for home. But fate had other plans: Mayr's time in the tropics was not yet over. Beginning in 1921 the AMNH had been exploring the islands of the South Pacific in what was later called "one of the most notable undertakings in the annals of ornithology":[67] the Whitney South Seas Expedition. Setting out from Tahiti in the seventy-five-ton schooner *France*, led by the inimitable Rollo Beck and staffed by a fluid crew of sailors, collectors, and scientists, the Whitney expedition roamed the South Pacific for the next decade, collecting birds, reptiles, amphibians, insects, and

97

anthropological artifacts for the museum. By 1928 the expedition had run into trouble, as Beck had developed health problems—both mental and physical—and had abandoned the ship in March of that year. A young Yale graduate, Hannibal Hamlin, had been forced to take over, but he lacked the necessary experience and skill. After twelve months Hamlin began to show signs of the psychological afflictions that had troubled Beck, and back in New York at the AMNH, Sanford, chairman of the bird department Frank Chapman, and curator of oceanic birds Robert Cushman Murphy started to despair. Sanford asked Stresemann what to do, and Stresemann suggested Mayr as a suitable candidate to take over. But Chapman was not so sure, although he was certain that Hamlin had to be removed:

> The Germans may have developed into admirable collectors, but they certainly are not the men to take charge of the Whitney Expedition, and I feel so strongly about the necessity of relieving Hamlin that if we are unable to find a man to replace him I think we should cable him to put the "France" out of commission and come home. If we do not take this step, I feel that we will assume a very grave responsibility.[68]

In the end, Mayr was hired as a compromise to take over Beck's ornithological collecting, and a couple of months later William F. Coultas relieved Hamlin as expedition leader, a position that Coultas kept until the expedition ended in 1932.

Despite his experiences with the challenges of expedition life in New Guinea, Mayr jumped at the chance to see and collect some more birds, although this meant he would not arrive back in Germany until

April 1930, two years after he had left. Mayr spent a good deal of his time enjoying the out-of-doors for the rest of his long life, but he never again went on a collecting expedition, devoting himself to museum and university research and the quest for ideas rather than specimens. The Whitney Expedition visited more than a thousand islands in all and shipped back to the AMNH forty thousand bird specimens and several tons of anthropological artifacts and other treasures. The collection was so large that Harry Payne Whitney, who had already funded the entire expedition, pledged an additional $750,000—serious money in those days—to build a new wing on the AMNH to house it. Mayr was to spend much of the next two decades in the AMNH's Whitney wing.

After the Whitney Expedition ended, Mayr returned first to his post in Berlin, but within a few months he was hired by the AMNH as a visiting research associate to work on the Whitney collections. He arrived in New York in January 1931. In the spring of 1932 Rothschild's collection had arrived at the AMNH, and they needed someone to curate it. Once again Mayr was the right man, in the right place, at the right time—another amazing coincidence that shaped both Mayr and much of ornithology for the rest of the twentieth century. Some have claimed that Mayr would never have blossomed in the rigid German system, where he would have remained in the long shadow of his mentor, Stresemann. In America he could set his own agenda and rise to whatever heights his considerable intellect and ambitions would allow. At the AMNH Murphy soon gave Mayr a full-time curatorship to focus on the South Pacific. The other curators already had their domains: Murphy for

South American seabirds, John Zimmer for Peru, James Chapin for Africa.

For the next twenty years Mayr focused his systematic work on South Pacific birds, discovering, defining, and redefining species, subspecies, and geographic variation in a long series of scientific papers,[69] and finally, in 2001, in a book with Jared Diamond, *The Birds of Northern Melanesia*.[70] As always Mayr was not just compiling but also thinking—about systematics, about evolution, about how science is done. Soon he was to publish the first of his pivotal books, *Systematics and the Origin of Species from the Viewpoint of a Zoologist* (Mayr 1942), a seminal treatise on the role of systematics in understanding organic evolution and the foundation for the "new systematics," an approach that followed Stresemann's in considering all aspects of a bird's biology to reconstruct its evolutionary history (see next section). Under Mayr's leadership the "new systematics" flourished through the 1940s, 1950s, and 1960s.

In 1953 Mayr, with Gorton Linsley and Robert Usinger, published *Methods and Principles of Systematic Zoology*,[71] the first ever attempt to codify how to actually do systematics, and thirty years later graduate students and practitioners were still using the book and subsequent editions as their bible. With the book, Mayr brought to systematics the practices of modern science, with (relatively) objective or at least clearly described methods, consistent concepts and philosophy, and reasonably repeatable procedures, finally bringing together all of the principles that emerged from Stresemann's and his own focus on populations, biogeography, and the nature of species. The year 1953 also saw Mayr move to Harvard and change his research emphasis from systematics to evolution.

By the early 1950s Mayr had laid the foundations for a modern, scientific approach to systematics embodied in his many publications on South Pacific birds and summarized in his two groundbreaking books mentioned above. He had also been instrumental in encouraging promising young ornithologists to pursue research in ornithology, much as Stresemann had done for him—Joe Hickey, Margaret Morse Nice, and Konrad Lorenz had all visited him, and by all accounts came away with renewed enthusiasm and perspective. But he was getting restless to establish a research group of his own, to have his own graduate students, and to focus his attention on the problems of evolution (chapter 2). Loyalty to Sanford, however, kept him at the AMNH until Sanford died in 1953. Upon Sanford's death Mayr moved immediately to Harvard, where he remained for the next fifty years, mentoring PhD students, working on evolutionary theory, the philosophy of biology, and, occasionally, systematics. Though his book with Diamond on the birds of Melanesia was fifty years in the making, it was worth the wait, incorporating systematics, ecology, evolution, and biogeography into what is probably now the best faunistic work ever produced on birds.

How did Mayr become such a towering figure of twentieth-century ornithology, biology, and science? Certainly there's no discounting his fertile intellect and his singleminded love of birds. Nor can we ignore his incredible luck and ability to capitalize on unexpected opportunities. Mayr was also an enthusiastic mentor, both confident—sometimes to the point of arrogance—and forward thinking, often seeing through problems with remarkable clarity. Though he spent a lifetime studying only birds, his

Members of the Department of Ornithology at the American Museum of Natural History in 1960 (left to right; approximate ages in parentheses). Front row—James Chapin (71), Robert Cushman Murphy (73), Dean Amadon (48), Jean Delacour (70), Charles O'Brien (ca. 55); back row: Robert Selander (33), Paul Slud (41), Charles Vaurie (54), Helen Hays (29), Wesley Lanyon (?), E. Thomas Gilliard (48), Stuart Keith (29), Eugene Eisenmann (54), Mary LeCroy (25).

knowledge of all branches of biology was legendary, and his mind was forever voyaging into new territory where he would stir up revolution. Throughout his long life he was slim, healthy, energetic, friendly, and compassionate, always aware of his place in history, but rarely too busy to encourage a young scientist showing promise or an established colleague with new ideas that challenged his own. Never one to back down from a good argument, Mayr, like many great scientists, had his share of failures, but it was his ability to rise above them that made him special (Haffer 2007b).

THE NEW SYSTEMATICS

By 1940 it was clear to many practitioners, and one keen observer, that systematics had undergone a major revolution after the two hundred years of discovering, describing, naming, and endlessly listing species (and by then subspecies) that followed the publication of Linnaeus's *Systema Naturae* in 1735. That keen observer was Julian Sorell Huxley, secretary of the Zoological Society of London, who, following in the giant footsteps of his grandfather, Thomas Henry Huxley, was one of Britain's leading scientists. Julian's contributions to zoology included his research in bird behavior (chapters 7–9), his encyclopedic book *The Science of Life* (1929–30) with G. P. and H. G. Wells, an Oscar-winning documentary film about Northern Gannets,[72] many popular books,[73] an extensive series of public lectures and radio talks,[74] and directorship of the London Zoo. Later he was even involved in the creation of UNESCO, where he served as its first director general (in 1946), and was famous for championing the work of the influential French philosopher Pierre Teilhard de Chardin and for his regular appearances on the panel of *Animal, Vegetable, Mineral*,[75] one of the BBC's first quiz shows. Huxley was arguably the world's best-known scientist by the middle of the twentieth century, dynamic, philandering, manic-depressive, prolific and talented as a writer,[76] and both knowledgeable and influential as an evolutionary biologist and humanist. He was particularly gifted in recognizing the winds of change in science and humanity and in portraying those changes in the most appealing and persuasive manner.

At the behest of the newly formed[77] (British) Association for the Study of Systematics in Relation to General Biology,[78] Huxley (1940) edited a book, *The New Systematics*, that brought together mostly, and deliberately, British scientists to write about the new developments in their field. As Huxley presciently recognized,

> the Committee is fully conscious of the somewhat presumptuous sound of the title it chose for the book. It would have been more accurate to call it Modem Problems in Systematics/ or Towards the New Systematics. For the new Systematics is not yet in being . . . it was felt that a good title goes a long way, and should be its own excuse. At least it will draw attention to the fact that a 'new Systematics', or at least a new attack on systematic problems, is an important need for biology.[79]

Not all were convinced, and Ernst Mayr later claimed that "there was little new systematics in that volume"[80]—and indeed there was precious little about birds. However, it was ornithology that had already initiated a new approach to systematics, and Mayr took it upon himself to provide both the bible and the commandments. Through the work of Mayr, Stresemann, Rensch, and their followers, the "new systematics" had already arrived in ornithology by the time Huxley wrote the title of his book. Mayr was mentioned only once in Huxley's book, in a figure caption, and birds only in passing— about thirty times with ten references—as brief examples in various chapters.

Just two years on the heels of Huxley's (1940) *The New Systematics*, Mayr (1942) published his first magnum opus, the classic *Systematics and the Origin of Species*, in which he both described and laid a foundation for a new systematics[81] brought about to a considerable degree by the recent work of ornithologists. In a way Stresemann had started the ball rolling with his emphasis on paying attention to the whole bird, but it was really Mayr, Rensch, and others who fully incorporated evolutionary biology and population thinking into the science of systematics. Systematists had certainly incorporated evolution at least since Haeckel, but it fell on Mayr and company to clearly define what a species is (chapter 2), to replace Linnaean and subsequent typological thinking with the notion that species actually vary morphologically in time and space, and to show that that variation has both genetic and environmental causes. You cannot look at a museum tray of Song Sparrows collected across North America and fail to realize that they are bigger in the west, darker on islands, and redder in the north. It is no surprise that studies of birds were instrumental in shaping the new systematics. By 1900 most species had been discovered and described; both private and public collections abounded, burgeoning with specimens; the trinomial revolution had uncovered myriad identifiable subspecies; and field studies had revealed much about dispersal as well as the extensive variation in ecology and behavior within and

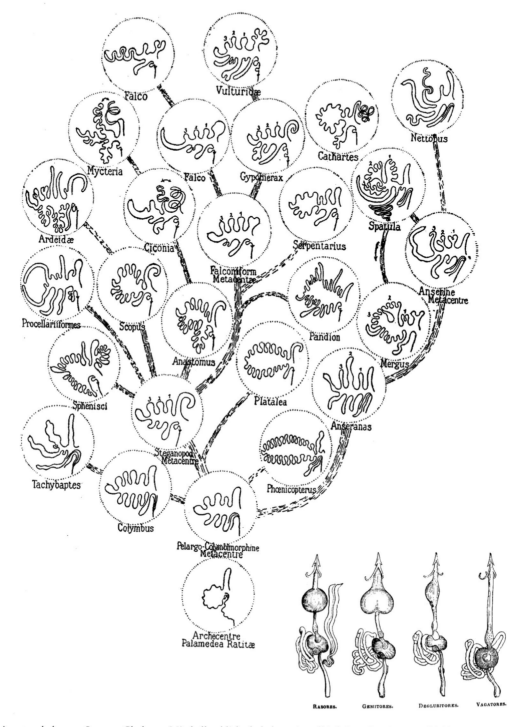

A gutsy phylogeny. In 1901, Chalmers Mitchell published phylogenies of birds based on intestinal folding patterns (in dotted circles here). Digestive tracts (including those in the inset from MacGillivray 1837: 99) had been used previously to infer phylogenetic relationships, albeit with limited success.

among species. On top of that, some of the best practitioners had already adopted objective methodologies and statistical techniques that were soon to be the hallmark of the future of systematics.

In the half century from 1920 to 1970 literally hundreds of publications on birds—from half-page descriptions of new subspecies to multivolume faunistic works—illustrate the gradual transformation of bird classification resulting from the new systematics. The early application of new philosophies and new methods can be seen in papers by Hartert and Stresemann and in the dozens of papers by Mayr in the 1930s and 1940s from his AMNH work. Many of these had pedantic titles, but they embodied the new philosophy, with taxonomic decisions based upon detailed measurements, descriptions, biogeography, and population ecology. Charles Vaurie's (1959, 1965) two-volume *The Birds of the Palearctic Fauna* is one example of the mature application of the new systematics. Here is Stresemann on reviewing Vaurie's (1959) volume on the Passeriformes:

> This book goes far beyond the usual scheme, as far as distributional, ecological and taxonomic information is concerned. Vaurie indicates the ecological niche of each species and the nature of its geographical variation, pointing out clines of color and size. He states the distinguishing characters of each subspecies and their relationships, and often calls attention to unsolved taxonomic problems at the species-subspecies level.... By grading the accepted subspecies into three categories of differentiation, marked by symbols, the author has introduced a novel and commendable method. The lowest category of subspecies—the slightly differentiated

local population and the intermediate stage in a cline—are placed in synonymy, marked with an asterisk.[82]

Mayr and Huxley felt that this revolution would transform systematics and it did, but it also played a major role in the decline of systematics as a field during the second half of the twentieth century, "reorienting several generations of taxonomists from pursuit of phylogenetic research to serving a supporting role to population genetics."[83] Mayr himself used his systematics research to build a foundation for the Modern Synthesis of evolutionary biology (chapter 2) of which he was a major architect. Little could they know that their new systematics would soon become old, and that what appeared in 1940 to be a relatively objective way of classifying birds and other organisms would soon be the focus of controversy. A stable taxonomy of at least the higher taxa of birds would take another half century to resolve—in a way that could not be conceived in the 1940s and with the sort of scientific rigor that Mayr could not have imagined when he wrote his *Principles*.

SYSTEMATICS WARS

At the British Ornithological Union's centenary conference held in Cambridge in March 1959, Huxley chaired the first session, devoted to systematics. Mayr gave the introduction: "And what was called the 'New Systematics' only a few years ago (Huxley 1940, Mayr 1942) is already beginning to look a little faded and shopworn. The replacements of the monotypic by the polytypic species, and of the morphological by the Biological Species Concept, are now so universally accepted that it is unnecessary to agitate any longer in their favour."[84] He was right, and he went on to suggest that two aspects of the new systematics were the wave of the future in ornithology and in general—population studies and phylogenetic systematics.

A few examples will illustrate the sorts of problems that had befallen avian systematics, symptomatic of the chaos that characterized museum work on most animal and plant taxa by mid-century. In the 1930s Alexander Wetmore published a comprehensive classification of birds listing twenty-five extant orders. Wetmore was head curator at the Smithsonian Institution and a recognized authority, but so was Stresemann, who listed forty orders in his own work published at the same time (Stresemann 1928). How could there be such a discrepancy? Arthur J. Cain, a frequent and acerbic critic of systematists, says it all:

> Probably more pure nonsense has been talked (and published) about phylogeny in birds than in any other group of animals. People have made the most astonishing assumptions about what must be primitive in given groups, and what must have given rise to what. One ornithologist tells us that the plumage is always more reliable in a group. Another says that the vomer, being inside the mouth, is less subject to adaptive modification and therefore more reliable for ancestral characters. . . . The whole of taxonomy is an attempt to extract as much information as possible from data of very various degrees of incompleteness; and it does no good to pretend that we can find out about a given group more than we can. . . . And since the "natural" classification sums up a great deal of information, it is

a most useful one. It would be a great thing if ornithologists would once again lead the taxonomic world by distinguishing clearly between natural and phylogenetic groups.[85]

The agents of change, however, were to come from entomology, not ornithology, and unbeknownst to Mayr and Cain when they wrote the words quoted above, the seeds for a new approach had already been sown by a relatively obscure German entomologist named Willi Hennig.

Despite the best efforts of Mayr and his (entomologist) coauthors to provide some consistency and rigor to systematics practices in their *Methods and Principles*, by the time that book was published, in 1953, the situation in systematics could best be described as confusing. That same year the newly minted journal *Systematic Zoology* published a paper on the systematics of termites by Clyde P. Stroud (1953) that heralded a brand-new approach based on statistical analysis of measured traits. Stroud was a graduate student at the University of Chicago, where one of his lab mates was the young Austrian Robert K. Sokal, who was also doing his PhD on termite systematics but was not at all (then) interested in maths and stats. No sooner had Sokal and Stroud agreed to collaborate on a research project in 1952 than Stroud died of cancer. But Sokal was hooked on systematics, and after beginning a faculty appointment at the University of Kansas in 1959, became increasingly critical of the "intuitive" (nonquantitative) approach taken by most taxonomists; as the philosopher David Hull said, "He went to graduate school to become a scientist, not an artist."[86] The eventual outcome was a book, *Numerical Taxonomy*, that Sokal wrote with Peter Sneath (Sneath and Sokal

1973). This was a comprehensive and at times highly technical text that defined the eponymous new field, a field that immediately attracted passionate adherents tired of the old subjective methodologies, embracing the potential for objective, repeatable, quantitative approaches to systematics. Not surprisingly, the ornithologists at the University of Kansas did most of the early work applying numerical taxonomy to the study of birds.

Ironically, Stroud's influential paper was followed immediately—in the same issue of the journal—by a rather scathing review of Mayr, Linsley, and Usinger's (1953) *Methods and Principles* by the ichthyologist Carl L. Hubbs:

> There is no sound weighing of the relative merits and defects of splitting and of lumping, merely a repeated indication of a strong bias toward lumping. Ecological speciation is largely denied or avoided, as in previous works by the senior author. Experimental methods are accorded little space or emphasis, comparative serology is only briefly treated, and other biochemical methods are barely mentioned. . . . The use of calculating machines, standard in some fields, is not mentioned. . . . [The reader] may think that some of the many examples drawn from ornithology are less effective than they would have been had they been taken from his field . . . for example in the statistical study of races and in the analysis of natural interspecific hybridization. . . . Such a systematist may feel a bit annoyed on being reminded so frequently of the completeness and pre-eminence of bird systematics.[87]

Hubbs had hit upon both the problems of the traditional approach (and the dominance of ornithologists) and the eventual solutions—biochemistry and "calculating machines." The first volley in what David Hull would later call the "Systematics Wars" had been fired—the 1960s and 1970s would see one of the most contentious periods in the history of any branch of science. While not everyone likes this term, to those of us looking in from the outside, systematists at the time seemed clearly at war, as egos and emotions ran high, and the field seemed to attract some of the most arrogant, opinionated, and downright nasty individuals who have ever called themselves scientists. It would take some serious psychological research to uncover the reasons for this.

Numerical taxonomists and traditional systematists would probably have had enough to argue about on their own, but the 1960s also saw the emergence of yet another brand-new approach to systematics, which Mayr called "cladistics." Cladistics was the brainchild of Willi Hennig, who wrote his pioneering book on the subject in German in 1950. In that book Hennig called his approach "phylogenetic systematics." This book was virtually unknown until 1965, when Hennig published a review paper outlining his principles, and especially in 1966 when his book was translated into English. Even then, the relative awkwardness of the prose made the book hard going and left some details open to various interpretations. Nonetheless, Hennig's basic thesis was simple enough: systematics should be based upon sound, logical principles resulting in repeatable and robust phylogenies derived from an understanding of branching patterns in the tree of life. He felt that shared derived characters—what he referred to as "synapomorphies"—were the key to defining where those branches joined.

Cladistics attracted some of the brightest—and most pugnacious—young scientists, drawn by the logic and objectivity of Hennig's principles. The main adherents, centered in the ichthyology department of the AMNH, were religious in their fervor and verging on inquisitional in their attacks on those who disagreed. David Hull (1988) provides a detailed exposé of the battles fought by the different factions that arose in the 1960s—traditional systematists, pheneticists, and cladists—but we still don't fully understand the underlying causes of these turf wars. Certainly this was a period of general upheaval and social revolution in Western society, and that may have had something to do with it.

How was ornithology involved in these revolutions, and how did it fare? Coincidentally, the University of Kansas, where numerical taxonomy was born, was also a major American center of ornithology at the time. Under the direction of Richard F. Johnston, the university's specimen collections were burgeoning, and many of the brightest young ornithologists went to graduate school there. The 1960s also saw ornithology emerge as a respectable vocation, and there were jobs to be had both in government and at the many new universities that were springing up all over the continent to serve the increased demand from the baby boom generation. Cornell, Berkeley, and Kansas had all distinguished themselves as centers of ornithological research and training, and their ornithological faculty—Josselyn Van Tyne, Charles Sibley, Arthur Allen, Joseph Grinnell, Alden Miller, and Richard Johnston—was famous. Kansas was the leading school for budding systematists, so it is not so surprising that many of the ornithologists who embraced numerical taxonomy got their start there.

The work of Gary Schnell, a Kansas graduate, provides some instructive examples. His study with Jerome Robins on relationships in the *Ammodramus* sparrows used forty-eight measurements of bones from a dozen species to assess how they might be related to one another (Robins and Schnell 1971). Possibly because their results were at odds with previous taxonomies, their work was largely ignored by the AOU, the body responsible for incorporating systematics work into the official American checklist of birds. Yet this work was a breakthrough because it showed for the first time how the interrelationships among bird species could be quantified objectively to make a highly informative phylogenetic tree. Previously, only crude trees could be drawn based on rather subjective classifications.

Birds also figured prominently in one of the great, and still ongoing, debates—how to formulate a useful and informative classification from a phylogenetic tree. Almost from the beginning it was clear that the clade of birds—the class Aves—was embedded within the clade of dinosaurs, not parallel to and independent of it, as we explored in chapter 1. That is, birds do not sit on their own, unique branch of the tree but are a branch within the dinosaur branches. Strictly speaking, birds are dinosaurs and should be classified within the dinosaur clade as one of its subgroups, rather than having independent status at the same level as the reptiles (chapter 1).

In the end, only a few studies of birds ever used the methods advocated by the early cladists and numerical taxonomists, but those approaches fostered a quieter revolution in systematics based on quantification, logic, and repeatability. As the botanist G. Ledyard Stebbins (1982) wisely noted, the resolution of many scientific controversies settles on what he called "modal themes," taking the best ideas from each side of the battle line. Stebbins recognized that many controversies—like the cursorial versus arboreal origins of bird flight controversy that we discussed in chapter 1 and the nature/nurture debate that engulfed ethology in the 1960s (chapter 7)—resolve themselves eventually, and often quietly, with the simple recognition that the truth lies somewhere in the middle, at the intersection of controversial claims. No wonder there is such fervor in these scientific wars: both armies "know" that they are right, because in a way each side is correct to some degree. The systematics wars found such a resolution, even though there remain many purists who feel the need to continue to battle, in the face of—or maybe even because of—the field moving on without them.

By the 1980s avian systematists, following the general approaches taken by scientists working on other animal taxa, had begun to move on, taking the most useful lessons from traditional systematics, phenetics, and cladistics with them, and forging a truly revolutionary approach to the construction of phylogenetic trees. There were some initial attempts, and successes (e.g., Robins and Schnell 1970; Cracraft 1974), at applying cladistic methodologies to the study of bird phylogenies, but by the 1980s that method was starting to give way to a more computational approach. Beginning first with parsimony methods, then moving to maximum likelihood analysis, the new techniques focused on quantitative methods to reconstruct phylogenies, using increasingly sophisticated and computer-intensive analyses. With first DNA markers, then DNA sequences, biologists finally had a quantitative measure that could be used to assess different models of

evolutionary change. At long last the sort of phylogenetic reconstruction pioneered (albeit feebly) by Darwin (1859) had come of age, no longer relying on but rather informing classifications. Indeed, the quantitative geneticist Joe Felsenstein (2004), at the University of Washington, has called this the "It-Doesn't-Matter-Very-Much" school of systematics, wherein phylogenetic reconstruction is the goal, and those interested in classifications are welcome (and encouraged) to pay attention to the resulting evolutionary trees, but whether they do or not is of little interest to most working scientists. Phylogenies have other uses, and these now pervade many areas of modern biology—as the basis for comparative analyses, as a focus for understanding evolutionary change, and as working hypotheses about evolutionary relationships. "Systematists voted with their feet to establish this school, . . ."[88] and the published literature on bird systematics since about 1980 supports this view.

Until the middle of the 1960s there were few published phylogenetic trees for birds. Darwin's (1859) attempt at a phylogeny of domestic pigeon breeds—the only phylogeny he ever published—was one of the first, and Konrad Lorenz's (1941, 1951–53) evolutionary tree of the ducks was constructed by mapping behavioral characters onto a tree, accounted for some degree of convergence in characters, and choosing the phylogeny that was the best fit to the available data. Most published phylogenetic trees for birds, however, were little more than drawings of relations among higher taxa based on fairly subjective analyses of relationships. It is possible—and reasonable—to make species' phylogenies from classifications, as Robins and Schnell (1971) did from Carl Eduard Hellmayr's (1938) and Robert Ridgway's (1901–47) classifications of *Ammodramus*, but those phylogenetic trees contain little information and nothing more than the classifications themselves. Phenetic and cladistic approaches changed all that by constructing trees that revealed a wealth of information—statistical support of relationships, lengths of branches, and the identification of unsolved problems (e.g., polytomies, where several branches emerge at the same point in a phylogeny)—that really cannot be usefully reproduced in the sorts of species lists and classifications that consumed the energies of Gadow, Hartert, Mayr, and dozens of other dedicated scientists for more than a century.

Advances in microcomputers, microbiology, and statistics, beginning in the 1980s, created a perfect storm of possibilities that revolutionized systematics and eventually produced a resolution to the problem of relationships among the higher categories of birds. Finally, today (2013) it can be said that we are beginning to see a fairly clear and reliable picture of the evolutionary relationships among the major groups of birds (Hackett et al. 2008). Darwin would be delighted.

SIBLEY WEAVES A TAPESTRY

It's late afternoon on a hot day in July 1986 when a tall, thin man in a brown suit strides through the new Ottawa Convention Centre, a sheaf of computer printouts tucked under his arm. He's trying to suppress a grin— smug, arrogant, determined, and maybe a little fearful. He is Charles Gald Sibley, and he is about to present the pinnacle of his life's work to the Nineteenth International Ornithological Congress in a most dramatic and memorable way, with a presentation that has become the stuff of legend in the ornithological community. He has been working

doggedly for more than a decade on a phylogeny of birds based on the analysis of DNA, and he has it here on a twenty-foot-long sheet of computer paper. He's going to tape it to the wall of the poster session for all to see—a "tapestry" of bird evolution and systematics.

The effect of Sibley's "tapestry" was electrifying: gaggles of ornithologists gathered around their favorite group of birds throughout the day, intrigued by both the surprises and the confirmation of ideas long held but little supported with evidence. The surprises were relatively few, but they were big, at least for the nonsystematists[89] in the crowd. For example, Sibley demonstrated that the passerine birds of Australia were largely descended from a corvid ancestor and were thus unrelated to the wrens, thrushes, and flycatchers that they resembled and were named after by the early ornithologists who first described them. Working with Jon Ahlquist, who managed the lab work, Sibley had compiled a phylogenetic tree of more than 1,100 species. This phylogeny supported an ancient separation of the birds into two clades, the Eoaves (ratites and tinamous, as well as the Galloanserae [quail, pheasants and waterfowl]) and the Neoaves (all the remaining extant species). As well, it suggested that the Neoaves comprised five major clades: Turnicae (button quail), Picae (woodpeckers), Coliae (colies), Coraciae (trogons, hornbills, rollers), and Passerae (all of the remaining birds, from herons, waders, hawks, owls, and pelicans to parrots, cuckoos, pigeons, and all of the passerines). Later, Sibley and Ahlquist (1990) moved the Galloanserae into Neoaves, but now (2013) these are considered to be the two sister clades in the Neognathae, and the Neognathae is the sister clade to the Paleognathae (Eoaves having been abandoned as a viable grouping of birds).

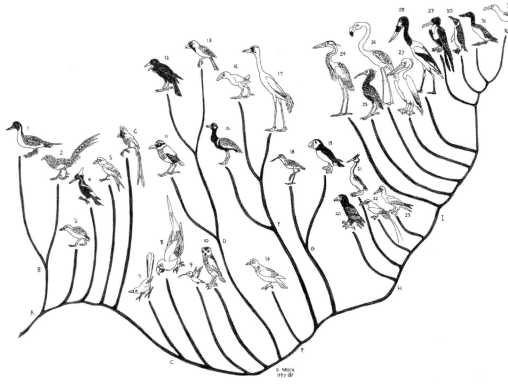

An artist's representation of the Sibley-Ahlquist tapestry.

Sibley grew up in California, and like so many ornithologists of his generation he was a keen birder, was inspired by the writings of John Burroughs and Ernest Seton Thompson, and got his professional start in a museum. His interests in avian biology led him to the University of California at Berkeley and the Museum of Vertebrate Zoology there, eventually to do a PhD with Alden Miller—a leading American professor of ornithology at the time. For his PhD Sibley studied hybridization in the towhees (genus *Pipilio*) on the central Mexican plateau, developing new analytical methods to create a hybrid index for quantifying plumage variation in zones of hybridization. His studies of hybridization led him, unwittingly, into the heart of avian systematics research.

Sibley began his academic career in 1948 at the University of Kansas, but by 1955 he had moved to Cornell as curator of birds, one of the most prestigious ornithological positions in the world. Several North American species hybridize in the Great Plains region, so they became one important focus of his work, and he often sent graduate students to Colorado, Kansas, Nebraska, and the Dakotas to collect specimens. In part to support a new PhD student, Paul Johnsgard, Sibley applied to the National Science Foundation to see if a recent technique involving paper electrophoresis could be used to reveal species-specific patterns in blood serum proteins—a potentially useful tool for assessing how much genetic material any hybrid individual had obtained from its two parental species. "Like most of the students advised by Charles, both then and subsequently, Paul stood in mortal fear of invoking his wrath, should he depart from the protocols carefully prescribed by Charles."[90] However, Johnsgard had read and was intrigued by a paper by Robert A. McCabe and Harold F. Deutsch (1952), outlining how egg-white proteins might vary among species. Without telling Sibley, Johnsgard collected some egg whites and analyzed their proteins while working on blood serum in the lab, and the results were stunning, especially in contrast to the blood serum results, which showed too much within-species variability. Sibley immediately recognized that egg-white proteins were a gold mine: "Almost overnight, he put aside his plans for using serum proteins to study the variation among hybrids, and began to lay plans for an electrophoretic study of egg-white protein variation in birds."[91] For the next fifteen years he immersed his lab in the study of these proteins, focusing especially on the relationships among the higher taxa of birds (orders and families). This was a significant change in career direction—from hybridization to systematics—that established Sibley as one of the founders of the new field of molecular systematics, one of the most vibrant fields in biology at the close of the twentieth century. We now know that the egg-white protein data were of little phylogenetic value, but they were Sibley's entrée into the world of molecular biology.

By 1965 Sibley had moved to Yale as professor of biology and curator of birds at the Peabody Museum, where he continued his egg-white protein work for a while, with Ahlquist still running the lab. Their egg-white protein results were published by Yale in two fat monographs (Sibley 1970; Sibley and Ahlquist 1972), but by then Sibley already knew that a different approach was needed:

"Its main contribution is seen by the authors as a review of the literature and a definition of the major problems. Most answers still lie ahead."[92] He was right, but at the time he thought it was just refinements in electrophoresis technology that were needed. By 1973, he realized that the analysis of DNA itself held greater promise: "We yearned for a single genetic mechanism, yielding clusters of related species, groups of related genera, and so on. Our first DNA data were so clear, so unambiguous, and so promising that any lingering doubts quickly disappeared.[93]

To facilitate this new direction in their research, Sibley perfected a "DNA machine" and the lab turned full bore to working on DNA-DNA hybridization, a method that reveals the similarity between strands of DNA from different individuals. For the most part, DNA hybridized well from individuals of the same species, less well from different species within the same genus, and so on up the hierarchy of classification. Thus a phylogeny could be constructed whose branching patterns reflected the recency of common ancestry, a major breakthrough for bird systematics, as we shall see in the next section. The culmination of the next decade of work in the Sibley lab was the "tapestry" and, later, Sibley and Ahlquist's (1990) impressive *Phylogeny and Classification of Birds: A Study in Molecular Evolution*, a thousand-page tome that summarized their DNA hybridization data and provided detailed accounts and a new classification of the entire class Aves. DNA was now the trait that would transform systematic biology, and Sibley had shown the value of thinking big.

Sibley was larger than life—brilliant, mercurial, arrogant, tyrannical at times, driven, and often difficult even to his closest associates:

He was a forceful man in figure and mind, highly articulate, extraordinarily well organised and energetic, and dominating in personality. Yet, he carried a chip on the shoulder, borne of a perception that he came from the wrong side of the tracks to the German establishment which, personified by Ernst Mayr, led systematic ornithology through nearly all the 20th century. . . . In argument he would bulldoze through, brooking no contradiction. Critics were baited with an acid tongue and, in fits of temper, he could be a cruel mimic. In short, lesser mortals were not tolerated easily and, as has been said by others, collegiate friends were few.[94]

He collaborated for almost thirty years with Jon Ahlquist on virtually all of his work on avian systematics, but they never became friends in the usual sense of the word (Ahlquist 1999).

One of Sibley's great talents[95] was getting people from around the world to collect and send him samples for his lab work. He was scrupulously careful in obtaining the necessary permits for these samples (Ahlquist 1999), keenly aware that some of the species he was interested in might be rare, endangered, or protected. Nonetheless, in 1973 he was charged under the US Lacey Act with illegally importing six egg specimens. Though he never contested this charge, and paid a huge personal, professional, and financial cost, there is some evidence that he had been framed: one of the specimens for which he was charged was labeled as the species *Torpis oocleptica*, which is not the name of any known species of bird, and translates roughly into "lazy egg collector." One of his many enemies may have perpetrated this fraud, but we'll probably never know. The upshot was that Sibley was forced to resign from the BOU and was not reappointed as director of the Peabody Museum. His reputation never fully recovered. The depression into which he sank began to ease only when the first results from DNA–DNA hybridization started to come through, showing the potential to reveal the holy grail of resolving the relations among the orders of birds—and more. Though Sibley's major contributions to avian systematics through his egg-white protein and DNA hybridization studies were soon superseded by more informative and more reliable DNA sequencing technologies, his books stand as a strong testament to the end of a long era in bird systematics research and mark the start of a new one.

DNA

Sibley would no doubt be both thrilled and more than a little jealous of the progress that has been made in the use of DNA technologies since 1990. We will see in chapters 8 and 9 how DNA profiling (fingerprinting) has revolutionized the studies of bird mating systems, sexual selection, and parental care; in chapter 10 how it has improved our understanding of population processes; and in chapter 11 how we can use DNA analysis in conservation biology. In the study of bird systematics, DNA sequencing been truly revolutionary, allowing us to finally realize the promise that Darwin initiated, and that Gadow, Hartert, Mayr, Wetmore, and so many others worked so hard, in vain, to achieve—a reliable, repeatable, informative method for constructing phylogenetic trees that show how species, genera, families, and orders of birds are related to one another, and to give as well a good indication as to when these various taxa diverged in geological time.

The first real breakthrough came in the 1970s and 1980s with the technological developments that were able to reveal the actual sequence of base pairs in a piece of DNA: DNA sequencing. Using this technology and the appropriate computational tools—tools that are still evolving to increasing sophistication and ease of use at reduced cost—we can infer the genetic relationships among populations and all higher taxonomic categories, we can estimate the age of recent common ancestors, and we can quantify numerous details about the genetic structure of populations. The first study to use DNA sequencing to infer a higher-level bird phylogeny was published in 1991 (Edwards et al. 1991), about a year after the chicken mitochondrial genome had been sequenced (Desjardins and Morais 1990). Thus began a flood of increasingly sophisticated work such that we now have well-established phylogenies of groups like the American wood warblers (Lovette et al. 2010). The focus is now on species and populations when constructing phylogenies, at a level of resolution probably not even imagined by early bird systematists who were often fairly content to guess at the relationships between genera and orders at best.

That long elusive holy grail of resolving the relationships among the higher taxonomic categories of birds (above the level of "order") has also been solved recently (Hackett et al. 2008) using DNA sequences 32,000 base pairs long, with a level of reliability and consistently as close to certainty as we can probably ever get. To do this the research team chose taxa to sample based on genetic distances revealed in Sibley's "tapestry,"[96] obtaining frozen tissue samples mainly from museum collections at the Field Museum of Natural History, Louisiana State University, the US National Museum of Natural History,

and to a lesser extent the AMNH. Thus, not only is this research a major breakthrough in the study of bird phylogenies, it also reaffirms the value of museums and their collections, long in decline due to lack of sufficient funding for both collection and storage of specimens. Increasingly, modern museums now store tissues in liquid nitrogen, along with skins and skeletons, an invaluable resource for all kinds of genetics research.

The Hackett et al. (2008) study highlights the power of modern sequencing analysis—a technology that improves every year—but also the necessity of large collaborative research groups with expertise in DNA technologies, bioinformatics, and ornithology, not to mention some serious research funding. While the systematists will no doubt continue to argue about the details of methodologies and analysis, the amazing progress that we have made since 1990, progress in no small part due to Charles Sibley's initiatives, suggests that we are now on the right track, finally.

DNA technologies have also led to the development of DNA barcoding—a technique by which a single sequence of DNA is used to identify any species (Hebert and Gregory 2005). In DNA barcoding, a 658 base-pair-long segment of the mitochondrial cytochrome-c oxidase subunit I (COI) gene is the usual sequence of choice for birds and most other animals, as it is relatively conserved within species and it can now be assessed with relative ease and low cost from a tiny blood or tissue sample. In birds, DNA barcoding has proven most useful for forensic work, the detection of a few cryptic species, and the preliminary screening of phylogenetic relations within families and genera. Thus, endangered and rare species can be readily identified in the pet trade and hunters' larders, bringing an important tool to enforcement and prosecution. Barcoding has also helped identify the possibility in a few genera of cryptic species, species that are genetically different but superficially virtually identical, especially when the time and expense for a more detailed DNA sequence analysis is prohibitive.

CODA

With the advent of DNA technologies we have, in a way, come full circle from the days of Darwin, Gadow, and Hartert, when morphological traits—bones, feather tracts, muscles, and the like—were used to evaluate relationships among species and higher taxa. Today, instead, we use computer analysis of DNA sequences to construct reliable phylogenies that then help us understand the evolution of those same morphological traits that interested the early systematists. Such modern phylogenies now also underpin our comparative analyses of migration patterns, sperm competition, parental care and sexual selection, to name just a few of the advances covered in other chapters. Because DNA sequences for many species are now readily available online or can be obtained from tissue samples sent to a fee-for-service lab, anyone with the appropriate software tools can reconstruct a phylogeny for themselves, one that is exactly repeatable by other scientists, using objective methods that are clearly described and understood. As of 2013, for example, there are about 400,000 DNA sequences of birds in the GenBank data set, representing about half of the known species of birds, and it probably won't be long before there is a DNA sequence available for every species. Future advances now seem to depend upon the development of ever-better computer programs to analyze those sequences. As Felsenstein (2004) cleverly noted, these advances have taken place in the virtual absence of considerations about classification systems, which are still more or less mired in the past. Phylogenies have instead taken their place as the foundation of systematics and the analysis of adaptation and evolution. Stresemann (1959) could never have guessed what the future had in store when he made the statement quoted at the head of this chapter. What he did clearly recognize, though, was that systematics provided a solid foundation for the study of birds.

BOX 3.1 Ernst Mayr

[This is a lightly edited excerpt from an interview with Mayr at age ninety-three conducted by Walter Bock in October 1997. The entire interview can be viewed at http://www.webof stories.com/play/99.]

While I was at the American Museum [of Natural History] I was interested not only in bird taxonomy, but also in other aspects of birds. For instance, I discovered that there was a geographic variation of degree of sexual dimorphism in several species of these island birds, so that on one island both male and female had the highly colored plumage of a male. And on another island the male was highly colored, the female had a drab protective coloration, and a third island—the same species—both male and female had the drab color of the female. So obviously this could not be due to the sex hormones as was at that time believed by everybody, it had to have another meaning.

So I got in touch with people working on the genetic determination of color in chickens at Storrs, Connecticut, and they in turn got me in contact with Professor L. C. Dunn at Columbia University, a geneticist who was interested in such questions. Dunn invited me to come to seminars, to Columbia University, which I did, and I became more and more friendly with the department there. Then Professor Dobzhansky arrived from California—from Pasadena, [where he] worked in T. H. Morgan's laboratory—to give the Jesup Lectures in 1936. And I invited him to come to the Museum and study my wonderful cases of geographic speciation of birds and he was quite excited about this. I was excited about his lectures, and we became very good friends.

In 1937, he published a pioneering book called *Genetics and the Origin of Species* which was the book that started the so-called evolutionary synthesis. I, myself, was invited to contribute one volume to that series in which Dobzhansky published his volume, and I published a volume on *Systematics and the Origin of Species*.

I always tell a little story about how Columbia University—when Professor Dunn was the chairman of that series—got the idea that I might be a suitable person to write this volume. There was a meeting in Columbus, Ohio, in 1939, at the Great Congress of the American Association for the Advancement of Science,

their annual meeting, and there was a symposium organized by Dobzhansky on speciation. The person who talked before me was a famous geneticist with the name of Sewall Wright who was famous for being a terrible speaker. There we were let into this huge auditorium, the biggest one that Ohio State had, and the platform was huge because their orchestras played there and the audience seated 3,000 people. In front of that platform was a lectern with a fixed microphone. Sewall Wright was the speaker just before me. He went to this microphone and he talked into it a little bit, but then after a very short time he left his place, he went to the other end, the back of the big platform where a series of blackboards were, because he's a mathematician and he had to put mathematical formula there. Nobody understood a word he said, and he talked to this blackboard all the time. Every once in a while he would go sort of halfway back to the lectern and talk to the audience. They still didn't hear him because he wasn't anywhere near the microphone. Then he overstayed his time. Anyhow, it was about as bad a lecture as you can imagine, however the people had heard the famous Sewall Wright and most of them left after that lecture, but those that stayed heard me.

I was glued to the lectern and the microphone, I had beautiful slides—this was in the days before Kodachromes, but we had an artist at the American Museum who was very good at coloring glass slides, and he colored these beautiful glass slides of geographic variation in my South Sea Island birds. I talked for only 25 minutes instead of, like Sewall Wright, about 45 minutes. After Sewall Wright's lecture, my lecture seemed to be a marvelous lecture. A very short time afterwards Professor Dunn came up to me and said, would I be willing to give some of the famous Jesup Lectures at Columbia University? And that was another one of these lucky accidents of my life, that Sewall Wright had spoken just before me.

Well, to go back to the Jesup Lectures—we were asked to talk about speciation. There were two people involved. I, to give two lectures on speciation in animals, and a botanist with the name of Edgar Anderson who gave two lectures on speciation in plants. Everything went fine except that Edgar Anderson was manic depressive and he got into a depressive state and didn't deliver his manuscript. So Columbia University Press came to me and they said could I expand my two lectures into a whole book, and well, I said, yes I would, and I did. And that is how my book *Systematics and the Origin of Species* originated.

That book had—I think I can rightly say—a very considerable impact because this area of evolutional biology had been totally ignored and neglected by the geneticists, and you go to the writings of the great population geneticists: Fisher, Haldane, Sewall Wright. Sewall Wright, incidentally, in that lecture in Columbus, Ohio, talked about evolutionary change within a lineage, but not about how species multiply, which was my problem.

I have been asked in later years quite often how I could have—in such a short time—written this book that is so rich in information and detail. Well, to begin with I, at that time, was a person with an extraordinary memory, I also was a voracious reader of the literature. The American Museum had a wonderful system that on Tuesday afternoon, every week, all the issues of the new journals that had come during the preceding week were laid out on a table and the staff could look at it and could put in a bid for getting that issue at the next week or the week following. So I knew not only the literature on birds, but all the other vertebrates, I read the insect literature, I read the literature on marine invertebrates, I read geology, I read anthropology. I had an enormous knowledge and I, apparently, more or less wrote the book off the top of my head with all that knowledge. And the remarkable thing was that, of course, my findings fitted perfectly well what the people that worked on butterflies had found, what the people working on snails had found. It didn't work so well on some other groups, but it was a spur to start working along new directions which people hadn't done up to that time.

Even in my own field in ornithology, most of the American ornithologists were still what we referred to as typologists; they described species as such entities, and if there was a little difference among species they made it a new genus. In my book, for instance, I make a list of 37 genera of American birds that were recognized as good genera in the official list of North American birds, published by the American Ornithologists' Union. Well, the official committee paid no attention to that at all, but in the course of time, so far as I know, all these 37 names that I said were not really valid, they all have been sunk into the synonymy, and a much more natural system has been adopted.

Birds are such a good stepping-stone to do researches in general biology because there is no other group of organisms that is as well known as are birds. Literally all the 9,500 or 10,000—depending how you split them— species that at present, exist, have been discovered, have been carefully described, their geographic variation is known, and so on and so forth. And so it is, birds are a marvelous stepping-stone in three directions: toward evolution, toward systematics, and toward biogeography.

CHAPTER 4

Ebb and Flow

The flow and ebb of the feathered tide has been sung by poets and discussed by
philosophers, has given rise to proverbs and entered into popular superstitions,
and yet we must say of it still that our "ignorance is immense."

—ALFRED NEWTON (1896), IN HIS *DICTIONARY OF BIRDS*, LAMENTING
HOW LITTLE WAS KNOWN ABOUT BIRD MIGRATION

THE GREAT RIDDLE

IN THE AUTUMN OF 1903 FIFTY-YEAR OLD William Eagle Clarke was ferried by tender from Blackwall in the Thames estuary twenty-one miles out into the North Sea to the Kentish Knock lightship, his home for the next month. Life on those lightships was "one of considerable hardship and discomfort,"[1] yet Clarke was excited by the prospect of what he might discover. He was one of a small group of ornithologists appointed by the British Association and directed by Alfred Newton to study bird migration. After several years of compiling information collected by others on migrant birds from around the British coast, Clarke felt that it was time to witness migration himself.

Because of the Kentish Knock lightship's position, Clarke suspected that it lay on an important migration route between Britain and the Continent—a suspicion that was amply verified during the following four weeks. On the morning of 11 October, for example, he witnessed "a conspicuous passage of Starlings, Skylarks, and Tree-Sparrows. By midday it had assumed the nature of a "rush," which was maintained without a break until 4 P.M. . . . So numerous were the Starlings . . . that when first observed in the distance they resembled dark clouds."[2] By midday the weather had deteriorated to become, in nautical speak, a "dirty day," with torrential rain and strong winds.

Nocturnal migration at Cape May Point, New Jersey, in late September looking to the southwest at about 1,000 feet up. The brightest object is Saturn, with the constellations of Sagittarius (lower left) and Scutum (upper right) clearly visible. The nocturnal migrants are Cape May Warblers, a Wood Thrush, a Common Yellowthroat, and a Black-and-white Warbler. Painting by Guy Tudor in the late 1950s.

"How the migrants braved such a passage was truly surprising. How they escaped becoming waterlogged in such a deluge of wind-driven rain was a mystery. Yet on they sped, hour after hour, never deviating for a moment from their course."[3] Clarke's subsequent book, *Studies in Bird Migration* (1912), which Alfred Newton encouraged him to write, made him an authority on a topic that had mystified everyone for generations.

Indeed, migration with its multiple facets was *the* great mystery of bird biology. Writing in 1895, Heinrich Gätke referred to it thus: "The winged traveller, speeding on his way during the darkness of night in unerring course over vast expanses of ocean, presents to the savants of our day as great a riddle as it did to the first observer in ages before the dawn of history."[4]

Today, nearly 120 years on, it is well known that many birds migrate vast distances between their summer and winter quarters. Nonetheless, the discovery in 2009—using sophisticated new technology discussed later—that Bar-tailed Godwits fly from one end of the Earth to the other, some 10,400 kilometers (6,500 miles) in a single, nonstop 175-hour flight, still left ornithologists awestruck. No less remarkable was the discovery in the 1980s of the subtle interplay between nature and nurture controlling migration behavior, and—still ongoing—the study of the extraordinary hierarchy of mechanisms that allow birds to navigate with such precision.

The study of migration has a long history, but our story starts with Alfred Newton at the very end of the nineteenth century. A Cambridge don, variously described as punctilious, honorable, obstinate, and high-spirited, Newton devoted his celibate life to the study of birds (Birkhead and Gallivan 2012). His most enduring publication was the monumental *Dictionary of Birds* (1893–96), in which, over several tightly packed pages, Newton provides a picture of what he called "migration . . . the greatest mystery which the whole animal kingdom presents—a mystery which . . . can be no more explained by the modern man of science than by the simple-minded savage or the poet or prophet of antiquity."[5] With extraordinary astuteness and typical Victorian expansiveness, Newton presents the facts as they were then known. You can sense his frustration—despite his synthesis of so much information, migration remained as mysterious as it had ever been.

Here then are the facts—accumulated over previous centuries—as they were known in Newton's day: most bird species—in temperate regions anyway—are migratory at least to some extent, the "flow and ebb of the feathered tide,"[6] Newton called it; the same species may be migratory in one part of its range and sedentary in another; some members of the same population may be migratory while others are not; migration is not an all-or-nothing phenomenon and certainly not a characteristic of a species; in the spring, males typically arrive before females; those populations of a particular species that migrate the longest distances have the longest wings—elegantly illustrated by populations of Willow Warblers and Northern Wheatears (Newton 1896).

Contentious points included the role of weather—was it the cause of migration or did it merely influence it? What about "migration routes" or "flyways," subjects that Newton said had been elevated to "superstitious importance"[7] by some ornithologists in the 1870s. He is referring here to Johan Axel Palmén, who was misled by the fact that migrants become concentrated on islands and spits like the Courish Spit, and hypothesized that they followed narrowly defined *Zugstrassen*—"migration streets"—for their entire journey (Palmén 1876). As soon became clear, migrating birds *are* concentrated at some points in their journey and they do sometimes follow geographical features, but there are no set routes in the way that Palmén and his followers envisaged (Thomson 1926).

In 1880 the British Association, at the urging of Newton,[8] established a migration committee comprising John Cordeaux and John A. Harvie-Brown and placed observers on lightships, in lighthouses, and at other locations to observe migration directly. William Eagle Clarke joined the committee in 1883, primarily to analyze the rapidly accumulating data.

On reading Clarke's account of the extraordinary mortality among migrant birds that he witnessed firsthand, Newton wrote to him saying, "The slaughter, if one may so call it, seems so indiscriminate; there can be scarcely room for Natural Selection to act."[9] Putting Newton straight, Clarke responded: "As to the waste of life, I am afraid Nature never contemplated lighthouse and lightship lanterns." Newton replied: "It does not seem to me that the destruction of life is due to lightships. They only enable one to see it.... The birds are evidently lost already.... In one way it is plain that Natural Selection does act. The birds that migrate successfully, and so carry on the species, must be of the best, any shortcoming must carry a fatal penalty."[10] Newton should have been smarter about this since Alfred Russel Wallace (1874) had previously spelled out

and with great lucidity the way that natural selection and migration might operate together.

The committee's mass of hard-won data helped to demolish Palmén's idea about migration routes but failed to answer any of the other main questions. As Alfred Newton said, a decade after the project ended, transforming this information into something intelligible was "still taxing the ingenuity and patience of Mr W. E. Clarke."[11]

The "most inexplicable part of the matter"[12] was how birds found their way. How did they know where to go? How did they navigate across continents or the oceans? "There was a time," Newton says,

> ...that what is called the 'homing' faculty of pigeons might furnish a clue, but Mr Tegetmeier[13] ... declare[s] that a knowledge of landmarks obtained by sight, and sight only, is the sense that directs these Birds, while sight alone can hardly be regarded as affording much aid to Birds ... which at one stretch transport themselves across the breadth of Europe, or even traverse more than a thousand miles of open ocean, to say nothing of those ... which perform their migrations mainly by night.[14]

Newton mentioned Alexander von Middendorff's ideas (see below) about a magnetic sense and rejected it, much as he dismissed August Weismann's idea of a "sixth sense" and the vague and unhelpful notion that migration is "an instinct"—in those days a catchall term for any aspect of behavior that no one understood. Reading Newton's account of migration today, what is most striking is that by the start of the twentieth century most of the key questions had been identified: Where do migrants go? What are

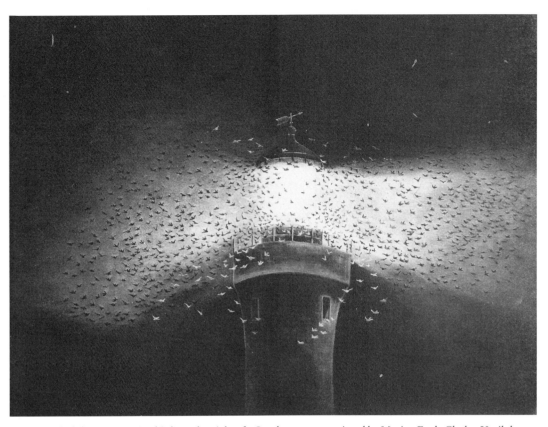

Eddystone lighthouse attracting birds on the night of 1 October 1901, as painted by Marian Eagle Clarke. Until the 1970s when the lighting was changed, lighthouses and lightships attracted, and killed, large numbers of nocturnal migrants.

the causes of migration? How did migration evolve? How do migrants find their way?

What happened next? The answer is, a great deal. Throughout the twentieth century and beyond, the phenomenon of migration has continued to fascinate and engage biologists more than perhaps any other aspect of bird behavior. Moreover, the complexity of migration has meant that it has required a wide variety of approaches. Such has been the volume of research on bird migration that throughout the twentieth century leading figures have felt obliged to provide state-of-the-art summaries, either in the form of papers in journals or entire books. For example, Arthur Landsborough Thomson's book *Problems of Bird-Migration* (1926) was intended as an update of what had been learned since Alfred Newton's day but also, as its title suggests, to "identify the problems [questions] relating to migration rather than an attempt, hopeless in the present state of knowledge, at their solution."[15] Interestingly Thomson, like Newton, also produced a monumental handbook, *A New Dictionary of Birds* (Thomson 1964). Opinions differ on how migration research developed and some, such as Don Farner (1955), felt that there were two main phases—pre- and post-1925—the watershed being the start of experimental studies, pioneered by William (Bill) Rowan. This division is slightly biased because Farner's interests were similar to Rowan's, very much focused on the causal aspects of migration (including its timing, which we discuss in more detail in chapter 6). Viewing migration more widely, a number of key experiments had already been conducted before 1925.

Like some of our predecessors (e.g., Gauthreaux 1996), we have structured this chapter around Tinbergen's four questions (chapter 7): evolutionary history, adaptive significance, ontogeny (development), and causation. We have done this because it seems to be the most pragmatic way to deal with the huge volume of migration research, but it is important to recognize that prior to the mid-1960s, before Tinbergen formulated his four questions, those questions were

Arthur Landsborough Thomson. Portrait by Sir Peter Scott in 1958 when Thomson was 67 or 68.

often confounded. Alfred Newton was certainly guilty of this. Thomson made a huge contribution, not only providing a succession of lucid summaries of what was known about migration (Thomson 1926, 1936a, b, 1964) but also, in the 1920s, being among the first to recognize the difference between proximate and ultimate explanations of behavior.

Proximate factors are the immediate causes of behavior, including environmental cues and internal, physiological processes, whereas ultimate factors are the selective forces shaping the behavior through natural or sexual selection. The addition of "evolutionary history" and "ontogeny" (development) as causes, by Tinbergen in the mid-1960s (chapter 7), helped to clarify things still further. Nonetheless, even with these clear distinctions, it is sometimes difficult with earlier studies to know what level of explanation the authors were dealing with, especially when the objectives of the study are unclear. While this might be frustrating, it is perhaps an inevitable part of science in an area of ongoing research.

The most recent comprehensive synthesis of bird migration research is provided by Ian Newton (2008) in his book *The Migration Ecology of Birds*. The fact that his "synthesis" comprises almost a thousand pages says a great deal about the volume of work that has been undertaken on this fascinating topic.

EVOLUTIONARY HISTORY

There are two main reasons why Alfred Newton struggled to make sense of migration, despite the large amount of descriptive information available at the beginning of the twentieth century. First, he was confused by the different types of movements birds exhibited—dispersal, irruptive movements, and there-and-back migration—and tried to explain all of these together. Here we focus mainly on there-and-back migration. Second, because Newton did not distinguish between different types of explanation, he muddled the evolutionary history of migration with its adaptive significance, as well as confusing the proximate and ultimate factors causing migration. As far as we can see, Newton barely considers the evolutionary history of migration, whereas for others this was the key question: How did migration evolve in a population of nonmigratory (i.e., sedentary) birds?

Wallace (1874) spelled out how migration might evolve through natural selection: "Those birds which do not leave the breeding area at the proper season will suffer, and ultimately become extinct."[16] This explanation encompasses both the evolutionary origin and the adaptive significance of migration, but he goes on to focus more explicitly on its origin. Imagine, he says, that originally the summer and winter ranges of a remote ancestor of the present species overlapped completely, "but by geological and climatic changes gradually diverged from each other."[17] In other words, migration evolved gradually in response to environmental changes until birds were commuting annually between their breeding and nonbreeding ranges.

The zoologist William Brooks had a simpler suggestion, saying that "the hypothesis of geological change seems gratuitous and unnecessary, since the known habits and instincts . . . of the birds . . . are . . . sufficient explanation."[18] In his view birds moved around in the breeding seasons in search of safe breeding sites, with selection favoring those that did so successfully, resulting in some cases in the eventual separation between breeding and nonbreeding areas. In terms of the origin of migration, rather than ultimate factors favoring it, Brooks's suggestion was both pragmatic and prescient.

Later, Thomson (1926) wrote, "Everything points to the conclusion that migration is a custom which forms part of the inheritance of the species and which is evoked to repeated activity by periodically recurring stimuli. For the origin of such an inborn custom one must obviously look to the past history of the races of migratory birds, and in this field no more than speculation is possible."[19]

Thomson acknowledged Wallace's idea that land movements and changes in climate, most notably the ice ages, might have played a role in the origin of migration. Specifically, he suggested that prior to the ice ages birds were nonmigratory, but were forced to migrate whenever the ice sheets spread south. Thomson also considers two other scenarios. The first is that an ice age pushes birds south, but after the ice recedes in later millennia there's an "inherited longing for an ancestral area"[20] and the birds return to their original breeding range. The second is that birds originally occupied what is now their winter range, but an increase in numbers resulted in competition for space and food, favoring individuals that moved elsewhere to what became the new summer range. Thomson is here alluding to Eliot Howard's idea that "territory" might be a crucial resource for which birds compete (chapter 10).

Writing in 1930, Ernst Mayr and Wilhelm Meise rejected the idea that migration evolved in response to glaciation (in the northern hemisphere), since birds in the southern hemisphere, where there had been no equivalent ice age, also migrate. Instead they suggested that migration evolved before the first glaciation. Mayr's views were almost certainly shaped by his PhD research (chapter 3) on the spread of the European Serin across central Europe during the nineteenth century. Originally confined to Mediterranean Europe, where it was nonmigratory, the species spread north and west beginning in the early 1800s, reaching Scandinavia by the 1960s. It is now known that all European Serin populations north of the Alps are partial or complete migrants (Berthold 1993). A similar change in migratory behavior occurred in the House Finch following its deliberate introduction from California, where it was nonmigratory, to northeastern North America, where within just thirty years a proportion of the new population became migratory (Able and Belthoff 1998).

A further example, albeit in the opposite direction, also occurred in the Fieldfare. In January 1937 deteriorating weather conditions resulted in a large group of Fieldfares migrating from southwestern Norway in search of better feeding. Normally the birds would have flown in a southwesterly direction toward Britain, but instead a storm carried them northward over Jan Mayen to eastern Greenland, from where—in round-the-clock darkness—some of the birds subsequently made their way to southern Greenland, where several were seen and shot. At least one bird appears to have reached Canada: in 1939 a mummified Fieldfare skin was found in the possession of an Inuit woman at Foxe Basin, Nunavut—the first record of the species in North America. There were only two previous records of Fieldfares from Greenland—single birds, both shot—but by 1944 that influx of birds in 1937 had resulted

in a small breeding population in the inner fjords in the southwest of the country. Remarkably, despite the Fieldfare's tendency to migrate, especially in response to poor weather conditions, the Greenland population was resident (Salomonsen 1951, 1979). That population persisted but was knocked back by heavy snow in the winter of 1966; the last breeding occurred in 1987, and the last bird was seen in 1990 (Boertmann 1994).

These examples illustrate a phenomenon that neither Alfred Newton nor his followers could have imagined: rather than being "fixed," migration is incredibly labile, with some species capable of very rapid changes in different aspects of migration (Newton 2008). But some ornithologists had anticipated this flexibility. Joseph Grinnell (1931), for example, felt that there was no need to invoke an ancient origin for migration:

> It would appear from this category of facts that the habit of migration is in most kinds of birds a perquisite easily and quickly taken on or put off. The physical equipment for migration is in nearly or quite adequate measure already possessed as a common attribute of birds, one of primordial standing. In the continual play of factors of existence consequent upon the continual changing of general and local environments, birds, by reason again of their endowment of motility, of sensitiveness, of extreme alertness, are able quickly to make adjustments. We may, indeed, say that the easiest thing they can do toward maintaining successful existence is to transfer their behavior from that of a population shifting annually to the condition of continual residence of their populations, subject only to daily and local circulation. And quite as easy is the reverse adaptation.[21]

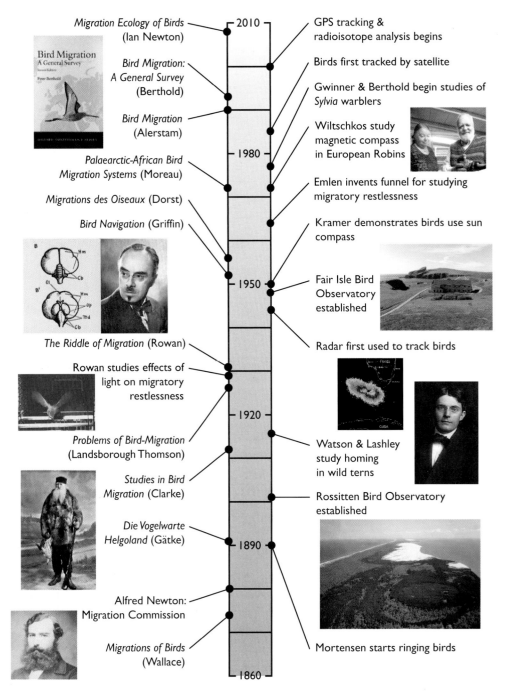

2010 — GPS tracking &
radioisotope analysis begins

Migration Ecology of Birds
(Ian Newton)

Birds first tracked by satellite

Bird Migration:
A General Survey
(Berthold)

Gwinner & Berthold begin studies of
Sylvia warblers

Bird Migration
(Alerstam)

Wiltschkos study
magnetic compass
in European Robins

1980

Palaearctic-African Bird
Migration Systems (Moreau)

Emlen invents funnel for studying
migratory restlessness

Migrations des Oiseaux (Dorst)

Kramer demonstrates birds use sun
compass

Bird Navigation (Griffin)

1950 — Fair Isle Bird
Observatory
established

The Riddle of Migration (Rowan)

Radar first used to track birds

Rowan studies effects of
light on migratory
restlessness

1920

Problems of Bird-Migration
(Landsborough Thomson)

Watson & Lashley
study homing
in wild terns

Studies in Bird
Migration (Clarke)

Rossitten Bird Observatory
established

Die Vogelwarte
Helgoland (Gätke)

1890

Alfred Newton:
Migration Commission

Migrations of Birds
(Wallace)

Mortensen starts ringing birds

1860

TIMELINE for MIGRATION. Left: Cover of Berthold (1993); Rowan's drawings of the brain of an American Crow; William Rowan; Eurasian Blackcap showing migratory restlessness; Heinrich Gätke in his shooting dress (painting in 1893 at age ca. 79); Alfred Newton. Right: Roswitha and Wolfgang Wiltschko; Fair Isle Bird Observatory in 2011; radar image of birds migrating off southwestern Florida during spring migration on 28 April 2002, showing as many as 16,000 birds per kilometer crossing over; John B. Watson at Johns Hopkins University, when he was in his 30s; Courish Spit, Prussia, the site of Rossitten Bird Observatory.

Although Thomson favored the idea that migration evolved gradually through natural selection, he also considered another scenario, based on the appearance of large numbers of exotic species well outside their normal range. In March 1888 reports were received from Russia of huge movements of Pallas's Sandgrouse. On that occasion large numbers of birds were seen moving westward away from the breeding areas in central Asia, with some reaching the east coast of Britain in May. This was a spectacular and exciting ornithological event, made even more extraordinary by the fact that some birds remained to breed in Britain.[22] Thomson described such invasions as "sudden attempts at extension of range due to some imperfectly known cause"[23] and wondered whether they could account for the almost spontaneous appearance of migration: "Had the new summer quarters proved more favourable, might not this species have suddenly changed from a resident to a regular migrant on any one or more of the occasions on which it has, so to speak, broken bounds? May not such occurrences . . . be the beginnings, or attempted beginnings of migration proper?"[24]

While Thomson was writing his migration book in the early 1920s, a battle was raging between the biometricians and the Mendelians over whether evolution occurred gradually by natural selection or by a single major shift (chapter 2; see also Provine 1971). Thomson doesn't mention it, but he must have been aware of this debate. The controversy lasted decades until it was realized, around 1930, that the two positions were actually end points along a continuum. In much the same way, Thomson's ideas about sandgrouse irruptions being the beginning of migration are simply a more extreme version of what

was later seen in the Eurasian Blackcap, where apparently random range extensions did indeed result in new migration behavior (see chapter 2).

More recently, ideas about the origins of long-distance migration have focused on three alternatives: that temperate-to-tropical migrants have evolved from (1) tropical-breeding birds that moved further north to breed, (2) temperate-breeding species that moved southward for the winter, or (3) mid-latitude-breeding species that moved both more northerly for breeding and more southerly for the winter. George Cox (1985) favored the latter scenario, proposing a stepping-stone model for Nearctic-Neotropical migrants in which the Mexican Plateau was the source region selecting for migration in species ancestrally breeding there. Another idea that emerged somewhat later was that there were single adaptive radiations within particular bird taxa, implying a strong phylogenetic component to migration (Rappole 1995). However, subsequent comparative studies using well-defined phylogenies show exactly the opposite: migration pops in and out of particular lineages very frequently and is extremely labile, as illustrated by the European Serin, Fieldfare, and House Finch reported above, and exemplified by the Eurasian Blackcap, as described in chapter 2.

It has been argued by some that the second of these scenarios—that migration emerged in the tropics—is more likely, based on the fact that many tropical species show altitudinal and geographical seasonal movements in search of resources. This idea has some support, for example from the comparative study by Douglas Levey and Gary Stiles (1992), which showed that among resident Costa Rican birds seasonal movements were most likely to occur among species with a fruit or nectar diet, or birds associated with dry habitats. These authors suggest that "long-distance migration to the Temperate Zone can be viewed as an evolutionary endpoint of this continuum."[25] A different view, originally expressed by Brooks (1898) and Grinnell (1931) and taken further by Peter Berthold (1999), is that most birds have the genetic programs for the suite of traits associated with migration, including navigational, physiological, and cognitive abilities, but in sedentary populations these are not "turned on." Once the costs and benefits of movement change, these genes are switched on and migration emerges very rapidly. Indeed, the tropical origin of migration may be a red herring, for, as Ian Newton has suggested, migration "is likely to evolve wherever the climate is seasonal, whether tropical or not."[26]

An arrow stork (*Pfeilstorch*); African arrows or spears in living White Storks found in Germany provided some of the first evidence of where these birds spent the winter.

COMING AND GOING

Before considering the second of our four questions about migration, some additional background information is necessary. The question about where migrating birds come from and go to has always been important. In a few cases the answer was obvious, as in the case of Common Quail observed flying over the Mediterranean between North Africa to Europe, occasionally landing on ships. In other instances there were tantalizing clues, such as the African tribal spear found in a live White Stork in Klutz, Germany, in 1822—one of several *Pfeilstorchs* ("arrow storks") subsequently recorded (see Vaughan 2009).

From the 1850s on observers made a concerted effort to witness migration across landmasses like Europe and Asia, enabling them to speculate about the breeding and wintering areas of particular species (Newton 1896). One of the most remarkable and undervalued pieces of research on this topic came from the amateur ornithologist Eliot Howard (chapter 7), who in the early 1900s plotted maps illustrating the summer and winter distributions (but not the migration routes) of different European warbler species. Remarkably, he provides no information on how he obtained information on their winter distributions in Africa, and we can only surmise that he must have gleaned it from the numerous accounts written by colonial administrators or from the specimens they collected. Certainly there were no ringing recoveries of European warblers from Africa until the 1970s.[27] The most remarkable feature of Howard's maps is just how accurate we now know them to be, based on hundreds or thousands of banding recoveries (see Wernham et al. 2002 for comparison).

The innovation of banding birds to better understand their movements is credited to the eccentric Danish schoolteacher Hans Christian Mortensen. Birds such as swans, herons, and hawks had been individually marked long before, with neck rings, claw and toe clipping, but in 1899 Mortensen was the first to mark birds—with leg rings—in a systematic way to study migration. He got the idea from reading about a Greater White-fronted Goose, marked with a brass neck ring, that escaped from a Dutch waterfowl collection in 1806 and was shot in 1835 in Poland. In 1890 Mortensen marked two adult Common Starlings near his home at Viborg, wrapping around one leg a narrow strip of zinc on which he had written in ink "Viborg 1890." Later Mortensen used aluminum rings—which had recently been introduced for marking chickens—first on a Red-breasted Merganser and soon after on young Common Starlings. Initially he had tried to stamp numbers onto the circular aluminum rings, but quickly gave that up and instead stamped flat strips of aluminum and then formed them into rings. In the summer of 1899 Mortensen used nest-box traps to catch and ring 165 starlings. The only recoveries were local, however, which Mortensen attributed to the inadequate inscription "Viborg" on the ring (along with a unique number). The next year he changed the lettering to "M" (for Mortensen) "Danmark" and obtained two recoveries, one from the Netherlands and one from Norway. Part of

his success was the result of making sure his scheme was well publicized.

Mortensen recognized that if his "ringing experiment" was to work, it had to be conducted on a large scale, and on species where the chances of recovery were relatively high. Ducks, which were trapped and shot in large numbers, seemed ideal. Accordingly, Mortensen visited a duck decoy (a duck trapping station) at Fanø on the North Sea coast of Denmark in 1907 and ringed 102 migrating Eurasian Teal. Some 20 percent of the banded birds were subsequently recovered, the first just 60 kilometers (37 miles) away a few days after release, but others were shot as far away as the Coto Doñana in southern Spain and the Bog of Allen, Ireland.

Fired up by the new information on bird movements revealed by banding, Mortensen marked a range of species, including starlings, White Storks, Grey Herons, and raptors. By the time he died in 1920, Mortensen and his colleagues had marked 5,000 individuals of 33 species. News of Mortensen's success spread rapidly, and very soon ringing schemes started up elsewhere: in Prussia (1903), Hungary (1908), Britain and the United States (1909), Yugoslavia (1910), Sweden (1911), Finland (1913), and Norway (1914).

In Prussia Johannes Thienemann started the world's first banding station-cum-bird observatory at Rossitten, now Rybachy, on the Courish Spit. As a schoolboy Thienemann became friends with Friedrich Lindner, a keen birder who in 1888 had explored the entire 97 kilometers (60 miles) of the Courish Spit. Lindner told Thienemann and a mutual friend, Kurt Floericke, about the remarkable bird life on the spit, and Floericke moved there in 1892, making a living by selling bird skins to museums and

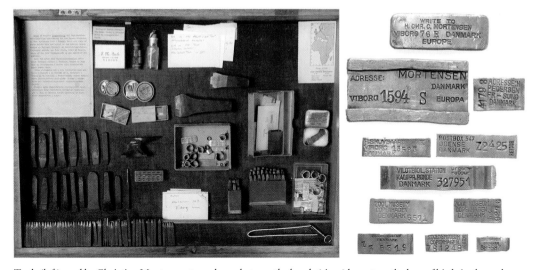

Tools (left) used by Christian Mortensen to make and stamp the bands (rings) he put on the legs of birds in the early 1900s.

collectors. Thienemann made his first visit in 1896 and fell in love with the place: "The rich bird life and regular passage of migrants . . . struck a chord in me which hitherto has been resting but waiting to be struck. I was seized by boundless enthusiasm. . . . "[28] A few days later, while drinking coffee in the garden, he shot a large bird flying overhead: it was a Pallid Harrier—the first he'd ever seen. What better place to establish a bird observatory?

When Floericke set off on a round-the-world adventure in 1897, Thienemann moved out to Rossitten. The coincidence of several events helped to ensure the success of Thienemann's banding station there, including: (1) the publication in 1891 of Heinrich Gätke's *Die Vogelwarte Helgoland* (published in English in 1895), a summary of his fifty years of migration studies on that German North Sea island; (2) the start of an ornithological center[29] in Budapest, Hungary, in 1893, where migration patterns of birds were recorded, and (3) the launch of Mortensen's banding scheme in 1899. The German Ornithological Society approved Thienemann's

plans for a ringing station at Rossitten, an extremely isolated village of just forty inhabitants in those days, with no roads and the nearest railway station 35 kilometers (22 miles) away. He became director in 1901 and was handed a detailed set of instructions regarding the type of information he was expected to collect. Thienemann stepped right into Floericke's shoes, acquiring not just his house but also his fiancée, Hertwig Hoffman (Vaughan 2009). Initially, Thienemann was forced to supplement his meager income by teaching but over time conditions—both in terms of communication and salary—improved, and by 1910 he was elevated to a professorship at the University of Königsberg as a result of his extraordinarily innovative migration studies.

Rossitten is located about halfway along the Courish Spit at a point where it narrows, concentrating migrant birds, of which there can be up to half a million per day. Initially, Thienemann conducted his research largely with a shotgun, collecting specimens both for identification and dissection. But in 1903 he also started banding, mainly the Hooded Crows that passed through in large numbers and had been trapped for centuries by the local people for food. The recoveries were not long in coming, and Thienemann was able to build up a picture of their movements, showing that they wintered on the north German plain and moved eastward in spring toward Finland (see Vaughan 2009).

Established in 1901, Rossitten was the first bird observatory and ringing station, followed by Heligoland (Germany) in 1910, the Sempach Bird Observatory (Switzerland) in 1924, Skokholm Island (Wales) in 1933, Isle of May (Scotland) in 1934, Cape Cod (USA) in 1939, Fair Isle (Scotland) in 1948,

and many others, each of which has a special story (Durman 1976). In 1905 Fair Isle was visited by William Eagle Clarke as part of his migration survey, and he soon recognized its importance. Word spread, and over the following decades birdwatchers came to Fair Isle to see the rarities for themselves. One visitor in 1935 was George Waterston, who started to make plans for an observatory there. The Second World War intervened; Waterston was captured and held as a POW, but after his release he purchased the island, opening the observatory and beginning to ring birds in 1948.[30]

Bird observatories collect records of the species seen, trapped, banded, and recovered and thus provide an important source of information on the routes and timing of migration of various species (Cornwallis 1959; Durman 1976). Combining data for all of the observatories in Britain, for example, it is possible to plot the northward progress of spring migrants. Observatories have also contributed to studies of morphology and molt, and more recently their records have been useful in monitoring population sizes (chapter 10). By providing opportunities for bird watchers to see birds in the hand, they can also serve to inspire young birders. However, it is generally acknowledged that the full scientific potential of bird observatories has not been fulfilled, largely because—with a few exceptions—most bird banders are not trained to synthesize, interpret, and write up their results for publication.

Even so, the introduction of bird banding, both at observatories and elsewhere by enthusiastic amateurs, resulted in a revolution in our understanding of bird migration at the start of the twentieth century. With so many birds banded, data accumulated

Johannes Thienemann (left, age ca. 66) and Oskar Heinroth (right, age ca. 58) in 1929.

rapidly, and it soon became possible to acquire a general picture for particular species. These were exciting times: the field was sufficiently promising that a journal, *Der Vogelzug*, devoted to bird migration—with Rudolf Drost and Ernst Schüz as editors—was launched in 1930, and enough recoveries of banded birds had accumulated by 1931 to allow Schüz and Hugo Weigold to produce an atlas of bird migration. Schüz and Weigold's atlas was the first of its kind, but its scope was limited, relying on only twenty-seven recoveries of banded birds. Following this, Gerhardt Zink produced four atlases of bird migration between 1973 and 1985, and together with Franz Bairlein, another in 1995. Other countries have followed suit, though slowly: the atlas for Britain was completed in 2002, presenting recovery data collected over a hundred years of bird ringing (Wernham et al. 2002),[31] and the atlas for Canada—four volumes published from 2000 to 2010—presenting banding and recovery data on more than 4 million birds banded in that country between 1955 and 1995.[32]

An improbable and little-known bird-banding venture was funded by the American military, which, initially concerned about wild birds transmitting the Japanese encephalitis virus to humans, decided that they needed to better understand the movements of birds in Southeast Asia. The project—known as the Migratory Animal Pathological Survey (MAPS)—started in 1963 and banded more than 1.1 million birds of 1,218 species in 18 countries. The 5,601 recoveries of 235 species provided the first information on bird movements within that region (McClure 1974). Because the project was funded by the military, China and the USSR assumed that the lead author, Elliott McClure,

was an agent of the enemy: "I had been on the [Communist hit] list since handbills were distributed in Tokyo in 1950 describing our work with birds and distorting it to be a study in biological warfare. We were supposed to be inoculating birds with viruses and freeing them to take infection to China and other lands."[33] His 1974 report on migration patterns is a classic of Asian bird studies.

In Europe banding has been remarkably successful, largely because it has been done on such a massive scale, with an estimated 115 million birds banded between 1900 and 2000.[34] In North America over 64 million birds have been banded since 1960.[35] However, placing bands on birds was much easier than trying to interpret the results that accrued from the subsequent few recoveries.[36] As researchers realized, the interpretation of banding results was usually problematic, not least because the data typically comprised just two localities per bird—one at the original banding site and the other where the bird was recovered. Usually, what the bird did between these two was completely unknown.

Technological advances have revolutionized our understanding of bird movements: first, with radio tracking in the 1960s and then with the advent of satellite tracking in the 1980s. Early progress in tracking birds was slow, mainly because the devices attached to the birds were bulky and could be used only on relatively large species: a Bald Eagle was the first bird to be fitted with a satellite transmitter, in July 1984, and the transmitter was a hefty 170 grams (6 ounces). The first satellites[37] were also not completely reliable and could not give data about bird movements at a very fine spatial scale. Since those days satellite tracking technology has greatly improved in terms of accuracy, transmitter size, and cost—satellite tags cost around £3,000 (= $4,750 US) each.

Starting at the turn of the twenty-first century, tracking devices began using GPS technology, which is both more reliable and allows for more fine-grained analysis. Geolocators, for example, cost around £50 (= $80 US) each and are small enough to put on small passerine birds. In 2010 thirty Aquatic Warblers—which weigh just 10 to 12 grams (0.35 to 0.4 ounces)—were fitted with geolocators weighing 0.6 grams (0.02 ounces) in the Ukraine; in 2011 four of these individuals, still with their geolocators, were recovered back in their Ukraine breeding site, having wintered in Senegal.[38]

The ongoing development of reliable miniature tracking devices will provide us with a new level of understanding about bird movements, and this approach has already allowed for intriguing new insights. For example, Yannis Vardanis and his colleagues (2011) followed seven adult Western Marsh Harriers on repeated migratory journeys between Sweden and West Africa from 2004 to 2009. They found that each bird was highly repeatable in its timing of migration, but there was considerable between-year variation in the route taken, suggesting an endogenous control of timing but route flexibility due to environmental factors. This shows the importance not just of plotting individual migratory routes but also of analyzing route repeatability at an individual level.

The development of radar (RAdio Detection And Ranging) to track warships and planes inadvertently also provided valuable information on bird migration. Developed during the Second World War, radar uses radio waves to determine the location, direction, and speed of objects by detecting the echoes from short pulses of radio waves. Said to have helped the Allies win the war, radar was not without its problems. "Throughout the war sudden and mysterious radar signals rushed combat men to battle stations, sent fighter planes on 'goose' chases, prompted lookouts to report unidentified aeroplanes diving into the sea, gave rise to several E-boat scares, started at least one invasion alarm, and tested the vocabulary of many skippers."[39] David Lack and the entomologist George Varley worked in the Army Operational Research Group on radar, and it was Varley who first recognized that those mysterious radar signals—clearly not coming from planes—were coming from birds. Lack and Varley identified Northern Gannets on a prototype coast-watching radar set near Dover in September 1941.

After higher powered transmitters were introduced in 1943, bird echoes became enough of a nuisance that Lack and Varley had to train radar operators to distinguish seabird echoes from ships (Lack and Varley 1945). Physicists were initially skeptical about these "bird echoes," and at one meeting they explained to Lack and Varley that the echoes could only have been caused by clouds of ions; a frustrated George Varley replied that he accepted their view, provided that the ions were wrapped in feathers (Lack 1958).

Even more powerful radar systems were developed after the Second World War, and it was the Swiss ornithologist Ernst Sutter (1957), using British equipment at the Zurich airport, who carried out the first systematic study showing that the "angels" often seen on radar while tracking planes were flocks of migrating birds. Radar technology provided much needed details about the volume, height, direction, and speed of migration, although it precluded the identification of species. Writing in 1958, David Lack commented that "future work may well show that radar will revolutionise the study of bird migration to an extent comparable to that exercised by the sound spectrograph on the study of bird song."[40] To some extent it did, and in the following decades radar studies of bird migration provided information on bird movements over huge areas of North America (Gauthreaux et al. 2003), as well as on flight speed, performance, and orientation (Bruderer and Bolt 2001; Alerstam et al. 2007).

THE ADAPTIVE SIGNIFICANCE
OF MIGRATION

Alfred Newton (1896) discussed two possible explanations for why birds migrate away from northern latitudes in winter—he suggested that they are in search of either food or warmer temperatures. Recognizing that birds are often capable of withstanding very low temperatures as long as they are well fed, he identified food as the main benefit of migrating. In this he agreed with Wallace's (1874) view of the way natural selection might favor migration. Newton believed that a shortage of food pushed birds south, where they then encroached on other birds, which in turn were also compelled to move south to avoid competition for food. In his own words: "As food grows scarce towards the end of summer . . . the individuals affected thereby seek it elsewhere; in this way they press upon the haunt of other individuals: these in like manner upon that of yet others."[41]

The journey north was trickier to explain, and on this point Newton says, "When we consider the return movement . . . doubt may be entertained whether scarcity of food

can be assigned as its sole or sufficient cause, and perhaps it would be safest not to come to any decision on this point."[42]

Part of the difficulty for Alfred Newton and his predecessors was, as we have seen, the confusion over proximate and ultimate factors. We now know that for most birds food is the ultimate factor determining migration, but rarely is it the proximate cue that initiates that migration (Newton 2008). Yet confusion was inevitable because a few species, like finches, waxwings, and nutcrackers often undertake huge southward movements (irruptions) in the northern hemisphere in response to a failure of their food supply.

By distinguishing between proximate and ultimate factors, Thomson (1926) recognized that food was the most likely ultimate factor favoring migration—a view consistent with the fact that most insectivorous species in temperate regions are migratory. Thomson also recognized that day length was important in providing enough time to forage, and that the shortening of daylight periods in the fall favored birds moving to lower latitudes for the winter months (Thomson 1926).

Assessing the hypothesis that food (or any other factor) is the single most important ultimate factor favoring migration has been difficult: it is almost impossible to test this idea experimentally, leading some ornithologists to suggest that discussing the adaptive significance of migration is pointless (Gauthreaux 1996; Berthold 1993). Comparative studies, however, show that latitude, diet, and habitat are important predictors of whether species are migratory or not (Newton 1995; Boyle and Conway 2007).

Another way to test the adaptive significance of migration is to compare individuals within populations (Schwabl 1983). Some bird populations are "partially migratory"—that is, they include both migratory and nonmigratory individuals—and this phenomenon has received much attention because it presents an opportunity to assess the relative costs and benefits of migrating or remaining resident within a species. Early hypotheses to account for the persistence of both "types" included Lack's (1943b) idea that the tendency to migrate varied between both populations and regions. There is now evidence from several studies—both empirical and theoretical—that socially dominant individuals are less likely to migrate (Smith and Nilsson 1987; Kokko 1999), and that migrating males have reduced breeding success when they return from their wintering grounds (Boyle et al. 2011).

With the advent of behavioral ecology in the mid-1970s, the idea of looking at the adaptive significance of migration by considering the costs and benefits of its various components, including flight speed, fuel reserves, stopover times, and so on became of interest (Alerstam and Lindstrom 1990). The aim was to use optimality modeling as a theoretical foundation to better understand the different aspects of migration, to generate predictions, and to identify new questions. For example, by adopting a particular route are migrating birds minimizing their total energy expenditure, the duration of their migration, or their risk of being preyed upon?

Perhaps the most obvious application of this approach is to predict fuel deposition—generally, the accumulation of fat—prior to and during migration. It is intuitively obvious that the greater the fuel load, the greater the energetic costs of flight, so that in terms of flight distance, there's a diminishing return. By making certain assumptions, the optimal fuel load can be predicted and then compared with observed values. For most passerines studied, the analysis of fuel loads suggested that birds acquired just the right amount of fuel to minimize the total duration of migration; for the European Robin, however, the results were more consistent with minimizing the total amount of energy expended during the trip (Alerstam 2011). Thomas Alerstam's analysis suggests that an optimization approach may help to explain why Bar-tailed Godwits make a single, long nonstop flight in their autumn migration between Alaska and New Zealand but make a two-step detour via Asia during the spring migration, when they are exploiting favorable winds or need to forage en route so that they arrive on their breeding grounds in good condition (Gill et al. 2009; Alerstam 2011). Alerstam is sanguine about the optimality approach, recognizing its strengths in identifying new questions but also aware that, like optimal foraging theory (chapter 8), "optimal migration theory" may have a short life. He also recognizes that new tracking technology will provide much-needed information about migration routes and stopovers that will enable researchers to test their optimality models quantitatively. Others are less optimistic; Ian Newton (2008: 614), for example, has warned of the practical difficulties of testing such models, arguing that birds probably migrate as rapidly as they can but are sometimes unavoidably delayed by unpredictable bad weather—something that optimal migration models are unable to include.

NATURE AND NURTURE OF MIGRATION

Almost from the start, the act of migration was assumed to be "instinctive," a view that goes back as far as Aristotle in the third century BC. Nicolas Venette, the extraordinarily perceptive author of *Traité du Rossignol* [A Treatise on the Nightingale], published in 1697, refers to the tendency of this species to migrate as an "instinct and inner guide that makes them fly . . . to the place where they want to go" (Birkhead and Charmantier 2013). Instinct is also implicit in Alfred Russel Wallace's assumption that migration is shaped by natural selection, and Darwin was forced to rely on instinct in his musings on bird migration: "It is a true instinct which leads the Brent Goose to try to escape northwards; but how the bird distinguishes north and south we know not. Nor do we know how a bird which starts in the night, as many do, to traverse the ocean, keeps its course as if provided with a compass."[43]

For centuries there had been debate about whether nature or nurture dictates the behavior of birds and other animals, including ourselves. As far as migration was concerned, there was a broad but uncritical consensus that it was entirely due to nature. A few ornithologists held the consciously contrary view, including Johann Palmén, who in the 1870s suggested that young birds learn their migration route by following older, experienced individuals. As evidence, he noted that it is usually young, inexperienced birds that get lost on migration, but Alfred Newton was quick to refute Palmén's logic, pointing out that the majority of birds do not migrate in flocks, thereby precluding "following." According to Alfred Newton, previous ornithologists, including Coenraad Temminck in the early 1800s and most clearly Heinrich Gätke on Heligoland, had reported that "the young and old always journey apart and most generally by different routes."[44] As Alfred

Newton says, the young birds "can have no 'experience,' and yet the greater number of them safely arrive at the haven where they would be." He argued that since many birds migrate at night in the dark, experience is irrelevant.[45] It is now known that in certain species, notably geese and cranes, young birds do learn migration routes by following their parents (Newton 2008), and it has been suggested that culturally transmitted migration routes can change more rapidly than those that are genetically influenced, exemplified by the recent changes exhibited by Barnacle Geese breeding and wintering in the Netherlands (chapter 11).

The assumption that migration was instinctive was also accepted by Thomson in the 1920s, but he was under no illusions regarding what was left to explain, referring to migration as " . . . an interesting example of instinctive behaviour. . . . To study the migration of birds is to investigate the nature of animal behaviour, and to do this is to probe the inmost mysteries and to ask the very meaning of Life itself."[46]

One of the first attempts to throw any light on the instinct idea was a set of experiments conducted by John Watson (who went on to found the field of behaviorism; chapter 7) in the early 1900s and later assisted by his postdoc Karl Lashley in 1913. In these experiments Watson took Sooty Terns and Brown Noddies far away from their breeding colony in the Dry Tortugas, 60 kilometers (37 miles) west of the Florida Keys, and reported whether they found their way home. Although the aim of the experiments was not to test whether homing ability was instinctive per se (and we will return to consider these experiments a little later), many of the birds did find their way back.

Watson's experiments were unique in that they were conducted on wild birds, but they were no different from the many "displacement experiments" that had been carried out on racing or homing pigeons for many years. However, the results were unexpected because, unlike pigeons that have to be trained to "home,"[47] these terns returned home with no training and did so over a featureless sea.

At around the same time, Thienemann in Prussia was wondering whether White Storks knew *instinctively* which direction they had to migrate. The White Stork was Thienemann's favorite species and one he considered to be the "predestined experimental bird,"[48] meaning that it was easily identified, large, and popular. By 1910 Thienemann had sufficient banding recoveries to be able to plot the stork's migration routes and wintering areas. Then, in the 1920s he capitalized on the stork's special status to undertake an ambitious experiment to answer the question of what first-year storks did if they were delayed from migrating (with experienced adults) at the normal time—in other words, did they know instinctively in which direction they should migrate? During 1926, 1927, and 1928 Thienemann hand reared a total of 140 young storks and released them several weeks after all other storks had migrated. By publicizing the experiment he obtained sufficient sightings and banding recoveries to show that these naive birds migrated in a southeasterly direction, exactly like the wild birds. So far so good. But as European banding recoveries accumulated, it became clear that storks from different parts of Europe migrated in different directions: those from Prussia went southeast but those breeding further west, in Germany, went southwest.

In the 1940s Ernst Schüz, Thienemann's successor at Rossitten, took advantage of this so-called "migratory divide" to conduct an ingenious experiment. Schüz took 754 nestling White Storks from East Prussia and sent them by train to be hand reared in western Germany—the other side of the so-called migratory divide—in order to see which direction they would migrate. Contrary to Thienemann's previous experiment—and therefore contrary to the notion that migratory direction was instinctive—these young birds migrated in a south*westerly* direction (Schüz 1949). The most likely explanation seems to be that the hand-reared birds simply followed the adult wild birds, their tendency was "to fly in a south-easterly direction was thus inhibited by the strong drive to join older individuals of their own species."[49] Schüz was persistent and devised a new experiment in which 144 nestling East Prussian storks were released on the western side of the "divide" in Essen, but crucially, after all other storks had migrated. This time they all migrated in a southeasterly direction, and as Ernst Mayr said, "The proof seems to be supplied that there was a genetic [i.e. instinctive or innate] disposition for flight in a south-east direction."[50]

The most comprehensive attempt to address the question of whether migration was innate was made in the 1970s by Peter Berthold and colleagues of the Max Planck Institute at Radolfzell in southern Germany. His results provided spectacular and convincing evidence for the role of genetics (Berthold and Querner 1981). Berthold crossed Eurasian Blackcaps from different populations in captivity to compare the migratory behavior of their offspring, using migratory restlessness (or *Zugunruhe*) as an index of migration

Hand-reared White Storks being transported during a navigation experiment, possibly conducted by E. Schüz in 1933.

direction and duration. For example, he crossed Eurasian Blackcaps from southern Germany, which typically migrate in a south-westerly direction, with those from the Canary Islands, which barely migrate, and then measured the migratory direction and duration of migration in their offspring. In each case—and he made many such crosses between several different populations—the migratory behavior of the offspring was intermediate between that of the parents. This suggested, first, that the migratory behavior is inherited, and second, because the behavior was intermediate between that of the two parents, that it is controlled by several genes. Had the behavior been controlled by a single gene, as with color in Gregor Mendel's peas, then the offspring would resemble one or other of the parents.

Subsequent experiments by Berthold and colleagues in Germany, using Eurasian Blackcaps and both Common and Black Redstarts, provided additional evidence for a genetic basis for migration. Estimates of the heritability of the direction and duration of migration, together with artificial selection, revealed that a migratory population could become fully sedentary within just three generations, and a sedentary population could become almost entirely migratory in just five or six generations (Berthold 1996). Migratory populations could also rapidly evolve a new migratory direction, as exemplified by the blackcaps flying northwest to winter in Britain each year. Such rapid microevolutionary change is possible because there is a substantial amount of variation in migration traits—a range of genotypes from resident through to migratory—within a single population.

The blackcaps from different populations that Berthold and others used for their experiments differed not only in migratory behavior but also in a number of other ways, including body mass, wing length, hyperphagy (the tendency to eat excessive amounts of food), and the propensity to lay down fat reserves. These traits also seemed to be under genetic control, and this suggested the existence of a "migratory syndrome"—a tightly integrated suite of genetically linked traits. While this seems to be true within individuals, there is no evidence that a migratory syndrome exists across genera or families. A test of this idea, which involved mapping these traits onto a phylogeny, showed that far from being a syndrome, traits associated with migration are extremely labile and have independently evolved many times (Berthold 2001). Within a single genus or family there is enormous variation in these traits, and that is because different species have evolved different ways to solve the same problems (Piersma et al. 2005).

Perhaps the most revealing experiment in terms of the roles of instinct and learning in migration was a study performed by the Dutch ornithologist Ab Perdeck in the 1940s and 1950s. After trapping some 11,000 Common Starlings during their autumn migration, he banded the birds and determined their age and sex. They were then displaced—by airplane—600 kilometers (373 miles) to the southeast, where they were released and allowed to continue their migration. It was already known from earlier banding studies where Dutch starlings normally wintered. What Perdeck was interested in was whether his displaced birds would adjust their migration and end up in their correct winter quarters, or whether "instinct" would simply dump them 600 kilometers (373 miles) southeast of the correct wintering area. The experiment was widely advertised, and Perdeck obtained over three

hundred recoveries of his marked birds. The results were intriguing: the young birds did winter 600 kilometers (373 miles) south of their normal wintering area, suggesting that they relied entirely on instinct to go in the right direction. In contrast, the adult birds compensated for the displacement, adjusted their migration, and ended up in the correct wintering area. Their ability to do this implied that as well as the clear instinctive or genetic effect, experience (learning) also had an important role in migration. Indeed, it was later shown that once a bird has undertaken its initial migration to its winter quarters, relying on instinct, it then learns the winter quarters' characteristics. It is this learning that enabled the older, experienced starlings to compensate for being displaced either experimentally or by unusual weather conditions (Perdeck 1958).

CAUSATION:
THE MECHANISMS OF MIGRATION

How birds find their way is a question that has mystified ornithologists for centuries. By 1900 there were lots of ideas but almost no evidence for any of them. It was well known, of course, that homing pigeons could—after some training in which they were given the opportunity to learn landmarks—find their way, but they did not actually migrate, and somehow homing and migration seemed to be separate phenomena. As Stresemann (1975) pointed out, the "old fundamental questions, what stimuli guide the migrating bird, and what provokes it to undertake the migration, remained unanswered until the method [of study] was radically changed. It had long been recognized that only tests would help, but because ornithologists, in

Heinroth's words, usually had 'an almost medieval horror of experiment' the fame to be won by the discoverer of new techniques eluded them."[51]

It was not until the early 1900s that the first experimental efforts were undertaken on wild birds to test their ability to find their way, and here we return to Watson and Lashley's work on the homing instinct of Brown Noddies and Sooty Terns in the Dry Tortugas.

Bird Key was a perfect location for displacement experiments. First, it was the most northerly breeding location of these otherwise tropical birds, so Watson could be sure that any locations where they released the terns further north would be new to the birds. Second, since Bird Key is the last stop between the Florida and Texas coasts, the birds could be sent hundreds of miles over open water, removing the possibility that they might use visual landmarks for navigation.

Watson captured the terns, marked them with oil paint—a different color for each cohort so their return could be documented—and transported them to the mainland. The birds were then released at various sites, including Cape Hatteras, North Carolina, over 1,400 kilometers (870 miles) to the northwest. To Watson's delight, most of the birds released from all sites returned, in some cases within a day: "The difficulty of explaining homing by current theories is seen to be great, but, while admitting this, we do not suggest the assumption of some new and mysterious sense. The task of explaining distant orientation is an experimental one, which must yield positive results as soon as proper methods are at hand."[52] This study provided the first conclusive evidence that wild birds displaced considerable distances could return home, and in some cases—over

wide expanses of water—without using landmarks.

Acknowledging that the pigeon was a good study species, Watson also pointed out that "the work on the homing pigeon has been carried out in too desultory a fashion and too much under the influence of particular theories to afford satisfactory material for hypothesis. Crucial experiments designed to bring out the facts as to what the untrained homing pigeon can do are lacking. Until such have been made speculation on the mechanism of return is useless."[53]

Watson abandoned his homing studies in favor of comparative psychology (where he had much greater control over what his subjects could do and experience), but it seems likely that had he persisted, he would have made a major contribution to understanding the orientation and navigation of birds. As it is, his subsequent behaviorist work eclipsed this early work on the terns, which remains underappreciated (chapter 7).

Throughout the early years of the twentieth century researchers became increasingly fascinated by migration, undoubtedly driven by the new technology of bird banding. Displacement studies (summarized by Thomson 1936b), pioneered by Watson and Lashley, continued to be popular both in Europe and North America. Werner Rüppell, who Stresemann (1975) described as "tireless and imaginative," conducted displacement studies on a large scale, enlisting the help of bird banders to capture nesting Common Starlings from all over Germany, releasing them from Berlin. Just as with Watson's terns, many of these ringed birds returned and with no obvious effect of either distance or direction on their ability to get home (Rüppell 1934). Rüppell went on to conduct similar displacement experiments with Eurasian

Manx Shearwater.

Wrynecks, Barn Swallows, Northern Goshawks, and Red-backed Shrikes—some birds returning to Germany from as far away as Madrid and Marseilles, a distance of 1,850 kilometers (1,150 miles) (Rüppell 1937). Huijbert Kluijver[54] (1935) similarly transported Common Starlings about 160 kilometers (100 miles) to the northeast of their breeding area at Wageningen in the Netherlands. Despite being anesthetized during part of their journey, they still managed to return (cited in Thomson 1936a). In Britain Ronald Lockley and David Lack undertook a modest displacement experiment in the summer of 1936. A Manx Shearwater (nicknamed Caroline by Lockley) was taken from Skokholm Island and released from south Devon, some 360 kilometers (224 miles) by sea from Skokholm, at 14:00 local time. Lockley checked the bird's breeding burrow at midnight that same day and was amazed to find Caroline incubating her egg—no more than nine hours and forty-five minutes after release, providing evidence that she knew where she was going (Lack and Lockley 1938). As Lockley said, their success encouraged further experimentation, and in due course a study was developed by Geoffrey Matthews, who arranged for releases from as far away as New York and Venice: the shearwaters still found their way back to Skokholm (Brooke 1990).

Summarizing the state of bird migration studies in the mid-1930s, Thomson urged researchers to conduct experiments, because "observational data constitute little more than the raw material for study."[55] He also

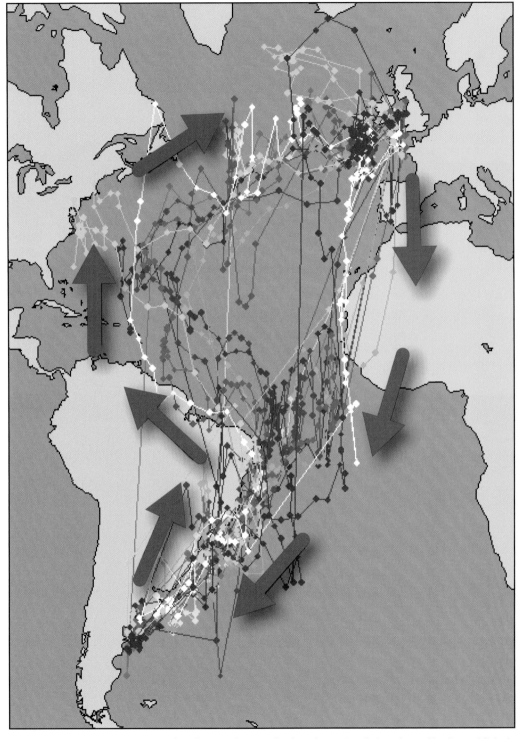

Migration route of Manx Shearwaters breeding on Skomer Island, Wales, to South America and back, established using geolocators. Different colors represent different individuals; large arrows show main routes.

recognized that the question of how birds found their way really comprised two problems: (1) what determines the goal? and (2) how do they find their way?

The studies reported above suggested that birds know which direction to fly (orientation) and also that they could find their way if displaced (navigation) either by a researcher or by bad weather. However, without knowing which route the birds took to get home after being displaced, it was impossible to know whether they found their way simply by chance. What was needed was information on the routes they took while migrating. Donald Griffin made the first attempt to obtain such information.

Born in New York in 1915, Griffin studied at Harvard, making bat migration and later bird navigation his main research focus. Colleagues told him that he had little hope of being taken seriously studying such "childish" topics. Luckily, however, Karl Lashley joined the Harvard faculty in 1935, and Griffin persuaded him to act as supervisor for his doctoral research (Griffin 1998). Griffin worked first on Leach's Storm Petrels and found that they could find their way home from up to 750 kilometers (466 miles) away, but not especially quickly. He then switched to American Herring Gulls and Common Terns as more convenient study species, and once again he found that the ability of these species to home was slower when they were displaced to unfamiliar territory. Trying to figure out what was going on was not going to be easy, but Griffin tactfully refrained from following Lashley's suggestion that he feed poison capsules to the birds so that after a predetermined time they would fall from the sky and reveal something about their route (Griffin 1998).

A breakthrough occurred when Griffin spoke to the yachtsman, aviator, and Harvard physiologist Alexander Forbes, who suggested that he could follow Griffin's gulls in his light airplane; their first trials were sufficiently encouraging that Griffin bought his own plane in 1941. After the war, he drafted new proposals for research that would help to clarify the sensory basis of navigation, and he settled on Northern Gannets as the ideal study species: large and white, they could easily be followed from the air, and their strictly pelagic nature meant that inland release points would almost certainly be unfamiliar. In 1947 Griffin, with his graduate student Ray Hock and Hock's wife, caught breeding gannets at Bonaventure Island in the Gulf of St. Lawrence and transported them to Caribou, Maine—200 kilometers (124 miles) inland from the Atlantic coast and 150 kilometers (93 miles) from the St. Lawrence River—where they were released and followed by Griffin in his three-seater Piper Super-Cruiser.

The gannets were followed for up to ten hours at a time over distances of 160 kilometers (100 miles). While most returned to their nests, they generally did not take the shortest possible route, suggesting to Griffin that they used random search until they spotted a familiar landmark. He later said, "This conclusion, which in retrospect seems so narrowly overconservative, was very much in keeping with the basic ideas on which I had been brought up at Harvard in the 1930s."[56] A strict ethos of reductionism was apparent in those days at Harvard: Loebian philosophy[57] and behaviorism dictated that behavior should be explained in the simplest possible terms (chapter 7). Griffin later recalled discussing the problem of bird navigation with his student friends and ridiculing the notion that birds might navigate using the sun or stars as compass mechanisms: "The possibility that birds might distinguish Polaris from other stars was so outlandish that I don't think anyone even dared to mention it."[58]

The homing ability of domestic pigeons was employed with great effect when they were used to carry messages during wartime, especially the First and Second World Wars. Fast and apparently indifferent to gunfire, certain pigeons famously conveyed information that saved the lives of troops; some birds were even awarded medals (McCafferty 2002). It is hardly surprising then that the military had an interest in bird navigation.

While Griffin (1944) conservatively suggested that birds used random searching and landmarks, other scientists preferred a more metaphysical explanation. Henry Yeagley, professor of physics at Pennsylvania State College, wondered whether birds used a combination of a magnetic sense (previously suggested by von Middendorff) and an ability to detect the Earth's Coriolis forces. He started training birds in May 1943 and carried out the first field test in November that year, using homing pigeons trained to navigate to Paoli in Pennsylvania. Officers of the US Army Signal Corps, including Major Otto Meyer, who was in charge of the Army Pigeon Service Agency, witnessed the experiments. In his first experiment Yeagley compared the homing ability of birds with either magnets or copper plates weighing 0.8 grams (0.03 ounces) attached to their wings. Of the "magnet" group, only two of ten birds returned the distance of 100 kilometers (62 miles), whereas eight of ten "copper plate" birds returned. These positive first results meant that the data and those from

five subsequent experiments were reported to the Army Service Forces alone. Only after the war was over was Yeagley able to publish his results. Yeagley's (1947) paper generated a lot of discussion at a special meeting at the Linnean Society in London in May 1948.

Independently, Swedish physicist Gustaf Ising had also suggested that birds might use Coriolis forces—possibly detected via the semicircular canals in the ear—to navigate (Ising 1945). The ecologist Howard Odum reported on the issue, identifying the question: "Is the superior navigation of birds possible because of their possible ability to orient to a Coriolis, magnetic, or other geophysical field of force in addition to keen powers of visual reference?"[59] After reviewing the evidence, Odum said that "the wandering and visual orientation theory is certainly part of the correct explanation. The magnetic theory is lacking in theory and upheld by experiments which for various detailed reasons need to be repeated. Even if valid magnetic effects exist, that they are anything but grossly inefficient has yet to be shown. The burden of proof still seems to lie with the proponents of the magnetic theory."[60]

When, in 1948, Griffin heard about Karl von Frisch's seminal experiments on honey bee communication, his response was "Good God, if mere insects communicate abstract information about distance and direction, where does that leave Loebean tropisms? If bees do something like that, how can I be so sure that homing birds simply search for familiar landmarks?"[61] Accordingly, in his next published overview of avian orientation in 1952, Griffin proposed a more sophisticated range of options: Type I orientation was the simple use of landmarks within a bird's familiar territory and undirected wandering in unfamiliar territory; Type II comprised a compass mechanism that enabled birds to fly in the same direction even when crossing unfamiliar territory ("one directional orientation"), and Type III allowed the bird to choose the correct direction to fly in even when placed in unfamiliar territory ("true navigation").

Gustav Kramer began his now famous studies on the physiological basis of bird orientation in 1945 at Heidelberg University, where he was based after finishing his war service. Two years later he moved to the Max Planck Institute for Marine Biology at Wilhelmshaven under Erich von Holst. Kramer exploited the fact that migrant birds held in captivity exhibit an agitated hopping, referred to as "migratory restlessness." The biological significance of this behavior was discovered by Nicolas Venette (1697) but is usually referred to by the German term *Zugunruhe*, since it is assumed (Berthold 1993) that it was discovered by the German farmer-cum-bird keeper Johann Andreas Naumann in the late 1700s (Birkhead and Charmantier 2013). Working with Common Starlings (diurnal migrants) in circular cages, Kramer noticed that their agitated hopping occurred in the direction in which they would normally migrate. He also noticed that if he obscured the sun, this "directionality" was lost (Kramer 1951). Allowing the sun to reappear, the birds' directionality reappeared too. More convincingly still, by using a mirror to change the apparent position of the sun he could make the birds reorient according to this "new sun." A sun compass! Strictly speaking, this is a sun azimuth compass, since it is based on the sun's position on the azimuth relative to the observer.[62] Possession of a sun compass also meant that birds must have a sense of time, to compensate for the sun's movements across the sky. Subsequent work by Kenneth Hoffman (1954) and Klaus Schmidt-Koenig (1990) confirmed that other birds also possess this ability.

Following his success with the sun compass studies, Kramer (1953) proposed a two-step navigational system, comprising a compass component and a map component, but left the latter mechanism unspecified. Both components, he suggested, were necessary for successful navigation: the compass for providing a direction to travel and the map for getting initial bearings. It is remarkable that, even though Kramer worked at the same time as von Frisch, and both used similar methods to test their sun compass ideas, neither knew of each other's work until much later.

At the same time as Kramer, Geoffrey Matthews—based at the Ornithological Field Station at Cambridge University and a colleague of ethologists Robert Hinde and Bill Thorpe (chapter 7)—was working on a different explanation for how birds use the sun (Matthews 1953). Matthews had become intrigued by the question of bird navigation after his own experience in the RAF during the Second World War, in which he had to visually locate tiny, remote atolls in the Indian Ocean. He decided to do a PhD[63] at Cambridge to explore bird navigation further and then stayed there for five years of postdoctoral work. Matthews's "sun arc hypothesis" proposed that birds could determine both direction and position from the sun—in other words, a compass bearing from its direction relative to their position but also positional information from the altitude of the sun in the sky. Since Matthews's "sun arc hypothesis" would, in principle, solve the mystery of bird navigation, it was very carefully scrutinized, and in the late 1960s Bill Keeton designed an ingenious test to distinguish between Kramer's and Matthews' hypotheses. His test revealed that Matthews was wrong (Keeton 1969); birds use the sun only as a compass, not for establishing their position.

A NOVEL ORIENTATION CAGE

What about nocturnal migrants? There was no evidence they used the moon (Berthold 1993), but in 1949 Kramer noticed that nocturnal migrants seemed to orient using the stars—an idea that only a few years earlier Griffin and his colleagues had considered preposterous. Kramer was just forty-nine in 1959 when he fell to his death while climbing to the nests of Rock Doves to get birds for his experiments. After Kramer's untimely death, Franz and Eleanor Sauer pursued the idea of a star compass and demonstrated that caged European warblers (genus *Sylvia*) oriented in the appropriate directions under a night sky. They were also the first to test birds in a planetarium, reporting that birds stopped moving when the planetarium sky corresponded with the sky of their normal migratory destination. Franz Sauer (1958) proposed that birds inherited an image of the night sky and were able to use this as both map and compass, allowing precise true navigation. Later work by Steve Emlen revealed that birds learned what the night sky looked like. Attempts to replicate Sauer's findings were unsuccessful, and there were criticisms of his statistical analyses. As a result, the idea of celestial navigation was controversial in the early 1960s. Others did find that caged birds under a night sky oriented

Gustav Kramer releasing pigeons in an experiment (photo date unknown).

their migratory restlessness in appropriate directions, but there were mixed results with respect to what happened on overcast nights—in some studies there was no change in orientation but in others the birds seemed to lose their sense of direction.

In 1962 Steve Emlen started work on his PhD at the University of Michigan. His thesis, "Experimental Analysis of Celestial Orientation in a Nocturnally Migrating Bird," was completed in 1966. Inspired by Kramer's and Sauer's previous studies of nocturnal migrants, Emlen realized early on that he needed a cheap and efficient way of recording the orientation of captive birds. His father, John (himself a well-known ornithologist), came up with a piece of apparatus—now known as the "Emlen funnel"—that consisted of a funnel of thick blotting paper, covered at the top with wire mesh and with an inkpad for a base. Birds were placed inside the cone, and the direction and intensity of their migratory restlessness could be quantified by the amount of ink they deposited on the sides of the funnel. There have been other designs of orientation cages, but the Emlen funnel has remained the most popular. Emlen employed his new device to test ideas centered around celestial navigation, using the Indigo Bunting as his model organism. This was an ideal species: a moderately long distance, nocturnal migrant (breeding in the eastern United States and

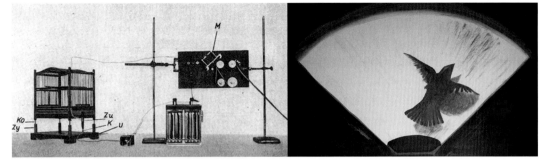

Changing technology to quantify *Zugunruhe* in captive migrants: a registration cage (left) from the 1920s (Wagner 1930) and an Emlen funnel (right) from the 1960s (e.g., Emlen et al. 1976).

wintering in the Bahamas, southern Mexico, and Central American south of Panama) and easily maintained in captivity (Emlen 1967).

The first experiments were carried out in an open field at the Edwin George Reserve of the University of Michigan (away from any towns that might produce a glow in the sky and bias the results—as pilot studies with other birds had shown). The planetarium tests were done at the Robert Longway Planetarium in Michigan—a dome 20 meters (66 feet) in diameter and 12 meters (39 feet) high. The planetarium allowed Emlen to present a wide array of different skies (of any latitude and longitude), providing precise control over what the bird saw. Emlen found that birds oriented in the direction appropriate to their migratory destination under both natural and artificial skies. Further, when he turned off the stars in the planetarium the orientation of birds became completely random. He also carried out experiments in which the skies were rotated according to local time to establish whether the star compass was a bicoordinate navigation system, but he found no evidence for this—birds tested under the clock-shifted skies did not change their orientations. Emlen's results disproved Sauer's hypothesis—at least in the Indigo Bunting—that the stars provided

both directional and positional information: it was a compass mechanism only.

Later, using a larger sample of birds, Emlen investigated the ontogeny of celestial orientation. His studies revealed that birds seemed to orient using only a portion of the night sky—specifically that region within 35 degrees around Polaris, the North Star. By manipulating the patterns of stars that young birds experienced, he also showed that young birds *learn* that the center of rotation of stars is north and that once they have learned this they do not need to see the constellations rotating around this point, but can get their bearings from the patterns of the stars in the sky. Birds that grew up without seeing any of the night sky before their first migration were unable to orient correctly. And birds that grew up under planetarium skies in which the star Betelgeuse was manipulated to be the center of celestial rotation later oriented to Betelgeuse as if it were north; only those raised under normal skies (in which the North Star is the center of celestial rotation) were able to orient normally on their autumn migration (Emlen 1970).

These studies provided conclusive evidence that various species of bird use the patterns of the stars to orient themselves, but there was no evidence that they possess a "map" of the stars—that is, they don't use the stars for true navigation. Neither was there any evidence that birds have any way of compensating for the movement of stars, but instead they learn the patterns relative to their reference point—the North Star.

One criticism of this work was that because the birds were caged, it was impossible to know how the results translated into their natural migratory flights. The next step was to follow free-living birds. After a number of early failures (where the birds simply chose not to fly under overcast conditions or after they had been clock-shifted), Emlen's wife, Natalia Demong, came up with a solution: a small cardboard box with a collapsible floor attached to a weather balloon that would transport the bird up to particular altitude. When the floor fell away the bird had no choice but to fly, and Emlen found that the majority of birds took off on a meaningful flight. A major boost came when NASA agreed to let them use NASA's tracking radars: the radar beam picked up the water content of the bird's body as it went up and then stayed on the bird once it was released. This allowed for real-time data on everything that the bird was doing, up to distances of 40 kilometers (25 miles). Emlen manipulated many cues available to his test subjects—for example, attaching magnets or forcing the birds to start flying at different altitudes so that they were above or below cloud cover or in the middle of strong crosswinds. He found that birds are very good meteorologists; they could shift to the altitude where the winds were most favorable, so long as they had access to landmarks below and the stars above. If they couldn't see the landmarks or stars, they were unable to compensate.

We had three tremendous seasons of getting as close as I've ever gotten to tracking birds that we had under our control for the first 20–40 kilometers of their migratory flights. We controlled the cues available to the birds by releasing them under varying meteorological conditions aloft and then monitoring the directional "flight decisions" the birds made in three-dimensional space. Obviously the systems are complex: the birds have multiple cues available to them, and they can integrate them in various ways, according to what's available. So it was a great ending to the work that I did in migration. Of course, researchers now are attaching miniature satellite tags to individual birds and tracking them, sometimes for their entire migratory journeys! So migration research has benefited tremendously from recent technological advances, but those were great experiments at the time.[64]

A SENSE OF SMELL, A SENSE OF DIRECTION

Although Griffin's (1952) classification of three types of orientation helped to define and guide the field, it was becoming clear by the 1970s that his proposed scheme was inadequate. Many studies did not fit into either Type I or II and so had to be classed as "true navigation" by default. The person responsible for setting a new direction to orientation studies was Bill Keeton. Encouraged by his parents, Keeton kept racing pigeons as a boy and became an astute observer of their behavior. He graduated from the University of Chicago, completed a master's in entomology at Virginia Polytechnic Institute, and entered Cornell in the fall of 1956 to study millipede systematics for a PhD. After obtaining his doctorate he joined the faculty at

Cornell in the Department of Entomology. Keeton was an outstanding teacher, and he turned his introductory biology course into a highly acclaimed textbook, *Biological Science*, first published in 1967. When we asked Wolfgang Wiltschko what Keeton was like, he took a deep breath and said that he had rarely met such a remarkable person, with such a tremendous ability to hold an audience. Keeton had heart valve problems, with a pacemaker installed at the age of thirty-seven, but as Wiltschko said, he seemed to compensate for his physical weakness by great mental facility.

Keeton's interest in pigeons and orientation continued and flourished. His colleague Steve Emlen said:

> I remember that when I first joined the faculty at Cornell, I would tease Bill, asking him facetiously: "What makes you think that a millipede taxonomist can solve the mysteries of bird orientation?" And he would get a twinkle in his eye and respond: "When I flew pigeons as a boy, they performed better than most of the results published in the homing journals. And furthermore, they homed under conditions when they shouldn't have been able to according to the literature." He was correct. By knowing the subtle nuances of pigeon behavior, and by spending time with his birds and observing them as individuals, Bill (and most serious pigeon racers) had achieved homing results superior to those reported by scientists.[65]

Keeton moved from the Department of Entomology to the new Department of Neurobiology and Behavior at Cornell in 1965 and there began his field studies of pigeon orientation. He soon discovered that his birds could orient even on overcast days,

Bill Keeton (photo in 1977 at age ca. 44).

indicating that in the sun's absence the birds had a backup system. Searching for this backup system, Keeton attached small magnets to his pigeons and found that their ability to orient under overcast skies was disrupted (Keeton 1971). Intriguingly, the magnets disrupted only the initial direction of departure, suggesting that there was yet another backup system.

Recognizing the existence of backup systems was a major conceptual breakthrough, because prior to this researchers had sought a single-cue explanation for orientation—that is, they thought that birds used either the sun or the stars or had a magnetic sense, rather than a hierarchy of fail-safes. As Emlen says in Keeton's obituary, once it was recognized that birds might use a hierarchy of cues to orient, the apparently conflicting results of previous experiments were viewed in a new light. This did not mean that there were no further conflicts—there were plenty—but Keeton's work caused a major shift in the way researchers thought about orientation and navigation in birds. Keeton's openness and willingness to engage with those whose results differed from his own provided a model for future researchers. His open attitude is nicely illustrated by a paper he coauthored with Floriano Papi, A. Irene Brown, and Silvano Benvenuti in 1978, which asked the question, Do American and Italian pigeons rely on different homing mechanisms? This demonstrates Keeton's progressiveness because the authors were on opposing sides of one of the most controversial topics to have come out of the study of pigeon homing: olfactory navigation. The outcome—see below—was an interesting one.

Among the many ideas at the beginning of the twentieth century about how migrating birds might find their way, the Russian-French physiologist Elias von Cyon suggested that birds had a special olfactory sense located in the "nose," distinct from the normal sense of smell. Von Cyon was director of the physiology laboratory at the University of St. Petersburg where Pavlov was a student, and his idea was that "the direction and temperature of the wind act upon the olfactory mucous membrane."[66] He made a modest attempt to test this, but with very few pigeons. In 1910 J. B. Watson conducted a slightly more extensive test in which he blocked the nostrils of some Brown Noddies with wax before transporting them up to 60 kilometers (37 miles) away, but they had no difficulty returning to their nest sites on Bird Key, thus providing no support for von Cyon's idea (Watson and Lashley (1915). The idea that olfaction might be used in navigation then fell out of favor for many years, the consensus being that birds had only a poor sense of smell.

Things changed in the 1960s, thanks largely to the work of Betsy Bang, Ken Stager, and Bernice Wenzel (Bang and Wenzel 1985; see also Birkhead 2012) who, through a combination of anatomical and behavioral work, demonstrated that the sense of smell in birds was much better developed than most ornithologists or physiologists cared to admit. But could birds use this sense of smell to find their way?

Hans Wallraff was one of the few researchers sympathetic to the idea of olfactory navigation. He noted that pigeons raised in aviaries open to the wind were able to orient but those screened from the wind could not. As he said, "I stated the involvement of unidentified 'atmospheric factors', but I did not believe that they might be chemical substances dispersed in the atmosphere. . . . I could not imagine in what way other airborne chemicals might provide positional information over such long distances. So I was closest to the deciding next step but did not make it. Fortunately, others were less burdened with preconceptions and made the deciding step."[67]

Those "others" included Floriano Papi, who provided the first strong evidence—in the early 1970s in Pisa, Italy—that a sense of smell was involved in homing. Papi found that pigeons whose olfactory nerve had been cut were much less effective at homing than control birds (Papi et al. 1971). For Wallraff this study was "the most stimulating event of the last years in the field of pigeon homing."[68] In 1972 Papi also offered a hypothesis on the mechanism of olfactory navigation. His "mosaic hypothesis" stated that there exists a patchwork of natural odors over the Earth's surface, and that young pigeons learn to recognize the specific odor of their loft, as well as "foreign" odors carried by the winds from nearby areas. Pigeons then form an "odor map," he suggested, that provides both positional and directional information. Papi's hypothesis came at essentially the same time as the importance of olfactory navigation was discovered, and this meant that the two ideas were linked together for many years, with criticisms against Papi's theory being taken as criticisms against the importance of olfactory navigation as a whole. "The result was some confusion and certainly contributed to the controversies of the last years."[69]

Some researchers, including Wallraff (1980), confirmed Papi's results, but others failed to do so, and this led to a long-lasting and often bitter controversy. Of Papi's results, Roswitha (Rosie) Wiltschko, Wolfgang's wife and collaborator, said that they "indicate an involvement of olfactory input in homing, but they do not prove that odours provide navigational information."[70]

The olfaction theory was "controversial from the beginning."[71] Some of the resistance almost certainly came from the persistent belief that birds have a poor sense of smell, but there were other issues with Papi's theory: "The initial results emerged suddenly, published in an obscure journal, by a research group which was at the time not one of the established players in the study of pigeon homing."[72]

Keeton was "impressed with the consistency of the results of Papi's group"[73] and was one of those who attempted to replicate the findings, but without success (Keeton 1974; Keeton and Brown 1976). Keeton's collaboration with the Italians, using Keeton's pigeons at Cornell, were so mixed that Keeton and Brown (US) and Papi and Benvenuti (Italy) wrote separate discussions to their joint paper, interpreting the findings somewhat differently (Papi et al. 1978). The Italian discussion starts: "The present results greatly reduce the discrepancy that seemed to exist . . . [between the American and Italian results]." The US discussion starts: "The present results do not remove the fundamental discrepancy between the previously published [American and Italian] results."[74] Such difference of opinion can be a normal part of science and identifies the need for further, better experiments.

In the 1980s the Wiltschkos attempted to resolve the olfactory map controversy, conducting their own experiments in Germany, America, and Italy (Wiltschko et al. 1987). When we talked to them, Wolfgang said this about the olfactory map: "I can tell you, in Italy it really works. . . . I've seen it, I was there! . . . in Ithaca [where Keeton worked in the United States] it does not work at all."[75] So what's going on? Rosie replied, "First of all, since the [olfactory] map is learned, different places might give different cues."[76] Wolfgang then interjected:

> And if you go to Arnino, the pigeon loft of Floriano Papi and you have a westerly wind you smell the sea and you smell the pines of the Ligurian pine forest. If you have a north-easterly wind you really smell the Apennines . . . and there is a good reason that pigeons might get different odours with different wind directions. If you go to Ithaca, New York, you have cornfields, cornfields, cornfields, in each direction. And if you are here in Frankfurt you have industry, industry, industry—except when you go to the east.[77]

Papi gave up working on pigeon orientation in the 1990s, believing that he had done all he could to demonstrate olfactory orientation and that the burden of proof rested with his detractors. One of his students, Anna Gagliardo, decided to continue the olfaction crusade by addressing one of the major criticisms: that by cutting the olfactory nerve to render the birds anosmic, Papi (and others) inadvertently disrupted the pigeons' magnetic sense as well. The crux of that critique was this: pigeons were thought to detect magnetic information with tiny particles of magnetite imbedded in the upper beak; this detection is then relayed to the brain through the ophthalmic branch of the trigeminal nerve, but this nerve is closely associated with the olfactory nerve in the upper beak area, so disrupting the olfactory sense, by applying local anesthetic to the olfactory membranes or sectioning nerves at the base of the beak, may also disrupt the relay of magnetic information (Mora et al. 2004). To test this, Gagliardo joined forces with the expert avian anatomist Martin Wild, from Auckland, New Zealand, and found no evidence that cutting the trigeminal nerve had any effect on homing ability. In contrast, the homing ability of birds whose *olfactory* nerve had been cut was much reduced (Gagliardo et al. 2009).

In rather striking contrast to all this, there was much less opposition when Gaby Nevitt suggested in the early 1990s that albatrosses and petrels use olfaction to locate both feeding areas and their breeding colonies. When we asked her about her "olfactory seascape" idea, she said:

> I think one of the reasons my work didn't ruffle as many feathers, perhaps, is because it flew under the radar. I came into this field having worked a lot in olfaction in fishes, but I had also worked in magnetoreception on spiny lobsters with one of my grad student friends, Ken Lohmann, so I had a foot in both camps. Also, Papi wasn't able to make as clear a link to a naturally occurring odorant, which I think would have helped his model. In my case, I got a lot

of interest from the ocean science community, especially people working with DMSP/DMS[78] with respect to global climate regulation. So maybe the work gained momentum by appealing to a different crowd.[79]

MAGNETIC CONTROVERSIES

The notion that birds might navigate by means of a magnetic sense has been described as an irresistible concept (Alerstam 2003). This idea emerged initially in the mid-1800s, when in the first account of the migration of Russian birds, Alexander von Middendorff (1859, cited in Berthold 1993) suggested that, like a compass needle, birds on their spring migration headed toward magnetic north. That idea was developed by French zoologist C. Viguier, who suggested in 1882 that the Earth's magnetic field provided not only a compass but also a map, allowing birds to identify their location. Henry Yeagley's experiments, conducted during the 1940s and described above, showed that attaching a small magnet to pigeons disrupted their homing ability and provided some of the only evidence for a magnetic sense (Odum 1948). Stresemann (1935, cited in Alerstam 2003) was sympathetic to the idea, but most ornithologists, including Thomson (1926, 1936a) and Griffin (1952), dismissed it.

In the late 1950s Franz Sauer reported that after dimming the light in his planetarium so the stars were no longer visible, the birds in his orientation cages simply went to sleep. In contrast, Friedrich Merkel, who was studying the physiology of migration, knew that even without access to the stars his birds remained active and continued to show migratory restlessness. Merkel and his student, Hans Fromme, wanted to check the birds' orientation and tested them in a round cage on the roof from which they could see the stars. The birds were indeed clearly oriented, but not in their normal migratory direction; instead, they were oriented toward the city of Frankfurt, where the sky was lighter. One evening Fromme covered the top of the orientation cage with a translucent cover to eliminate the glow in the night sky from Frankfurt—and with it, of course, the stars. The birds now oriented in their migratory direction, showing that the stars were not essential and raising the question of whether the birds were using a magnetic sense. Determined to solve the problem, Merkel and Fromme went off to the physics department and managed to get a Helmholtz coil, a device that could create a strong magnetic field and alter the apparent direction of north around the cage. Having reversed the magnetic field around their caged birds using the Helmholtz coil, the birds seemed disoriented, and Merkel and Fromme concluded that there was no magnetic sense.

At this time—the late 1950s—Merkel's department had a steel vault for conducting experiments in extraterrestrial biology. This was the Sputnik age, and there was great interest in sending animals into space. Fromme had already observed that inside the vault a bird's ability to orient disappeared. Knowing that radio waves could not penetrate the walls of the vault, he wondered whether birds might use radio waves to orient. Radio astronomy was a new area of physics, and when Wolfgang Wiltschko began his studies with Merkel, the idea that birds might use radio signals from the stars was extremely exciting. To test the idea, Wiltschko kept a group of European Robins for several days in the steel vault (to isolate them from radio signals

and desynchronize their stellar clock). On placing a single bird inside an orientation cage within the vault on 12 October 1963, Wiltschko was amazed to see the bird clearly orienting toward the south. Wiltschko later recalled that of the several hundred birds he and Rosie tested, this single individual showed the strongest, most obvious preference for its migratory direction.

Since the magnetic field was only weakened, not totally obstructed inside the vault, and all other cues had been removed, these results suggested that the birds oriented with the help of the magnetic field. Apparently they needed several days to adjust to intensities that differed from the local magnetic field. The reason why this result differed from previous trials was that the magnetic field created by the Helmholtz coil had been too strong for spontaneous orientation. Effectively, robins can use a magnetic field to orient themselves only if its intensity is similar to what they are used to. A stronger or weaker field simply causes the birds to be disoriented.

When Wiltschko experimentally shifted magnetic north, keeping the magnetic intensity similar to that of the local geomagnetic field, the birds changed their orientation accordingly. The results of these groundbreaking experiments were first presented at a meeting of German zoologists in Jena in 1965 and a year later at the International Ornithological Congress (IOC) in Oxford. The response was one of polite disbelief. As Rosie Wiltschko told us, "At that time people found it hard to imagine a mere bird possessing such a sophisticated sense."[80] However, it was at that IOC meeting that the Wiltschkos met Emlen, who was interested in their results, despite not finding any evidence of a magnetic sense, as reported in his own recently completed PhD thesis. The meeting with Emlen resulted in the Wiltschkos going to Cornell University, where they met Bill Keeton.

In 1972 the Wiltschkos completed an experiment that allowed them to identify the magnetic compass of birds as an inclination compass,[81] and the paper sailed into *Science*; Wolfgang recalled that it was the easiest paper he ever published (Wiltschko and Wiltschko 1972). Controversy persisted, however, as not everyone was convinced. The result was a collaboration with Emlen in which everything they could think of was carefully controlled. The result: another paper in *Science* (Emlen et al. 1976). Magnetic orientation in birds had come of age.

As the Wiltschkos said when we spoke to them, one of the major difficulties with orientation studies was that the standard scientific approach of experimentally manipulating factors one at a time simply did not work. This was because the orientation system has so much redundancy built into it: "On a clear day a pigeon will use a sun compass, but if it is overcast, it will use its magnetic compass. From these results, one cannot conclude that the sun is not involved. Not finding evidence for a particular cue doesn't mean that the cue is not playing any role—it simply shows that it is not used in the test situation."[82]

Despite increasing evidence for a magnetic sense, there was still a major obstacle to overcome. There was no obvious sense organ capable of detecting the Earth's magnetic field, so the biophysical basis for magnetic orientation remained elusive. The first glimmers of a resolution came in the form of two very different hypotheses, which presented themselves at almost the same time.

In 1979 Charles Walcott, James Gould, and Joe Kirschvink published an article, "Pigeons Have Magnets," in *Science*, reporting the existence of microscopic particles of magnetite next to the skull and raising the possibility that these may have a role in detecting a magnetic field. Further studies confirmed and extended this finding. For example, it is possible to alter the magnetization of the particles by treating birds with brief, high-intensity pulses; this treatment has clear effects on orientation ability (e.g., Munro et al. 1997).

It had also been shown in the 1970s that certain chemical reactions were influenced by magnetic fields and induced by light, prompting Klaus Schulten to speculate that birds might literally "see" the Earth's magnetic field (Schulten et al. 1978). The theoretical aspect of this research was pursued by Thorsten Ritz and colleagues, who proposed an influential model to account for light-dependent magnetic orientation (Ritz et al. 2000). Empirical studies support the idea. The Wiltschkos demonstrated that magnetic compass orientation was affected by the wavelength of light, suggesting that the compass may have its basis in the visual system (Wiltschko and Wiltschko 1998). From experiments in which they covered one eye with a frosted "contact lens," they also showed that pigeons were able to "home" more successfully with their right rather than their left eye, suggesting that the right eye might be crucial in detecting the Earth's magnetic field. To test this they placed European Robins in orientation cages inside a magnetic coil (as before), but this time also with each eye covered in turn. With both eyes uncovered or just the right eye, the birds continued to orient appropriately, but when

Some of the most spectacular of all migrations revealed by satellite tagging: Bar-tailed Godwits (top) migrate from their breeding grounds in Alaska to wintering areas in New Zealand in a single, nonstop nine-day flight (yellow lines, right); blue lines show the route taken by the subspecies breeding in the New Siberian Islands to northern Australia. The northward spring migration (left) for both subspecies goes via China.

only the left eye was uncovered, the birds were unable to orient (Wiltschko et al. 2002).

Further experiments using opaque, frosted lenses—which allowed the bird to detect the light but prevented them from seeing a clear image—revealed that it was not light per se but the ability to see a clear image of the landscape with the right eye that provided the magnetic sense (Stapput et al. 2010). Subsequent research questioned the plausibility of this, presenting evidence that the magnetic sense occurs in both eyes (Hein et al. 2011). The explanation was provided by the Wiltschkos and their collaborators, who demonstrated that while this was true in young robins, as they mature the sense becomes fixed in the right eye (Gehring et al. 2012).

In summary—and this is very much ongoing research—there is at present (2013) unequivocal behavioral evidence that birds can detect and use the Earth's magnetic field

153

to help them find their way. The physiological mechanism, however, remains elusive. Migratory birds appear to have at least two magnetic senses: (1) a light-dependent "compass," located in the eye(s), and (2) a magnetic sense linked with the trigeminal nerve, located in the upper mandible. Another possibility is (3) that the magnetic sense is located in the inner ear (Wu and Dickman 2012). The practical problems of distinguishing between these ideas is considerable because of what Henrik Mouritsen and Peter Hore (2012) call "the chronic 'diseases' of magnetic-sense research— the difficulty of independent replication and the lack of truly blind protocols."

CODA

The study of migration—in all its various forms—has preoccupied researchers for longer and more consistently than almost any other area of ornithology. By comparison, "instinct" (chapter 7) has a forty-year (1930s–1970s) history, and behavioral ecology (chapters 8 and 9) has a life span that runs from the mid-1970s to the present. Migration research, on the other hand, reaches back well before 1900 and continues to this day. The feats accomplished by certain migratory birds—like Arctic Terns, Bar-tailed Godwits, and certain hummingbirds—continue to amaze us, as does the array of adaptations to migration. The hierarchies of fail-safe mechanisms, and the intimate links with other aspects of birds' life histories, are all truly remarkable.

Perhaps one of the most exciting discoveries about migration is that it is not a discrete species-specific phenomenon. It now seems that all individuals of most bird species of mid- to high latitudes have the potential to be migratory; they all possess most of the traits—such as the ability to navigate, to lay down fuel supplies, and so on—required for migration. As a result, migratory behavior can evolve very rapidly, as demonstrated by studies of Eurasian Blackcaps. We should not assume, however, that all aspects of migration are adaptive. When behavioral ecology first emerged, it was pointed out that some traits are simply consequences of others and hence should not be called adaptations (chapter 8). The migration routes of some bird species appear to be the consequence of gradual changes in geographical distribution and hence are unlikely to be the most economic route. The best of several known examples (see Newton 2008) is the Northern Wheatear. As this species spread eastward across Europe into Siberia, the birds breeding in Siberia continued to migrate the long distance to sub-Saharan Africa—as their European ancestors do—rather than minimizing the costs of travel and wintering in southern Asia.

Whereas once it was assumed that there was a particular migration syndrome, with the various adaptations, including morphology, fuel physiology, navigational ability, and so on all tightly genetically linked, it now appears that these traits can evolve independently, according to the various circumstances birds experience.

Like many other areas of research during the twentieth century, the study of migration has been heavily dependent upon technology: bird banding (early 1900s), orientation cages and radar (1940s–1960s), and satellite tracking (1980s), each of which has had a revolutionary effect on the types of information collected and its interpretations.

The miniaturization of tracking devices that began in the 1990s is currently producing a further revolution in migration studies, by revealing both the routes and stopover locations of individual birds.

As we wrote this account we talked to a number of ornithologists actively engaged in migration and orientation research and attended conference presentations where these issues were discussed. The anatomy and mechanisms responsible for the magnetic sense in birds continues to be hotly debated, with different factions vigorously defending their views. It is remarkable just how difficult this area of research is. With no magnetic sense ourselves, it is extraordinarily difficult to design appropriate experiments and even more difficult to imagine what the confounding factors in such experiments might be. Who would have imagined, for example that radio waves can disrupt a bird's magnetic sense?

BOX 4.1 Peter Berthold

At age ten, in the countryside around my hometown of Zittau [Germany], I encountered many kinds of bird, some of which—Common Kingfisher, Eurasian Bullfinch, Long-tailed Tit, and partridges—fascinated me so much that I never lost my enthusiasm for the world of birds. This fascination was probably a genetic predisposition, as my parents and grandparents were very close to nature, and it was directly stimulated by binoculars my grandfather had salvaged from the First World War. A European Serin was the first caged bird given to me, and it awakened my enduring passion for keeping and breeding birds. And as a ten-year-old I even started capturing birds illegally—simply in order to take a close look at them before setting them free again. Thus I was granted a crucial experience: catching a ringed Great Tit, with a ring bearing the label "Vogelwarte Radolfzell"! I was terribly alarmed: evidently there existed an institute that officially captured birds and ringed them for scientific purposes—I had to learn more about this!

As fate would have it, in 1953 my mother moved with me to Württemberg, where my father had settled after his time as prisoner of war in England. From there it was just seventy kilometers to Schloss Möggingen, the site of the "Vogelwarte Radolfzell." In 1954 I inspected the presumed "mecca" for ornithologists while on a cycling tour and began to plan how I could become linked to the "leading lights" of bird science who were working there. The successful strategy was joining the German Ornithological Society in 1955 and in the same year visiting (by bicycle) its annual meeting in Frankfurt, where for the first time I encountered Konrad Lorenz, Erwin Stresemann, Heinz Sielmann, Ernst Schüz—who later mentored my doctorate—and many others. Above all, I obtained permission to carry out scientific bird-ringing in summer. That was the beginning of my scientific ornithological career, which increasingly tended to focus on questions related to bird migration.

As a youth, my professional goals at first oscillated widely, from whale fisher through teacher to psychotherapist, but by the time I had qualified for university admission it had become clear that only a proper university education could get me into the true heaven of ornithological research. So I studied biology, chemistry, and geography, at the University of Tübingen. There, in 1959–64, I completed the most rapid course of study in the postwar period—altogether only ten semesters—inspired by my desire to soon begin independent ornithological research. To monitor my PhD studies I chose an external "doctor father," Ernst Schüz, who at the time was working in Stuttgart but had formerly been leader of the Vogelwarte Rossitten (after Johannes Thienemann). A nice thing about this collaboration was the way it started. When I asked Schüz to mentor my dissertation, at first he thought it wouldn't work, because he had never had a PhD student before. I responded by saying that I might be his last chance to do so; he laughed—and took that chance! He assigned me as a thesis topic the relationships between migratory and reproductive behavior and physiology in starlings, the experimental work to be done at the Vogelwarte Radolfzell. The dissertation itself was something like squaring the circle: in the same spring 150 juvenile starlings from Holland, Finland, and southern Germany were raised, and then they were sheltered during the winter of 1962–63, the coldest for a century. But it worked—and in 1963 they all brooded very early, almost simultaneously, and thus demonstrated that the timing of breeding depends strongly on the migratory period (which had been suppressed in the aviaries).

After these initial studies on starlings and some others, my colleague Ebo Gwinner and I decided to start a cooperative long-term *Sylvia* warbler program in which we would study as many migration features as possible. That was possible beginning in 1967, when I returned as postdoc to the Vogelwarte, where I then proceeded to work until today, as scientist, local head, director, and now as emeritus—altogether fifty-eight years, a Vogelwarte record.

From the findings of strongly fixed patterns of *Zugunruhe* and the rigid migration control by circannual rhythms, it soon became clear to me that genes most likely would play a key role in the control of migration. With respect to the many questions regarding partial migration, David Lack had already foreseen that the problem could only be resolved with the help of very large-scale and difficult experiments. Fortunately, we found the Eurasian Blackcap a suitable species for large-scale breeding and selection experiments, so that with my Radolfzell group we could initiate two new research fields: experimental migration genetics and experimental microevolutionary studies. With more than 2,000 blackcaps bred in our aviaries and several other species, we were able to demonstrate more and more the general role that genetic factors play in the control of avian migration.

I consider that my "big five" important contributions are the first-time findings that: (1) genes are immediately involved in all essential elements of migration, such as whether to migrate or not, timing and distance, physiology and orientation; (2) partial migration as the most common migratory behavior acts as a turntable for extremely rapid microevolutionary changes from migratoriness to residency and vice versa; (3) probably all bird populations are equipped with genes for migratory and nonmigratory behavior (potential partial migration, inherited from preavian ancestors), compiled in a novel migration theory; (4) central European birds in general and migrants in particular are on the decline (demonstrated already in the 1960s) and that global climate change strongly affects birds in general and especially migrants (in the 1980s); (5) restoration (rebuilding) of natural habitats in devastated areas is the only way to overcome the loss of species richness, i.e., the biodiversity crisis (current experiments). Even as a schoolboy in the 1950s I felt driven to accompany my early research projects by nature-protection activities, which today occupy first place for me, like a new profession.

Unsurprisingly, no other institute anywhere in the world has undertaken such extremely expensive and difficult breeding and selection experiments, and hence when I acquired emeritus status and "Blackcap City" was closed, the entire research field fell back asleep. My successors in the Institute, fascinated by the most modern minitransmitters and telemetry, have followed the general trend and returned to performing field observations, although at a level above that during the period of "pure observations" in the nineteenth century. So it probably won't take long before ingenious experiments explain, for instance, which genes control which elements of migration, which migratory syndromes are jointly inherited and selected, but also how true navigation is controlled, etc. Thus we would have the first brief demonstration, for example, that migratory restlessness of Common Rosefinches (which migrate at night) can be inherited even in domestic Atlantic Canaries—a starting point for all kinds of genetic studies even in smaller laboratories. Well, eventually more obsessives will get to work.

BOX 4.2 Steve Emlen

I had the good fortune to grow up in a biological family. By this, I mean that my father was a biology professor at the University of Wisconsin, where he and his students studied the behavior and ecology of birds and mammals. My parents constantly urged my two brothers and me to "get out into nature," which we did by hiking, bicycling, canoeing, and camping in many wild places. I developed a passion for bird watching while still in primary school and joined the Junior Audubon Society. My two best friends throughout high school were also avid birders. We kept lists of the birds we saw and planned many "expeditions" to

find species to add to our lists. A favorite was bicycling to a peninsula on Lake Mendota, where we monitored the arrivals of warblers during spring migration. The male plumages of American wood warblers are extremely brilliant, and we kept track of the order in which species arrived during the three weeks when arrivals were most intense. Another favorite destination was the Horicon Marsh, a national wildlife refuge where we witnessed the migration stopovers of tens of thousands of Canadian geese.

I attended Swarthmore College in Philadelphia, Pennsylvania, a small liberal arts school where the faculty acted as true mentors. During those years, I spent two summers at a biological field station high in the Rocky Mountains. The first year I pursued my first independent research project—with results that were less than impressive. I described changes in the population densities of several species of mice, but I had not thought enough about what patterns I expected, nor why. The following summer I returned to the same station as a research assistant for one of my professors, Ken Rawson, who was studying the homing abilities of one of "my" mouse species. From Ken I learned the importance of hypothesis-driven research—how to design the collection of data to test and potentially reject specific hypotheses—in this case, about the cues mice use in homing. The contrast between the two projects was a revelation. Good science, I realized, requires explicit hypotheses, and science progresses when the data collected allow one to distinguish among alternative explanations.

After Swarthmore, I pursued graduate studies at the University of Michigan in Ann Arbor. It was the early 1960s, and the modern disciplines of behavioral ecology and sociobiology were just emerging. John H. Crook and David Lack were suggesting that the mating systems and social structures of animal societies could be predicted, in part, from the ecologies of where and how the organisms lived. William D. Hamilton (and later Robert L. Trivers) were demonstrating the importance of including kin—thus measuring inclusive fitness rather than individual fitness—when studying the costs and benefits of the behavioral decisions of individuals.

I found my graduate studies so exciting that much of my own research since has centered on behavioral, ecological, and sociobiological questions. Specifically, I have focused on (1) why certain bird species live in multi-generational family-based groups; (2) how their living with relatives allows us to predict many patterns of avian family dynamics; and (3) whether such predictions are useful in better understanding dysfunctional dynamics in our own human family interactions.

For my doctoral thesis, however, I focused on bird migration. I asked how migratory birds find their way while traveling thousands of miles between their breeding and wintering grounds, and especially what cues they use for direction finding at night, when ground landmarks are less visible and presumably less useful. I built upon the prior studies of Gustav Kramer and Franz Sauer, German ornithologists who had placed migrating birds into circular "orientation cages" and observed their behavior.

To succeed, I needed three things: (1) a recognizable behavior that indicated the bird's motivation to migrate; (2) an orientation cage that easily recorded the directional preferences of the bird inside; and (3) the ability to control the cues the bird might use to find its migratory direction. At the times of spring and fall migration, nocturnal bird migrants become extremely active at night. This spontaneous nocturnal activity (termed *Zugunruhe*, or "travel unrest," by the German scientists) became my measure of a bird's migratory motivation. If a bird wasn't motivated, it was inactive. In collaboration with my father, I invented a funnel-shaped orientation cage in which a bird recorded its own preferred takeoff directions by leaving ink footprints as it jumped up the side walls of the funnel, only to slide back to the bottom, which contained a small inkpad. The record of spontaneous "takeoffs" left on the funnel cone provided information on the bird's ability (or inability) to orient in its migratory direction given the cues available to it. I performed many of my experiments in a planetarium, where I could control the (artificial) starry night sky. Through several series of experiments, both outside under the natural night sky and inside the planetarium where I altered various aspects of the celestial world, I was able to demonstrate that indigo buntings use stars to determine their migratory flight directions in both spring and autumn. Specifically, the buntings recognize spatial patterns of stars in the northern sky and use them to determine the direction of geographic north—analogous to the way we humans use the Big Dipper to find the North Star. The birds' star compass is elegant and simple and does not require time calibration to compensate for the Earth's rotation: spatial patterns among stars remain constant whether the constellation is rising in the east or setting in the west.

After grad school I accepted a position at Cornell University, where I have happily remained ever since. At first I continued caged-migrant studies, testing ever more complex hypotheses about celestial orientation. Later, my wife Natalia and I collaborated with the National Aeronautics and Space Administration to study free-flying migrants. We used powerful NASA tracking radars in Virginia to follow individual birds that we released aloft under various conditions. We now know that migrating birds can use many sources of information to select and/or maintain their flight directions: natural selection has produced migrants that can use multiple, sometimes redundant, types of information. As the individual migrant matures, it learns to integrate information from sources such as star patterns, the sun, polarized light at sunset, geomagnetic fields, and meteorological patterns. If one system fails or is temporarily unavailable, the bird can often switch to a backup.

I have always viewed science as being like a detective game. Our curiosity drives us to seek solutions to nature's mysteries. To do so, we must use the tools of observation and experimentation to "interrogate" our subjects and get them to reveal clues. Some clues will be misleading, and only replication and new experimentation can separate the false from the true. Each clue points the way to another, in a never-ending chase for knowledge. To the scientist, nothing is more intellectually stimulating than being on this chase. I feel very fortunate to have been able to be a scientific detective for most of my life.

CHAPTER 5

Ecological Adaptations for Breeding

The contents of the Ibis in recent years were sad indeed. Now there
is a revival with the younger men taking things over.

—LETTER FROM ERNST MAYR TO ERWIN STRESEMANN ON 22 DECEMBER 1945,
WITH REFERENCE TO WORK BY PEOPLE LIKE DAVID
LACK REVITALIZING ORNITHOLOGY

HARDY'S SWIFT

SWIFTS ARE AMONG THE MOST EXTREME birds—superbly adapted for high-speed life in the air. Short-legged (hence their family name *Apodidae*, meaning "footless"), long winged, and highly streamlined, swifts were once known as "devil birds." In his book *The Inner Bird*, Gary Kaiser describes how a 100-gram (3.5-ounce) White-collared Swift he caught in the Andes "lay stiffly in my hand, more like a model plane than a living creature."[1] The complete antithesis of a typical soft and fluffy small bird, swifts are fast flyers that feed, sleep, and copulate on the wing.

In 1946 David Lack and his wife, Elizabeth, started a study of Common Swifts breeding beneath the thatch of cottages in villages near Oxford. The following year they discovered a better study site in the tower of

Oxford University Museum. Built in 1855 and funded partly by the sale of Bibles, it was here that Thomas Huxley in 1860 trounced the Bishop of Oxford over Darwin's idea of natural selection. Some sixteen to thirty pairs of swifts nested in the tower, entering through the ventilation flues in the steeply pitched roof and making their simple nests among the beams and rafters. For the Lacks, access to the tower was an arduous succession of stairways, culminating in a long wooden ladder that projected vertically deep into the roof space. One of us (TRB) visited this colony in the 1980s, and although used to climbing sea-bird cliffs, found the vertical ladder an unnerving experience.[2]

By positioning nest-boxes inside the ventilation openings, the Lacks had unprecedented

European species that played a special role in the study of breeding ecology (clockwise from top left): Great Tit, Common Redstart, European Pied Flycatcher, Tawny Owl, Grey Heron, Common Starling. Paintings by Eric Ennion ca. 1960.

The tower of the Oxford University Museum (right), where David and Elizabeth Lack studied Common Swifts in the 1940s.

access to the birds and their nests, enabling them to make a detailed study of their breeding biology. Many of the results are presented in Lack's 1956 book *Swifts in a Tower*. Among other things, the Lacks demonstrated that such extreme specialization to an aerial lifestyle isn't without its costs. Typically, Common Swifts lay a clutch of two or three eggs, but sometimes only one, and very occasionally four. The number of eggs laid and the number of chicks reared depends upon the weather. In good summers, when aerial arthropods are abundant, both clutch sizes and breeding success are good, but in wet, cool summers, the opposite is true and chicks starve to death. Food availability—as we will see—was the key to both the swifts', and David Lack's, success.

During the 1940 and 1950s, the Zoology Department in Oxford was run by Alister Hardy, famous for his studies of marine plankton but increasingly interested in aerial plankton—the invertebrates on which swifts feed. Lack and Hardy were good friends and had much in common, including a strong religious belief. Lack, however, kept his Christianity and evolutionary thoughts in "water-tight compartments," while Hardy attempted—unsuccessfully—to bring the two together in a single "harmonious scheme" (Hardy 1973). In 1963, when Hardy (who by then had been knighted) retired, Lack used the occasion to name a new species of swift in his honor. Hardy's swift (*Apus durus*), Lack said, was a high-altitude specialist, feeding on the aerial plankton above 1,000 meters (3,280 feet); an ecological replacement for the Common Swift. It was a spoof, of course, reflecting Lack's sense of fun, but some people took the account of Hardy's swift seriously.[3]

Many who knew David Lack during the 1960s, in the decade or so before he died, remember him as balding, bespectacled, of medium height and somewhat serious in manner. To students attending the annual EGI conference he could seem intimidating, for he was intolerant of those who turned up late and irritated by unprofessional behavior. On one occasion, when a cocksure student speaker attempted to turn his presentation into a comedy act, Lack summarily dismissed him from the stage.[4] Lack's schoolteacher tactics—remnants of his first career—belied the fact that he was also a kind, gentle, and committed family man. Compared with many of his colleagues, he considered his life to have been relatively uninteresting, in the sense that it consisted of little other than his ornithological work. Always industrious, he was exceptionally focused. He went to bed early, and few who knew him dared to telephone after eight p.m. Accounts of his younger life, at Dartington for example, when he was in his twenties, reveal, perhaps not surprisingly, a more carefree, easy-going individual, epitomized by photographs of him on Bear Island and the Galápagos in the 1930s.

BEGINNINGS

Born on 16 July 1910, David Lack had a privileged, if emotionally bereft, childhood, described in a candid memoir that he wrote in the mid-1960s.[5] His father was a leading ear, nose, and throat surgeon and his mother, Kathleen, daughter of Lieutenant Colonel McNeil Rind of the Indian army, had been on the stage with a touring company. The family lived in a large house in Devonshire Place, London, with seven servants and a chauffeur. Lack and his three siblings were brought up by a "succession of nurses, each of whom left when we had become emotionally dependent on her."[6] Lack's interest in birds started early and blossomed through his teenage years. Obviously an able pupil, his experience at an unpleasant boarding school made him vow that if he became a schoolmaster, he would be a kind one—which indeed he was. He later attended Gresham's School in Norfolk, where he was much happier and where his interest in birds was encouraged. At the age of eighteen

he was awarded a school prize for his study of nesting European Nightjars, discovering that the species has two broods each year and not one as the textbooks claimed.

It was reading the semipopular books of William Pycraft—a biologist at the Natural History Museum in London—that started him thinking in an evolutionary way. Lack had made it clear that he did not want to be a doctor, so his parents thought a career as a chartered accountant might be appropriate. Instead, he decided to study zoology at Magdalen College, Cambridge—Alfred Newton's college and the birthplace of the BOU seventy years earlier. Lack described Cambridge as "a spring awakening after the winter of a public school,"[7] not because of his undergraduate lectures, which he found dull, but because of the ornithological opportunities it provided. He continued to conduct detailed field studies and read the then meager literature on bird behavior, including books by Edmund Selous (chapters 7 and 9), as well as those on evolution by J. B. S. Haldane.

After becoming president of the Cambridge Bird Club and joining an expedition to St. Kilda in 1931, Lack planned his own expedition to the Arctic with Colin Bertram the next year. On the advice of the explorer Tom Longstaff, they went to Bear Island, off northern Norway. As Lack later said, in Longstaff's view "the ideal number for an expedition is one, and the party should consist of the minimum number above that to attain one's ends, which has been fully borne out by my own experiences."[8] In 1933, his last year at Cambridge, Lack spent several weeks with Bertram and Brian Roberts in East Greenland. Before they set off Julian Huxley (chapter 2) came to speak at the Cambridge Bird Club and told Lack about a new coeducational school at Dartington Hall in Devon that, on the suggestion of Huxley, was looking to employ an ecologist. Lack jumped at the opportunity and started later that year.

Lack was already an established ornithologist by the time he began teaching at Dartington, with two nightjar papers and a prominent and perceptive, but aggressive, attack on the increasingly popular concept of territoriality, which he wrote with his father (Lack and Lack 1933).[9]

Dartington was not a conventional school—pupils decided whether or not to attend lessons—and the children loved it; so did Lack. Many years later, children's author Eva Ibbotson singled out Lack as her most inspiring teacher:

> His lessons were so clear. . . . He helped me to see biology as a fascinating, rich subject. He was also a gifted artist and very musical. I remember going with a few others . . . to see him in his room after he had broken his leg. David was sitting on his bed wearing pyjamas, playing the guitar. You couldn't get away with that now, but at the time this easy informality seemed perfectly innocent as well as natural. There was a boiled egg on a plate by the side of the bed, which he ate whole. He then asked us to guess why he was doing this, eventually explaining that while he normally left the shell he was consuming it this time because his broken leg needed all the calcium it could get.[10]

In his second winter he started trapping and color-ringing European Robins around the school, mainly for the benefit of the children, but his own interest grew and eventually resulted in a four-year investigation. Lack's study was inspired by James P.

Burkitt, a civil engineer from County Fermanagh in Ireland, who started ringing robins in 1922. Being color blind, Burkitt did not use colored rings but instead applied different combinations of metal rings to the birds' legs. Burkitt plotted the territory boundaries of his robins and was the first to estimate the *average* life span of any bird from his ringed individuals. His results were published in *British Birds* between 1924 and 1926, but "the British ornithological community paid little attention to Burkitt's innovative study of the life history of the robin until the 1930s."[11] Burkitt did not pursue his research further and was surprised by the interest that ornithologists (eventually) took in his work.

Lack recalled in his memoir how, one beautiful March morning at Dartington, all

David Lack (age ca. 22) as a Cambridge University undergraduate on an expedition to Bear Island in 1932.

of his thirteen-year old pupils opted not to attend his class. Freed of any responsibility, Lack went to watch his robins instead. The following week he assumed the same would happen and went straight to the woods, leaving his pupils frustrated. After that they were scrupulously punctual, and they "gleefully told me that I would have to teach them that day and could not go out for Robins."[12] The main themes of his robin research were behavior—which he later lost interest in—and survival, which became a core feature of his subsequent research (chapter 10). His findings were summarized in *The Life of the Robin* (1943), an ornithological classic and one of the first popular monographs on a single bird species.

Lack went directly from university graduate to schoolmaster—he never did a PhD. Nonetheless, he received advice and support from a number of key people, including Eliot Howard, who suggested focusing on a single species, which is why Lack decided to concentrate on the robin while at Dartington.[13] Julian Huxley became his unofficial supervisor for the robin work, encouraging him to conduct what, at the time, was an extremely innovative study of behavior and ecology. Huxley also encouraged Lack to experience the tropics, suggesting he visit Reg Moreau in Tanzania; Lack did this during the summer vacation of 1934, broadening his knowledge and making a lifelong friend (see below). The following summer saw him in North America, where he met Joseph Grinnell in California and Ernst Mayr at the AMNH. Mayr in particular became an influential friend.

Such was the nature of Dartington School that in 1938 Lack was given a year-long sabbatical, setting off with several others for the Galápagos Islands. Once again Huxley provided encouragement and smoothed the way. It was a tough trip both socially and meterologically, but one that had an important effect on ornithology. Lack's resulting book, *Darwin's Finches* (1947a), not only resulted in his election to the Royal Society, it shaped the rest of his career and much of ornithology to come (chapter 2).

Returning to Europe just in time for the start of the Second World War, Lack spent much of the next few years in Orkney, enjoyably working on radar and using the opportunity to look at bird migration (chapter 4). It also allowed him to complete his robin book. In 1945, soon after the end of the war, Bernard Tucker, a zoologist at Oxford, suggested that Lack apply for the directorship of the Edward Grey Institute. This was one of a very few ornithological jobs, and there were others interested, including H. N. (Mick) Southern, and James Fisher, the latter championed by Max Nicholson. Fisher was a keen birder and effective publicist for ornithology, but not a scientist. Once he knew that Lack had applied, Southern pulled out, and fortunately for ornithology Lack was appointed.

To put Lack's EGI appointment in perspective we need to backtrack a little. In 1927–28 Max Nicholson, of the Oxford Ornithological Society, founded a group called the Oxford Bird Census, which among other things initiated the annual census of heronries (chapter 10). The success of that census encouraged Nicholson and Tucker to create a permanent center for ornithology. They appointed Wilfred "W. B." Alexander as director of the census, which evolved into the British Trust for Ornithology in 1932. In 1938 Oxford University backed the scheme to commemorate their late chancellor, Lord

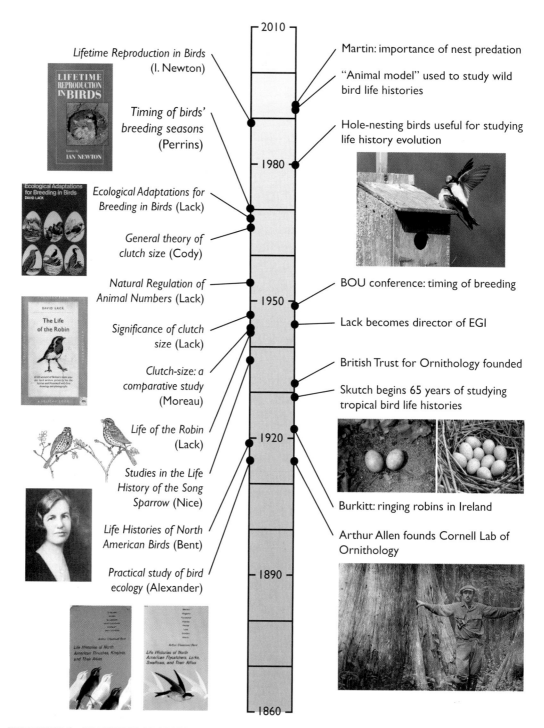

Lifetime Reproduction in Birds
(I. Newton)

Timing of birds'
breeding seasons
(Perrins)

Ecological Adaptations for
Breeding in Birds (Lack)

General theory of
clutch size (Cody)

Natural Regulation of
Animal Numbers (Lack)

Significance of clutch
size (Lack)

Clutch-size: a
comparative study
(Moreau)

Life of the Robin
(Lack)

Studies in the Life
History of the Song
Sparrow (Nice)

Life Histories of North
American Birds (Bent)

Practical study of bird
ecology (Alexander)

2010

1980

1950

1920

1890

1860

Martin: importance of nest predation

"Animal model" used to study wild
bird life histories

Hole-nesting birds useful for studying
life history evolution

BOU conference: timing of breeding

Lack becomes director of EGI

British Trust for Ornithology founded

Skutch begins 65 years of studying
tropical bird life histories

Burkitt: ringing robins in Ireland

Arthur Allen founds Cornell Lab of
Ornithology

TIMELINE for BREEDING ADAPTATIONS. Left: Covers of Newton (1989), Lack (1968), and Lack (1944); Song Sparrows; Margaret Morse Nice; covers of two of Bent's *Life Histories*. Right: Tree Swallows copulating on a nest-box; clutches of Fiery-necked Nightjar (left) and American Coot (right); Arthur A. Allen looking for Ivory-billed Woodpeckers in Florida in 1924 at age ca. 39.

Grey of Fallodon, forming the Edward Grey Institute of Field Ornithology, with Alexander as director and sole member. Alexander retired in 1945 but remained as librarian, his personal collection of bird books providing the foundation for what was to become one of the most comprehensive ornithological libraries in the world—the Alexander Library, named after him. When Lack took over as director in 1945, he was only thirty-five but already had three books[14] and fifty publications to his name.

Moving to Oxford—and primed by his earlier experience studying robins—Lack focused his research on the population biology of birds and on the evolution of two life history traits: clutch size and the timing of breeding. Somewhat reluctantly, he switched from robins to Great Tits (chapter 10), but it was a clever move, and the following two decades were spectacularly productive. Even though this success earned Lack the reputation for bringing about a revolution in field ornithology in Britain during the 1940s and 1950s (Johnson 2004; Haffer 2007a), the need for a field-based ornithology had been championed previously by a number of individuals.

Among these was Henry "Harry" Witherby, best known today for initiating bird ringing in Britain (chapter 4), and for starting the journal *British Birds* (started in 1907) and writing the *Handbook of British Birds* (first published 1938–41). Witherby also tried to introduce ecology into ornithology. In 1914 at a British Ornithologists' Club dinner, he asked Horace G. Alexander (Wilfred's brother) if he would write a piece for *British Birds* to introduce the approach, and "A Practical Study of Bird Ecology" appeared the following year.[15] Witherby was delighted and

wrote to Alexander, urging him to follow up his article with more detail, suggesting further topics for investigation, such as what could be learned by studying a flock of tits:

> Observe how it is built up of families: mark the trees it chiefly affects: the chief food: whether that food depends upon those particular trees: whether the food is difficult to find: whether it is more easily found by a flock than singly: whether the flock is an advantage against enemies if any: etc, etc. . . . I had a dream to work out flocks of Tits in that sort of way years ago but drifted off it chiefly because I am not enough in the country. . . . Anyone who could work out such a problem with the aid of entomology and botany would beat G. W. [Gilbert White]. He was a saint in ornithology far far in front of his time but why should we not beat him—Howard for example might if he were not so infernally theoretical.[16]

Alexander never capitalized on Witherby's ideas, blaming the war: "The slaughter of innocents in Flanders dried me up, and I had little heart for such efforts. To go out and forget for a few hours, as one watched birds, was very comforting; but serious ornithological work, whether in the field or in the study, was too difficult."[17]

The second champion was Captain John Kennedy, who wrote to the editor of *Ibis* in 1924 urging him to make it more readable and interesting: "On looking over several of the later volumes . . . I am struck by the fact that many of the articles are devoid of interest to the ordinary student of ornithology. In particular, the long list of skins, unaccompanied by field-notes upon the habits of the birds, which so frequently appear in our pages, can surely possess no value for anyone."[18] Claud Ticehurst, one of Alfred Newton's disciples and a staunch BOU member, answered Kennedy's letter, expressing surprise at its sentiment. While he accepted that there might be a place for more "chatty" articles about birds, that place was not within *Ibis*: ". . . do not let us forget that the Union is a scientific body, and that we must keep its Journal a scientific publication in the eyes of the world, as it has been for over sixty years."[19]

The third champion was the twenty-two-year-old Bill Thorpe (later a major figure in ethology; chapter 7), who wrote to support Kennedy's proposal. Recently elected to the BOU, Thorpe cautiously suggested that "more space might be devoted to articles dealing with the relation of ornithology to other branches of biology and to the more general biological problems of the day. . . . At present the tendency seems to be to deal with ornithology as if it were a subject apart, having no connection with other branches of science."[20]

Fourth was Max Nicholson. While not trained as a scientist—he read history at Oxford—Nicholson was a keen bird watcher with a genuine interest in the scientific aspects of ornithology. Inspired by the likes of W. H. Hudson, Edmund Selous, Eliot Howard, and Julian Huxley (chapter 7), Nicholson championed their ideas on the behavior, ecology, and evolution of birds in a series of books, including *Birds in England* (1926) and *How Birds Live* (1927). He was under no illusions about the scale of inertia in field ornithology, writing in 1927 that "bird ecology . . . is in a hopelessly backward state, and shows no sign of improvement."[21] Among other things, he advocated measuring—quantifying—the breeding success of birds to understand how populations are regulated, and it was his efforts that galvanized the Oxford Ornithological Society into action, beginning practical work in this direction.

Reading Nicholson's books now, one might wonder why he hasn't had more credit as an architect of the new field-based ornithology. Intriguingly, David Lack wrote in his memoir that "unlike most of my contemporaries, I was unimpressed by E. M. Nicholson's *How Birds Live*, partly owing to its style and perhaps because it reached me too late."[22] We can see what Lack meant: although Nicholson was writing for a general readership, his text reads at times rather carelessly. Nicholson was a man in a hurry and, as his later role in conservation demonstrated, a mover and shaker, and these books were written with the explicit intention of shaking up the museum-based ornithological establishment.

It is less clear what Lack meant when he said that Nicholson's work reached him "too late." Lack graduated in 1933, the same year that he and his father wrote their ferocious critique of Howard's territory theory. Perhaps it was Nicholson's enthusiasm for Howard's idea that David Lack reacted against. Or maybe it was their conflict over who should run the Edward Grey Institute and its precise role. As we said above, the EGI was the brainchild of Nicholson and Tucker. When Lack took over as director, his remit was to make the EGI a "national recording centre for the field ornithology done by amateurs."[23] He chose to ignore this, focusing instead on mainstream biology and was often critical of the unscientific work of amateur ornithologists. Lack's focus on the biological aspects of birds irritated Nicholson and his protégé James Fisher, but it helped to make ornithology a respectable scientific discipline (Johnson 2004).

By the 1940s the inertia in UK ornithology—epitomized by the BOU—was colossal. In 1941 Lack was incensed when Ticehurst, the editor of *Ibis*, summarily rejected a paper by Oxford zoologist John Baker on the timing of birds' breeding seasons. Lack wrote to Mayr to complain, and Mayr's reply was a rallying cry for a new ornithology: "There are a number of good young men in England who could change all this radically, but they never seem to have made an attempt to oust the old guard. If you can round up sufficient members . . . you will not have any trouble instituting a new deal.[24] It was precisely because Ticehurst and many others like him in the BOU were so conservative that the "old guard" continued to dominate UK ornithology for a full twenty years after a "new deal" had been struck on the continent as a result of Erwin Stresemann's vision (chapter 2; see Haffer 2004a).[25]

Ticehurst's reign as the dreary, old-school editor of *Ibis* finally came to an end when he died, in 1941. The following year Lack became assistant editor and things began to change. As Moreau (1970) later commented, Lack's appointment was ". . . a remarkable gesture of hope and faith in those dark days."[26] Change was in the air, and *Ibis* started to resemble the German and American journals in its content. As director of the Edward Grey Institute, Lack's influence extended beyond what was published in *Ibis*, and he was able to direct the course of ornithology more generally. By 1947, when Moreau—now based in the EGI—took over as *Ibis* editor, the "revolution" was complete, and even those clinging to the last vestiges of the old ornithology were forced to admit

defeat. Moreau again: "In the space of a very few years a revolutionary change occurred, as *Ibis* contributions came to reflect the fact that ornithology was no longer inbred and isolated from the main currents of biological science."[27]

Lack's new institute was an inspiration: a center of ornithological science that attracted research students and visiting scientists from all over the world. In a stroke of genius, Lack started an annual student conference in 1947, providing a venue for young ornithologists from Oxford and elsewhere to meet. Just about everyone in Britain who subsequently became a professional ornithologist cut their teeth at those meetings. The current aim of the conference is for students to "present their work in a constructive atmosphere,"[28] but nonetheless these meetings were—and still are—sometimes threatening for students. Typically numbering twenty-five to a hundred students, the conferences were regularly attended by a number of senior scientists—including Ian Newton, Niko Tinbergen, and Mike Cullen—who had the useful if unnerving knack of asking exactly the right question. Lack's old friend, Arthur Cain, professor of evolution at Liverpool, was a regular and enthusiastic attendee who, unlike Lack, was keen to join the students in the pub in the evening. At a conference in the early 1970s, Cain started a conversation with students by saying, "Did you know that there's a bird of prey that lives entirely on fruit?" In his commanding voice, he then told the entire pub about it.[29]

Lack's enthusiasm for research students meant that the EGI became a major training ground for ornithologists. Between 1947 and 1973 he supervised nineteen DPhil students,[30] and several of them in turn—notably

Robert Hinde, Chris Perrins, Ian Newton, and Peter R. Evans—have trained their own students. As a result, the Lack academic dynasty is vast, contributing to his enduring reputation.

By today's spoon-feeding, bureaucratic standards, Lack offered little in the way of formal supervision. His approach was largely species oriented, and as Ian Newton told us of Lack:

> Supervision was absolutely minimal, if you had a specific question and he could answer it, he would do briefly but there was never any long discussion on where your work should go. You were left on your own to do that, to work out everything you did basically, but of course we used to work together as students and that obviously was a great help. . . . I think most of the stimulus I got or we got from David Lack was coffee table conversations, but they were almost all centered on whatever David Lack was working on at the time. I don't want to criticize him for that because we learned an awful lot from that and during the time I was at the EGI he wrote three books: *Population Studies of Birds*, *Ecological Isolation in Birds*, and *Ecological Adaptations for Breeding in Birds* and that was a very, very great book. He'd always be working on a book and he'd come down and talk about whatever he was working on at the time, and that was very educative . . . because first of all we learned about the subject, but the real interest for me was seeing how he thought about problems and seeing how he looked at problems. And he looked at everything from an evolutionary point of view and that was I think one of the main lessons I got from David Lack.[31]

Newton knew that Lack had read his thesis in detail because he included some of the information, on Eurasian Bullfinches, in his 1966 book *Population Studies of Birds*, adding that Lack was

> very helpful when I wanted to go on. I wanted to continue working on bullfinches and he put in for a grant for me from the Agricultural Research Council, and I can remember this because this was about the longest time I ever spent in his company. It must have been about 50 minutes or an hour and he just sat down with me in his room and he typed the whole grant application, there and then, the whole lot and sent it off that day. We heard very quickly that it had been successful and I'd got the three year postdoc to carry on working on bullfinches. That was how I came to work for six years on finches in Oxford.[32]

ECOLOGICAL ADAPTATIONS FOR BREEDING

Lack's approach to ornithology helped to start a new wave of field studies of birds in Britain, along with research by like-minded individuals in North America and elsewhere. If one looks back at copies of *Ibis* and *The Auk* from the 1950s, 1960s, and 1970s, it is striking how many papers have titles like "The Breeding Biology of. . . ." These largely descriptive papers now seem rather pedestrian, but they were the building blocks upon which David Lack would create his new ornithology; one of his great strengths was the ability to synthesize huge amounts of information and in so doing see the big picture in a way few others could.

By the mid-1960s Lack felt the need to summarize this wealth of information and started to plan a book that covered both population biology and reproductive biology. Essentially this was to be a book on the

Members of the Edward Grey Institute in January 1955. Back row (left to right; approximate ages in parentheses): Jim Lockie (31), R. Gale (?), Dennis Owen (24), Lance Richdale (55), A. M. Richdale (?), W. J. L. (Bill) Sladen (?), L. Mc-Cartan (?); middle row: David Snow (31), Reg Moreau (58), W. B. Alexander (70), David Lack (45), John Gibb (36), Monica Betts (?); front row: Peter Walters Davies (27), Gordon Orians (23), Derek Summers (?).

evolution of avian life histories: longevity, age of first breeding, dispersion (colonial versus solitary nesting), reproductive success, timing of breeding, reproductive effort (clutch size, egg size), and chick growth rates. He soon realized the magnitude of the task, and it was Chris Perrins's wife, Mary, in 1964 working as EGI librarian and helping David Lack with the references, who suggested he break it up into two books. In due course these became *Population Studies of Birds* (1966) and *Ecological Adaptations for Breeding in Birds* (1968).

Possibly regretting his tirade against Wynne-Edwards and group selection in *Population Studies* (which we discuss in chapter 10), or his failure to convince Wynne-Edwards

of the fallacy of group selection, Lack adopted a more positive strategy in *Ecological Adaptations*, using it as "an attempt to build a positive case on the other side."[33] This was evidently an uplifting experience, and Ian Newton described Lack as being in a "state of feverish excitement"[34] about his book.

Ecological Adaptations was inspired largely by John Hurrell Crook's comparative studies of weaverbirds, studies that elegantly identified the links between nesting dispersion (colonial versus solitary nesting), diet, and mating systems (Crook 1964). Crook's approach to mating systems, developed by Lack in *Ecological Adaptations*, eventually became an important component of behavioral ecology (which we discuss in chapters 8

Plate by Robert Gillmor from Lack's *Ecological Adaptations for Breeding in Birds* (1968), illustrating the convergent evolution between American blackbirds (icterids) (left) and African weaverbirds (right). Gillmor illustrated many of Lack's books.

and 9); here we focus on the second half of Lack's book—on clutch size, egg size, chick growth, and the timing of breeding.

A number of people (starting with Charles Darwin and including the pioneering ethologists Konrad Lorenz, Oskar Heinroth, and Charles Whitman) had used the comparative method to address evolutionary questions, but Crook did so in a particularly effective way, alerting Lack to the possibility that he might employ the same technique to explain many different life history traits across birds as a whole. Moreover, Lack realized that a comparative approach was the only way

certain types of question could be addressed: "It is hard to evaluate ecological adaptations because one normally finds in nature only the successful products of evolution and not the failures."[35] One couldn't, for example, force Rooks to nest solitarily in order to work out why they usually nest in colonies; nor could one force Common Blackbirds to breed without a territory to understand why all males of this species defend a territory.

Comparisons between species provide a powerful method for addressing such questions, but Lack was also aware of a major pitfall: the potential lack of statistical independence among closely related species. The essence of the comparative approach is to assume that two species show similar life history traits because they are subject to similar selection pressures. But as Lack was well aware, two species can also exhibit similar traits simply because they share a common ancestry. To minimize this "phylogenetic effect," Lack based his comparisons not on species but on families or subfamilies, the assumption being that the further back in the phylogenetic tree one goes, the weaker the phylogenetic effect is likely to be. This in turn required that taxonomic relations were correctly identified, which in many cases was not true (chapter 3). Finally, Lack attempted to quantify his results, but at the same time he avoided cluttering the text with statistics. While this lack of formal statistical analysis made for easier reading, it caused consternation among some ornithologists (see below). As his son, Peter, said, "He more or less avoided formal statistics all his life and, although in later work he often compiled quantitative data in support of his theories, many of these had started out as hunches based on his natural history knowledge. To the annoyance and frustration of his scientific colleagues, these hunches too often proved correct."[36]

Lack decided to focus on the *ecological* adaptations that govern breeding, rather than behavioral or physiological adaptations (which we discuss in chapter 6). His main aim was to provide a comprehensive overview of the breeding biology of birds, as a way to understand why there are differences in traits like egg size, clutch size, and chick growth rates, and how these traits were shaped by the bird's environment: "I consider . . . that all these features are adaptations, or the result of adaptations, evolved through natural selection, which enable each species to raise, on average, the greatest possible number of young, and my aim is to interpret the nature of these adaptations, particularly by means of comparisons between different groups of species."[37]

We should pause—briefly—to ask where all the information that Lack amassed on avian life history traits came from. The answer is that it had accumulated over centuries, reused and refined as errors became identified and rectified. The accumulation of knowledge about life histories of birds is similar to that of many other areas of ornithology. At the beginning of the twentieth century most ornithologists assumed that everything there was to know about the ecology and behavior of birds was already known and available in handbooks like that of the German ornithologist Johann Friedrich Naumann, first published between 1820 and 1860. Remarkably, Naumann's volumes *Naturgeschichte der Land- und Wasser-Vögel des nördlichen Deutschlands* were reprinted between 1897 and 1905 (Haffer 2001) as if they were the last word in ornithology and no

new knowledge had been accumulated in the previous half century. Sadly, this was probably true of field ornithology: Jürgen Haffer (2007a) describes the period between 1850 and 1920 in Europe as being at a "temporary standstill." It was really Oskar Heinroth (1917, 1930) who breached the walls of this intellectual inertia, identifying the limitations of Naumann's work and attempting to fill some of the gaps. Heinroth did this through a project of monumental proportions, systematically hand rearing over three hundred species of European birds and documenting their incubation periods, growth rates, and development (Heinroth 1922; Heinroth and Heinroth 1924–34).

Stresemann in turn encouraged an interest in the breeding biology of birds among German-speaking ornithologists during the 1920s through both his *Aves* volumes (1927–34) and his editorship of the *Journal für Ornithologie* (JfO). He actively encouraged his colleagues to collect field data—which they did—resulting in the JfO moving far ahead of *Ibis* and *The Auk*, both of which continued to focus mainly on systematics. The difference in attitude toward life history studies from 1920 to 1940 between central Europe on the one hand and Britain and North America on the other is striking. Margaret Morse Nice—Mrs. Nice, as she was known—for one, found it easier to publish her life history work on the Song Sparrow in Germany than in her native United States.[38]

The development of ornithology in North America during the twentieth century was rather different from that in Europe. Perhaps the most striking contrast was the teaching of "ornithology" in North American universities from the early decades of the century. At the University of California at Berkeley, Joseph Grinnell was giving lectures on birds as early as 1909. In 1935, Lack visited California and met Grinnell at the university. He was amazed to learn that ornithology was taught, and he concluded that "the university's academic standards must be very inferior."[39] Presumably he considered it better to teach general principles—ecology and evolution—rather than taxon-specific material.

On the other side of the country, Arthur A. Allen had founded the Laboratory of Ornithology at Cornell University in 1915,[40] after completing his PhD in 1911. Allen's thesis research—published in 1914 as *The Red-winged Blackbird: A Study in the Ecology of a Cattail Marsh*—was pioneering because it broke away from the tradition of faunistic and museum studies. Frank Chapman (chapters 3 and 11) called it "the best, most significant biography which has thus far been prepared of any American bird."[41]

Allen's contribution to the development of American ornithology was far reaching. As well as his own studies, such as that on the sexual displays of Ruffed Grouse, Allen was among the first to take color photographs of birds—many of which were reproduced in *National Geographic*. He gave many public lectures, developed techniques to record and analyze bird song, and was responsible for numerous undergraduate courses and PhD students. Like Lack, he created a vast dynasty of ornithologists who took his ideas and developed them across the country; by 1933 no fewer than thirty-six colleges in the United States were giving undergraduate courses in ornithology. One of his former doctoral students, Olin Pettingill (1968), said that Allen's approach was very open minded: he did not prescribe projects to his students,

nor did he try to influence them to work in a particular field. "They could undertake faunal, life-history, taxonomic, population, or behavioural studies—any studies so long as the objectives were scholarly."[42]

In 1957, with the financial assistance of amateur birder and businessman Lyman Stuart, the Laboratory of Ornithology acquired its own building, based in Sapsucker Woods, Ithaca. In 2003 the Lab moved to quarters within the Imogene Powers Johnson Center for Birds and Biodiversity, a futuristic building of wood and glass that acts both as research station and visitor center, thereby fulfilling Allen's desire to advance both ornithological research and public engagement.

Even with these early advances, some individuals, like Allen's student Herbert Friedmann, were frustrated that the AOU was holding things back. Writing to Ernst Mayr in 1932, Friedmann said that "the A.O.U. was hopeless as far as science was concerned, and that the 'Auk' was as incapable of progress as its equally dead name-sake."[43] He continued: "If it were only possible to get A.O.U. people to study the change in the J f Ornithologie since Stresemann took charge, it might be possible to talk to them, but it isn't."[44] In a letter to Grinnell in 1935, Friedmann continued the theme:

> The tendency toward senility in the A.O.U. is a very noticeable one, and one that I think will take some time (and some deaths) to overcome completely. It is largely due to the fact that it is so much easier to do over and over again the obvious, superficial things that were the main problems in ornithology 50 years ago, and so many of our members are either set in their ways of thinking, or too

impressed by the gray-beards of a past day, to do the harder but necessary deeper ornithology of the present.[45]

Friedmann, Mayr, and other "reformers" were particularly incensed when "Mrs. Nice" was passed over for election as a fellow to the AOU in 1935 and the aging Edward Preble elected in her place.[46]

In those days the second major difference between ornithology in North America and Europe was the lack of "domination" by a key individual in North America. Stresemann in Germany (1920s–40s) and Lack in Britain (1940s–70s) were extremely important in promoting many aspects of ornithology (Mayr 1975: 366). In North America no one individual can be credited with bringing about a paradigm change; instead, a number of people were simply working toward a common goal, complemented by the development of museums, universities, wildlife departments, and ornithological societies. One possible exception is Mayr himself, yet for all his contributions (see Haffer 2004b, 2007b), he had few research students before the 1950s, so his impact on ornithology in this respect seems less than that of Stresemann and Lack.

In terms of field ornithology, there were some key publications, including Arthur Cleveland Bent's momentous twenty-one-volume *Life Histories of North American Birds*, published between 1919 and 1968 (completed after his death). Bent's volumes provided a useful reference point for research programs dealing with North American birds until 2002, when the *Birds of North America* project was completed and available online.

Other important early figures in North America included Samuel Prentiss Baldwin

and Margaret Morse Nice. Using bands supplied by the American Bird Banding Association, which was formed in 1909, Baldwin started banding birds caught accidentally in sparrow traps at his home in Ohio in 1914.[47] Up to this point, bird banders had concentrated largely on banding nestlings, with data coming from the chance recovery of dead birds. Baldwin instead concentrated on trapping and retrapping adult birds. Publication of his first data in 1919 aroused considerable interest, and in Baldwin's obituary Charles Kendeigh (1940) wrote: "Doubtless this paper will go down in the annals of ornithology as one of the classic publications of all time in this science. It opened a whole new field for ornithological endeavor."[48] By 1937 Baldwin had banded 21,862 birds at his home, half of which were House Wrens. Using his marked birds, Baldwin documented the "marriage relations" and the incidence of divorce in the wrens (e.g., Baldwin 1921). Despite the pioneering nature of this work, Baldwin remains little known.

Nice worked on the feeding preferences of Northern Bobwhite for her master's project at Clark University, but her studies were halted when she married, so she never received a PhD. She banded her first Song Sparrow—a male that she named Uno—on 26 March 1928; it was the start of an extraordinarily detailed study that continued for fourteen years. She found it difficult to publish her observations in *The Auk*, because they did not accept articles longer than ten pages. On meeting Nice, Stresemann was sufficiently impressed that he had her work translated and published in two parts in the *Journal für Ornithologie* in 1933 and 1934. Mayr wrote to her full of praise: "I consider your Song Sparrow work the finest piece of life-history work, ever done."[49] Like Mayr, Nice was keen to integrate Stresemann's vision into North American ornithology, which she achieved in part by translating and reviewing, in the journal *Bird-Banding*, literally thousands of ornithological publications coming out of Europe. Her impact on ornithology is all the more remarkable given that she was raising five children at the time.[50]

In summary, then, for the first sixty years of the twentieth century, ornithology in North America progressed on a much broader front than in Europe. North America was ahead of Europe by including ornithology as part of undergraduate degrees between 1910 and the late 1930s, but similar to Britain in focusing largely on faunistics and classification. As with Britain, serious change happened in the 1940s, with the introduction of "new blood" into the national unions and their journals and a subsequent change in focus to field ornithology. By the time the university systems of North America and Britain expanded in the 1960s, ornithology was functioning in a similar and productive way in both regions.

Thanks to these various efforts, English-speaking ornithologists eventually caught up with—and overtook—Germany such that from the mid-1940s *Ibis*, *The Auk*, and *The Condor* published what seemed like an ever-increasing number of life history studies. Indeed, in America the journal *Bird-Banding*[51] was established in 1930 as an outlet exclusively for life history studies. With Nice as associate editor from 1935, *Bird-Banding* also published detailed reviews of books and papers coming out of Europe, facilitating the spread of ideas about field ornithology. It was through these descriptive studies of breeding ecology that Lack was able to create his

magisterial synthesis of avian ecology, *Ecological Adaptations for Breeding in Birds* (1968).

When we started this book we conducted a survey among senior ornithologists, asking them which books they felt were the most influential in the twentieth century. Lack's *Ecological Adaptations* (1968) was the one most frequently listed. Because of its iconic position, it provides a convenient standpoint from which to look both back and forward at the development of life history studies. Here, we focus on two topics—clutch size and the timing of breeding—that Lack made his own in the 1940s but which have continued to intrigue and fascinate ornithologists and other ecologists ever since.

CLUTCH SIZE

Lack (1947b) is usually credited with being the first to ask (and answer) why different species of birds produce average clutch sizes that range from one to twelve eggs. Yet, as is usually the case with scientific breakthroughs, Lack's ideas did not emerge de novo; rather, he developed his hypotheses from the thoughts and writings of others—hence the expression "standing on the shoulders of giants." In terms of clutch size, however, unbeknownst to Lack (or indeed to almost anyone else), the correct explanation—as we will see—had been published thirty years previously.

Since the earliest times it had been suggested that there was a physiological limit to the number of eggs a bird could lay. This idea was dispelled in the seventeenth century when Martin Lister discovered that by daily removing each egg from a Barn Swallow's nest as soon as it was laid, the female continued to lay—in this case nineteen eggs in as many days (Birkhead 2008: 355). Species like the Barn Swallow that continue to lay as successive eggs are removed are referred to as "indeterminate" layers. But not all birds respond in this way, and those that lay only a fixed number of eggs per clutch are "determinate layers" and may be physiologically limited in how many eggs they can lay. Another seventeenth-century idea was that clutch size was determined by how many eggs a bird could cover and incubate properly, but this too was dispelled by the simple expediency of adding an egg to completed clutches and seeing no reduction in hatching success (e.g., Wallander and Andersson 2002). The third explanation—Lack's idea— was that clutch size was shaped by natural selection and the clutch size for each species was that which resulted in the greatest number of surviving offspring.

Lack got this idea in the 1940s, when as assistant editor for *Ibis* he read a paper submitted by his friend Reg Moreau on clutch size in African birds: "It came to me in a flash that the clutch-size of nidicolous [altricial] birds must have been evolved in relation to the number of young which they can feed and raise."[52] Moreau's paper on clutch size had its origins in nineteenth-century oology, when collectors realized that birds in the tropics typically lay smaller clutches than those in temperate regions (Moreau 1944). The first to attempt to make sense of this pattern was Erwin Stresemann (1927–34), who suggested that birds in the tropics produce more clutches in a season to compensate for smaller clutches—a seemingly sensible idea but one that was later shown to be incorrect. Another suggestion was that birds breeding at higher latitudes enjoyed longer day lengths, allowing more

Reg Moreau (photo in 1963 at age ca. 66).

foraging time and thereby enabling parents to rear larger broods (Hesse 1922; Hesse et al. 1937).

Reg Moreau was a late developer. He left school at seventeen with no formal scientific training and no real interest in birds. Seeking a career in the British Civil Service, Moreau was sent from England to the Cairo branch of the Army Audit Office, and it was there that he met the entomologist Carrington B. Williams, who fostered Moreau's scientific development and encouraged his interest in birds. That Moreau needed some distraction is hardly surprising, since his job, which comprised little more than counting blankets and lumps of cheese, was excruciatingly dull. In 1928 Moreau and his wife, Winnie, moved to Amani, Tanganyika,[53] where the Colonial Office was establishing an agricultural station. Nestled in the shadow of the Usumbara mountain range, this is an area of cloud forest that averages 200 centimeters (79 inches) of rain per year. Moreau and his family were plagued by dysentery and malaria, but his scientific career flourished, due to the tremendous biological diversity of the cloud forest: "Indeed in that enchanted time and place anything one gave attention to showed something "new"—undescribed

Southern Ground Hornbills. Moreau conducted a comparative study of hornbills while working in Africa, intrigued by their unusual breeding system and extreme brood hierarchies.

birds, nests and eggs, breeding seasons, weights, incubation and nestling periods, behaviour at the nest, the inter-relations of allied species . . ."[54]

Becoming increasingly intrigued by the life history traits of birds, Moreau was particularly fascinated by hornbills, birds whose specialized breeding biology goes something like this: at the beginning of the breeding season a pair of hornbills chooses a tree cavity in which to nest; both members of the pair then collect little balls of wet mud, which they plaster on the sides of the cavity

entrance; when only a small gap remains, the female squeezes inside and continues to plaster over the remaining gap with mud provided to her by the male. Eventually the female is sealed inside the cavity, as only a tiny slit remains through which the male passes food and she ejects dung and other waste. She can stay in the cavity for up to four months, during which time she undergoes a complete molt while looking after the young. Early ornithologists were intrigued by this behavior, convinced that the male "locked up" his partner against her will.

Moreau conducted a comparative study of sixteen of the twenty-six species of African hornbill, and he was struck by the extreme variation in egg-laying intervals, which resulted in clear within-brood hierarchies (Moreau 1937). Inspired by his initial findings, he began a much wider ranging comparison, focusing on clutch size and exploiting the vast amount of information amassed by generations of oologists.

Moreau retired early due to poor health and returned to England, where he made a surprisingly rapid recovery and was given a position by Lack at the EGI in 1949, with complete freedom to do what he liked. Known for his "dynamic, sometimes violent and extremely amusing conversation,"[55] he became the assiduous editor of *Ibis* and, eventually, president of the BOU. He was an exceptional editor and could "reduce a manuscript to half its original length, losing nothing but surplus verbiage, but gaining precision and clarity."[56] His book *The Bird Faunas of Africa and Its Islands*, published in 1966, was the first important account of the biogeography and history of African birds. Moreau is best remembered now for the *Palaearctic-African Bird Migration Systems*,[57] completed on his deathbed and published posthumously in 1972.

Moreau's (1944) monumental comparison of clutch size in African and European birds was inspired by earlier studies, including those by Richard Hesse and Bernhard Rensch in the late 1930s (Hesse et al. 1937; Rensch 1938). Ably assisted by Winnie, who helped him to compile close to four thousand records by trawling the literature of several journals, Moreau confirmed that birds in equatorial Africa produced smaller clutches than those in southern Africa and that clutch sizes in southern Africa were between a half and a third those of related species in Britain. His explanation was that British birds produce larger clutches because they experience higher adult mortality, as a result of either enduring harsher winter conditions or because of the perils of migration. In other words, Moreau's explanation was based on the assumption that clutch sizes are adjusted—through natural selection—to balance the level of mortality that adult birds experience. But as Lack suggested,

> This plausible idea rests on a mistaken view of both population balance and natural selection. Clutch-size could be adjusted to the mortality and achieve population balance only if it were much lower at high than at low population densities, which is not the case. Further, natural selection operates on the survival-rate of the offspring of each individual or genotype. If one type of individual lays more eggs than another and the difference is hereditary, then the more fecund type must come to predominate over the other.[58]

It was this flaw in Moreau's reasoning that allowed Lack to see, in a flash, the correct explanation: clutch size evolves through natural selection, "to correspond to the largest number of young for which the parents can, on average, provide enough food."[59] Lack also recognized that the idea of Richard Hesse and colleagues (1937) about day length and the amount of time available for parents to forage for their young elegantly accounted for latitudinal trends in clutch size. Food was the critical factor—something that Moreau had not realized.

Lack's ideas on clutch size were first published in 1947, as a three-part series of papers in *Ibis*.[60] But his solution to the clutch size problem had been anticipated over thirty years earlier by the virtually unknown British ornithologist Eric B. Dunlop. Noticing the high incidence of starvation among late-hatching Rook chicks, Dunlop suggested that if the parents had laid fewer eggs they would have left more surviving offspring (Dunlop 1913). Rob Magrath (1991), who rediscovered Dunlop's paper, wondered whether Lack's solution should be called "Dunlop's solution." It shouldn't, for although Dunlop was described as a "promising young ornithologist," there is no evidence that he recognized the general significance of his comment before he died, at the age of thirty, in 1917 during the First World War (Anonymous 1917).

Lack's clutch size hypothesis provoked a vigorous response from Alexander Skutch, who had spent many years studying birds in Costa Rica (chapter 8) and simply could not accept that the tropics were less productive than temperate regions. Skutch (1949) agreed with parts of Lack's hypothesis and accepted that it could not be dismissed, but he believed that it was only relevant in temperate regions, where there was dramatic winter mortality. In the tropics, Skutch argued, populations were more constant, with low adult mortality, so there was no need for large clutch sizes to maintain population sizes. He therefore subscribed to Moreau's point that clutch size had evolved to match the mortality rate—an idea that Lack vigorously rejected. Skutch was an ardent group selectionist, so it was hardly surprising that Lack was reluctant to consider his ideas seriously. Unlike Moreau, however, who acknowledged the logic of Lack's argument, Skutch stuck to his erroneous group selection views, but, as we'll see, he was correct about productivity and the idea that food

was unlikely to be limiting for birds in the tropics.

As Lack knew, his hypothesis rested on a number of assumptions, including the fact that clutch size must be heritable. What he did not seem to know was that this had already been demonstrated in poultry.[61] In the 1940s and for several decades thereafter, the links between poultry research and ornithology were almost nonexistent, and so it is hardly surprising that Lack did not use this information to bolster his hypothesis. Even by the mid-1970s, when the heritability of clutch size was first demonstrated in a wild bird—the Great Tit—there was no reference to the earlier poultry results (Perrins and Jones 1974). One reason why heritability estimates from poultry studies was ignored may be because they usually came from birds in which the degree of inbreeding and artificial selection were unknown and so were unlikely to be representative of wild birds (Boag and Van Noordwijk 1987). When we asked Peter Jones and Chris Perrins about their landmark study, they told us that they couldn't recall whose suggestion it was to address the question, commenting that Lack did not appear particularly interested in their idea.[62] To be fair, Lack was sick by this time (he died in March 1973), but even so it seems that Lack took it for granted that the traits he assumed to be under selection—like clutch size and timing of breeding—would be heritable.

Given its significance, the Perrins and Jones (1974) paper[63] hardly seems to have received the credit or citations it deserves. There are several possible explanations. The first is that in the early 1970s not many ornithologists were interested in avian genetics—with a few notable exceptions such as Fred

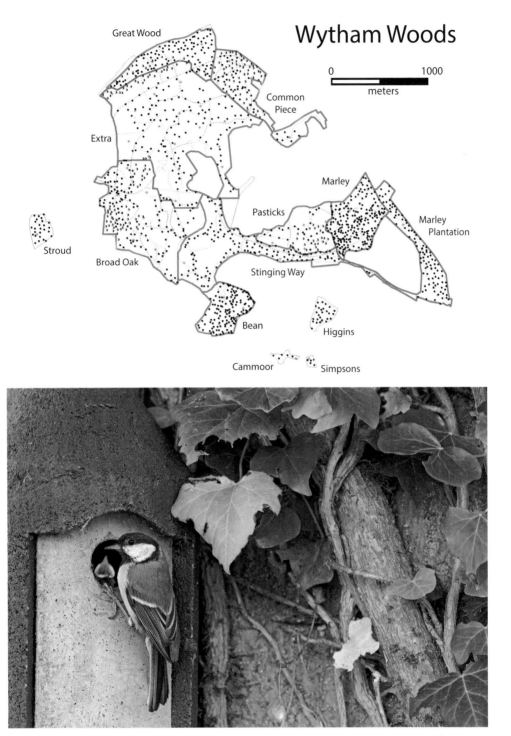

Map of Wytham Woods near Oxford, showing the location of nest-boxes (upper), many of which are occupied by Great Tits (lower) during an exceptional long-term study.

Cooke, who told us, "It was a key paper but not widely recognized as such. I suspect part of the reason may have been that the authors were not primarily geneticists."[64] A second reason is that the paper was published as a short note rather than as a "proper" paper. While publishing this paper in *The Condor* may have seemed appropriate in the 1970s, a result of equivalent significance today would be published more prominently, partly because there are more journals, and partly because self-promotion is so much more important today. It also seems likely that the more extensive work by Arie van Noordwijk on avian genetics, which first started to appear in the 1980s, superseded and eclipsed that of Perrins and Jones (Boag and Van Noordwijk 1987).

Although animal breeders had been estimating the heritability of various traits since the 1940s (e.g., Falconer 1960), the first attempts to test Fisher's Fundamental Theorem (1930) that the heritability of life history traits is low,[65] occurred only in the mid-1980s. In 1986 Lars Gustafsson analyzed data from his study of Collared Flycatchers on the island of Gotland in the Baltic Sea. Using data collected over a five-year period, he calculated the heritability of morphological characters, fitness components, and lifetime reproductive success, and showed—as Fisher predicted—that there was an inverse correlation between the heritability of a trait and its influence on fitness (Gustafsson 1986). Birds proved to be excellent for such research since—as we will see—several long-term studies were beginning to provide data on lifetime reproductive success (see below). Starting around the year 2000, ornithologists, ecologists, and behavioral ecologists began to apply the "animal model"—widely used by animal breeders (Sheldon et al. 2003; Kruuk 2004) to estimate genetic parameters and to explore the genetic aspects of senescence, inbreeding depression, and gene flow (Charmantier et al. 2006).

As far as Lack was concerned, the crucial test of his hypothesis was that the commonest clutch size was that which resulted in the survival of the greatest number of offspring. A key prediction was that individuals that laid *more* than the average number of eggs would leave fewer descendants than others because they were less able to provide each chick with sufficient food. The first evidence that this was true came from Lack's study of swifts—the body mass of chicks in larger broods was lower than that in smaller broods. If the body mass of a chick at fledging was an index of its body condition, and hence its likelihood of survival—as turned out to be correct—then this was consistent with Lack's hypothesis.

Lack also checked and confirmed his hypothesis using data from Common Starlings collected by Swiss researchers at Vogelwarte Sempach, the Swiss Ornithological Institute. Lack visited Sempach in 1946, and with the help of the director Albert Schifferli, he went through the mass of records to extract the necessary data. He then conducted endless calculations by hand (there were no convenient calculators or computers in the 1940s) to establish whether average-size broods resulted in the greatest number of surviving offspring. They did (Lack 1948).[66]

Others were also keen to test Lack's clutch size idea. The Finnish ornithologist Lars von Haartman[67] was interested in all aspects of life histories and adaptations for breeding. Importantly, he analyzed data on polygamous birds to distinguish between Lack's individual selection hypothesis of clutch size and the group selection hypothesis of Skutch (Von Haartman 1954). His argument was that under individual selection polygamous (actually polygynous) birds would have smaller clutches than monogamous species because the offspring of polygynous species are provisioned only by the female. In contrast, the group selection hypothesis predicted no difference in clutch size—because clutch size was predicted to have evolved to match adult mortality, rather than what the parent(s) can rear. The results revealed little difference in clutch size between birds adopting the two mating systems—that is, no evidence for Lack's theory. Von Haartman concluded that "in theorizing about the ways of natural selection we should be indeed very careful, and I do not wish to claim that the question of adaptation of clutch size is in any way solved by these findings."[68] He pointed out that Lack's theory—which he called "interindividual selection"—could not easily explain cases like the Eurasian Penduline Tit, which lays relatively large clutches, even though the female alone cares for young: "Either the nesting place of *Remiz* must be a 'Schlaraffenland,' where food is unrestricted, or the *Remiz* female is the most skilful bird in the world at finding food, or the clutch size is wholly unadapted. Neither of these explanations seems very attractive."[69]

In the mid-1960s Martin Cody—then at the University of Pennsylvania and now at UCLA—revisited the clash between Lack and Skutch over clutch size, pointing out that the differences they observed could be due to the types of environment in which each author's birds lived: temperate England and the tropics respectively (Cody 1966). Lack proposed that food influenced clutch

Martin Cody (photo in 1969 at age 28).

size by determining how many young could be reared; Skutch could not accept that food was limiting in the tropics and felt that nest predation might be the selective force favoring a smaller clutch size there. Cody demonstrated that both Lack and Skutch could be right: the genotype that leaves the most descendants might be determined by features (food and predators) of the particular environment in which they breed. Cody's paper—one of several that helped to refine Lack's ideas—became a citation classic, and he later commented that his proposal was "simple, logical, general, realistic and in general accord with the evidence; I like to think that this explains why the paper has been cited so often."[70]

Remarkably, it was not until the 1980s that other ornithologists started to question some of Lack's life history ideas, including his assertion that food was the single most important evolutionary factor. Among the first to raise doubts was Sven Nilsson (1984), who found that losses to predators were much greater among Great Tits nesting in natural cavities than in nest-boxes. This suggested that, since most of Lack's conclusions came from nest-box studies, the role of predation may have been underestimated. It was also pointed out that if researchers typically removed nest material from nest-boxes each year, the role of ectoparasites might also be underestimated (Møller 1989). A final issue was that most of Lack's conclusions were drawn not only from nest-box studies but also from highly modified, temperate woodlands where predator numbers were unnaturally low. Was it appropriate to generalize across birds as a whole from such a limited worldview?

In the 1990s Tom Martin started to amass data—both from the literature and from his own field studies in many different regions of the world—to reexamine the role of predation. His results were clear: breeding failure was much more likely to result from nest predation than starvation (Martin 1993, 1995). Ironically, Martin's research also exploded Lack's one concession that predation might be an important selective force accounting for the larger clutch sizes and longer nestling periods of cavity-nesting birds. Lack argued that birds breeding in cavities are safer from predators and hence can afford to produce larger clutches and have chicks in the nest for longer (Lack 1948, 1954). When Martin tested this idea with his own data, he found that it was not true. The best explanation for the larger clutch size in cavity nesters was the "limited breeding hypothesis," which states that when cavities are scarce, birds should invest more in breeding opportunities when they arise. Simplifying a complex story, Martin's data suggest that limited breeding opportunities, rather than nest predation, is the best explanation for larger clutch size in cavity nesters.

These results emphasize the importance of examining new hypotheses rather than blindly accepting longstanding dogmas. Indeed, although nest predation appears to be less important to life histories of cavity-nesting birds, nest predation apparently influences life history traits among many groups of birds for which food limitation has traditionally been considered to be more important.[71] Martin's work, with its wide geographic coverage and enormous data sets, has revealed, perhaps more than anything else, the extraordinary complexity of the relationships between different life history traits.

No Such Thing as a Free Lunch

In 1966 evolutionary biologist George Williams transformed the study of life histories by introducing the concept of "residual reproductive value." Dividing reproduction into current and future effort, he formalized the old idea of a trade-off between survival and reproduction. Williams's (1966b) paper, "Natural Selection, the Costs of Reproduction, and a Refinement of Lack's Principle," was to become particularly influential. In it he pointed out that "for species with low annual mortality rates . . . the parent's lifetime reproductive success may demand a low reproductive effort in a given season, much lower than that which would maximize reproduction in that season."[72] Put another way, Williams realized that Lack's clutch-size argument was erroneously based on the outcomes of individual breeding seasons, rather than on lifetime reproductive success. With this new viewpoint came the potential to address some of the empirical anomalies, perhaps the most striking of which was the ability of some birds to rear artificially enlarged broods.

Most attempts to test Lack's idea involved manipulations of clutch size—either adding or (more rarely) removing eggs or chicks. If Lack was right, then artificially increasing or decreasing brood sizes should result in lower breeding success. In many cases it didn't, yet, remarkably, ornithologists continued to accept Lack's hypothesis, possibly because—as Torbjörn Fagerström suggested—theory often carries more weight than empirical data. When Herman Klomp (1970) reviewed manipulation studies, he showed that in most cases the results were consistent with Lack's hypothesis, but the anomalies that did exist revealed that the observed clutch size was always smaller than that which the parents could experimentally rear—that is, smaller than the most productive one. This was consistent with Williams's (1966b) idea of a trade-off between reproductive effort (i.e., clutch size) and survival that maximized *lifetime* success rather than success in any one breeding season; presumably birds who raised artificially enlarged broods would suffer poorer future survival and reproduction. This came to be known as "the cost of reproduction."

In the 1970s Eric Charnov and John Krebs produced a simple model of optimal clutch size that focused on individual fitness, following Williams, rather than the "most productive clutch," as Lack had done. They argued that rearing offspring is costly in terms of adult survival; larger broods were more costly to rear because the parents had to work harder. If the costs—increased energy expenditure and predation risk—rise very steeply with increasing clutch or brood size, it pays parents to lay fewer eggs. The Charnov and Krebs (1974) model neatly accounted for observed clutch sizes often being smaller than the most productive clutch size each year: by investing less in reproduction in any single season, the adults live longer. In other words, there's a trade-off between reproduction and survival, and natural selection favors the optimal solution, the clutch size that results in the greatest number of surviving offspring (Stearns 1976).

The first test of the idea that there is a trade-off between reproductive effort and subsequent adult survival was conducted by Conny Askenmo (1979), on a nest-box population of European Pied Flycatchers in southwestern Sweden. Males whose broods had been experimentally enlarged were significantly less likely to return in subsequent seasons than controls, suggesting that the greater effort associated with an enlarged brood reduced their lifetime fitness. Nadav Nur (1984) conducted a more detailed study on Eurasian Blue Tits in Wytham Woods, Oxford. Over three breeding seasons (1978–80), he manipulated brood size (to three, six, nine, twelve, or fifteen chicks) and measured both adult weight (mass) loss during the chick-rearing phase and subsequent adult survival. For females (but not males) both body mass and survival decreased with brood size, as predicted (Nur 1984; see also Gustafsson and Sutherland 1988).

For many species, the trade-off between reproduction and survival seemed like a sufficient explanation for the disparity between the most common clutch size and the number of offspring that the adults could successfully rear. However, in the 1990s Pat Monaghan at Glasgow University wondered if an additional factor might be involved. She had been looking at the energetic cost of producing eggs—which up until then had been considered negligible. Using Lesser Black-backed Gulls as a study species, Monaghan and colleagues demonstrated that increased egg production also carries fitness costs—in years when laying costs increased, parents were less successful at rearing their young

(Monaghan et al. 1998)—and she wondered whether females lay fewer eggs than they can actually rear to increase the effort they put into rearing their current brood. That is, rather than there being a trade-off between clutches—between reproduction and survival—the trade-off might occur within clutches.

The problem—as Monaghan recognized—with the numerous brood manipulation experiments was that they failed to mimic the true situation. Simply giving a pair of birds one or more additional eggs ignores the physiological cost of actually producing those eggs. Accordingly, Monaghan and Vicky Heaney undertook a brood manipulation experiment on Common Terns that incorporated the full cost of producing, incubating, and rearing an extra egg. By removing the first laid egg (for twelve hours), they induced the female terns to lay an additional egg. The experiment comprised three treatments: (1) full cost—as just described; (2) free egg—females given one more egg (from another pair) than they actually laid; and (3) free chick—females given an additional chick (from another pair). The birds in group 1 paid the full cost of producing and incubating an additional egg and rearing the extra chick; birds in group 2 paid the cost of incubating (but not producing) the extra egg and rearing the chick; while birds in group 3 paid the cost only of rearing an additional chick.

Terns in group 3 had the greatest success, followed by those in group 2 and then the birds in group 1. The terns in group 1 provisioned their chicks at a lower rate, resulting in slower growth and higher mortality. This was a clear demonstration that producing and incubating an additional egg was energetically demanding and that cost was reflected in poorer performance during chick rearing. Heaney and Monaghan (1995) concluded that all brood manipulation studies that involved adding extra eggs or chicks to broods were misleading because they assumed that the main cost of reproduction occurs between—rather than within—clutches: "These findings cast doubt on the evidence that the most common clutch size is less than the Lack value [i.e., the number of young the parents can adequately provision]."[73] By identifying the within-clutch trade-off between the costs of egg production and chick rearing, Heaney and Monaghan provided another crucial refinement to Lack's original hypothesis.

Lifetime Reproductive Success

Inspired by George Williams's insight that the most appropriate currency for understanding the evolution of life history traits was lifetime reproductive success rather than success in a single season, Ian Newton was the first to measure it in birds, in the 1980s. Newton had been running a long-term population study of the Eurasian Sparrowhawk in southern Scotland since the 1970s. After following the same birds for several successive seasons, he realized he had enough information to estimate their lifetime reproductive success:

That for me was one of the highlights of the study because at that time nobody had recorded lifetime reproductive success in a bird. . . . I'm sure that kind of data existed on two or even three tit populations but nobody had thought about looking at it from that angle. . . . Tim Clutton-Brock looked at this in deer at the same time. We obviously had the same idea completely independently and he published a book on this which . . . I found helpful when I was getting my stuff together for publication.[74]

Newton went on to edit a book on lifetime reproductive success in birds (Newton 1989):

> . . . people used to say "oh that's fantastic, you're looking at natural selection in action," because some individuals were of course very much more productive than others, but . . . most of it I think was largely a chance thing because . . . some birds are bound to be more productive than others just because they happen to live longer, which could have little to do with their genetics. . . . In the Sparrow Hawk it turned out that 5% of each cohort of birds produced half the young produced by that entire cohort, so . . . some birds are much more productive than others.[75]

TIMING OF BREEDING

From the earliest times it was obvious to country people—especially bird catchers, who typically took young birds to hand-rear—that different bird species reproduced at different times of year. John Ray, in the seventeenth century, mentions striking differences in the timing of breeding of different bird species and, indeed, anticipated Lack's explanation by 250 years (Birkhead 2008). Egg collectors in the nineteenth century were also attuned to birds' breeding seasons—and were far more knowledgeable about this than any ornithologist of the day. Indeed, when Lack came to discuss birds' breeding seasons in the 1940s, some of his best data came from the egg collector Arthur Whitaker, who had been searching for nests in northern England for forty years (1904–46).

One of the discoveries to emerge from the activities of egg collectors and other naturalists was the relationship between latitude and the timing of breeding. Using their data, Oxford zoologist John Baker (1938) showed that in the northern hemisphere breeding started twenty to thirty days later for every 10-degree increase in latitude. Baker also made the important distinction between what he called "proximate" (immediate causes) and "ultimate" (evolutionary causes) factors controlling the timing of breeding seasons. He suggested that the proximate factors in boreal and temperate regions were day length and temperature, whereas in the tropics they were rainfall or what he called the "intensity of sunlight."

Probably as a result of Lack's interest, the BOU organized a conference on birds' breeding seasons in November 1949, and the papers from that meeting made a special issue of *Ibis* the following year. In his introduction, Arthur Landsborough Thomson (1950) provides a useful overview, summarizing geographical variation, differences among bird families, and reiterating Baker's distinction between proximate and ultimate factors.

Lack's focus was on the ultimate factors determining breeding seasons—the adaptive significance of breeding at a particular time—inspired, in part, by his earlier work on European Robins. Remarkably, European Robins have been recorded breeding in every month of the year in Britain, but Lack (1950) was quick to notice that those pairs breeding during the normal breeding season were much more likely (55 percent) to fledge young than those breeding at all other times

European Robin.

(14 percent). Lack's main message was that breeding seasons have evolved through natural selection, so that young are in the nest when food for that particular species is most abundant. Support for his hypothesis came from John Gibb's study of Eurasian Blue Tits in Wytham Woods, Oxford, where he demonstrated that the seasonal availability of caterpillars (the tits' main food)—measured by the density of the caterpillars' droppings (frass)—peaked at the time when the birds had chicks in the nest. In other words, food is the ultimate factor determining the timing of breeding (Gibb 1954). Embracing the work of William Rowan and others, who

demonstrated that day length was an important proximate factor (chapter 6), Lack recognized that different species (or even different subspecies) have evolved physiological differences in their response to day length: "The proximate factors do not have significance in themselves, but only in relation to natural selection and successful raising of a family. That is why very different proximate factors may be important in different species of birds. The ornithologist is concerned with these factors in so far as they relate to the life of the bird in its natural habitat."[76]

John Gibb, David Lack's field assistant at the EGI, was succeeded in 1951 by Dennis

Owen, who in turn was succeeded by Chris Perrins in 1957. Born in 1935 and educated at Charterhouse School and Queen Mary College, London, Perrins first imagined he might be an entomologist. At school he became interested in birds as a result of David Lack initiating a nest-box study of tits nearby at Alice Holt. While he was an undergraduate, Perrins ringed birds with Kenneth Williamson on Fair Isle and with Bert Axel at Dungeness, both key birding spots in Britain. In the first year of his degree Perrins saw a notice advertising the EGI's annual student conference. He attended for each of his three undergraduate years, after which David Lack offered him a position as field assistant. The rest, as Perrins said to us, "is history,"[77] in the sense that following Lack's death in 1973, Perrins took over as director of the EGI and remained in that position until his retirement in 2002.

Perrins completed his DPhil in 1963 on the survival of Great Tits in Wytham Woods, recognizing that an important factor influencing the survival of young birds was the date of hatching. His parallel study of Manx Shearwaters on Skokholm Island, Wales, also revealed that the earliest breeders had the highest breeding success and their young had the greatest chance of surviving to breeding age (Perrins 1965, 1966). This raised the question of why all individuals did not breed early. Perrins's (1970) explanation for this apparent anomaly was that some individuals are constrained by a lack of food early in the season, forcing them to breed later than is best for producing the optimal brood size.[78]

One of the first to test Perrins's idea was Hans Källander (1974), who provided Great Tits in Sweden with additional food throughout the winter. Although the extra food resulted in earlier egg-laying, it did so only up to a point—about five days— suggesting that factors other than food might also be important. Many subsequent food-supplementation studies were conducted in an effort to understand the timing of birds' breeding seasons, only a few of which had much effect on clutch size, possibly because most of these studies provided extra carbohydrate—in the form of sunflower seeds and peanuts—rather than protein, a nutrient much more likely to be limiting at that time of year (Meijer and Drent 1999).

Perrins's (1970) idea caused Lack to modify his conclusions regarding timing: "The breeding season evolved by each species is that which results in its leaving most offspring, but it has been evolved in relation to the needs of the laying female as well as those of the young."[79] As we have seen, given that egg production is energetically costly and may constrain the onset of breeding, clutch size and the timing of breeding are even more intimately associated than Lack had imagined: if optimal clutch size is determined only by costs incurred during the nestling phase, we would not expect clutch size to be sensitive to conditions at the time of laying, unless these foretell conditions later in the season. If the fitness costs of egg production are dependent on conditions at the time of laying, clutch-size decisions become inextricably linked with decisions on the timing of reproduction.[80]

Perrins's observation, that breeding success declined through the breeding season, sparked a number of studies to identify the cause. There were two main ideas: (1) late-breeding birds were young and inexperienced, and hence less competent at rearing offspring; and (2) environmental conditions deteriorated through the breeding season (Brinkhof et al. 1993; Nilsson et al. 1999). Researchers attempted to distinguish between these hypotheses through a variety of manipulations, such as advancing the timing of laying through supplementary feeding, or delaying breeding by forcing birds to lay replacement eggs, but it was eventually concluded that "clean" manipulations of the timing of breeding are impossible, since manipulation of any sort introduces other, potentially confounding factors. For example, in the case of birds laying replacement eggs or clutches, forming a second egg or clutch subjects these individuals to additional energetic costs (as shown by Monaghan and Heaney, above). Overall, it was concluded that both parental quality and date per se were equally important in explaining the seasonal decline in breeding success (Verhulst and Nilsson 2008).

LIFE HISTORIES

Lack concluded that the breeding habits and other features discussed in his book *Ecological Adaptations for Breeding in Birds* have "evolved through natural selection so that . . . the birds concerned produce, on average, the greatest possible number of surviving young. . . . The main environmental factors concerned in this evolution are the availability of food, especially for the young and to a lesser extent for the laying female, and the risk of predation on eggs, young, and parents."[81]

For many ornithologists, *Ecological Adaptations* was ahead of its time and inspirational—a bold attempt at some kind of ornithological synthesis. Like many bold ventures, it was one that attracted both approval and disapproval. The most critical

review appeared in *Science* from the pen of Martin Cody (1969). As one of Robert MacArthur's PhD students, Cody had visited Lack in Oxford in 1962. Acknowledging the value of Lack's synthesis, Cody identified what he considered to be the book's two serious limitations: Lack's methods and his views on clutch size.

Lack's approach was to identify correlations between different life history traits, and from these infer their adaptive significance. Cody himself excelled at identifying patterns in nature, but he disliked Lack's associations, describing them as confusing, inconsistent, and having a "will-o'-the-wisp quality."[82] Moreover, Cody felt that the associations between nesting dispersion, pair bonds, and food that were so clear in John Hurrell Crook's weaverbirds—the inspiration and motivation for Lack's book—simply did not hold up so well for birds in general. He also felt that Lack's preferred method of focusing on *pair-wise* associations sidestepped much of the complexity, and while this made for a neater story, did little to further understanding of life history traits. As for clutch size, Cody accused Lack of ignoring the refinements provided by various ornithologists that dealt with the anomalies. Cody's final blow was to suggest that Lack had added little to the basic ideas on life histories set out by Lamont Cole fourteen years earlier (Cole 1954). Intriguingly, although it is now apparent that Cody's assessment of *Ecological Adaptations* was accurate, his review did little to dampen enthusiasm for the book.

AN ALLURING GLIMPSE OF GENERALITY

Life history studies of birds began with the observations of Skutch, Moreau, and Lack.

Their work, especially that of Lack, helped later researchers to develop the idea that suites of traits tended to co-occur, epitomized by the live-fast-die-young strategy shown by species like Common Quail and the Zebra Finch, and the live-slow-die-old strategy exemplified by albatrosses and certain tropical passerines. In reality there is a continuum, one that became known as the "r-K continuum," with short-lived species at the "r" end and long-lived species at the "K" end (MacArthur and Wilson 1967; Pianka 1970). The basic idea was that different environments resulted in the evolution of optimal combinations of life history traits.

The study of life histories developed mainly using species, like fruit flies, that were easier experimental subjects than birds, focusing on the evolution of different combinations of traits and seeking to explain variation in the age at sexual maturity, longevity, clutch size, egg size, and the development of offspring. Steve Stearns was dismissive of Lack's type of comparative approach, saying that it provided only "alluring glimpses of generality."[83] Instead, he favored what he called an empirical, predictive approach (Stearns 1977).

Another strategy, and one adopted by Rudi Drent and Serge Daan—who were both disciples of Lack—advocated a shift to the study of mechanisms, in particular energetics and body condition. Drent and Daan were interested in the ways in which birds adjust body condition, fat and protein reserves, and reproductive effort to maximize their fitness, using techniques like doubly labeled water[84] to measure energy use (Drent and Daan 1980). In addition, as we saw earlier, there were important advances in our understanding of the genetic bases of life history traits beginning in the 1980s.

NEW METHODS—NEW INSIGHT?

Some thirty years after the publication of *Ecological Adaptations*, new methods for conducting comparative studies were developed by Paul Harvey and Mark Pagel (1991). Using these new comparative methods, together with ongoing refinements in avian phylogeny (chapter 3) and an ever-increasing volume of life history data, Peter Bennett and Ian Owens attempted to update Lack's study, revisiting all the questions (and more) that Lack (1968) had identified. After controlling for phylogenetic effects, many of the patterns Lack had identified intuitively, and often with rather limited data, were still evident (Bennett and Owens 2002). Overall, however, Bennett and Owens's conclusions were very different from Lack's. Detecting strong phylogenetic effects, Bennett and Owens argued that this was evidence that traits such as clutch size, egg size, and duration of incubation were nonadaptive and that Lack's interpretation was therefore flawed. But as Rob Freckleton (2009) later pointed out, this isn't true. Simply because a trait or combination of traits has a strong phylogenetic signal does not mean that they are not adaptive: "It could be that new species live in similar environments to their ancestors, and evolve traits that allow them to exploit these environments. The traits would show phylogenetic signal, but obviously be adapted to the conditions in which the species find themselves."[85]

In contrast to Lack's (1968) attractive and accessible volume, Bennett and Owens's *Evolutionary Ecology of Birds* (2002) was unashamedly businesslike, reflecting a shift toward harder-nosed science—and a harder-nosed publisher keen to save space. The book received mixed reviews, including one from Arie van Noordwijk (2002), who said:

In summary, this book is a progress report. Almost 35 years have passed since Lack's epic work was published. Nearly all of the progress in our understanding of avian diversity has been made in the past decade. This book provides a good record of this progress but I hope and expect that it will be outdated a decade from now. Meanwhile, it provides a useful overview of the strengths and weaknesses of the comparative method, and it certainly stimulates thinking about how we can learn more about variation in avian life histories and mating systems. Part of this stimulation is achieved through annoying the reader. Whether or not this was intended doesn't really matter—it works.[86]

STRONG ADVOCACY

In 2000 Robert Ricklefs reflected on the role of Lack, Moreau, and Skutch as pioneers in the study of life histories. He reaffirmed Lack's dominant role, describing him as "a visionary who held his beliefs strongly and argued them effectively. For the most part his insights were brilliant and, projected through his strong advocacy and productivity, fostered the most influential work in ecology in the middle third of the twentieth century."[87] With the benefit of fifty years of hindsight, he also points out that Lack's view was narrow, and others have commented on Lack's preference for single-factor explanations—in this case food. Subsequent studies have shown that those other factors—disease and predation—that Lack considered to be relatively unimportant are as significant as food. In addition, some of the ideas described by Lack in simple terms have proved to be much more complex.

During his life David Lack received many honors. Those he most cherished included election to the Royal Society (1951), election as president of the International Ornithological Congress in 1966—he later told his son Andrew that when he first heard the news he was too excited to sleep—and the Royal Society's Darwin Medal in 1972. Extremely modest, Lack wrote his "memoir" only reluctantly—saying his life was too dull—and did so only to encourage Reg Moreau to write his own. Unlike Darwin, David Lack never reflected in print on the reasons for his success. His son Andrew told us that "he simply did not talk about these things. He regarded his own life as uninteresting. He worked with great speed and concentration. Everyone who knew him knew that. He could be fairly scathing of lazy people, or those he perceived as lazy. He did not regard himself as out of the ordinary in intellectual ability but saw dedication and concentration as vital."[88]

And here is Ernst Mayr, a significant if subtle influence in Lack's life:

> I have known only few people with such deep moral convictions as David Lack. He applied very high standards to his own work and was not inclined to condone shoddiness, superficiality and lack of sincerity in others. This did not always go well with those who preferred to compromise in favour of temporary expediency.... His intolerance of shoddy thinking did not mean that he was intolerant of disagreement.... David and I had many disagreements but they never disturbed our close friendship.... David with his intense enthusiasm and dedication tended to adopt single factor interpretations.... For me phenomena such as territory, clutch size, niche partitioning or species number on an island were rather the phenotypic compromise between several selection pressures.[89]

On the centenary of Lack's birth, on 16 July 2010, the EGI held a one-day conference in his honor. In a concluding talk, Ian Newton captured what he and many others consider to be key features that made Lack so successful as a biologist: (1) he was an astute observer whose drive and intelligence enabled him to see the evolutionary significance of a wide variety of facts; (2) he recognized that birds comprised a tractable study system; (3) he was unusual in his thinking, especially in terms of evolution and individual selection; (4) he was lucky in his timing—there were relatively few ornithologists in the 1940s and 1950s and funding was relatively good; and (5) he was a good communicator—writing with simplicity, brevity, and clarity enabled him to promote his ideas in an engaging and inspirational manner.

In appraising the roles of Skutch, Moreau, and Lack in the development of avian life history studies, Ricklefs (2000) emphasized how Moreau (1944) clearly "understood not only the principle of evolutionary optimization of clutch size to maximize individual fitness, but also that many considerations might lead to an optimum clutch size smaller than that set by the food supply."[90] Lack thought that Moreau was arguing that clutch size (and reproductive rate) was adjusted to balance the mortality rate. He wasn't. As Ricklefs says, Moreau merely stated that "reproduction and adult mortality must be balanced in a population whose size remained constant."[91] Whereas Lack said that "if clutch-size is inherited and if other things are equal, those individuals laying larger clutches will come to predominate in the population over those laying smaller clutches."[92] The key to coming to grips with the misunderstanding between Lack and Moreau is Lack's phrase "if other things are equal." In Lack's view they were; in Moreau's view they might not be. For Lack, food was the only factor affecting clutch size; for Moreau, predation might also be important. Moreau (1944) suggested that nest predation could favor smaller brood sizes, a view that Skutch (1949) felt might account for the smaller clutch size of tropical birds: "A possible advantage of small broods and infrequent parental visits to the nest is the smaller likelihood of betraying its position to enemies."[93]

Lack's narrower viewpoint lumped Moreau together with Stresemann and Rensch, whose views on clutch size Moreau considered "teleological," with "unsupported speculations." Lack can perhaps be forgiven for this because Moreau was sometimes ambiguous and sloppy in his terminology, implying group selection thinking. What is surprising is that Lack and Moreau worked almost side by side for so long without (apparently) resolving their differences. Summing up their contributions, Ricklefs said that Lack was a visionary driven by his own convictions, whereas Moreau was more reflective and more amenable to considering alternative viewpoints: "Hindsight . . . offers a sobering view of our progress in understanding the diversification of life histories, and how science works in general."[94] Lack, Moreau, and Skutch were pioneers in this field, but Lack carried the day: "a glorious victory for certain powerful concepts and a sad loss for untested alternatives."[95]

CODA

Almost seventy years after Lack first formulated his life history hypotheses, these topics continue to intrigue and challenge biologists.

Why birds breed when they do, and why they lay a certain clutch size, seem at first sight to be relatively simple questions, and one might naively imagine that they require only simple answers. This is now very clearly not the case. Nature is often complex, and many life history traits—including clutch size and timing of breeding—are influenced by a wide range of factors and often in subtle ways. As more data and different ways of thinking about biological problems emerged from the mid-1960s onward, it became apparent that Lack's original views required modification. As we have seen, such modifications and refinements have continued to occur. Lack, we suspect, would have been delighted by these improvements and excited by the idea—and evidence—that there are various trade-offs.[96]

It is difficult to know how excited Lack would have been by Tom Martin's work, which has shown that the interrelationships between different life history traits are complex and that identifying the targets of selection is difficult. One reason for Lack's success was that he kept things simple; this was partly the way his mind worked, but it may also have been a consequence of the limited amount of data he had at his disposal. It is hardly surprising that with the enormous volumes of data available to Martin, a more complicated picture emerges. Complexity is probably a biological reality, but we crave simple solutions, and as far as life histories are concerned, simple solutions might not exist.

BOX 5.1 Robert Ricklefs

I grew up close to nature—at the end of the street, so to speak, with the woods right out the back door. I have always felt comfortable in nature. My father enjoyed camping and fishing, and we spent memorable days in the countryside in, what was then, uncrowded northern California. When I was nine, my family moved to the Monterey Peninsula, where my father, a lifelong educator, founded a private college preparatory school for boys. Naturally, I attended. The disadvantages of being the headmaster's son were outweighed by the eclectic faculty that a private school could hire (and a fledgling one could afford), including retired military officers and graduate students at nearby Hopkins Marine Station. It was a teacher in fifth or sixth grade, Margaret Moody, who introduced me to bird watching—I might have been eleven at the time—when she took a few students to the mouth of the Carmel River with a spotting scope. I never recovered from that experience!

Margaret introduced me to the local Audubon Society, and I became a rather fanatic, although not exceptionally skilled, birder, looking forward to weekend bird trips and enjoying a certain status as a young boy with a serious interest in birds. I remember especially Laidlaw Williams, a transplanted easterner who studied the life histories of local birds, his paper in *The Condor* on the Brewer's Blackbird being something of a classic. John Davis, director of the University of California's Hastings Natural History Reservation in Carmel Valley, also was a beloved mentor. With the Pacific coast a ten-minute walk from our house, I particularly enjoyed shorebirds, but I also developed a naturalist's interest in seashore life and could easily have become a marine biologist. The legacy of Ed "Doc" Ricketts of John Steinbeck's *Cannery Row* was still fresh in the area; the local sardine canning industry hadn't yet fully collapsed.

I entered Stanford University in 1959. This was shortly after the Russians had sent Sputnik orbiting around the globe, and students were flocking to engineering and the nascent space race. I was caught up in this craze for a couple of semesters until I realized that I liked biology (and biologists) much more than physics, chemistry, and mathematics. My undergraduate years preceded the rise of universities as research powerhouses, and our access to professors was truly amazing. The department botanist, Richard Holm, provided an inspiring foundation in classical evolutionary biology, which was complemented by courses in geology and paleontology. I particularly enjoyed a directed study on the Miocene environment of coastal California—my first real research project—and nearly became a paleontologist. I also remember my first exposure to ecology in a course taught by marine biologist Don Abbott, using Eugene Odum's *Fundamentals of Ecology* text. This would have a lasting impact on my career.

Although in the early 1960s most biology majors were going on to medical school, I never questioned that I would continue my training in some area of natural history. When it came time to think about graduate schools, I asked recently hired Assistant Professor Paul Ehrlich where I should apply. In his mind, my only real choice was to work with Robert MacArthur at the University of Pennsylvania. I had not heard of MacArthur, but I followed Paul's advice, was accepted into the graduate program, and headed east to graduate school. This was the most amazing experience. MacArthur and E. O. Wilson had just published their "Equilibrium Theory of Insular Zoogeography" in *Evolution*, and MacArthur was playing a central role in the resurgence of community ecology and the emerging field of evolutionary ecology. Along with fellow grad students Mike Rosenzweig, Martin Cody, and Henry Hespenheide, we met in MacArthur's office every day for lunch, with frequent visitors including Evelyn Hutchinson, Richard Levins, Egbert Leigh, and Larry Slobodkin, among others. The conversations were mind boggling.

Because of MacArthur's interest in island biogeography, and the presence of James Bond—the expert on West Indian ornithology (and, yes, the namesake of Ian Fleming's hero)—at the Academy of Natural Sciences downtown, I began a project on Caribbean birds inspired by Wilson's ideas about taxon cycles drawn from the distributions of Melanesian ants. However, MacArthur viewed such historical analysis as natural history without testable hypotheses, and thus unsuitable to a doctoral dissertation, so islands were pushed to the back burner. My first summer as a graduate student was spent in the

Sonoran Desert in Arizona, where I stumbled through some projects (best left undetailed) until fellow bird watcher Will Russell steered me toward the nesting biology of some of the local birds. This led to my first paper, "Brood Reduction in the Curve-billed Thrasher," published in 1965 in *The Condor*. What an accomplishment! I became hooked on breeding biology, life histories, and the physiological ecology of desert birds, particularly the Cactus Wren. I convinced a fellow grad student, Reed Hainsworth—who worked on the neurobiology and physiology of temperature regulation in rats—to accompany me to the field in 1966. Reed introduced me to experimental science, and I introduced him to birds.

In 1965, MacArthur moved to Princeton University, but I remained at Penn, where it was my good fortune to be taken under the wing of avian ethologist W. John Smith. Along with others in the department, John got me through my doctoral dissertation defense and off on a postdoctoral year at the Smithsonian Tropical Research Institute in Panama, with mentors Neal Smith, Stan Rand, and Mike Robinson. My only other experience in the tropics had been an eye-opening month on Jamaica collecting plants (mostly bird watching on my part) with Henry Hespenheide. Like many others, I was hooked on the tropics, and diversity has remained a theme in much of my work.

I joined the faculty at Penn as an assistant professor in 1968, advancing through the ranks there but moving to the University of Missouri–St. Louis in 1995 so that my wife, Susanne, and I could have positions in the same department. Her subsequent move to the University of Munich turned us into transatlantic commuters but also gave me the opportunity to interact more with European colleagues. I won't dwell on my academic career over the years, which has been very rewarding as a life experience. People—colleagues, postdocs, students—have been most important to me, and I would be remiss not to mention some important early influences.

On returning to Penn, I found that Frank Gill had recently been hired as curator of birds at the Academy of Natural Sciences, and we enjoyed many years together. In 1969, in Arizona, I ran across George Cox, who was a professor at San Diego State University. He took a sabbatical year with me at Penn and we revived the West Indian taxon cycle idea. However, some ecologists took a dim view of the concept at the time, and I shifted my efforts to seabird growth and development to test some of David Lack's ideas about slow growth in pelagic seabirds. This interest led to fieldwork on many remote islands, from Antarctica and Bird Island, South Georgia, with the British Antarctic Survey (and John Croxall and Peter Prince), through the tropical Pacific, often with friends and colleagues Ralph and Betty Anne Schreiber, to the North Atlantic and many seasons of research on Leach's Storm Petrel at Bowdoin College's Kent Island Biological Station, including stimulating collaborations with Joe Williams and Matthias Starck.

The 1980s was a bleak period for community ecology, but the challenge of understanding patterns in biodiversity continued to tug at me. When the distinguished limnologist Ruth Patrick invited me to write an article on ecological communities for *Science* (1987), I sought a balance between local and regional processes in shaping patterns of diversity. This seemed to strike a harmonious chord with many ecologists, and soon I was editing a book with Dolph Schluter on historical and geographical perspectives on communities, which I believe helped to integrate ecology, biogeography, and evolution. At the same time, DNA sequencing was becoming widely available, and I teamed up with Eldredge Bermingham to confirm my earlier insights with George Cox about taxon cycles, using phylogeographic approaches. At about the same time, gerontologist Caleb Finch contacted me

about teaming up on a demographic analysis of aging in birds, bringing me back to life histories and comparative biology.

I look back on this kaleidoscope of influences and ideas with some amazement, glad that my curiosity about life has not dimmed, and happy to have grown up in natural history and science with encouragement to think broadly about nature from some absolutely amazing people.

BOX 5.2 Kate Lessells

I don't know where the enthusiasm for birds came from, but when asked aged eight what I would like as a pet, I plumped for a pair of Budgerigars. The small cage that they lived in was quickly superseded on my ninth birthday by a (secondhand) outdoor aviary, more Budgerigars, and eventually a pair of Cockatiels, all of which were allowed to breed in the summer months. The Budgerigars also prompted the purchase of the *Observer's Book of Birds*,

and in the wake of my disappointment that the Budgerigar didn't figure, I set about trying to identify some of the birds in the garden. Through the Young Ornithologists' Club I got to know Tom Kittle, who trained me to ring birds, and I subsequently became involved in the Wash Wader Ringing Group, run with gusto by Clive Minton. Having fallen in with the wader crowd, I spent the summer holiday between school and university traveling to Morocco in a Land Rover and mist-netting waders in salt pans on the Atlantic seaboard, and I was later involved in wader-catching trips to Norway and prerevolutionary Iran.

The book that inspired me when I was at school was David Lack's *Natural Regulation of Animal Numbers*, chosen more or less at random (although the puce cover of the paperback must have helped) from the shelves of Foyles bookshop. It introduced me to the idea that life history traits—in the case of birds, clutch size—have been shaped by natural selection in the same way as other biological traits, and was also responsible—along with the then current popularization of Niko Tinbergen's work on gulls—for me choosing to study zoology at Oxford. As it turned out, Lack died between me obtaining a place at Oxford and going up as an undergraduate, and Tinbergen retired at the end of my first year. Looking back, this was the changing of the guard in which population ecology and animal behavior morphed into behavioral ecology. Being at Oxford in the '70s and early '80s was akin to being Obelix dropped into the cauldron of magic potion, only this time it was a heady mixture of behavior, ecology, and evolution. I have never quite managed to escape the lure of behavioral ecology.

I stayed on in Oxford to carry out a DPhil under Chris Perrins's supervision at the Edward Grey Institute of Field Ornithology. I was interested in clutch size in birds that don't feed their young: if, as Lack suggested, the clutch size of many bird species is determined by their ability to feed their offspring, what happens in bird species without parental provisioning? I wanted to carry out experiments manipulating brood size at hatching, so I chose to work on Canada Geese: they were plentiful to the point of being regarded as pests in some areas, and Clive Minton (again) organized an annual roundup of molting Canada geese and their young in the Midlands so I could follow the survival of my web-tagged goslings. In contrast, it was not so easy to study the behavior of Canada Goose families, and I was lucky to have the opportunity to study Snow Geese at Fred Cooke's La Pérouse Bay study colony in my third summer of goose fieldwork. My DPhil work was followed by a postdoc while a prize fellow (the first woman fellow) at Magdalen College, Oxford. The original aim was to study clutch size in Pied Avocets in the salt pans of the Camargue as a contrast to the geese, but I ended up working on Kentish Plovers instead, which turned out to have a fascinating mating system: they

are one of a handful of bird species in which either sex of parent can leave their partner to care for the brood and remate to produce another family.

After nine years I fledged from Oxford and had a three-year interlude from birds at Imperial College at Silwood Park as Mike Hassell's postdoc. The main aim was to study the population dynamics of a small seed-eating beetle, but I managed to infiltrate some behavioral ecology into the work while also learning how relatively simple mathematical models can be used to answer ecological and evolutionary questions.

From Imperial I moved to a lectureship at the University of Sheffield, taking the beetles with me but also moving back to birds. John Krebs and Mark Avery had started a study on helping at the nest in European Bee-eaters a few years before, and with the help of Terry Burke, I took this over at the stage at which we were carrying out one of the first (minisatellite) DNA fingerprinting studies. The DNA profiles allowed us to determine the degree of relationship between helpers and the nestlings they aided, in particular to show that they were not ever the parents of these chicks but at the same time were (almost) never unrelated. Both of these negatives would have been impossible to demonstrate without the molecular data. My closest colleague in Sheffield was Tim Birkhead, who was moving into sperm competition from his earlier seabird studies, and he got me involved in mathematically modeling the process. Our models enabled us to identify which mechanisms were most likely to account for the patterns of sperm precedence observed.

The molecular work led to the offer of a job from Arie van Noordwijk at the Netherlands

Institute of Ecology. The institute was about to build its first molecular lab, and Arie was willing to appoint a permanent molecular assistant, which was an attractive proposition compared to the need to gain a series of short-term grants. We developed RAPD markers for sexing and were one of the first groups to be able to sex newly hatched chicks. Oxford is known for its long-term studies on Great and Blue Tits, but the Netherlands Institute of Ecology hosts an even older study (which Lack emulated in starting the Wytham study). Beginning with the sex ratio of their nestlings, I began studying various aspects of parental investment in the Great Tits. I'm particularly interested in the role of various evolutionary conflicts of interest within the family. The Great Tits' ready use of nest-boxes makes them an amenable study system, and mathematical modeling provides an important complement in understanding the evolutionary outcome of sexual and parent-offspring conflicts of interest. Data collected as part of our experimental studies have uncovered a remarkable change over the years in the relative contribution of the sexes to parental provisioning. This forms one piece of the overall picture of how the birds are responding to environmental change and shows that, despite being one of the best-studied species in terms of parental care, much remains to be discovered and understood.

CHAPTER 6

Form and Function

Ornithologists . . . became leaders in various branches of biology, ranging
from the new systematics and speciation research to endocrinology
and behavior. . . . It would be of interest to the historian to take a closer
look at the progress in various areas of ornithological research.

—ERNST MAYR, IN STRESEMANN'S (1975: 382) *ORNITHOLOGY:
FROM ARISTOTLE TO THE PRESENT*, EXPLAINING WHY BIRDS ARE
IDEAL STUDY ORGANISMS IN MANY FIELDS OF BIOLOGY

EXTRAORDINARY ANATOMIES

IN NOVEMBER 1998 KEVIN MCCRACKEN traveled from Louisiana State University to Argentina to collect specimens for a study of duck systematics. Impressed by the size of the male Lake Duck's cloacal swelling, Mc-Cracken dissected out the phallus from the little pouch in which it is stored when not in use and was overwhelmed by its size and bizarre spiny appearance. At 20 centimeters (8 inches) long, the penis of this bird is larger than that known in any other duck. Describing his findings in *The Auk*, McCracken speculated that the Lake Duck's phenomenal phallus may have evolved through sexual selection and might serve to remove the sperm of rival males from within the female's reproductive tract (McCracken 2000). The paper excited considerable attention both from the media and other scientists.

But, as McCracken told us:

In these specimens, the phalluses were not everted and I was still very curious what the naturally everted phallus looked like in the live (or dead) duck . . . the Auk paper photos were from phalluses that had been dissected and turned inside out. . . . Three years later, in April 2001 I shot [another] Lake Duck . . . at some distance with a rifle and had to wait for it to blow across the lake. It was raining pretty hard and we just stayed in the car for a while. We didn't want to draw too much attention either as we didn't know whose farm we were on. . . . When I first picked it up I thought I had blown out a part of its intestine. But then to my surprise I realized we had a fully everted phallus and I ran back to the car to tell my soon to be graduate student Rob Wilson the good news.[1]

Razorbills and Common Murres underwater. Painting by David Miller.

The everted phallus on this specimen measured 42.5 centimeters (16.7 inches)—longer than the duck itself and longer than the phallus of any other bird, including the ostrich—a record that merited a further publication (McCracken et al. 2001).

It has been known since Aristotle's time that male waterfowl are unusual among birds in possessing a phallus. From Aristotle to the Renaissance, there was little more than passing interest in duck penises, but by the seventeenth century anatomists such as Claude Perrault (1680; cited in Cole 1944), William Harvey (Whitteridge 1971), and others were excited by the discovery that the males of several birds, notably ostriches, other ratites, and screamers, also possess a penis (King and McLelland 1981).

The most detailed account comes in 1914 from the German anatomist Walther Leibe, who described the manner in which the mallard's non-erect penis lies coiled and invaginated (that is, turned inside out) within a thin peritoneal sac adjacent to the cloaca and how, on becoming erect, it is folded outward—like the finger of a glove, albeit a spiral-shaped one.[2] Leibe assumed that erection of the phallus occurs inside the cloaca of the female, but because of the speed with which copulation occurs, he doubted whether the phallus is always fully erect during mating (King 1981).

Why only certain groups of birds possess a penis has puzzled ornithologists for a long time (Montgomerie and Briskie 2007). Within waterfowl, however, the relative size of the phallus seems to be explained by the intensity of postcopulatory sexual selection (chapter 9). A comparative study revealed that phallus size varied markedly across different species of waterfowl, with relative phallus size covarying with relative testis

Argentine Lake Duck with its extraordinarily long penis extended.

size (Coker et al. 2002). Since relative testis size reflects the intensity of sperm competition, Christopher Coker and colleagues (2002) concluded that a relatively long and often spiny phallus had evolved in response to sperm competition. Waterfowl are renowned for the incidence of forced extra-pair copulation (Huxley 1912a; McKinney et al. 1983)—copulations forced on the female by males other than their mate—thus, sperm competition is expected to be particularly intense in some species in this group.

The focus of both McCracken's and Coker's waterfowl work was on males, but one of us (TRB) wondered what the implications of a relatively enormous phallus were for females, and together with Patricia Brennan initiated a comparative study of female reproductive tracts. The results were completely unexpected. In all previously studied birds the vagina is a simple, tubelike structure, but in certain waterfowl the vagina is extraordinarily complex, with up to three blind-ending side branches and a spiral structure adjacent to the uterus. In all species, the vagina spiraled in a clockwise direction, whereas the erect phallus always spiraled counterclockwise. Waterfowl species in which the male has a long phallus have a correspondingly complex vagina—assumed to have coevolved through an arms race—allowing females to retain some control over fertilization in the face of forced inseminations (Brennan et al. 2007). Brennan and colleagues (2010) then conducted some experiments to see, first, how insemination was achieved and, second, how the spiral structure of the vagina might affect insemination. High-speed video recording and a transparent (glass) model vagina confirmed Leibe's idea that eversion occurs within the female's vagina and does so explosively. As the male's cloaca touches the female's, the phallus is everted into the vagina within 0.36 seconds. Confronted, however, by a vagina spiraling in the opposite direction to the penis, penetration was almost completely prevented (Brennan et al. 2010). Obviously, such experiments are relatively crude, but they provide additional insight not only into the process of insemination in ducks but also into the coevolution of male and female reproductive anatomy.

This brief account of history of waterfowl genitalia is similar to the history of avian anatomy as a whole; characterized by successive phases of description, speculation about function and, eventually, by experimental tests of hypotheses.

Descriptive knowledge of avian anatomy accumulated slowly, beginning mainly in the Renaissance (Cole 1944). Indeed, one of the oldest branches of ornithology is the study of external and internal anatomy, a field that reached its peak during the nineteenth century. As Jürgen Haffer (2001) said, the aim of these anatomical studies was twofold, to establish a "natural system" of classification and to understand the function of particular structures and organs. Walter Bock and Gerd von Wahlert (1965) reflected on this: "The studies done in this great period [nineteenth century] of classical comparative anatomy constitute one of the major contributions, to date, by morphologists to evolutionary theory."[3] However, they also felt that the tendency for researchers to consider "form" independently of "function" was a serious obstacle when using anatomy to reveal evolutionary relationships.

The nineteenth century heyday of avian anatomy was celebrated in the monumental volumes by Max Fürbringer (1888) and Hans Gadow (Gadow and Selenka 1891). Following their publication, it was almost as if the last word on the topic had been said, and there was nothing new to discover; by the beginning of the twentieth century the study of avian anatomy had virtually ground to a halt (Haffer 2001, 2006b). As Charles Sibley (1955) noted, "Descriptive anatomy languished somewhat"[4] after the turn of the century, despite the efforts of a few twentieth-century stalwarts like Robert Shufeldt, William Beecher, Dietrich Starck, William Pycraft, Robert Storer, and Harvey Fisher. Indeed, by the mid-1950s researchers were lamenting the fact that anatomical studies were out of favor. Here is Harvey

Fisher: "The 'modern' trend in biological sciences seems all too often to imply that 'anatomy as such' may be overlooked in the evolution of the 'better and more accepted' avenues of approach to biological problems."[5]

Given the general lack of interest in descriptive anatomy during the mid-twentieth century, it is remarkable that Anthony S. King and John McLelland were able to persuade Academic Press to publish their edited *Form and Function in Birds* (1979–89), a collection of papers in four volumes, which at the time seemed as though *they* might be almost the last word in avian anatomy. Much of the anatomical (and physiological) work in these volumes had its origins in studies of poultry and was thus driven by commercial interests. Sadly, little of this vast store of information made it across the conceptual gulf that kept poultry biologists and ornithologists apart, yet studies of poultry anatomy (and physiology)—for example, in reproduction—later became the foundation, if not the inspiration, for much subsequent ornithology, exemplified by the study of avian phalluses described above. Moreover, as that small but exciting body of work on waterfowl genitalia illustrates, structural and "mechanistic" aspects of avian biology are once again back on ornithologists' research agendas, especially when linked with other areas of biology, such as sexual selection theory.

COMPARATIVE PHYSIOLOGY

Research on avian form and function exhibits clear trajectories through the twentieth century. The study of form was at its peak in the nineteenth century and declined rapidly through the first few decades of the twentieth century. The study of function, or comparative physiology, began only in the 1930s, and it peaked initially between the 1950s and the 1970s (Dawson 1995; Haffer 2006a).

Physiology—the study of how the body works—covers a huge range of topics, including body systems, metabolism, the senses, neurobiology, and endocrinology.[6] Prior to 1930 most physiological research was conducted on mammals, "with particular emphasis on human physiology and its relationship to medicine. By comparison, the physiology of birds [was] neglected."[7] So wrote Paul D. Sturkie in his book *Avian Physiology* (1954), the first single volume to deal with the physiology of birds. One reviewer (Branion 1966) described Sturkie's book as a "gift from the gods," emphasizing the need for such a volume. The book dealt "mainly with the chicken, the duck and the pigeon, because most of the research has been conducted on these species and they represent species of economic importance to man."[8] Subsequent editions, edited (rather than written) by Sturkie—and later others—incorporated increasing amounts of information on nondomestic species.

The rise of avian physiology in the 1930s (e.g., Stresemann 1927–34) was part of the general increase in research on comparative physiology, which in turn was driven by an increase in the experimental approach to physiology and a bias toward medical research. The subsequent decline in avian physiology in the 1970s was the result of several factors. First, there was a perception that most basic physiological problems had already been solved. Second, the field of physiology had become more biochemical, and later more molecular, and it was here that most research funding was eventually

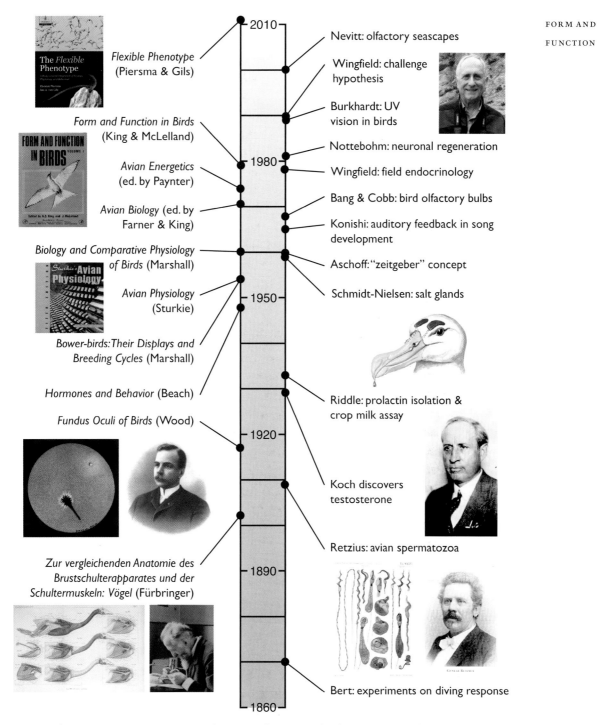

TIMELINE for FORM AND FUNCTION. Left: Covers of Piersma and Gils (2011), King and McLelland (1979), and Sturkie (1954); fundus of a kiwi, by Casey Wood; Casey Wood; musculature of a Greylag Goose by Fürbringer; Max Fürbringer examining a specimen. Right: John Wingfield; location of salt glands in an albatross; Oscar Riddle; bird spermatozoa by Retzius; Gustav Retzius.

directed. Physiology continues to thrive as a discipline within biomedical or molecular biology departments at universities, but mainly through research at a molecular level so that "everyone these days is a molecular biologist, a molecular geneticist or is so specialised they would not recognize an animal if they saw one."[9] In contrast, in zoology departments in Britain and North America, comparative physiology has all but disappeared from most undergraduate programs, and research funding for traditional, lab-based comparative physiology is difficult to obtain (Wagner and Paterson 2011).[10]

William Dawson's (1995) overview of avian physiology up until the mid-1990s, published in *The Condor*, identified five major areas of physiological research, which in descending order of abundance in that journal were (1) annual cycles (20 percent of *The Condor* papers); (2) temperature regulation (15 percent); (3) energetics (10 percent); (4) water and electrolyte balance (less than 10 percent); and (5) eggs and embryos (less than 10 percent). This ignores—just as it does for the study of "form" or anatomy—the mountain of information collected by poultry researchers since the 1950s. Ernst Mayr may have identified the reason why the work of poultry biologists (reviewed by Sturkie 1954 and King and McLelland 1979–89) has been largely ignored by ornithologists: he said this research failed to produce any "biologically significant generalisation."[11] While true, this is slightly unfair, since the motivation of the two groups of scientists is so different: one's is commercial, the other's is academic. Another possibility is that ornithologists—naively—felt that domesticated birds under strong artificial selection had little relevance to the study of wild birds.

It is impossible for us to cover the entire breadth of avian physiology in this chapter. Instead we focus on three main topics, two of which Mayr (1975: 385–86) identified as especially "interesting and important": (1) physiological adaptations to harsh environments and (2) the annual cycles of birds. The study of annual cycles is unusual and of special significance, since it is one of the few areas of comparative physiology where research has continued to flourish to the present day. The third topic we cover here—sensory physiology—is a long-neglected area of research that started to develop only during the 1970s and is now, we feel, bristling with potential.

PHYSIOLOGY IN HARSH ENVIRONMENTS

In 1670 the chemist and physicist Robert Boyle tested the commonly held belief that "Nature have furnished ducks and other waterfowl with a peculiar structure of some vessels about the heart, to enable them, when they have occasion to dive, to forebear for a pretty while respiring under water without prejudice."[12] With rather less concern for his birds than is usual (or allowed) today, Boyle experimented by attaching weights to the birds' feet to see how long they would survive when submerged. Boyle reported that ducks seemed to survive for longer than hens. This result was confirmed by the French physiologist Paul Bert two centuries later, in 1870, when he recorded that ducks held underwater survived for fifteen minutes whereas hens lasted only three minutes—a difference that appeared to be linked to a difference in physiology. Bert noticed that when the ducks were submerged their heart rate slowed dramatically—a phenomenon known as "bradycardia" and now known to be typical of all diving vertebrates—and also that ducks have more blood than hens and hence can hold more oxygen in their bloodstream (Bert 1870, quoted in Andersen 1966). He then conducted an ingenious but brutal test, bleeding ducks until they had the same blood volume as a hen and then comparing how long they survived submersion. The difference between the two species disappeared, and Bert concluded that the greater blood volume—and concomitantly larger oxygen store—was what allowed ducks to remain submerged for so long. Ingenious, yes, but as we'll see, incorrect.

It was Bert's student, Charles Richet, who proved him wrong, showing that bradycardia, not blood volume, accounted for the superior survival of submerged ducks. Cutting the vagus nerve prevented the slowing of the heart, and as a result the submerged ducks rapidly expired. Richet also recognized that bradycardia was a reflex, triggered by the duck's head coming in contact with water (Richet 1894). The study of diving birds and mammals was picked up again in the 1930s by Laurence Irving and his students, including Per Scholander, whose 1940 monograph was described by his colleague Harold Andersen as "the most comprehensive and important single effort in the field of diving physiology."[13] Scholander showed that the diving reflex also occurs in humans. During the 1930s, Scholander worked on seals and diving birds in Greenland and Spitzbergen, intrigued by the way they were able to avoid the "diver's disease"—the bends, or decompression sickness, that humans experience after coming to the surface too quickly when diving to similar depths. To explore this phenomenon he devised ways to continuously

record both oxygen consumption and carbon dioxide and lactic acid production in animals underwater. The results were remarkable: while oxygen levels in the blood of submerged animals were higher than in nondiving species, they were not sufficient to allow aerobic respiration. A decrease in aerobic respiration was accompanied by bradycardia and selective vasoconstriction, with blood diverted away from the peripheral circulation toward the vital organs—the heart and brain—such that the muscles started to function anaerobically, resulting in an "oxygen debt" and a buildup of lactic acid. Because Scholander and Irving arrived independently at similar conclusions at around the same time (in the late 1930s and early 1940s), this physiological response to diving is often referred to as the "Scholander-Irving classic response."

When Scholander (and others) observed the relatively short dives of free-living Macaroni, King, and Gentoo Penguins in Antarctica, he realized that they were diving within their aerobic capacity and thus that the "classic response" could not be the whole story. Technological advances in the 1970s allowed researchers to use radio transmitters fitted with sensors to relay information about heart rate, blood flow, and blood pressure, confirming that voluntary dives were much shorter than forced dives, that bradycardia was either absent or very low, and that blood pH changed little during the dives; the birds remained within their aerobic limits. Michael Millard and colleagues (1973) used radiotelemetry to measure cardiovascular responses to natural dives in Gentoo and Adelie Penguins and suggested that diving physiology was "a composite of diving and exercise response."[14] That is, the birds were balancing their physiological responses between the classic bradycardia response and their typical response to exercise in air by an increased heart rate, increased blood perfusion to the muscles, and increased oxygen consumption. Millard and colleagues suggested that during short, natural dives the balance was tipped toward the exercise response (with the possible exception of the Emperor Penguin, which naturally dives for long periods), but if birds were *forcibly* submerged, the balance tipped more toward the classic response—that is, diving birds had a safety mechanism that enabled them to stay underwater if necessary. Subsequent work on Tufted Ducks and other species provided strong support for this idea (e.g., Woakes and Butler 1983).

Gerald Kooyman started to work with Scholander in the 1960s, miniaturizing devices developed for whales and seals for use on birds, to see how deep they dived. Recorders placed on Emperor Penguins (body mass 25 kilograms or 55 pounds) at Cape Crozier in the Antarctic revealed the astonishing fact that some individuals dived to depths of 265 meters (869 feet) and could remained submerged for as long as eighteen minutes—far beyond their aerobic limits. Later studies of this species recorded a remarkable dive of 564 meters (1,850 feet) and—in a different individual—a duration of twenty-seven minutes (Sato et al. 2011). Even the much smaller Thick-billed Murre (body mass 1 kilogram or 2.2 pounds) has been recorded as deep as 210 meters (689 feet) (Croll et al. 1992). Technological advances continued with the production of smaller devices, for use on smaller species, and devices that can record dive profiles for species like Adelie Penguins and European Shags. More recently, small digital cameras—"crittercams"—attached to larger species, such as Emperor Penguins, allow researchers to actually see what birds are seeing when they dive (e.g., Takahashi et al. 2004).

Instruments that allowed internal measurements were a huge step forward. Pat Butler and Tony Woakes first pioneered implantable data loggers in the late 1970s, using Tufted Ducks and Common Pochards (1979) and Barnacle Geese (1980). Unlike previous loggers attached to the external body surface, these were placed inside the abdominal cavity through a small incision in the brood patch, so that reliable measurements of temperature, heart rate, and respiratory frequency could be taken. As the technology has improved, these implanted loggers have become much more sophisticated, enabling researchers to monitor dive depth and duration (e.g., Green et al. 2003).

As Kooyman told us:

Using the Emperor Penguin as the premier model, numerous investigators with a variety of devices and protocols, have refined the concept of the Aerobic Diving Limit (ADL), which initially was based on the rise of peak lactate at the end of dives of a consistent duration. There has been much speculating about the cardiovascular controls necessary to manage the oxygen stores of the Emperor Penguin. At one point the problem seemed impossibly complex until Paul Ponganis and his students . . . devised the instruments and procedures necessary to attack the problem of oxygen management in free diving birds. In a series of papers they have shown that myoglobin is the key component of the store, an extremely reduced heart rate is critical to sequestering the store, and an exceptional tolerance of the brain and heart to low oxygen levels is essential for full utilization of the store. The orchestration of all these and other elements have resulted in a breakthrough to our understanding of how birds make their exceptional dives.[15]

Life at sea generates another physiological challenge: an excess of salt. Seawater contains about 3 percent salt, but the body fluids of birds and other vertebrates are only about 1 percent salt, so ingesting seawater results in the osmotic loss of water and thus dehydration. We humans are particularly susceptible because our kidneys are relatively inefficient, especially compared to those of whales and seals, which can produce very concentrated urine. Birds' kidneys are also incapable of dealing completely with ingested saltwater, so marine birds have evolved an additional way of handling the problem, whereby excess salt is excreted through nasal glands. These glands, which are located in a depression in the skull above each eye's orbit, were first described in 1665 (Peaker and Linzell 1975), but their function remained unknown until the 1950s. The first account in English came from the British surgeon George Bennett: "In 1828 when dissecting the head of the Albatross, I observed embedded in a bony cavity, situated immediately over the orbit, a gland." He goes on to describe its structure, finishing by saying, "An interesting subject next for inquiry is, what this gland secretes, and what is its use in that situation? which at present cannot be answered."[16]

In his *Dictionary* Alfred Newton says that the function of the nasal glands is to "moisten and cleanse the mucous lining of the nasal cavities."[17] It was also known in Newton's day that while these glands occur in all birds, they are larger in marine species,

promoting the ingenious explanation that the glands had a protective function. By the 1930s it was quite clear that " . . . from the fact that the glands as a rule are better developed in marine than in freshwater and terrestrial species, it seems that an important function of the gland is to protect the lining of the nasal cavity from the effects of sea water."[18]

The true function of nasal glands—or "salt glands," as they called them—was discovered by Knut Schmidt-Nielsen, Carl Barker Jorgensen, and Humio Osaki in Double-crested Cormorants in the 1950s. They asked how seabirds function without fresh water: do they drink seawater as ornithologists usually assume (the answer was yes), and if so how do they cope? Their work was part of a rise in studies of salt and water metabolism fueled by the newly discovered flame photometer method of measuring salt concentrations in the 1950s.

Suspecting that the nasal glands were involved in the excretion of salt, Schmidt-Nielsen and colleagues started in the most obvious way, making a Great Black-backed Gull ingest (via a tube) a large volume—134 milliliters (4.5 ounces)—of seawater and recording what was excreted from the salt glands. Within three hours the gull had eliminated 90 percent of the salt through its salt glands, expelling drips and dribbles of concentrated (5 percent) salt solution far more rapidly than was possible to process through the kidneys (Schmidt-Nielsen et al. 1957). Further studies on a range of other bird species revealed that, in contrast to the kidneys which function continuously, the salt glands do so only intermittently, when there is salt to excrete (Schmidt-Nielsen 1960). Salt glands occur in all birds but are functional only in marine species and a small number of terrestrial birds.[19] They also occur in reptiles, notably marine iguanas (Peaker and Linzell 1975).

It has been recognized for millennia that the vision of birds, especially that of falcons and eagles, is much better than our own. In the 1940s the ophthalmologist André Rochon-Duvigneaud (1943) coined the phrase "a wing guided by an eye" to capture what he considered to be the essence of being a bird. The basic anatomy of the vertebrate eye was described in the seventeenth century, part of the broader investigation into the nature of light itself by Johannes Kepler, René Descartes, and others. Also in the 1600s, William Harvey recognized that the excellent vision of birds was the result of their relatively large eyes compared with those of mammals—the larger the image projected onto the retina at the back of the eye the better, just as with a TV screen. The insight of Renaissance researchers was often remarkable. The Irish philosopher William Molyneux, for example, provided an answer to something that had puzzled Descartes: why it is that we see an image "the right way up" when it is so obviously projected upside down on the retina. Molyneux provides the answer by saying, "The eye is only the organ or instrument, 'tis the soul that sees by means of the brain."[20] In other words, while the eye records the image, it is the brain that interprets it. Research in the twentieth century confirmed this.

The excellent vision of diurnal raptors is accounted for by a very high density of light receptors—rods and cones—in the retina, but especially cones. They also have two fovea, a fact discovered by another ophthalmologist, Casey Albert Wood. During the first quarter of the twentieth century, Wood, who was based at the University of Illinois, was among the most eminent of eye specialists. He was also interested in birds, and with the same ophthalmoloscope he used on his human patients, he made some fundamental discoveries about avian vision, including the presence of two fovea in raptors and a linear, horizontal fovea in certain seabirds (Wood 1917).

The basis for color vision in humans was first investigated in the 1600s with Isaac Newton's studies of the properties of light and Leeuwenhoek's microscopic studies of the retina. But it was not until the early 1800s that the properties of rods and cones began to be studied and theories of color vision developed. In combination, our red-, green-, and blue-sensitive cones provide the full spectrum of color visible to us. These same three cone types also occur in birds, and for a long time it was assumed that their visual system was much the same as our own with respect to the colors they can see. Then, in the 1970s, UV-sensitive cone types were discovered in birds. It had been known since the 1880s that insects could detect UV light, but it was assumed that this was a private communication channel inaccessible to predators such as birds. Current evidence suggests that UV sensitivity occurs across a range of animal taxa, but because we lack the ability we have always assumed it to be rare. Once it was known that pigeons and hummingbirds were sensitive to UV light (Wright 1972; Huth and Burkhardt 1972), UV vision was soon found to be widespread in birds (Bennett and Cuthill 1994) and that it is used to choose food and mates (chapter 9). Some fruits reflect UV light—as does the urine of small mammals, which allows raptors to track them—and UV reflectance can play an important role in mate choice (Hill and McGraw 2006). The way birds see each other's colors and the way birds use ambient light to enhance their own colors, especially in relation to mate choice, resulted not only in renewed interest in the anatomy and physiology of vision but also in the ultrastructure of feathers (chapter 1).

One of the most remarkable twentieth-century discoveries relating to vision is the fact that birds appear to use their left and right eyes for different tasks. Pierre Broca, a physician, was the first to discover that the two sides of the human brain process different types of information, but it was assumed that this "lateralization" was unique to humans. In the 1970s Fernando Nottebohm discovered that it wasn't.

This particular story starts with Fred Koch and his student Laura McGee, who discovered and identified the hormone testosterone in bulls' testicles in 1927 (McGee 1927). Just eight years later biochemists had synthesized testosterone, and dishonest canary breeders were injecting it into female canaries—which do not normally sing—in order to pass them off as singing males. The fact that testosterone did indeed cause females to burst into song—for a few days—was later confirmed experimentally (Baldwin et al. 1940).

What did testosterone do to a female canary to make it sing? The answer to this question, which simultaneously addressed the more fundamental question of what triggers the springtime song in male Atlantic Canaries (and other birds), emerged from the extraordinary work of Fernando

Nottebohm in the late 1960s. Born in Buenos Aires in 1940, Nottebohm studied zoology at UC Berkeley, where he met Peter Marler, a British-born ethologist (chapter 7) who had moved to America in the 1950s. Nottebohm conducted his final year honors project with Marler—his "scientific father," as Nottebohm called him—studying the way the rooster syrinx produces its characteristic crowing.

Encouraged by his undergraduate results, Nottebohm remained with Marler to conduct his PhD research, looking in more detail at the way the syrinx worked. As part of his PhD studies, Nottebohm spent a year (1964–65) in Cambridge with Bill Thorpe (while Marler was in Africa studying monkeys). While there he conducted an experiment designed to understand how, once a Common Chaffinch had acquired its song, that song remained so stable. In performing this experiment, which involved cutting the nerves that run from the right and left sides of the syrinx to the right and left sides of the brain, respectively,[21] Nottebohm made a remarkable discovery. If the nerves on the left-hand side were cut, certain elements in the birds' songs were lost or modified. In contrast, cutting those on the right had almost no effect (Nottebohm 1971). What this meant became clear only later, when Nottebohm—now studying Atlantic Canaries and White-crowned Sparrows—showed that song production required input mainly from the left side of the brain (Nottebohm and Arnold 1976). This was exactly the same pattern found with human speech—an extraordinary and exciting discovery.

Flexibility is another feature of the human brain—if the left side is damaged the right side can sometimes compensate. Nottebohm asked whether the same was true in birds, and by cutting the twelfth cranial nerve—the hypoglossal—leading to the syrinx in chaffinches at different ages, he was able to show that the right side of the brain compensates if this operation is performed while the bird is young and has not yet acquired its song. The birds are later able to sing normally (Nottebohm 1971, 1976).

Nottebohm also demonstrated a subtle link between the testes, testosterone, the syrinx, and song. While in Cambridge with Thorpe, he removed the testes from a young male chaffinch whose song was still incomplete. The bird stopped singing, as predicted. Remarkably, however, when Nottebohm injected the bird with testosterone two years later—long after a wild chaffinch would have finalized its song—this bird began to sing. The implications of this experiment were enormous: it meant that the critical period for song learning—a sensitive period (chapter 7)—was determined by the bird's neural development rather than by its age, as everyone had previously assumed (Nottebohm 1969).

After his year in Cambridge with Thorpe, Nottebohm returned to the United States to complete his PhD, after which, in 1967, he took an assistant professor position at Rockefeller University. Over the next few years he continued mapping the bird's brain, identifying two major regions responsible for song: the higher vocal center (HVC), essential for song learning, and the robustus archistriatalis (RA), responsible for song production (Nottebohm et al. 1976). Effectively, as he told us, "RA is the pit where the orchestra is and the programme in the HVC tells the musicians what to play during every millisecond of song."[22]

212   During this work it became obvious to Nottebohm that there are major differences between the brains of male and female birds. As he cut slices of brain for histological examination, he could tell the difference between male and female brains with his naked eye: holding the microscope slide up to the light it was obvious that both the HVC and the RA were larger in males (Nottebohm and Arnold 1976). It was this greater volume of neuronal tissue in the male brain that enabled males to acquire and perform their songs.

Perhaps the single most important of Nottebohm's remarkable series of discoveries was that neurons in the canary brain are able to replace themselves—a process called "neurogenesis." This had been suggested in the 1960s by Joseph Altman but, with little evidence to support his claim, the idea was dropped.[23] In a series of pioneering experiments in the 1980s, Nottebohm discovered that male canaries' brains changed throughout the year, losing old neurons and gaining new ones. "What my lab did was to provide incontrovertible evidence that new neurons were formed and added to existing circuits of the song system."[24] Delving further, Nottebohm and his researchers found that these changes in the brain tracked changes in testis size and testosterone production. This also explained why female canaries given a shot of testosterone were able to start singing: they too showed neurogenesis in response to the hormone (Nottebohm 1981). Prior to the 1980s, it was widely assumed that the number of neurons in the brain was fixed and that they were irreplaceable: that once a neuron was damaged or died, it had gone forever.

Nottebohm's research had far-reaching implications, not just for ornithology but for the entire field of neurobiology, overturning the long-held belief that neurons were incapable of growth and regeneration. "What began with the song of a small bird had changed an entire paradigm in neuroscience."[25] These findings created initial debate over whether the same phenomena could occur in mammals' brains, with mammalian neurobiologists skeptical that Nottebohm's findings had any relevance to their work. Rather, the evidence for avian neurogenesis was viewed as an exotic specialization related both to the necessity for flying creatures to have light cerebrums and to their seasonal cycles of singing.[26] Subsequent research revealed that neurogenesis also occurs in human brains, sparking an entire new field of biomedical research: if the secret of neuron regeneration could be discovered, this could revolutionize research into the neurological disorders of humans.[27]

Over the following decades, it became clear that lateralization was not restricted to song acquisition: day-old domestic fowl chicks, for example, were found to use their right eye for close-up activities like feeding, while the left eye scans for predators (Rogers and Anson 1979). In tool-using New Caledonian Crows (chapter 8), individuals show near-exclusive individual preferences for the side of the beak against which they hold a tool (Weir et al. 2004).

Remarkably, the more lateralized an individual, the more proficient it is at particular tasks. For example, it has long been known that parrots are either right or left footed, and the more biased an individual parrot is with respect to which foot it uses to hold food, the better it is at solving puzzles posed by researchers (Magat and Brown 2009). Lateralization in birds was assumed to be genetic in origin, but in the 1980s Lesley Rogers showed that it was environmental, dictated by which eye was exposed to light in

the embryo. During normal development in the egg, one eye faces inward and receives no light, while the other faces outward and receives some light through the shell—enough to establish lateralization. Rogers revealed that by gently turning the embryo's head through 180 degrees, she could reverse the direction of lateralization (Rogers 1982).

Influenced by the notion of "a wing guided by an eye," vision has come to dominate our view of how birds perceive the world, at the expense of research on almost all other senses. A review of avian senses by Jerry Pumphrey in 1948 said as much, and with only a few exceptions not much has changed since then. There has been remarkably little work on the sense of touch, for example, even though many bird species spend a large amount of time allopreening. The recent discovery of pressure sensors in beaks of wading birds and kiwis (Piersma et al. 1998; Cunningham et al. 2007) suggests that there is still much to learn about the sense of touch in birds. Similarly, taste in most birds has been almost ignored, despite some excellent work on the Mallard by Herman Berkhoudt (1985).

Pumphrey referred to avian hearing as an "unjustly neglected field for experiment and observation."[28] In an innovative attempt to rectify this, he used some descriptive information collected by Gustav Retzius in the late 1800s on the cochlea—the inner ear—of birds and revealed that the length of the cochlea was positively associated with the degree of what he called "musicality" in birds. In other words, the extent to which birds rely on song and hearing in their day-to-day lives is reflected by the relative size of the basilar membrane inside the cochlea. The fact that the hair cells on the basilar membrane in our own cochlea—the cells responsible for the detection of sound—are not replaced when they die is one reason most people's hearing deteriorates as they age. Deafness is especially likely in rock musicians and their fans who have enjoyed, or endured, loud music—our hair cells are extraordinarily sensitive and easily damaged. The discovery in the late 1980s that, unlike us, birds do replace the hair cells in their ears has stimulated much new research in hearing, albeit in humans rather than birds, so far with no success in finding the genetic bases for hair cell replacement (Salvi et al. 2008).

The sense of hearing in birds is intimately related to the development of song, as demonstrated by another of Marler's pioneering students, Masakazu (Mark) Konishi. Born in Japan in 1933, Konishi ended up as an undergraduate at Berkeley in 1958, where he worked on juncos in Marler's lab. He remained at Berkeley for his PhD, with the aim of using experiments to test specific hypotheses about song learning. Knowing that deaf humans had difficulty in speaking normally, Konishi wondered whether the same was true in birds. As he started to work his way through the rather limited literature on hearing in birds, he came across the work of Johann Schwartzkopff,[29] director of the Allgemeine Zoologie (Institute for General Zoology) in Bochum, Germany. Schwartzkopff had conducted some of the first studies of hearing in birds, and in the 1940s he developed a technique for surgically removing the cochlea—the inner ear—from live birds. During his PhD research, Schwartzkopff noted how the normally flutelike calls of Eurasian Bullfinches became "shriller" after deafening, but with no access to a sonograph machine he was unable to quantify this effect.

Removal of the cochlea is a difficult operation because it lies deeply embedded in the bones of the skull, but Konishi mastered the technique, allowing him to test his idea that auditory feedback was essential for birds to develop a normal song. He started on cockerels but found that deafening had no effect on their crowing. When he performed the same experiment on White-crowned Sparrows, however, the results were dramatic. With young birds, whose song was incomplete, deafening prevented song acquisition, suggesting that auditory feedback was an essential aspect of song acquisition. With adult birds, whose song was complete, deafening had no such effect (Konishi 1965).

Not only do young birds have to learn their song, usually from a conspecific male tutor, they also have to be able to monitor how the song they produce compares to the song of their tutor, a song stored in their brain. Preventing young birds from making that comparison by deafening them means that the birds are incapable of correcting any errors they make while learning. Konishi's remarkable results identified the importance of sensory feedback for the appropriate development and maintenance of several types of behavior, in addition to singing.

Pursuing his interest in the neurophysiology of hearing, Konishi made other breakthroughs. After hearing Bob Payne talk about his thesis research on prey capture in Western Barn Owls in 1963, Konishi started breeding a pair of these birds at Princeton in 1966. In 1975 he took up a full professorship at Caltech and was able to take twenty-one owls with him. Barn owls have slightly asymmetric ears—the right one is positioned higher on the owl's head than the left ear —and this allows accurate sound localization. Collaborations with Jack Petti-grew and Eric Knudsen led to a number of significant findings, including their now classic study of auditory receptor fields, from which they were able to produce a "map" showing how space is represented in the owl's brain (Knudsen et al. 1977; Knudsen and Konishi 1978).[30]

Other than vision, the one sense that is now better understood today than it was a century ago is olfaction. It has been a long, slow journey, and many people continue to find it difficult to accept that many birds have a good sense of smell. Ornithologists were misled initially in the late 1700s by John James Audubon, who on the basis of some rather inept experiments—compounded by his misidentification of the two American vulture species—vehemently rejected the idea that birds had a sense of smell. In contrast, Richard Owen, Darwin's nemesis, relying on some of his admirable anatomical studies in the 1830s, declared that kiwis and New World vultures, with their enormous olfactory bulbs, *must* have a good sense of smell. In the 1920s John Gurney, weighing up the evidence for and against olfaction in birds, concluded simply that it was "perplexing" and pleaded for more research. It was not until the 1960s that a breakthrough occurred.

Betsy Bang was a medical illustrator working for her husband studying respiratory disease in birds. She noticed that the nasal conchae of certain birds were much more complex than in others and suggested that the elaborate conchae in the kiwis, Turkey Vulture, and Black-footed Albatross reflected an olfactory lifestyle (Bang 1960). Soon after, she teamed up with the eminent neurobiologist Stanley Cobb to conduct a

landmark study of the olfactory bulb size across a wide range of bird species (Bang and Cobb 1968). They devised a simple index of relative bulb size, showing that it was enormous (index ~30) in species in which there was anecdotal evidence that olfaction might be important, like the kiwis, Turkey Vulture, and albatrosses. In contrast, it was small (index ~3) in songbirds. Bang later joined the physiologist Bernice Wenzel to provide a comprehensive, and positive, review of olfaction in the mid-1980s in King and McLelland's *Form and Function in Birds* (Bang and Wenzel 1985).

Despite the best efforts of Bang, Cobb, and Wenzel, most ornithologists continued to struggle with the idea that smell might be important in the lives of birds. This is one reason why Floriano Papi's idea that birds use smell as an aid in navigation had such a rocky reception when it was first mooted in the 1970s (chapter 4). Change came from an unlikely quarter.

Whalers and other seafarers had long suspected that Procellariiformes could smell cetacean carcasses. Some, like Captain J. W. Collins in the 1880s, had even conducted their own experiments to make the point. On throwing bits of liver out on the sea where no birds were visible, he noted that petrels and shearwaters appeared soon after, flying upwind "out of the fog, flying backward and forward across the vessel's wake, seemingly working up the scent until the gloating pieces of liver were reached."[31]

In 1992 Gaby Nevitt started her olfactory research career, studying salmon. There was already some scientific evidence for Captain Collins's conjecture, but the break came when Nevitt met atmospheric scientist Tim Bates on an Antarctic research vessel near

Elephant Island southeast of Tierra del Fuego, Argentina. Bates was studying a substance called dimethyl sulphide, DMS for short, released into the atmosphere from the bodies of phytoplankton as they were consumed by zooplankton. Nevitt realized that DMS might be what she and the albatrosses were looking for. And indeed it was: when she released DMS onto the ocean the birds came (Nevitt 2000). DMS provided an olfactory seascape, and Nevitt provided convincing evidence that this is how certain seabirds (Procellariiformes) find their food.

That birds possess an olfactory sense is clear, but much remains to be done. In principle it would be possible to use fMRI to screen a large number of species to assess their neurological responses to a suite of odors. It is also now possible to use tomography to obtain volumetric measures of olfactory bulb size, as Jeremy Corfield and colleagues (2008) have done for the kiwi. Indeed, the entire field of avian sensory biology—linked with molecular and behavioral studies—seems ready for a new era of research (Birkhead 2012).

ANNUAL CYCLES

A "rude, rugged and improbable academic,"[32] Jock Marshall helped to construct the foundations of what has become the study of avian breeding cycles.[33] Born in Sydney, Australia, in 1911, Alan John Marshall acquired his nickname as a boy, because he aspired to be a jockey, but as it turned out he was too tall. At school he was considered "totally undisciplined," preferring riding and shooting to academic work. "I was a monumentally lazy little bastard, did almost nothing but lie around reading."[34] Two of the books he

read were given to him by his father in reaction to his mother sending him to Sunday school: Haeckel's *The Riddle of the Universe* and Darwin's *Origin of Species*. They made a deep impression and helped to consolidate Marshall's interest in natural history. They did little to temper his behavior, however, and Marshall was eventually expelled from school. Just before his sixteenth birthday an accident resulted in the amputation of his left arm just below the shoulder: a defective safety catch caused his shotgun to explode as he was poking it among firewood. Jock was depressed by his disability, and hoping to help him, his mother wrote to the ornithologist Neville Cayley, author of one of the first field guides to Australian birds,[35] to ask how she might encourage her son's interest in natural history. Cayley passed the letter to his friend Alec Chisholm, fellow of the Australian Zoological Society, who on meeting Marshall was sufficiently impressed by his knowledge of natural history that he introduced him to other fellows of the AZS at the Australian Museum.

For the first time in his life Marshall had a goal. He began working at the museum, mainly on birds and mammals and soon learned to achieve with one hand what others could do with two. Colleagues were astounded by his ability to dissect birds, and in response Marshall quipped, "It isn't necessary to have two arms—the weight and space would be far better taken up by an extra penis!"[36]

In 1930 a group of Harvard-based biologists arrived in Sydney en route to New South Wales and Queensland. Marshall persuaded them to let him tag along, reinforcing his taste for fieldwork and allowing him to put his field skills to good use. The following year he went to Queensland on his own, his single arm forcing him to use a sawed-off shotgun for collecting specimens. Typically unkempt and armed with this dubious weapon, he was reported to the police, who locked him up in the local jail until the museum vouchsafed his authenticity.

In 1934 Marshall was able to join an Oxford University expedition to the New Hebrides (Vanuatu), and it was on this trip that he met Tom Harrison and the expedition leader, Oxford academic John Baker. The aim of the fieldwork was to investigate the reproductive physiology of animals breeding in one of the most uniform, nonseasonal environments in the world (Baker 1938; see also chapter 5).

Jock Marshall (photo in 1937 at age 26).

The Vanuatu expedition was a life-changing experience. Writing to Alec Chisholm from there on 9 May 1934, Marshall said:

It's fascinating work—measuring every aspect of the climate (including ultra-violet light) in that it may be expected to affect animal life. Each month between the 10th and the 20th I have to dissect and preserve the gonads of 30 fruit-bats, 50 insectivorous bats, 60 lizards, 30 *Pachycephala* (*pectoralis*) [Golden Whistler], and 30 *Trichoglossus* [lorikeets] . . . Also I have been studying, in collaboration with Tom Harrisson, the moult rhythm of about 30 species, and also the food, ecology etc. of many others. . . .[37]

There were more expeditions: New Guinea in 1936 and Spitzbergen in 1937, followed by a study of Satin Bowerbirds closer to home, an investigation rudely interrupted by the start of the Second World War. Despite his disability, Marshall saw active service in New Guinea, where he commanded a small reconnaissance unit (nicknamed Jockforce), operating behind Japanese lines and using carrier pigeons to relay information back to headquarters. When he wasn't fighting, Marshall watched and collected different bowerbird species, preserving their gonads in gin or whisky. Once the war was over and now in his mid-thirties, he signed up for a biology degree at the University of Sydney, where he continued his bowerbird studies.

The Satin Bowerbird's extraordinary bower-building behavior was first described by John Gould in his *Birds of Australia* in the 1840s, and it immediately captured the public imagination. In the 1880s George Romanes pronounced that "these curious bowers are merely sporting-places in which the sexes meet."[38] For Romanes, bower building was mere "recreation" or "play," and because of this he considered bowerbirds to be more intelligent than other birds. It was an attractive thought, and one that for many seemed to account for the extraordinary behaviors exhibited by these birds. Marshall, on the other hand, considered the idea to be utter nonsense: "These . . . [behavioral] attributes have caused a voluminous popular literature to spring up about the family, most of it marred by anthropomorphic generalisation and all of it unsupported by experimental evidence."[39] With financial assistance from the then newly formed Association for the Study of Animal Behavior (chapter 7), he set out to test the recreation hypothesis by performing what he called a histophysiological study of the bird's annual cycle. Essentially Marshall used the size and internal (histological) state of the gonads as an assay of the Satin Bowerbird's reproductive state. To do this he shot and dissected birds, preparing their gonads for microscopic examination. He even castrated a small number of males, bringing their sexual behavior, including bower building, to an abrupt halt, but rapidly restoring it by an injection of testosterone.

Drawing all this together in his book *Bower-birds: Their Displays and Breeding Cycles* (1954), Marshall concluded that the gonad was the primary organ of periodicity. In a paper read at the Society of Endocrinology in 1955 he developed this idea, writing that "the most plausible view of the regulation of sexual periodicity is to think of the gonad as a sort of cog-wheel which is seasonally engaged, so to speak, by various environmental teeth, which differ in combination and significance from species to species."[40]

After finishing at Sydney, Marshall went to Oxford in 1947 to work with John Baker for a DPhil on arctic birds. The fieldwork required another expedition, this time to Jan Mayen, a small volcanic island between Iceland, Norway, and Greenland. Previous expeditions to this remote island had reported extensive nonbreeding among both seabirds and landbirds, and Marshall was keen to investigate. Marshall and his colleagues, however, found little evidence for nonbreeding among Northern Fulmars—a species that Vero Wynne-Edwards (1939) had previously suggested might breed only once every three years (chapter 10)—but widespread nonbreeding among Glaucous Gulls. Extensive collection and dissection allowed Marshall to compare the gonads of breeding and nonbreeding gulls and to describe the "seasonal secretory activity of the avian testes, and to postulate [for the first time] an endocrinological role for lipids in the post-nuptial semeniferous tubules."[41] It is now obvious that Marshall would have had no way of knowing whether his "nonbreeding" gulls were sexually mature birds taking a year off from breeding or immature birds (in adult plumage) yet to start breeding.

In 1949 Marshall obtained a position as reader in zoology and comparative anatomy at St. Bartholomew's Medical College, then part of the University of London. It may seem strange that an avian reproductive biologist should obtain a position at a hospital, but in those days would-be medics received training in basic science, and Marshall's role was to teach zoology to these "premeds." An ex-student remembered going to see him one day in 1959. While they were talking, Marshall spotted a Common Wood Pigeon in a plane tree outside the window. Taking a

Male Satin Bowerbird picking up a decoration in front of its bower.

12-bore shotgun from a cupboard, he opened the window and shot the pigeon, saying, "Be a good chap and nip out and collect that bird."[42] He did.

Independent of the wood pigeon incident, Marshall left Barts the following year to take the chair of zoology and comparative physiology at the newly formed Monash University in Melbourne. Prior to this he had applied for a similar position at Reading University in Britain but was unsuccessful. He later discovered that John Baker's letter of "support" had highlighted Marshall's limitations as a scientist and alluded to his acrimonious divorce; it seems that Baker sabotaged him, but it is not known why he did this. The fact that he felt the need to mention Marshall's divorce suggests that Baker may have been rather prudish. Another

possibility is that Baker still considered Marshall to be his "pupil" and was reluctant to consider him an equal. Or it could simply be that Baker was coldly and devastatingly objective; he had once described himself as "sincerity 100%: tact nil."[43] Whatever the reason, Marshall was deeply hurt by what he felt was Baker's betrayal.[44]

Back in his native Australia, Professor Marshall was a force to be reckoned with. As one of his ex-PhD students told us:

Monash was built on the grounds of a former mental hospital that was bereft of most of its vegetation. Like most Australian cities, Melbourne was landscaped primarily with exotic vegetation of European origin. The initial plans for Monash called for a continuation of this approach. Jock, however, convinced the

219

university's senate to recommend that ONLY native vegetation be used at Monash, his rationale being that Monash, as an AUSTRALIAN university should reflect Australian values in as many areas as possible, including its landscaping. If you have visited Monash's main campus at Clayton, you will appreciate Jock's foresight and lasting contribution.

In the northwest corner of the original grounds there was a 12 ha remnant stand of dry sclerophyl eucalyptus surrounding a small pond. The master plan for this area was to remove the forest, fill in the pond, and use the site for student residences. Jock was furious at this proposal, especially given the university's decision to retain and use native vegetation in its landscaping. Nevertheless, Jock lost this battle in senate. Plans went ahead, the surveyors moved in, and the machinery was assembled to begin clearing the site for construction. Under the cover of darkness, the day before construction was to start, Jock, with the assistance of his lab manager, removed all the surveyor's stakes. Construction was obviously delayed and Jock eventually won his battle. The area became the Department of Zoology's animal holding facility with pens and aviaries for research animals. The reserve survives to this day as the Jock Marshall Reserve.[45]

In 1960, the first volume of Marshall's *Biology and Comparative Physiology of Birds* was published (the second in 1961). With contributions from twenty-three leading researchers and Marshall as editor, this was a landmark volume providing a much-needed synthesis.

MINUS FORTY-SEVEN CELSIUS

For those of us located well away from the equator, the seasons we know as spring, summer, autumn (fall), and winter mark the Earth's annual journey around the sun and are caused by the 23.5-degree tilt of the Earth on its axis. In temperate regions most birds start to breed in spring, and prior to the twentieth century it was widely assumed that this was because the summer is warmer than the winter. In the mid-1700s efforts to industrialize the poultry business focused on getting hens to lay eggs throughout the year—and especially during winter—by warming up their winter accommodation. This did not work, and in seeking an explanation, poultry breeders went off down the wrong—but not totally unrelated—route. They had assumed that hens failed to lay in winter because they had not molted, so attempts were then made to devise ways to encourage hens to molt. Breeders reasoned that if the hens molted they would start laying in response to the increased temperature.

It is clear that the people raising poultry did not talk to those who caught and kept small birds like finches and canaries. Bird catchers had known since at least the 1500s that breeding cycles could be altered by manipulating the amount of light the birds received. Plunging a bird into darkness in May—at the height of the breeding season—precipitated an early winter for the birds. They called it "stopping"—stopping the light. As a result, the birds molted several months earlier than usual, and as the keeper increased the light, the bird "emerged" in full song as though it was spring in August or September (Birkhead 2008).

Because "stopping" was apparently unknown among the scientific community of the early 1900s, it was still widely assumed that temperature rather than light was the trigger for breeding in the spring. Some experiments by Gustave Loisel (1900; cited in

Allender 1936), for example, on springtime gonad development in the House Sparrow presumed that temperature was the cause.

At a meeting of the Edinburgh Natural History Society in the early 1900s, physiologist Edward Schäfer gave a talk "On the Incidence of Daylight as a Determining Factor in Bird Migration." His basic message was that by moving south birds were searching for longer days (implying that light was the ultimate factor). By saying that "the incidence of the proportion of light to darkness is a constant factor, and might even be conceived to be operative in exciting the migratory instinct into activity,"[46] he unwittingly identified day length as the proximate cue.

The real turning point came in 1920 with the work of two botanists—Wight Garner and Harry Allard—interested in what triggered flowering in tobacco plants. By altering day length, just as bird keepers had done, they discovered they could manipulate events in a plant's annual cycle pretty much at will.

William Rowan, a zoologist at the University of Alberta, Canada, in the 1920s, reached the same conclusion, apparently independently. Brilliant, eccentric, and difficult to work with, Rowan operated very much on his own and had little access to the scientific literature; like almost everyone else he was oblivious to the bird catchers' habit of "stopping." Interested in birds since childhood, Rowan gradually recognized that the only cue reliable enough from year to year to initiate migration was day length—temperature and barometric pressure were simply too variable. In 1924 he came across a paper by Gustave Eifrig in *The Auk* on the role of photoperiod (day length) on migration. Eifrig publicized Garner and Allard's finding and proposed that it was equally applicable to seasonal timing in birds; he suggested that spring migration in birds occurred in response to the growth of their gonads, which in turn was triggered by photoperiod. Eifrig's paper was like a bomb, rousing Rowan from his inertia and forcing him to start putting his ideas to the test. Rowan was convinced that Eifrig was wrong: it was not that light triggered gonad growth and that this caused migration, he thought, but rather light that initiated gonad growth and migration independently. The difference was subtle but, to Rowan, crucial.

Setting up the experiments was a nightmare. Rowan's head of department, Henry Tory, disapproved and felt that zoological research should be conducted only in the laboratory and that Rowan should focus his efforts on applied biology, rather than blue skies ideas. As a result, Rowan had to fund and build the aviaries himself on his own property and conduct the research in his own time. The idea was to provide some birds (juncos) with artificial light—from a 75-watt light bulb—in winter and then to establish how this affected their migratory behavior (Rowan 1929). If they were affected by the light, then even in midwinter the birds should head north on being released, as though it was spring. The effect of the artificial illumination was dramatic, and by Christmas—at temperatures as low as –47°C (–53°F), the juncos came into full breeding condition, with enlarged testes and in full song. Releasing them, however, was a disaster. Instead of flying off in a direction that Rowan could measure, the birds merely shot into the undergrowth, rendering this crucial part of the study inconclusive.

Undaunted, Rowan repeated this ambitious experiment using American Crows, whose tails he dyed bright yellow prior to release so they could be more easily tracked (Rowan 1930; 1932). Exploiting the power of the media and offering rewards for either sightings or the bodies of shot crows, Rowan received numerous reports of his experimental birds, confirming his hypothesis. Birds stimulated by light in winter developed their gonads and migrated northward; the control birds went south or remained in the area. Photoperiod was the key to the riddle of migration.

Rowan assumed that light resulted in an increase in the birds' "wakefulness" and that it was this wakefulness, rather than light per se, that was the cue to migration. After a lot of unpleasant experiments in which birds were kept awake inside rotating drums, it was clear to Rowan that his "wakefulness" idea was wrong.

Rowan summarized his findings and ideas in *The Riddle of Migration*, published in 1931. Written for a general audience, the book was a great success. Nonetheless, Rowan was concerned by the difficulty of getting his results published in the scientific literature. In May 1931 he wrote to his friend Reg Moreau in Africa: "I'm afraid my efforts are threatening to become too highbrow for ornithology. I see Witmer Stone [editor of *The Auk*] didn't even bother to notice my attempt with crows in the Auk."[47] Moreau, who admired Rowan's work, recognized that Rowan was addressing general biological problems rather than ornithological ones. Moreau wrote back saying that he was "painfully conscious of the narrowness and amateurishness of so much ornithological work,"[48] deploring the fact that few ornithologists bothered to consider the wider implications of their studies—a deficiency epitomized by Witmer Stone. It was precisely because Moreau could identify

important and general biological issues that he was so effective when he later became editor of *Ibis* (chapter 5). In the end Rowan published his studies in the Proceedings of both the Boston Natural History Society and the US National Academy of Sciences.

Rowan's biographer, Marianne Ainley (1993), sums him up as one of the most renowned Canadian scientists of the first half of the twentieth century: " . . . an intuitive, creative and highly energetic scientist who favored an interdisciplinary approach to research. In some ways Rowan was similar to Marshall in resisting scientific authority and steering his own, innovative but rocky course that embraced both field and laboratory work."[49] Marshall later also applauded Rowan, saying that "Rowan's . . . experimental work will remain for all time a landmark in the literature." He continued: "With great determination and ingenuity he [Rowan] devised and performed vital experiments in an era before the planned experiment became a widely recognized tool for attacks on ecological problems."[50]

Yet while Marshall praised Rowan, he did not agree with all of his assumptions, in particular with Rowan's near-exclusive focus on photoperiod as the key to birds' annual cycles. Marshall wrote: "It is a matter of regret that Rowan's great advance has tended to obscure the obvious fact that not only day length, but many and various factors are influential in the control of the breeding seasons (including migration) of vertebrate animals."[51] He points out that while the virtue of day length is its regularity, it can also be a destructive force if birds respond to it automatically. As is well known, Marshall says, in temperate regions birds breed later in cool springs than in warm ones, suggesting that photoperiod cannot be the only cue they use to time their breeding. Marshall was also well aware that, in his native Australia, environmental conditions suitable for breeding were often unpredictable, rendering photoperiod a completely inappropriate cue. Through a series of ingenious aviary experiments in which he simulated rainfall, aridity, and the availability of fresh vegetation, Marshall identified the cues used by arid-zone birds, like the (wild) Zebra Finch, to time their breeding. He also showed, by examining their gonads, that birds could respond very rapidly as soon as conditions were suitable. Working with Dom Serventy, Marshall also found that Zebra Finches kept in the dark continued to produce sperm, indicating an internal rhythm of sperm production (Marshall and Serventy 1957).

It is now recognized that while photoperiod is the main driver for seasonal events, many other factors, including temperature, food availability, nest-site availability, nest material, and social interactions all influence whether a bird comes into breeding condition. How else, one can ask, would they have the flexibility to respond adaptively to variation in conditions (Williams 2012)? As Brian Follett explained,

> Photoperiod seems to act strategically, a least in most birds away from the tropics, in setting up the system and shaping the main events of the year—entering refractoriness in early summer (post breeding), breaking refractoriness in late autumn as short days recur. Then the bird is photoresponsive and will grow its gonads as days lengthen after the winter solstice—they start before actually. The "other factors" are of immense importance tactically (especially in the female) since they bring the final stages to fruition. Ecologists view the other factors as being critical, physiologists tend to place more emphasis upon the strategic picture. Both are equally correct methinks![52]

The idea—promoted by Marshall—that the testes drive the annual cycle in birds was as old as Aristotle. It had been known since the third century BC that if deprived of their testicles males became less aggressive and more female-like. In the 1840s, Arnold Berthold, professor of physiology at the University of Göttingen, found that if he reimplanted testes inside the body of castrated cockerels, the birds returned to their former, aggressive, sexually enthusiastic state:

> So far as voice, sexual urge, belligerence, and growth of comb and wattles are concerned, such birds remained true cockerels. Since, however, transplanted testes are no longer connected with their original innervation, and since no specific secretory nerves are present, it follows that the results in question are determined by the productive functions of the testes, i.e., by their action on the blood stream, and then by corresponding reaction of the blood on the entire organism . . . of which . . . the nervous system represents a considerable part.[53]

The key to understanding what was going on was the recognition that the reimplanted testes could have their effect *only* through the bloodstream: chemical messengers.[54] Excited by this idea, seventy-two-year-old Charles Brown-Séquard started to inject himself with the crushed testes of guinea pigs, dogs, and monkeys, reporting an instant and startling rejuvenating effect. After publishing his findings in *The Lancet* in 1889,

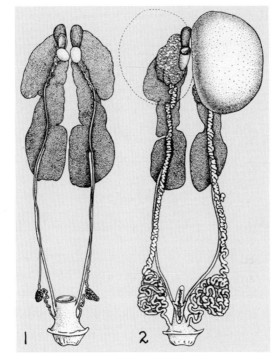

The dramatic seasonal changes in the size of birds' gonads is shown here by comparing the male House Sparrow in winter (left) and spring (right).

"organotherapy," as he called it, became extremely popular. Now recognized to be a placebo effect, Brown-Séquard's results caused a surge of interest in revealing the testicular chemical. Once testosterone was identified, its effects on male birds were found to be various and far reaching, including stimulation of sexual and aggressive behavior, spermatogenesis, development of secondary sexual characters, reduced immunity, reduced survival, and interference with both the pair bond and paternal care.

In the late 1920s, Oscar Riddle and colleagues at the Carnegie Station for Experimental Evolution at Cold Spring Harbor isolated a relatively pure form of a new hormone secreted by the anterior pituitary in the brain. They named it "prolactin" because it stimulated milk production (lactation). In 1928 two German researchers had induced lactation in rabbits by injecting them with extracts from the anterior pituitary gland, and in 1930 Riddle's colleague George Corner followed this up, finding that extracts from the anterior pituitary caused lactation in spayed rabbits that had never ovulated. Intrigued by Corner's work (but unaware of the German study), Riddle made a purified form of the hormone from the frozen pituitaries of cattle and sheep, which he then injected into pigeons (Riddle et al. 1932). As one of Whitman's students, Riddle knew that pigeons fed their young on "crop milk," but he was amazed to discover that his purified prolactin stimulated the pigeon's crop to thicken and start producing milk. Riddle et al. (1933) developed a procedure using the response of the pigeon crop—involving the laborious injection of thousands of birds—as an assay for prolactin, a procedure used until the development of radioimmunoassays in the 1960s (Smith 2004).

HORMONES AND BEHAVIOR

The 1930s saw the start of a surge of hormone research in both birds and mammals. The main proponents were Frank Beach, Richard Whalen, and William Young, who were interested in the links between the brain, hormones, and behavior. Together they launched a series of informal conferences called West Coast Sex Meetings, where they and chosen participants shared ideas relating to sexual behavior. The key to the success of these meetings was that participants were not allowed to talk about published results. As the field flourished, the meetings morphed first into the

Conference on Reproductive Behavior and then in 1997 became the Society for Behavioral Neuroendocrinology.

Beach has been described as "arguably the premier psychobiologist of his generation."[55] Though he examined mating and parental behaviors from a physiological stance, he remained dedicated to a holistic view of behavior, including the ecological and evolutionary factors involved. "Physiological analysis can quickly become reductionist; in Beach's hands, by contrast, it was integrative."[56] He was a firm believer in experimentation and excelled at this, although in a "seat-of-the-pants, follow-your-nose kind of experimentation, rather than one based on sophisticated mathematical analysis or elaborate equipment."[57] His book *Hormones and Behavior* (1948) defined the field, drawing together all the evidence from the first half of the century and providing a theoretical basis for the developing field to follow.

By the 1960s a picture was emerging. The hypothalamo-pituitary gland complex in the bird's head was the control center. Referred to as a neuroendocrine organ, the hypothalamus is part of the brain and contains many ancient neural circuits (including those for circadian clocks, eating, drinking, and thermoregulation) but also produces a range of neurosecretions. It is activated both by external factors, like light, and by its own internal clock or clocks (see below). Once stimulated, the hypothalamus releases neurohormones that then, via the pituitary (another endocrine gland), encourage the gonads to grow and to start producing both gametes and their own hormones. The testosterone released from the testes in turn affects behavior like aggression, singing, and courtship, and male courtship then affects the female's endocrine system and her sexual behavior.

The way male courtship influences female reproductive activities takes us back to the late nineteenth century and to Charles Whitman again. A student of the famous Rudolf Leuckart, who was in turn a student of Rudolf Wagner's, Whitman was one of the most influential biologists of his day. Like his predecessors, he was fascinated by reproduction, but it was not until the 1890s, when he was in his fifties and professor of zoology at the University of Chicago, that he developed a special interest in birds. He became enamored of pigeons, and with his students he worked on several aspects of their biology, including behavior, genetics, and reproductive physiology. Whitman never got around to writing up much of his own research, but his students, including Wallace Craig, Eugene Harper, and Oscar Riddle, all excelled.

It was Craig who helped to solve the longstanding mystery of what triggered reproduction in female pigeons. Female domestic fowl are happy to lay eggs without ever seeing a male, providing there is sufficient light (a fact discovered in the 1800s and widely used in commercial operations by 1920). The same is not true of pigeons.

In 1911 Craig demonstrated experimentally that it was the sight of a male that stimulated ovulation in female pigeons. This result later helped to make sense of a discovery, made in the 1930s, that while male House Sparrows will come into breeding condition and start to produce sperm in response to increasing day length, females seemed reproductively indifferent to day length (Riley and Witschi 1938). It turned out that, like pigeons, female House Sparrows need male courtship to complete the development of their reproductive organs. This also made sense of numerous anecdotal field observations showing that males typically arrive back from migration ahead of females and in full breeding condition, whereas females arrive later and require several days for their reproductive system to develop.

In the 1960s Danny Lehrman (chapter 7) decided to look in detail at the ways in which behavior and hormones interact during a bird's breeding cycle. Lehrman had conducted his PhD on parental care in captive ring doves (African Collared Doves) with Theodore Schneirla at New York University before joining the faculty at Rutgers University, Newark in 1950, four years before completing his PhD. The secret to Lehrman's subsequent success was combining ethological and psychological approaches to explore the importance of external and internal factors controlling behavior. Building on the work of previous pigeon researchers such as Whitman, Craig, and Matthews in the 1930s, Lehrman teased apart the underlying causes of the ring dove's breeding cycle: courtship, nest building, egg laying, incubation, and chick rearing (e.g., Lehrman 1964). Newly formed pairs of ring doves presented with a nest containing eggs rarely performed incubation behavior. If, however, the birds had been together for five to seven days and had laid eggs, they always incubated the eggs, suggesting that behaviors, such as courtship and nest building were necessary prerequisites to incubation. However, injecting the hormone progesterone, which had been isolated from pigs in 1929,[58] into newly formed pairs presented with a nest resulted in immediate incubation behavior. Lehrman's experiments were among the first to explore the interactions between external

stimuli, hormones, and behavior; Robert Hinde (chapter 7) conducted similar types of experiments on canaries at the same time. Lehrman showed that the level of steroid hormone in the blood can change a bird's response to a behavior, such as courtship, and that this response can then affect the endocrine system—that is, that hormones not only influence behavior but can be influenced by behavior.

Marler later said that Whitman's efforts to link behavior with endocrinology around 1900 largely fell flat, and except for Wallace Craig's studies "they remained dormant, until they burst forth, phoenix-like, in ethology, and the integrative research programs of people like Hinde and Lehrman. You can reasonably regard Lehrman's magisterial 1961 review of parental behavior in W. C. Young's *Sex and Internal Secretions*, the bible of the time, as the culmination of a scientific progression that began as much as a century before, in the pigeon lofts of Charles Otis Whitman."[59]

BIOLOGICAL CLOCKS

Donald Farner's decision to study the control of avian breeding cycles in the 1940s marked the beginning of three decades of extraordinary science in his lab. Born in 1915, Farner studied zoology first at Hamline University in Minnesota and then at the University of Wisconsin for his PhD on the digestive physiology of chickens. On joining the US Navy's Medical Services Corps in 1943, he studied arthropod-borne diseases but became increasingly interested in ornithology. After leaving the navy he worked at the Universities of Kansas and Colorado before settling at Washington

State University in Pullman for the period 1947–65, thereafter taking up a professorship at the University of Washington. Over several summers he had worked as a park naturalist at Crater Lake and had written a book on the birds there. Once at Washington Farner decided that his future lay in ornithology. The White-crowned Sparrow—a migrant subspecies that wintered in the Snake River Canyon just twenty miles south of Pullman—was the perfect study species to explore the photoperiodic control of the annual cycle. Farner's starting point was a state-of-the-art paper that he prepared for the 1948 AOU meeting in Omaha, Nebraska, summarizing what was known and what was still unknown (Farner 1950).

The first evidence that migration might be controlled by hormones came from a study by Helmuth Otto Wagner, director of the Bremen Museum in Germany, who had elicited migratory restlessness in Common Whitethroats by injecting them with thyroxin (Wagner 1930). A few years later Thomas Bissonnette (1937), noticing that starling gonads started to regress in June while the days were still getting longer, speculated (correctly) that there was an internal rhythm controlled by the anterior lobe of the pituitary gland. Gregorius van Oordt (1928), who studied plumage and gonad development in Red Knots and Ruddy Turnstones, also suggested that migration might be timed by an "internal physiologic rhythm" (Farner 1986).

Based on these ideas, and much else, Farner embarked on a study to identify the internal factors initiating migration. His eventual success guaranteed his position as a leading figure in the study of avian photoperiodism and reproductive biology.[60] As well

as excelling as a researcher, Farner was also a remarkable teacher: "Although he might sometimes grouse about 'the little buggers,' he took his classroom duties seriously."[61] Among his other successes was the *Avian Biology* series, which he edited with James King and dedicated to the memory of Jock Marshall (e.g., Farner and King 1971). Part of Farner's influence was the result of working with an international network of collaborators, including Brian Follett.

Follett studied biological chemistry as an undergraduate at Bristol, remaining there for a PhD in pharmacology. Laboratory rats and rabbits were his study species. In 1962 Follett's colleague Hans Heller organized a neurosecretion meeting in Bristol and invited Farner from Washington and Andreas Oksche from Geissen: "They told a remarkable story—remarkable that is to the white rodent people who dominated the meeting—that in seasonally breeding birds one could drive the entire gonadal cycle (in males anyway) by altering photoperiod. This opened up (to me) the possibility of sorting out the neural and neuroendocrine circuits controlling higher vertebrate reproduction and so I approached Don Farner."[62]

On completing his PhD, Follett and his wife went to Washington State, where his postdoctoral project with Farner was to assess the role of pituitary gland hormones on the breeding cycle of the White-crowned Sparrow. As a nonbirder, Follett was horrified to discover that Farner expected him to go out into the field and collect his own birds. The idea was to record the pituitary hormones through the bird's breeding cycle. The amounts of hormone involved were minute, meaning that large numbers of birds would have to be sacrificed simply to have enough hormone to measure. Even then, the assays were crude.

The breakthrough came in the late 1960s with the development of a new technique: radioimmunoassay, developed by Sol Berson and Rosalyn Yalow for measuring human insulin.[63] Radioimmunoassay allowed researchers to tag the hormones they were interested in with radioactive iodine, which was easily measured. Follett, with his old friend Frank Cunningham at Reading and Colin Scanes, a graduate student, collected vast numbers of chicken pituitary glands, extracted the luteinizing hormone (LH) gonadotropin, raised antibodies to them in rabbits, and developed by 1971 a radioimmunoassay for avian LH. As Follett says, "The day we perfected the assay was remarkable. In one leap we had increased our ability to measure LH 200,000 times. From needing the extract from three pituitary glands (and therefore also very dead donors!) we now only needed 10 microlitres of blood, a volume that can be taken easily and repeatably from living small finches or in my case mainly Japanese Quail."[64]

Two years later, in 1973, Follett, together with Philip Mattocks and Farner, introduced the technique to laboratory White-crowned Sparrows. The first major discovery was that White-crowned Sparrows use a circadian rhythm to measure day length and activate gonadotropin secretion. It was the increased sensitivity that was the key because the birds had to be exposed to only a single pulse of light given once to increase LH release (Follett et al. 1974).

Rowan and Marshall had previously assumed that breeding, molt, and migration were controlled by the gonads, but by Follett's day it was increasingly obvious that the

control center was in the head. In the late 1960s Follett's colleague Mike Menaker performed an ingeniously simple experiment that extended Benoit's (1930) pioneering experiments, which had suggested that birds were able to perceive light directly through the skull. Menaker tested this idea more directly by plucking the feathers from the tops of the heads of House Sparrows and painting the skin black: they did not come into breeding condition, because light was unable to penetrate the paint (Menaker et al. 1970).

The next question was where in the brain the photoreceptors were located. Fiber optics were unknown in the 1970s, but Follett found a clever way of illuminating different parts of the brain. On a trip to Japan he had seen billboards glowing in the dark and realized that the paint used to create these luminescent ads might provide the answer. He ordered some of the special paint, which was delivered to him in due course at the zoology department in Bangor, Wales. Handing over the package, the postman said, "I don't know what's in this, but it's warm."[65] Of course it was! The luminescent paint contained radioactive promethium. By placing tiny spots of this paint on needles and seeing which areas of the brain responded to these pinpricks of light by secreting hormones, Follett and his colleagues were able to show that the photoreceptor lies in the hypothalamus.

Further work, in which they burned out tiny areas of the brain and cut lesions between different regions, allowed Follett and his coworkers to identify two or three distinct areas of the brain essential for the photoperiodic response. They asked themselves: Was one area the photoreceptor, one the clock, and one the neurosecretory pathway? They did not know, and indeed, when we talked to

Brian Follett in October 2011, twenty years after he retired from research, it is still not known. Even with the use of fiber optics to stimulate different parts of the brain, the precise location of the photoreceptor(s) remains a mystery, but as time passes it does appear as if the bird's brain contains a number of photoreceptors—possibly each attached to a different physiological function. This is common across other vertebrates but not in mammals, where the only photoreceptors lie in the eye.

Farner and Follett's work was mainly on captive birds, and they assumed that the patterns they observed were the same as those in wild birds. It was John Wingfield who challenged this and in the 1970s launched the new era of field endocrinology. Born in Derbyshire, England, Wingfield was interested from an early age in the comings and goings of birds and also in physiology; he dissected any dead animal that he came across. He studied at the University of Sheffield whose zoology department, in the 1960s and 1970s, was composed largely of endocrinologists and was the endocrine center of Britain. Yet while enthused by comparative endocrinology, Wingfield was a natural historian at

heart, and he hankered after some training in animal behavior.

After completing his degree in 1970, Wingfield applied to do a PhD in Jimmy Dodd's department at the University of Bangor, Wales, where Brian Follett had recently been appointed. Follett was unable to accommodate Wingfield, so he opted to work instead with Andrew Grimm on the endocrinology of fish. It proved to be good training, and Wingfield developed a way of simultaneously obtaining estimates of several different steroid hormones—including testosterone, estradiol, progesterone, corticosterone, and luteinizing hormone (LH)—from a single blood sample. This was a major step forward, and it put him very much in demand when he looked for a postdoctoral position; he had offers from Menaker (Texas), Al Meier (Louisiana), and Farner (by then at the University of Washington). Encouraged by Follett, he chose Farner and moved to Seattle in 1973. Farner more or less greeted Wingfield with the words "Well, let's see this assay then!"[66]

Wingfield's first task was to add the steroid hormone immunoassays to that of LH, which had been put in place some months earlier. He was also finally able to work on wild birds. When spring came he decided that the returning White-crowned Sparrow population would be a good system in which to try both his multisteroid and Follett's LH assays. Up to this point only captive birds had been sampled, with one exception: Temple (1974) obtained blood samples from wild Common Starlings, but he did so by shooting them. Wingfield, being a birder, wanted to catch wild birds, take a blood sample for hormone analysis, and release the bird so it could be resampled later in its breeding cycle. Proposing this unusual approach to

Brian Follett (age 72) in conversation with Mike Menaker (age 78) at a conference in Oxford in 2011.

Farner, Wingfield said that "Don listened to me in his polite way with one eyebrow raised and puffing on his ever-present pipe."[67] Fortunately, Farner could see the potential and "to his credit, and after my infernal pestering, he gave me a chance."[68]

It was a chance that paid off. By following individually color-ringed birds whose precise stage in the breeding cycle was known, Wingfield was able to demonstrate the dynamic nature of hormone secretion (see Wingfield and Farner 1976). Prior to this, birds' annual cycles had been lumped into crude categories, such as early breeding, incubating, or feeding young, and hormone levels within each category were averaged across time. The fine scale detail and fluctuations revealed by Wingfield's new approach was so much more informative than anything that had gone before. He recalls Farner's reaction:

> He came into my office and he had his pipe, his ever-present pipe, and he looked at the stuff on my table. You could see the temporal patterns of hormones completely different from what we were expecting and from anything that had been collected in the lab, even though they were mature birds. . . . And I remember him standing there and looking at this and he looked at me and said "handsome data John, handsome data."[69]

Field endocrinology had begun.

The gulf that separated ultimate and proximate approaches to behavior was at its widest in the 1970s. Typically, ornithologists (and other biologists) did one or the other, not both. Luckily, Peter Marler, then at the Rockefeller Field Station in Millbrook, New York, could see the advantages of combining the two. Starting in 1981, Wingfield spent a productive six years in Marler's group.

Wingfield wanted to follow up on some of the pioneering work of Adam Watson (chapter 10), who in the late 1960s had looked at the behavioral consequences of testosterone implants on wild male red grouse. The effect was dramatic: Watson's testosterone-enhanced grouse became more aggressive, obtained a larger territory, and in some cases even acquired a second female (see Watson and Parr 1981).

Wingfield was puzzled. If only a slight increase in testosterone gave males such an advantage, why didn't selection favor all males having more testosterone? The answer, it was later discovered, was because testosterone carries costs. Bengt Silverin, working on European Pied Flycatchers in Sweden, was the first to show that increased testosterone levels reduced male parental care (Silverin 1980). Bob Hegner and Wingfield (1986, 1987) demonstrated the same effect in House Sparrows: males with experimentally elevated levels of testosterone neglected their young and had reduced breeding success. Wingfield then demonstrated—in what he called the "challenge hypothesis"—that male birds *manage* their testosterone levels in a way that enables them to exhibit both aggression *and* paternal care. Most of the time, circulating testosterone levels are low, elevated only in response to a challenge, such as by a territory intruder or during competition over a sexually receptive female (Wingfield et al. 1990).

In the mid-1980s Ellen Ketterson and her husband, Val Nolan, visited Wingfield's lab to learn his techniques so they could apply them to the Dark-eyed Juncos they were studying. The aim was to use hormone implants (mainly testosterone) to establish the endocrine basis of life history trade-offs, in what they refer to as "phenotype engineering" (Ketterson and Nolan 1992; Ketterson et al. 1996).

In a series of elegant experiments, Ketterson found that male juncos whose testosterone levels were artificially elevated invested more in reproduction—mating, seeking additional partners, siring more offspring—but less into parental behavior and, significantly, less into self-maintenance. As a result testosterone-enhanced males survived less well. Ketterson and her colleagues, graduate students, and postdocs have continued to study the way natural selection operates on traits like hormone levels. As Trevor Price has commented, "Dr. Ketterson's work is close to being unique in that it couples results from long-term study with experimental manipulation. . . . [Her recent research on the junco system is] perhaps the model study of the way in which links between physiology and fitness are being made in natural populations."[70]

AN INTERNAL CALENDAR

But is a circadian-based photoperiodic clock the only way for birds (and other creatures) to measure time and build a calendar? Fascinatingly, no. In 1960, just before taking up his post at Monash University, Jock Marshall obtained convincing evidence for the existence of some kind of internal calendar regulating migration and breeding. Together with seabird biologist Dom Serventy, he studied the Short-tailed Shearwater, the Australian "Muttonbird": "The astonishing regularity of their Australian landfall and subsequent egg-laying seemed to suggest [this] might be timed by some astronomical constant, possibly decreasing day-lengths."[71] To establish whether an external cue, like day length, or an internal

clock was responsible for this remarkable time keeping, Marshall and Serventy captured shearwaters and kept them for several months under either a natural or an artificial (twelve hours light, twelve hours dark) light regime, before dissecting them to examine their gonads. By today's standards it was a crude experiment—many birds died and the final sample size was small—but the results showed clearly that regardless of the conditions the birds experienced, their breeding rhythm remained unaltered, suggesting the existence of an internal circannual clock.

Marshall and Serventy's study was inspired by the work of Jürgen Aschoff, now acknowledged as the founder of the study of biological rhythms. Born in 1913, the son of a famous pathologist, Aschoff studied medicine at Bonn in Germany, then moved to the University of Göttingen to work with Hermann Rein on the physiological basis of thermoregulation. In 1952 Rein moved to the new Max Planck Institute for Medical Research at Heidelberg, taking Aschoff with him. After noticing a spontaneous twenty-four-hour cycle of heat loss in his own body, Aschoff wondered about the physiological basis of this rhythm. He was aware of Gustav Kramer's work from 1950 on the sun compass (chapter 4), which implied that birds had a sense of time. He also knew of studies, like those cited above, that demonstrated persistent activity rhythms in animals kept under conditions of constant light or dark, but it was not known whether or not these rhythms were innate.

By raising birds and mice under constant conditions, Aschoff confirmed the innate nature of these activity rhythms. He also noticed that chaffinches kept in the dark exhibited a daily cycle of activity, but he saw that rather being twenty-four hours long it was closer to twenty-two hours, gradually "drifting" over time. This is a free-running activity rhythm, as there is no environmental cue that would maintain a twenty-two-hour cycle. It also means that in order to maintain a twenty-four-hour rhythm, some kind of timekeeper was required (Aschoff 1960).

In 1958 Aschoff was offered a position by Konrad Lorenz and Erich von Holst at their newly formed Max Planck Institute for Behavioral Physiology at Andechs/Seewiesen, near Munich, to study biological rhythms. His research "led to a new conceptual view of the synchronization of circadian rhythms. It postulated an innate biological oscillator which under natural conditions is synchronized with the Earth's rotation by the response to a 'zeitgeber'—a word Aschoff coined and contributed to the English language."[72] A "zeitgeber" is a timekeeper, and the best timekeeper or synchronizer of endogenous rhythms is, of course, the light-dark cycle. "As early as 1955, Aschoff forecasted the existence of endogenous circannual timers in annual reproduction."[73]

The circadian rhythms of birds we now know are controlled by a number of neural clocks located in the hypothalamus, the retina, and the pineal gland. This is a complex system, much more so than in mammals, where the central clocks are located in only the hypothalamus. It seems likely, although far from proven, that the various clocks within the bird's brain are coordinated by the hormone melatonin. The pineal gland and the retina both secrete melatonin in a cyclic manner, peaking in the hours of darkness in quail and weaverbirds (Ralph et al. 1967). Suzanna Gaston and Mike Menaker (1968) provided the first evidence that the pineal gland was important by showing that

House Sparrows kept in darkness, with the nerves to their pineal gland cut, no longer exhibited a circadian rhythm. In the 1970s Fred Turek and colleagues (1976) conducted the reciprocal experiment and showed that melatonin implants in House Sparrows had marked effects on their circadian rhythms.

One of Aschoff's star students was Ebo Gwinner. Encouraged by Hans Löhrl, one of the best known German ornithologists of the 1960s, Gwinner chose not to go into the family locksmith business but to become a zoologist instead. He started studying biology in 1958 and soon focused on the social behavior of hand-reared Northern Ravens. At the beginning his supervisor was Gustav Kramer, but Kramer died in an accident the following year (chapter 4) and Gwinner completed his PhD under the inspiring supervision of Aschoff, Löhrl, and Konrad Lorenz. During this time, Gwinner came under the spell of the ethological approach, an attachment he maintained even though he later enthusiastically embraced new research paradigms.

In 1964 Aschoff invited Gwinner to join him as a postdoc at the institute in Andechs/ Seewiesen. Aschoff's group had already made pioneering discoveries about the circadian rhythms that underlie daily behavior in a range of taxa including humans, insects, mammals, and birds. Gwinner started his own circadian rhythm experiments on birds, showing (among other things) that birds could synchronize their endogenous clocks by song. He also began to study the behavioral shifts of captive migrants during the migration season, becoming ever more intrigued by the annual timing of migration and reproduction.

For birds at high latitudes the circadian-based photoperiodic clock offers the best solution, but what about those species that winter in the tropics where the regular changes in day length are slight? Willow Warblers are a good example, and in the mid-1960s Gwinner set about examining this by taking young birds before their first migration and keeping them under constant 12:12 light:dark conditions for several years. Despite the absence of any environmental cues, the Willow Warblers maintained their annual pattern of migratory restlessness and other seasonal processes, confirming the existence of an internal circannual clock.

Gwinner then extended this work in two ways: he added a comparative aspect by including Wood Warblers, Common Chiffchaffs, and Western Bonelli's Warblers to the study, and he complemented the experiments on captive birds with those living under more natural conditions. In an ambitious experiment, he transported warblers to their wintering grounds in the Congo at the very start of their migratory restlessness. If they were using environmental cues, they should stop "migrating" once they were in the wintering area. They didn't, and on the basis of this Gwinner suggested that warblers possess an internal (endogenous) program that carries them in a particular direction at a particular speed until the drive runs out, leaving the birds in their wintering area. As further evidence for an inherited program, Gwinner found that the different leaf warbler species retained their species-specific patterns of migratory restlessness. Making use of Konrad Lorenz's setup, Gwinner put the idea of inherited programs to yet another test: he analyzed the egg-laying dates of the different goose species kept at the institute in Andechs/Seewiesen, and of their hybrids, confirming his prediction that timing was

under genetic control. His remarkable suite of studies (e.g., see Gwinner and Helm 2003) revolutionized the way ornithologists thought about endogenous programs, resulting in Gwinner receiving in 1973 the first Erwin Stresemann prize of the Deutsche Ornithologen-Gesellschaft, the German Ornithological Society.

In 1968 the Vogelwarte Radolfzell, a ringing and research station near Konstanz in southern Germany, became part of Aschoff's institute in which Gwinner worked. Backed by Aschoff, Gwinner was able to initiate a large-scale project there on comparative migration biology of *Sylvia* warblers, in collaboration with Peter Berthold (chapter 4). That project involved the simultaneous study of migrating warblers at three different stopover sites and extensive work with captive birds. Gwinner continued to oversee this project formally in his capacity as head of the Vogelwarte Radolfzell from 1979 until 1998, but as it turned out he spent only a few years actually stationed in Radolfzell.

Seeking further inspiration, Gwinner set out to the United States in 1969 to work at the two leading labs in his field of interest, first with Don Farner. Although he found the intense work regime in Farner's lab a "harrowing experience," he later emulated Farner by "keeping abreast of novel techniques and incorporating them smoothly in his own work."[74] He then went to Stanford University to work with Colin Pittendrigh, who, together with Aschoff, is considered the founder of modern chronobiology. Gwinner returned to Germany in 1971 to continue his work on the way endogenous programs guide migrating passerines between their winter and summer quarters. As well as looking at circannual rhythms, he continued to investigate circadian rhythms, leaving his mark in many areas, such as by demonstrating the role that melatonin and the pineal play in measuring day length, by showing the effect of testosterone on circadian clocks, by exploring the oscillator underlying *Zugunruhe* (chapter 4), and by elaborating the interplay between different pacemakers for regulating avian circadian clocks (Gwinner and Benzinger 1978).

In 1991 Ebo Gwinner was made director of the Max Planck Institute for Behavioral Physiology, with its official seat in Seewiesen and local research centers in Andechs and Radolfzell. Following this upgrade, the institute experienced another phase of expansion. Existing laboratories for endocrinology and energetics were complemented by work on neurobiology, tissue culture, genetics, and molecular biology. In 1998 Gwinner and Berthold, respectively, were made directors of these two centers, and in 2004 these became the Max Planck Institute for Ornithology, which now resides in both Seewiesen and Radolfzell.

Gwinner died in September 2004 at age sixty-five as the result of an aggressive cancer. In their obituary, Rudi Drent and Serge Daan refer to Gwinner's relationship with Peter Berthold, who he had known since their student days in the 1950s:

> The extended collaboration of these "scientific twins" had a stimulating but also corrective influence as the two investigators kept each other on their toes. . . . In approach and temperament the two differed, but in many ways their efforts were complementary and we must acknowledge the wisdom of the Max Planck philosophy in allowing the two sister departments an existence on their own providing the ideal environment for an element of healthy rivalry. In retrospect the three constellations (the Wiltschkos at Frankfurt . . . Gwinner . . . and Berthold . . .) interacted to provide a series of breakthroughs that underlined the special power of imaginative experimentation with hand-raised birds on a scale never witnessed before.[75]

In his own obituary of Gwinner, Peter Berthold says that when he and Gwinner disagreed they took themselves off to a nearby monastery to sort out their differences, agreeing to speak "with one voice" to the Max Planck Institutes. Gwinner's longtime collaborator Barbara Helm told us that the relationship between Gwinner and Berthold was "complex" and that their respective ideas emerged gradually and particularly from their "regular, famous sessions in local pubs." She said,

> Overall, Ebo Gwinner was an intellectual leader and mentor, with wide academic interests and a great love of chronobiology. He constantly sought conceptual challenges and new techniques to meet them, while Peter, being more of a "just-do-it" man, took to study migration biology at full speed. I believe that the main idea of a comparative breeding

Ebo Gwinner (photo in 2004 at age 65).

program was Ebo's, arising directly from his earlier comparisons of leaf warblers, goose hybrids and *Sylvia* populations, and from his exposure to Lorenz' ideas of inborn instinct, but it was Peter who implemented it on an unexperienced scale. The two men were in many ways complementary: Ebo was deeply involved in the international scientific community and fostered an enthusiastic research group, while Peter was a public figure that attracted great interest in ornithology and bird migration.[76]

The Max Planck Institute at Seewiesen focuses now on behavioral neurobiology and on the evolution of mate choice and parental care, and Radolfzell continues its focus on the ecological and genetic aspects of migration.[77]

CODA

With the exception of work on avian breeding cycles, the study of avian physiology has declined in popularity since the 1970s.[78] Thus, funding for physiological research is currently difficult to obtain unless it incorporates molecular techniques. One consequence of this is that avian physiologists have cherry picked a handful of tangential but trendy topics and have abandoned the big questions. As Tony Williams reflects:

> My own view is that the "age-old, questions" ... such as why clutch size, or parental effort varies among individuals, and the consequences of this variation, have not remained intractable simply because of a dearth of non-genomic approaches. Rather we have stopped focusing on certain fundamental questions even though they are unresolved, and which

can be tackled with traditional, experimental approaches (albeit enhanced and aided by genomics knowledge and techniques). Identifying the genes or DNA sequences underlying variation in life-history traits is an important goal (Ellegren and Sheldon 2008), but unveiling the physiological mechanisms linking genotypes to phenotypes, and ultimately to fitness, in natural populations is even more important.[79]

While this chapter may give the impression that we know a great deal about the reproductive physiology of birds, most of what we know relates only to males; many fundamental aspects of female reproduction remain unexplored. A striking example is that we know almost nothing about what controls the speed with which females come into breeding condition. A male passerine—like the juncos studied by Rowan—typically takes six or seven weeks to develop full-sized testes and assume breeding condition, but a female can go from being reproductively quiescent—with a threadlike oviduct and tiny ovary—to an egg-laying machine in just six or seven days.

Historically, the ease with which male birds can be brought into breeding condition by manipulating their photoperiod has left the study of females in the dark. As a consequence, the study of avian breeding cycles has effectively been a study of male physiology and anatomy. Even after Danny Lehrman and others in the 1960s and 1970s showed that male courtship was essential for bringing females into breeding condition, the way that physiological, metabolic, energetic, and hormonal mechanisms regulate female reproductive cycles remains largely unknown. Farner and Follett (1966)

235

recognized this gap in our knowledge in the 1960s, and it says little about progress that forty years later Gregory Ball and Ellen Ketterson have said the same thing (Ball and Ketterson 2008).

In truth, a great deal *is* known about the reproductive physiology of female birds, based on extremely detailed research on the domestic fowl. Indeed, many of the basic principles of female reproduction, such as the physiological regulation of ovulation, were discovered in domestic fowl but remain known only for that species and a handful of other poultry. However, domestic fowl, turkeys, and ducks differ from wild birds in general and from passerines in particular in a number of important respects, and the kinds of questions asked by field ornithologists are often very different from those asked by poultry researchers. For example, under extreme artificial selection and constant environmental conditions, female poultry very obviously do not require male courtship to bring them into reproductive condition; they do not produce discrete clutches, and if their eggs are removed they usually remain in breeding condition and lay continuously throughout their adult lives. Understandably, poultry biologists have little interest in the way physiological mechanisms influence lifetime reproductive success or the way different individuals respond to environmental conditions.

One of our aims has been to provide a sense of our current state of knowledge, but as researchers we also consider it important to identify what we still need to know and hence to identify new research opportunities. The "molecular revolution" has led to many new discoveries, but by leaving classical comparative physiology behind, we risk overlooking the bigger picture—that is, the whole bird in its environment. Luckily there are exceptions, including two rather different areas of research that we feel have enormous potential. The first is the use of various imaging technologies to explore brain form and function, exemplified by a recent study of American Crows. Using PET (positron emission tomography) to assess brain activity, John Marzluff and colleagues (2012) confirmed that crows use the same visual sensory system as humans (and other mammals) to recognize human faces (see page 427). The potential of brain scanning technology is considerable. The second is the integration of different aspects of physiology with ecology and behavior exemplified by Theunis Piersma and Jan A. van Gils (2011). This innovative approach, inspired by the work of Rudi Drent in the 1980s, has simultaneously increased our understanding of the biology of migratory shorebirds and indicated a worthwhile direction for future form and function studies.

BOX 6.1 Fernando Nottebohm

I was born in 1940 in Buenos Aires. My father was a rancher. My mother, Amelia Grant Menzies, enjoyed nature and encouraged my curiosity. From my earliest age I was fascinated by birds, those in the garden at my family's home in B.A., those at the ranch and birds at the

B.A. zoo. As a boy I collected bird eggs. My mother showed me how to empty them and how to take only freshly laid ones. My mother would also drive me to a lake in the Palermo Park in Buenos Aires. It had lots of ducks. We would take bread and feed the ducks. It gave

me enormous pleasure. Then she would take me to the zoo. I brought with me a long stick with sticky glue at one end. I would reach into the cages and collect lovely feathers. One summer I came back from holidays, and when I opened the box that held my collection of feathers, I found only the rachis of feathers. Moths had eaten the rest. I collected no more feathers. I also had a collection of stones and a box with iron filings that I harvested from the beach in Mar del Plata. I would take a magnet to the beach and move it through dry sand. Iron filings jumped to it, like long black hairs. I felt this form of iron must be of interest and valuable, so it too went into a box, along with ingots of molten lead soldiers. Other boxes held my collection of stamps. I collected stamps from the British Empire. Stamps from the colonies often had lovely birds. As a child I also kept birds — pigeons of various breeds and canaries. The latter I caught in the garden. They had, undoubtedly, escaped from previous owners.

I had mediocre biology teachers in high school and found little inspiration in them. I was raised Catholic, but by age sixteen religion had triggered in me antireligiosity, and that encouraged me to ask questions. High on my list was the issue of the seeming dichotomy between brain and mind. I resolved it fast. It had to be all matter, and matter acted in predictable ways. So mind was what happened in our brains and we were probably just spectators. For some reason, though, we felt responsible for the events that happened in our brain, and consciousness gave rise to the notion of free will. By age seventeen I decided I was a determinist and that God had not created people, but the other way around. A book I read at that time, *The History of Philosophy* by Will Durant, influenced my thinking, but I liked even more reading *Far Away and Long Ago*, by William Hudson, about his boyhood in the pampas, among the birds he loved.

At age eighteen I was totally fascinated by nature and the problem of origins: the origin of the world, of people, of matter, time, and space, of language, and, more generally, of vocal learning. When it came time to go to college, I went to the United States, to the University of Nebraska, in Lincoln, to study agriculture. The plan was to become a modern rancher and return to work with my father, whom I loved. But I found crops and cattle very boring, and after a year I switched from Lincoln's School of Agriculture to the zoology department at Berkeley. There I met Peter Marler. He had arrived a few years earlier from the University of Cambridge, after two PhDs, one in botany and one in zoology, where he had studied chaffinch song as a student of William H. Thorpe, a Cambridge professor. Thorpe had written the first detailed account of song learning in any songbird, and the monograph he wrote in the English journal *Ibis* in 1958, "The Learning of Song Patterns by Birds, with Special Reference to the Song of the Chaffinch, *Fringilla coelebs*" became an instant classic. When Marler moved to Berkeley in the late 1950s, he turned his attention to a local songbird, the White-crowned Sparrow, that like the chaffinch also showed local song dialects.

Peter Marler was my scientific father. I was inordinately fond of the man and loved the way he did science. He avoided dogmas and just asked questions. All he wanted was answers, not a particular answer. I found, later, that some people put great importance on hypothesis and that one should approach one's work as a constant process of hypothesis testing. I prefer, like Marler, to just ask good questions. Careful observation and relentless asking of questions leads to the truth, and one avoids biasing one's thinking by premature allegiance to any one explanation. In later years some of my grant proposals sent to NIMH were turned down for not being "hypothesis driven." I shrugged and thought to myself, "Insects." Plato and Marler advanced their knowledge by asking questions. That was good enough for me.

At the end of my undergraduate training at Berkeley I did an honors project in Marler's lab. Under his guidance I constructed a transparent chamber in which I placed the syrinx of a rooster and ran air through it. To my delight, it produced sounds with the frequency of a crowing cock, and using a stroboscopic light, I was able to show that the frequency of vibration of the tympaniform membranes corresponded to that of the fundamental of the sound produced. That got me interested in the syrinx, and later, during my thesis work, still under Marler, I de-enervated the right or left half of the syrinx of chaffinches. To my delight, I found that chaffinches were left-handed singers. Seeing this evidence of a relation between handedness and vocal learning convinced me that my future lay in working out how the brain of birds acquired its song. For this I had to discover the "song system," and many other things followed. Oh the joy of it all! But there was a dark side. I did not like doing surgery on my little wards. There was nothing in it for them. I was never able to resolve this dilemma, and in my mind I accept that though I was a successful scientist, I was also a criminal. If there is a God and God is a chaffinch, there will be hell to pay. But before then, I still have a few years to work on the origins of things.

BOX 6.2 Ellen Ketterson

All of my professional life I have studied one species of bird—the Dark-eyed Junco or snowbird—pursuing interests in migration, mating systems, hormones, behavior, and evolution. Each successive research question began with some episode in which I wondered, Why do juncos do what they do?

Why, for example, do females make longer migrations than males? Why, when you remove a male from his territory, do some of the males that replace him care for his offspring, while others do not? Why do males differ in their testosterone profiles? If males could be made to have higher testosterone, would that influence how long they lived or how many offspring they produced? Most recently, I have asked why juncos have colonized new environments in Southern California that differ greatly from their ancestral montane environments? And why do juncos look so different from place to place, yet still manage to interbreed? How many species of juncos are there?

My junco journey has taken me to sites all over North and Middle America and brought me in contact with ornithologists around the world. I have found enormous pleasure in the sense of connection that birds have provided to other like-minded people and also to the natural world. I also feel especially lucky to have played a hand in the training of nearly forty PhD and postdoctoral students. Each of these people has enriched my life and contributed to our knowledge of bird behavior and evolution.

I was born in Orange, New Jersey, in 1945 at the close of the Second World War, the third child, second daughter, of Lois Meadows and John Boyd Ketterson. I was strongly shaped by my mother's belief in the importance of learning. She was the daughter of southern Baptist missionaries to China at the beginning of the twentieth century, and she believed in sharing, tolerance, and other virtues. I was not able to assimilate all her beliefs, but along with my husband, Val Nolan, she and my sister Emily were the strongest influences in my life.

My brother John was clearly a factor in my becoming an ornithologist. John is eleven years older than I and a physicist at Northwestern University. But as a teen he was an avid bird watcher who spent his time outdoors. At one point he decided that his little sister should learn the birds in his bird book before she could read, which I apparently did.

I attended Indiana University in Bloomington, Indiana, for all of my university degrees. As a PhD student in the early 1970s I studied ecology under Val Nolan Jr. and pursued postdoctoral research in ecophysiology with James R. King at Washington State University. After a short stint as assistant professor at Bowling Green State University in Ohio, I returned to Indiana, where I have been on the faculty ever since.

In 1980 I married Val Nolan, and we were academic and life partners until he died in 2008. Almost every decision I make on a daily basis is influenced by the time spent with Val. During my first academic leave, I was also lucky to work with John Wingfield at Rockefeller University.

Many of the ornithologists I admired—Glen Woolfenden, Frank Gill, Bert Murray, Joe Jehl—were boy birders who became professionals after they learned it was possible to earn a living studying birds. Despite my early exposure to pictures of birds, my path was different—I forgot about birds early on and was encouraged by my mother to be interested in flowers with no particular goal other than to be an educated woman. I grew up at a time when career opportunities for women were narrower, and my expectations were shaped by my mother's advice. Thus I went to college expecting to find a husband who would set the agenda, and I would follow.

But I also had scholarly ambitions, and after reading Konrad Lorenz's *On Aggression* and *King Solomon's Ring*, I decided to study animal behavior. This desire led me to Val, who was both a law professor and an ornithologist who had taught himself biology. Val's monograph, *Ecology and Behavior of the Prairie Warbler*, was awarded a Brewster Medal by the AOU. Owing to his enormous intellectual capacity and determination, and to some wise administrators, his accomplishments in bird biology led to his being appointed professor of both law and zoology. Val took that opportunity to found an ornithological dynasty at Indiana University—he advised seventeen PhD students who have made lasting contributions to our understanding of birds, including Charles Thompson, Ken Yasukawa, and Dan Cristol.

I was Val's third or fourth student, and I briefly studied Prairie Warbler vocalizations. But the question that fired my curiosity was why junco populations in Indiana consisted of more males than females during winter. Val had banded juncos for many years and knew that the sex ratio was nearly equal during migration but male biased in winter. Where were the females? I sought them by sampling populations in the field, which entailed its own challenges, such as running out of gas on my first solo trip to Alabama, where I caught only a single junco (a female). In time I found the females in the southern United States. It turned out that juncos exhibit differential migration and females make longer migrations than males. This pattern is now known to typify many species of bird, and the question interests me still. Newly developed methods in ecoimmunology and our ability to track free-living birds promise to answer this and other long-standing questions in migration biology.

My interest in evolutionary biology was sparked by our studies of hormones and male parental behavior in the junco. In the late 1980s Val and I began a series of studies that we later referred to as "phenotypic engineering with hormones." We used hormone implants to "create" male juncos that invested less in parental behavior, but the "ah ha moment" came when we realized that we had altered more than one thing—males treated with testosterone sang more, had larger home ranges, and were more attractive to females. So we committed to learning how natural and sexual selection would act on suites of traits influenced by testosterone. Pursuit of this goal involved spending our summers, from 1983 to 2001, living in a small cabin at Mountain Lake Biological Station in Virginia, with all the joys of working around the clock with students and field assistants. I still study the juncos at Mountain Lake, and our nearly thirty-year effort continues to yield insights into annual variation in extra-pair mating, the impact of changing temperatures on breeding phenology, and the reasons for variation in life histories.

The common theme in all my research has been organismal biology in relation to ecology, evolution, and behavior. For me it is essential to study birds in the wild . . . to follow where the bird leads . . . and to conduct research in collaboration with others who make the research better and your life more meaningful.

BOX 6.3 Theunis Piersma

I grew up in a small village in the countryside of Friesland, a rural province in northern Netherlands. My father was a country vet, and at home we were surrounded by animals. I kept rabbits, goats, chickens, and parakeets and other birds, and I was especially fond of our cats. The village greengrocer took me to find lapwing eggs, the church caretaker took me out to check nest-boxes. I was the only naturalist child in our village but otherwise did not stand out. Family vacations took us to the Mediterranean, where I spent much time snorkeling, shell collecting, and watching lizards and ants. High school provided incentives to be creative in theater and pop bands, to wander in the countryside photographing places and people, and to slowly become aware of the power of the seasons: how landscapes and bird communities show profound and repeated change.

Fired by this youthful enthusiasm for biology in general, and sea life in particular, the choice for the University of Groningen was easy. With a tradition in ethology and animal ecology, it is close to the Wadden Sea and well known for its links with the Netherlands Institute for Sea Research–NIOZ, an attractive scientific bastion. During spring and summer, weekends back home were spent studying the roosting of shorebirds along Lake IJsselmeer. In Groningen I soon joined a group of shorebird enthusiasts, and with them I started counting waders and working on ecological studies in the Wadden Sea. The disappearance in autumn of so many shorebirds kindled questions about their migrations. I soon found quiet, understated support in Rudi Drent, who was to become my thesis supervisor. He always mentored us when working out of the Zoological Laboratory as undergraduates or unemployed enthusiasts.

Early 1980 found four of us on an expedition to the Banc d'Arguin in Mauritania, a region with large but previously uncounted numbers of shorebirds. Financed by Prins Bernhard Cultuurfonds and the magazine *Nieuwe Revu*, this turned into an adventure that I only just survived, as I contracted a rare viral disease and had to be flown out by helicopter. We counted more than 2 million waders and found high feeding densities on pretty scarce benthic resources, and somehow we managed to write up the results promptly. This first experience with migratory shorebirds on tropical mudflats soon led to further adventures, in the springs of 1981 and 1982, in coastal Morocco, where we studied the northward return migration. These exciting studies were the prelude to follow-up expeditions, in 1985, 1986, and 1988, to the Banc d'Arguin that yielded a rich scientific harvest.

The organization of these expeditions brought me in contact with the Wader Study

Group, then a group of British professional and amateur shorebird workers that was beginning to reach out internationally. Soon I was involved in the group, and my first voluntary tasks were the writing of a paper on an international cooperative study of northward migrating Red Knots and the compilation of data on breeding wader populations in Europe. At this time I also initiated projects with amateur wader-ringers back in Friesland. I carried out most of these activities independently of the formal requirements for a biology master's degree. In the "official" part I studied the circannual body mass changes and energetics of captive shorebirds, communication in Little Gulls, and body composition, molt, and diet of Great Crested Grebes on Lake IJsselmeer, a study that I continued during eighteen months of civil service. I also worked with Leo Zwarts to come to grips with approaches to study interactions between shorebirds and benthic food stocks. It became clear to me that traditional approaches to the study of shorebird predation—mainly consisting of correlating benthic biomass with shorebird densities at various scales—were unlikely to help us understand the structuring principles of these communities. Instead, further experiments unraveling the mechanisms of interactions between benthic prey and shorebird predators (i.e., habitat selection and foraging decisions of the predator in the context of recruitment processes, growth, and predator-avoidance behavior on the part of the prey) would probably yield some insights.

What I ended up looking for was an easy-to-study system with a long-distance migrant shorebird eating benthic prey. It took me several years of unemployment to find the system (and the resources!) to start that PhD project. During the intervening four years, I expanded my horizons with respect to waders and intertidal benthos by conducting fieldwork in Korea, doing body compositional analyses of birds, and, perhaps most important, learning some limnology (during my civil service). The latter work—on the interactions between populations of Great Crested Grebes and their pelagic smelt prey, with perch and pikeperch as competing fish predators—showed me the possibilities of large-scale ecological research.

For my PhD work I settled on a shorebird that I was familiar with through the Wader Study Group—the Red Knot. This species migrates long distances, shows intraspecific population structure of recognizable morphs with different migration routes and annual schedules, and eats only hard-shelled molluscs in soft sediments in coastal nonbreeding habitats and only surface-living arthropods on its tundra breeding grounds. Intriguingly, but I did not know this when choosing Red Knots, the birds go through great changes with respect to levels of nutrient and energy stores, digestive capacity, and sustained working power.

I guess I never looked back. The PhD project was productive, and after graduation I landed the best job that I could dream of, as ornithological researcher at NIOZ. With the prey choice and nutritional demands of this shorebird predator pinned down, and the availability of its main bivalve prey partly determined, it seemed time to expand the scope of the studies and to look at the ongoing evolutionary interactions between shorebirds and their intertidal benthic prey. A PIONIER grant of the Netherlands Organisation for Scientific Research in 1996 made my activities come full circle. Having started as a biology-interested shell collector in my hopeful teens, I was coming back to molluscs as a professional marine biologist looking at them through the

Red Knots.

eyes of their avian predators. The PIONIER grant enabled me to bring together in a single team various young specialists sharing a research vision. The gradual loss of good study areas in the Dutch Wadden Sea at this time was highly disturbing, with declining populations of predators and prey attributable to ongoing industrial shellfishing.

Through my contact with amateur wader catchers I had learned about nonmarine (freshwater) waders. This may have yielded my most gratifying insight: seeing an exciting ecological contrast between marine and freshwater species of shorebirds and the possibility that (genetically based or expressed) immunocompetence and habitat-determined disease factors may mechanistically and evolutionarily explain this difference. Testing these ideas would fall largely outside the scope of my research remit (which was constrained to marine ecology), but a partial move in 2003 toward the position of animal ecology

professor at the University of Groningen took care of this. The key word since that move has been "comparison," with respect to phenotypes (including immunological aspects) and vital statistics such as recruitment, survival, and dispersal—between populations, between species, between habitats, between seasons. In 2006, together with Allan J. Baker and other friends, I started the Global Flyway Network, a worldwide consortium of shorebird scientists devoted to long-term studies of movements and demography. This enterprise is rapidly gaining relevance in a world with so much habitat destruction. One way to fight the cynicism that comes so easily when confronted with rampant loss of beautiful places is to look at cases of habitat destruction as informative and interesting experiments. We try to learn from the fates of individually marked birds in well-described ecological contexts and hope that the stories that result help avoid further losses.

©Robert Bateman
1979

CHAPTER 7

The Study of Instinct

This observation, like all other observations of animal behaviour, however trivial
they may seem to be, gives rise to the question upon which the scientific study
of behaviour, or ethology, is based: Why does the animal behave as it does?

NIKO TINBERGEN (1951), IN THE PREFACE TO HIS BOOK *THE STUDY OF INSTINCT*

BEHAVIORAL RESPONSES

IT IS LATE AFTERNOON ON AN EARLY WINTER evening; a flock of Common Starlings is foraging busily in a grassy field. A Peregrine Falcon appears high in the sky, and the starlings crouch in fear. As the raptor glides away, apparently uninterested, the starlings resume feeding in the grass. They eventually take flight in an untidy flock and head toward their roost. Suddenly the peregrine is back, and in an instant the starlings bunch together in flight, weaving and undulating in unison as the raptor swoops at them. The peregrine is trying to break up the flock and isolate a victim, but the tactic fails. The starlings disappear into their reed bed roost; the falcon searches elsewhere.

In his landmark book, *The Study of Instinct*, Niko Tinbergen (1951) used his observations on the response of birds to the presence of an aerial predator to draw some general lessons about bird behavior. The most important of these was that the escape response is instinctive, triggered by the raptor's distinctive silhouette.

Tinbergen's conclusions emerged from his observations and experiments, including some conducted in 1936 with Konrad Lorenz, in which they recorded the reactions of young (and hence behaviorally naive) domestic turkeys, ducks, and geese to a model bird "flying" overhead (Lorenz 1939; Tinbergen 1939). While flawed in certain respects, as we discuss later, these and other experiments were important in shaping Tinbergen's and Lorenz's developing ideas. Their bird model was a cardboard silhouette, shaped like a T, with a short "neck," a long "tail," and symmetrically shaped wings. Remarkably, when

Adjacent pairs of nesting Black-legged Kittiwakes greet their partners. Painting by Robert Bateman in 1979.

the model was flown with the short end first, the young birds crouched in fear, but if it passed over in the opposite direction, with the long end first, the young birds ignored it. Tinbergen realized immediately that the young birds' interpretation of the silhouette must have been based on the direction of movement—a short neck and long tail saying "predator" or "hawk" and a long neck and short tail saying "no predator" or "goose" (Tinbergen 1948).

Tinbergen and Lorenz's experiments with model predators were not without precedent. Sixty years earlier, in the late 1870s, English zoologist Douglas Spalding reported how his twelve-day-old domestic fowl chicks " . . . while running about beside me, gave the peculiar chirr whereby they announce the approach of danger. I looked up and behold a sparrow-hawk [presumably a Common Kestrel] was hovering at great height overhead."[1] As Philip Gray describes it, "Spalding's next step was to take his tamed and pinioned hawk and throw it over a hen and her brood of week-old chicks. On the sounding of a call from the hen the chicks immediately ran and hid while the hen itself attacked the hawk when it landed."[2]

In a similar vein, the German ornithologists Oskar and Magdalena Heinroth commented in the 1920s that young birds sometimes exhibited the escape reaction in response to the sudden appearance of Common Swifts, whose silhouette resembles a predator (Heinroth and Heinroth 1924–34). That idea was first tested in the 1930s—before Lorenz and Tinbergen conducted their experiments—by German ornithologist Friedrich Goethe, who exposed Western Capercaillie chicks to differently shaped silhouettes and noted that the more

closely those shapes resembled a predator, the greater the escape response (Goethe 1940; cited in Schleidt et al. 2011). Tinbergen and Lorenz were also aware of the work by another German, Heinrich Krätzig—a promising young researcher killed in the last days of the Second World War—who tested hawk-goose silhouettes with Willow Ptarmigan chicks. It was he who first noted how a T-shaped model presented as a "hawk" elicited an escape response, while the same model presented in the "goose" direction did not (Krätzig 1940).

As Tinbergen recognized, the behavior of the young birds could be interpreted at several different levels. First, their ability to identify a predator seemed to be innate, or instinctive—not something that a young bird had to learn. Second, through ingenious forward-and-backward experiments with the T-shaped silhouette, he identified the cause of the prey's evasive response: only the short neck and long tail released the escape behavior. Third, Tinbergen considered the adaptive significance of the behavior, recognizing that by bunching tightly together in flight, starlings reduced the chance of attack. He sometimes called this the "immediate function" of behavior, meaning its advantage in terms of survival.

Niko Tinbergen was one of the founders of ethology—the study of the behavior of animals in their natural environment—a field that is now called simply "animal behavior." Ethology's main focus, initially at least, was on instinct and the causation of behavior. Tinbergen's genius, however, was to see that there were also other questions that could be asked about behavior, and it was this insight, among other things, that set him apart from those who simply watched and wrote about

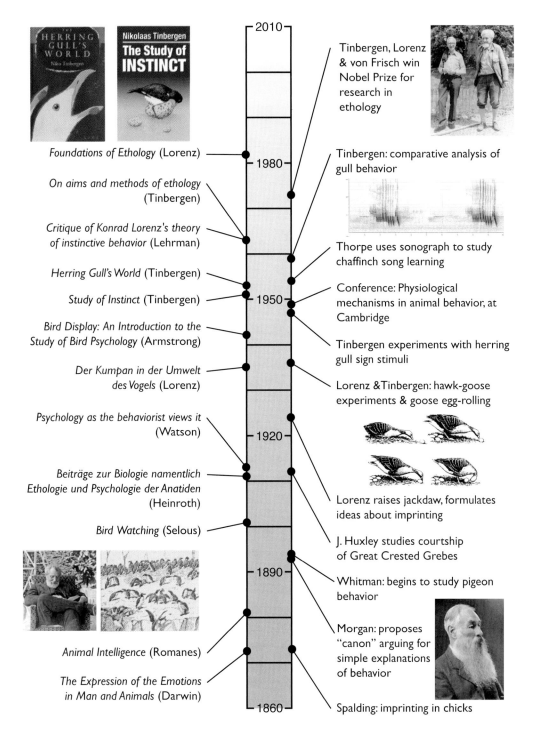

2010

Tinbergen, Lorenz
& von Frisch win
Nobel Prize for
research in
ethology

Foundations of Ethology (Lorenz)

1980

Tinbergen: comparative analysis of
gull behavior

On aims and methods of ethology
(Tinbergen)

Critique of Konrad Lorenz's theory
of instinctive behavior (Lehrman)

Thorpe uses sonograph to study
chaffinch song learning

Herring Gull's World (Tinbergen)

Study of Instinct (Tinbergen)

1950

Conference: Physiological
mechanisms in animal behavior, at
Cambridge

Bird Display: An Introduction to the
Study of Bird Psychology (Armstrong)

Tinbergen experiments with herring
gull sign stimuli

Der Kumpan in der Umwelt
des Vogels (Lorenz)

Lorenz &Tinbergen: hawk-goose
experiments & goose egg-rolling

Psychology as the behaviorist views it
(Watson)

1920

Beiträge zur Biologie namentlich
Ethologie und Psychologie der Anatiden
(Heinroth)

Lorenz raises jackdaw, formulates
ideas about imprinting

Bird Watching (Selous)

J. Huxley studies courtship
of Great Crested Grebes

1890

Whitman: begins to study pigeon
behavior

Morgan: proposes
"canon" arguing for
simple explanations
of behavior

Animal Intelligence (Romanes)

The Expression of the Emotions
in Man and Animals (Darwin)

1860

Spalding: imprinting in chicks

TIMELINE for ETHOLOGY. Left: Covers of Tinbergen's books *The Herring Gull's World* (1953b) and *The Study of Instinct* (1951); Edmund Selous; detail from drawing of a Common Murre colony in Selous's *Bird Watching*. Right: Tinbergen and Lorenz at Madingley; sonogram of a Common Chaffinch song; goose retrieving egg; Conwy Lloyd Morgan.

Niko Tinbergen (probably in the 1960s when he was in his late 50s or early 60s).

birds behaving. Few ideas in science are truly original, and Tinbergen like many others built on the ideas of his predecessors. To see who these were and how Tinbergen rose to such eminence, we need to go back to the beginning of the nineteenth century and the very beginnings of field ornithology.

ETHOLOGY BEGINS

Charles Darwin can be considered the father of the modern study of animal behavior because of his books *The Descent of Man* (1871) and *The Expression of the Emotions in Man and Animals* (1872). By showing that the differences between humans and other animals were only in degree, Darwin paved the way for the comparative study of behavior. His information about nonhumans was based mainly on "facts" and observations of domesticated or captive animals, most often sent to him by colleagues, including

his friend, George Romanes, who used the same approach in his book *Animal Intelligence* (1881). An exceptionally competent comparative physiologist, Romanes—like Darwin—took many of the reports that he received from his correspondents at face value. However, these reports were often biased by superstition, religion, and other preconceived notions and really lacked the objectivity necessary to form sound conclusions about animal behavior. Today Romanes's book, the first real treatise on "comparative psychology," is remembered as merely a collection of anecdotes, including several relating to "clever birds" (see chapter 8).

What Darwin started in the mid-1800s was by the turn of the twentieth century becoming the scientific study of animal behavior. The most notable advances occurred in Europe, North America, and Russia. In Britain, Darwin's evolutionary ideas provided the inspiration for four key figures, now considered the forefathers of bird behavior research: Edmund Selous, Frederick Kirkman, Eliot Howard, and Julian Huxley.

Early in the 1900s Edmund Selous made groundbreaking contributions to the study of bird behavior by watching birds in the field. Selous was an "amateur," operating outside the museum-based research—then considered to be "scientific ornithology"—whose sole objective was to understand the taxonomy and geographic distribution of birds (chapter 3). For museum ornithologists, the very idea of spending time watching bird behavior in the field was eccentric, and they were "united in a common hatred and contempt for the field-naturalists,"[3] typified by Selous and his followers. For his part Selous (1901b) was consumed with loathing for the museum men:

I must confess that I once belonged to this great, poor army of killers, though happily, a bad shot, a most fatigable collector, and a poor half-hearted bungler, generally. But now that I have watched birds closely, the killing of them seems to me as something monstrous and horrible; and, for every one that I have shot, or even only shot at and missed, I hate myself with an increasing hatred . . . the pleasure that belongs to observation and inference is, really, far greater than that which attends any kind of skill or dexterity, even when death and pain add their zest to the latter. Let anyone who has an eye and a brain (but especially the latter), lay down the gun and take up the glasses for a week, a day, even for an hour, if he is lucky, and he will never wish to change back again. He will soon come to regard the killing of birds as not only brutal, but dreadfully silly, and his gun and cartridges, once so dear, will be to him, hereafter, as the toys of childhood are to the grown man.[4]

Over his lifetime Selous studied a wide range of species, including Ruffs, Rooks, and Black Grouse, described in his detailed "observational diaries" in the journal *Zoologist* and in a series of popular books. Retiring and socially awkward, Selous lived in the shadow of his famous brother Frederick, the big game hunter. Perhaps Edmund's scorn for the professional ornithologists was a reaction against his macho brother. Whatever his motivation, Selous was a superb observer, and his first book, *Bird Watching* (1901b), received particular praise from W. Warde Fowler in the *Saturday Review*: "Reading Mr Selous's book I feel that if I were beginning life again, I would give all my spare time to watching as he has watched. He has taken a new departure, and needs to

be supplemented and tested. There is a wide field in front of the beginner who will follow in his footsteps."[5] Among his many "firsts" was Selous's description in 1902 of the bizarre precopulatory display of the Dunnock (not published until much later: Selous 1933). But his florid writing style meant that it was—and still is—often difficult to identify his many discoveries. This style, his reputation for being "difficult," and his disdain of the professional ornithologists of the day, meant that few considered Selous a key player in scientific ornithology during his lifetime.[6] Only much later did David Lack (1959) include him among the most influential ornithologists of the twentieth century.

While Selous wanted nothing to do with professional ornithology, another British ornithologist, Frederick Kirkman, worked hard to try and bridge the divide between field and museum studies. Less confrontational in his writings than Selous, Kirkman contributed to ethology in several ways: through his long-term studies of the Black-headed Gull (which was published before some of Tinbergen's work), his involvement with the Institute for the Study of Animal Behaviour (of which he was a founding member in 1936) in London, and his work as editor of the *British Bird Book* (1910–13). The aim of that multivolume work was to describe the habits of all British bird species, with hundreds of photographs and paintings showing birds behaving in their natural environment rather than in the conventional "perched-on-a twig" pose that was typical of bird books until then. Selous was an early, invited contributor to that series of books, but Kirkman removed him from the project after Selous expressed his disgust of the egg collectors and "museum men" in some of his accounts. As Kirkman said in a letter to Selous: "Were it not for his [the egg collector's] labours, the egg descriptions in the Book could not have been written; to cast reflections on him in a book which profits by his collections is an absurdity which I feel I could not be expected to foresee that anyone would perpetrate."[7]

The third English pioneer was Eliot Howard, who took the observation of bird behavior to a different level as he tried to understand the minds of birds. A successful businessman from Worcestershire, Howard got up at dawn each day to study the behavior of birds in the field before going to work. He is now best remembered for introducing—in 1920—the concept of territory in bird life, a "discovery" that changed the course of ornithology (chapter 10). Howard's ideas on the avian mind were laid out in 1929 in his *An Introduction to the Study of Bird Behaviour*. In that undertaking, he was mentored by the comparative psychologist Conwy Lloyd Morgan at University College Bristol (who made his own contribution to the study of bird behavior; see below). Struggling to find a publisher, Howard told Morgan of his difficulties, to which Morgan replied:

The plague of it is that we have to face the fact that our interpretation of behaviour is not only not popular but is <u>unpopular</u>. In these matters what <u>we</u> think to be a scientific interpretation is distasteful to the great majority of people. What they like, and what they want, is <u>not</u> a scientific interpretation (in our sense) but—well something else with a strong flavour of poetic mythology radically anthropomorphic [underlining in the original].[8]

Eventually Cambridge University Press took the book and produced a large-format, beautifully illustrated work consolidating Howard's material on the breeding behavior of British birds. The scientific community responded with glowing reviews, both in America: "This book . . . is, in the opinion of the present reviewer, the most important and significant interpretive study of avian life ever made in any country"[9]; and in Britain: "It forms a contribution to general biology of which ornithology may well be proud."[10] Yet by today's standards Howard's writing seems obtuse. For example:

Yours truly,

Jos. Kirkman.

P.S. Selous has been taken off the British Bird Book. He has been a terrible nuisance to us.

250 Postscript on letter dated 30 Octobewr 1910 from Frederick Kirkman to Eliot Howard, complaining about Edmund Selous.

Eliot Howard (date of photo unknown) and two illustrations of Chaffinch behavior from his books: fighting (right), from *Territory in Bird Life* (1920) illustrated by Henrik Grönvold, and courting (left), from *An Introduction to the Study of Bird Behaviour* (1929) drawn by George Lodge.

We can take any one reaction and study it by itself and learn something about mechanisms, but we can learn nothing about its utility or value, for that depends upon other reactions which minister to it and to which it ministers. One reaction in itself is neither more or less important than another; each forms a portion of the environment for others; each is sensitive to the modification of others—they form a constellation, and somewhere in the organization of the living bird they have a common structural link.[11]

Selous, Kirkman, and Howard provided the groundwork, but it was Julian Huxley who blazed the way for the new science of animal behavior, just as he did with evolution (chapter 2) and systematics (chapter

3). Huxley excelled as a synthesizer and popularizer and liked nothing better than championing causes, just like his famous grandfather. Ethology was one of his causes, and his study of the courtship behavior of the Great Crested Grebe,[12] published in 1914 (chapter 9), was a landmark in the early history of animal behavior research. Huxley promoted ethology tirelessly through public lectures, films, books, and, eventually, in his capacity as first president of the Institute for the Study of Animal Behaviour[13] established in 1936.

BEHAVIOR AS PHYSIOLOGY

In Russia the study of animal behavior emerged very differently—underpinned by a strong physiological ethos. During the late 1800s, Ivan Pavlov began his famous work on the conditioned reflex, training dogs to associate the ringing of a bell with the appearance of food.[14] After a short training period the dogs would salivate after merely hearing the bell. The fact that a reflex (salivating), could be modified, or "conditioned," by training, was a breakthrough in the study of behavior.[15] Prior to Pavlov's work, reflexes were considered to be simple and nonmodifiable, requiring no conscious thought and neatly accounting for behaviors—such as that escape response of birds on seeing a raptor. Pavlov's student, Nicolas Popov, showed that the conditioned response was not restricted to mammals but also occurred in pigeons (see Razran 1933). Pavlov's groundbreaking work revealed that reflexes could account for changes in behavior—that is, for learning—and this was to have profound impacts on the study of animal behavior, partly in Britain but mainly in the United States and in particular through the work of John Watson (below).

In Britain Eliot Howard's mentor, Lloyd Morgan—one of T. H. Huxley's students and later professor of psychology at what became Bristol University—preferred Pavlov's physiological approach to bird behavior over the "softer" anecdotal approach adopted by the likes of Romanes: "I felt, as no doubt he [Romanes] did, that not on such anecdotal foundations could a science of comparative psychology be built."[16] While Morgan was enthusiastic about the field study of bird behavior, he was doubtful that ornithologists were the ones to do it: "The trouble here is that so many of them love the drama of bird life more than the science thereof."[17]

Morgan himself made important contributions to the study of bird behavior, in

both his empirical work on the development of chick behavior and his theoretical ideas about migration (Adler 1973). An outstanding scientist he was the first psychologist to be made a fellow of the Royal Society—Morgan insisted on defining concepts and terminology, encouraged the replication of experiments, invented the term "trial-and-error learning," and perhaps is best known today for his "canon": "In no case may we interpret an action as the outcome of the exercise of a higher psychical faculty, if it can be interpreted as the outcome of the exercise of one which stands lower in the psychological scale."[18] In other words, apparently intelligent and insightful behavior may have a simpler explanation (see Seibt and Wickler 2006). Thus behaviors like string-pulling by European Goldfinches—a trick, popular with bird keepers since the Middle Ages, in which the bird pulls up a small bucket of water in order to drink—might, if we adopt Morgan's canon, be explained by simple trial-and-error learning.

In the United States Edward Thorndike, a psychologist at Columbia University, referred to Morgan as "the sanest writer on comparative psychology."[19] In his early days at Harvard, Thorndike had carried out tests of maze learning with chicks, and found that, once chicks had "discovered" the right way out of a maze, they did so faster on subsequent trials and with fewer "useless" behaviors. When Thorndike moved to Columbia, he began experiments with cats and dogs using his famous "puzzle box"—a device that allowed him to quantify and control his subjects' behavior. Thorndike's book *Animal Intelligence* (1911) introduced a new technique for comparative psychologists now known as "operant conditioning," in which the chances of a behavior occurring is manipulated by linking it with either a reward or a punishment. He called it his "law of effect": behavior that is followed by "satisfaction" will become linked to a particular situation and will then happen again when the situation reoccurs.

John Watson, whom we met in chapter 4, rejected Thorndike's "law of effect" as being unnecessary and hedonistic, because he felt that "satisfaction" was an unknowable and untestable trait. Instead, he argued that the intensity of a stimulus, and how recently it had occurred, were sufficient to account for learning.

In 1908 Watson moved from the University of Chicago, where he was an instructor, to a full professorship at Johns Hopkins University, where he began his investigations into the behavior of seabirds in Florida. Much of that research was motivated by his interest in the homing ability of birds, but they also represent some of the earliest experimental studies (some of them in collaboration with Karl Lashley) of the behavior of free-living birds (Todd and Morris 1986).

A report of his fieldwork in Florida, published in 1908, described the breeding biology of Sooty Terns and Brown Noddies in detail, providing comprehensive life history data as well as the earliest "ethograms" (Todd and Morris 1986). In addition, Watson also carried out ingenious experiments, asking whether, for example, marking eggs or changing them for fake ones would affect parental behavior (it didn't, possibly because the terns were unable to recognize their own eggs). He also found that if he placed an egg in an empty nest he could elicit the full range of incubation behavior by the bird—suggesting that to begin incubation a bird merely needed the stimulus of an egg in its nest.

In reviewing Watson's report for *The Auk*, Joel Asaph Allen (1909) said that "such a minute and detailed study, conducted with scientific exactness, of the activities of any species of wild bird has doubtless never before been made, and is hence of the highest interest as a contribution to the life histories of the two species here under investigation, aside from its value from the psychologic side."[20] He also said, "Dr. Watson's paper . . . is noteworthy from the double viewpoint of ornithology and psychology."[21] Much later Lorenz complained that "if J. B. Watson had only once reared a young bird in isolation, he would have never asserted that all complicated behaviour patterns were conditioned."[22]

Despite this promising start with field studies, Watson was frustrated by his lack of control over the birds' behavior—there were simply too many variables—and he eventually swapped the field for the laboratory. In a landmark series of lectures at Columbia University in 1913, Watson delivered a paper titled "Psychology as the Behaviorist Views It," and with this launched the new field of "behaviorism"—a field characterized by its rigid, mechanistic view of behavior:

> Psychology as the behaviorist views it is a purely objective experimental branch of natural science. Its theoretical goal is the prediction and control of behavior. Introspection forms no essential part of its methods. . . . The behaviorist . . . recognizes no dividing line between man and brute. The behavior of man, with all of its refinement and complexity, forms only a part of the behaviorist's total scheme of investigation.[23]

Behaviorists assumed that animals functioned independently of the environment in which they had evolved, and as a result were best studied in empty cages with maximum experimental control. Indeed, in contrast to Selous, Kirkman, Howard, and Huxley, Watson and his followers completely ignored Darwin's ideas: in Watson's rigorous, mechanistic world, nurture was everything. His overtly psychological approach to animal behavior gained credibility among his early critics when in 1916 he incorporated Pavlov's ideas to provide a physiological basis for his ideas.

Behaviorism went on to dominate the study of animal behavior in America for decades, but, ironically, Watson's own career in animal behavior came to an abrupt end. In 1920 his wife Mary discovered that he was having an affair with his graduate student, Rosalie Rayner. Mary learned about the affair during a dinner party held by Rosalie's wealthy family. Following a hunch, Mary excused herself from the dinner table to go to the bathroom, but instead she searched through Rosalie's bedroom and found incriminating letters from her husband. Making copies of the letters, she gave Watson an ultimatum. He chose Rosalie and endured a highly acrimonious divorce and a very large settlement to Mary. Resigning from Johns Hopkins University at the age of forty-two, Watson eventually took his newly developed theories of human behavior to the J. Walter Thompson advertising agency in New York, made a fortune, and never returned to academic research.

Watson had many disciples, the most enthusiastic of whom was Burrhus F. Skinner, who took the subject to new extremes in the 1930s in the form of "radical behaviorism."

Skinner obtained a PhD in experimental psychology at Harvard in 1931, and like Watson he believed that *all* behaviors in both humans and nonhumans were shaped via reward or punishment. Again like Watson, Skinner also believed that only by having complete control over the environments experienced by experimental animals could psychologists draw meaningful conclusions. While still a graduate student, he designed his famous (infamous to many) "Skinner box" to achieve that control over everything (sound, smell, visual and tactile cues), while simultaneously reinforcing—by reward or punishment—the production of only one kind of behavior. This approach engendered great enthusiasm among behaviorists but was deplored by field ethologists. By rejecting the notion of consciousness or emotions—partly because he considered them too difficult to study—Skinner alienated those who considered humans to have rather more going on in their heads than conditioned responses. Thus, Skinner assumed a baby's brain to be a blank slate and thought that children acquired language entirely through learning. Later, ethologists' studies of the way birds acquire their "language," or songs, was to have a profound effect on the way psychologists subsequently viewed language acquisition in humans.

After completing his PhD and postdoctoral research at Harvard, Skinner moved to the University of Minnesota in 1938 as assistant professor of psychology. One day in 1939, after Warsaw was laid to waste by German aircraft bombing, Skinner was on a train thinking about the threat of airborne attack and how it might be averted. As he stared out of the window he was inspired by an unlikely source. "I saw a flock of birds lifting and wheeling in formation as they flew alongside the train. Suddenly I saw them as 'devices' with excellent vision and extraordinary maneuverability. Could they not guide a missile? Was the answer to the problem waiting for me in my own backyard?"[24] Skinner immediately set to work on the project with the support of General Mills, the food manufacturer. Then in 1943 the National Defense Research Council awarded Skinner $25,000 when they finally recognized the promise in what he was doing. Pigeons have excellent vision and could be trained to reliably direct missiles to within 6 meters (20 feet) of their target. Harnessed in snug jackets inside the cone of the missile, they were trained to peck at a screen on which was a projected image of the target; they were reinforced for pecking the target and could track the target in this way for impressive amounts of time. Yet after the one-year grant expired, Skinner was unable to convince those in authority that "Project Pigeon" was worth pursuing, so it was abandoned in 1944. To the physical engineers involved in the project, the idea of trusting a mere pigeon to guide a bomb seemed ridiculous, even though the pigeons provided a better degree of accuracy than anything else they had tried.

Skinner returned to Harvard in 1948 as a professor in the psychology department, where he remained until his retirement in 1974. Most of his work involved mammals, including humans, but he also continued to use pigeons in his studies—in one remarkable case training them to play Ping-Pong—demonstrating not only his skill as an experimenter, but also the power of operant conditioning (Skinner 1962). Skinner's pigeon work alerted reproductive biologists and experimental psychologists to the value of the pigeon as a "model organism." Darwin had started that ball rolling, but Skinner was responsible for making the pigeon central to research in psychology (Montgomerie 2009).

NATURALISTS AGAINST "BEHAVIORISM"

Behaviorism might have been an objective way to study behavior in a controlled environment, but for European researchers interested in bird behavior in the wild it had no biological relevance, and for this reason they detested it. Selous, Kirkman, Howard, and Huxley were passionate about birds—and about nature in general—and could not have been more different from

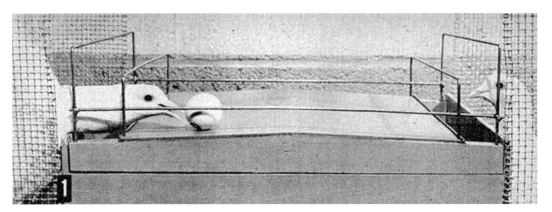

Pigeons playing Ping-Pong, evidence of Burrhus Skinner's skill at training animals thorough reinforcement.

the apparently unsentimental behaviorists. However, being emotionally involved with their subjects, and lacking any real conceptual framework, the field naturalists found themselves at the opposite extreme, struggling to develop an objective way to study bird behavior and often interpreting that behavior from a human perspective. As Herbert Friedmann wrote in his review of Howard's (1929) book: "The plain truth of the matter is that until very recently the great majority of bird-students did not know how to think in the cold, dispassionate manner so essential to scientific progress. They were bird-lovers, not bird-students."[25] Friedmann insightfully called his review "The New Study of Bird Behavior."

While Selous was watching Dunnocks and Howard was chasing warblers in England, Oskar Heinroth and his wife, Magdalena, were studying the courtship behavior of ducks in Germany. Heinroth had been fascinated by birds from an early age; he "learned to walk in the family hen house where he watched birds and mimicked their calls."[26] Like many other ornithologists of the day, Heinroth started out studying medicine but then switched to zoology. On return from an expedition to the Bismarck Archipelago in 1900–1901, Heinroth began working at the Berlin Zoo. In 1913 he became director of the aquarium, remaining until his death in 1945. It was the ten-volume *Brehm's Thierleben* [Brehm's Life of Animals], published in 1876–79, that inspired Heinroth to undertake a comparative study of waterfowl courtship behavior, presenting his initial results at the fifth International Ornithological Congress (IOC) in Berlin in 1910. In what was to be a key moment for the early study of animal behavior, Heinroth suggested that

courtship displays could be studied in the same comparative way that had been applied to the study of anatomy, similarities and differences between species reflecting their evolutionary history. The enormous significance of Heinroth's ingenious idea was not appreciated until much later, when Konrad Lorenz developed it and to some extent took ownership of it. As Erwin Stresemann said, Heinroth's study "was the first move toward bridging the gulf between biology and systematics; the two great branches of ornithological research, separated for centuries by 'fashion', began to approach each other and to touch—a most important process."[27]

For twenty years Oskar and Magdalena (who undertook the lion's share of the research) worked long hours each spring hand-rearing almost every species of European bird—308 species in total, from Goldcrests to Bearded Vultures. Their aim was to document and describe the development of the birds' instinctive behaviors. That was a monumental undertaking, and the results were eventually published in four volumes as *Die Vögel Mitteleuropas* (Heinroth and Heinroth 1924–34), a superbly illustrated encyclopedia of avian growth and behavior. In a review following the publication of the first volume, it was described as "one of the most important contributions to the life history of European birds."[28] Some of their hand-reared birds subsequently bred in the Heinroths' home, including a pair of European Nightjars that was so tame the birds regularly landed on the heads of dinner guests. Incredibly, the nightjars nested on the pelt of a wild boar on the dining room floor, producing several broods in two successive seasons.

Magdalena Heinroth died of adhesion ileus (or a blockage of the bowel) in 1932; she

Oskar Heinroth with young Tawny Owls (left) and his first wife, Magdalena, with a Ural Owl (right). These are two of the three hundred or so species the Heinroths hand reared to document the birds' behavioral and physical development (photos in 1920s).

was just forty-nine. Oskar subsequently remarried, and his second wife, Käthe—who survived him when he died in 1945—later became director of the Berlin Zoo. In his obituary of Oskar Heinroth, Stresemann referred to him as "one of the greatest ornithologists of our generation,"[29] adding that the "publicity of an author and lecturer but rarely corresponds with his worth as a promoter of science. Heinroth, however, like a few other pioneers . . . helped to open an immense new field of research—the comparative study of animal behavior."[30]

Heinroth was a major influence on Konrad Lorenz, who grew up at Altenberg in Austria surrounded by animals and encouraged by his nurse, Resi Führinger, who had a "green thumb" for rearing animals. It was this same nurse who read to him Selma Lagerlöf's (1906) *Nils Holgersson* [The Wonderful Adventures of Nils], which fueled Lorenz's desire to become a wild goose:

As a very little boy, I loved owls and I was quite determined to become an owl . . . when I realized that owls could not swim, they lost my esteem. My yearning for universality drove me to want to become an animal that could fly and swim and sit on trees. A photograph of a Hawaiian goose sitting on a branch induced me to choose a Sandwich Goose as my life's ideal. Very soon it dawned upon me that I could not *become* a goose, and from then on I desperately wanted at least to *have* one and when my mother obstructed this, because geese are too damaging in a garden, I settled for a duck.[31]

At the age of six he chose a tiny duckling from a nearby farmer, and his childhood friend and future wife, Gretl, got one the next day, from the same farmer. Both children spent the summer living with their ducks and ducklings, learning about the repertoire of actions and calls that Lorenz later

Konrad Lorenz followed by two imprinted goslings (photo in 1961 at age ca. 58).

called the duck's "ethogram." One result of these early experiences was the discovery of imprinting, not only of the ducks on the young children but of Lorenz onto ducks: "My undying love for ducks is a good illustration of the fundamental irreversibility of the imprinting process."[32] Despite this early interest in and aptitude for the study of behavior, Lorenz decided, after seeing a fossil *Archaeopteryx* (chapter 1) when he was ten, that he should become a paleontologist.

To satisfy his father, however, Lorenz studied medicine, first at Columbia University in New York City (where he had been sent in an effort to dissipate his affection for Gretl) in 1922, then in Vienna, where he qualified to practice medicine in 1928. In Vienna he was taught and inspired by Ferdinand Hochstetter, an enthusiastic proponent

of the comparative method for both anatomy and embryology. Lorenz quickly realized that these two subjects offered a better access to the problems of evolution than did paleontology and that the comparative method was as applicable to behavioral patterns as it was to anatomical structures. He also attended seminars given by the psychologist Karl Bühler,[33] who encouraged Lorenz to assess critically what were then the two main schools of thought about behavior. On the one hand, there were behaviorists like Watson and Skinner, who rejected all involvement of the mind and studied simple "stimulus-response" contingencies. On the other hand were the "vitalists," who believed that behaviors were caused by an immeasurable, internal "life force." These ideas were hugely significant for Lorenz: "I suffered a really shattering disillusion: none of these people really knew animals. . . . I felt crushed by the amount of work that was still to be done and that obviously devolved on a new branch of science that, I felt, was more or less my own responsibility."[34]

Although the study of medicine was his main task in Vienna, birds were never far from his thoughts. In 1925, shortly after purchasing a young Western Jackdaw that he named Jock, Lorenz was given Heinroth's *Die Vögel Mitteleuropas* by his close friend, Bernhard Hellman. It was a decisive moment: "I realized in a flash that this man knew everything about animal behavior. . . . Here, at last, was a scientist who was also an expert!"[35] Lorenz kept detailed notes on Jock's behavior, discovering sexual imprinting (see below). Gretl—by then his fiancée—and Hellman sent a copy of Lorenz's jackdaw notes to Heinroth, who was sufficiently impressed that he encouraged Lorenz to write up his observations as a formal paper for the *Journal für Ornithologie*, whose editor was Erwin Stresemann. As we have seen (chapters 2 and 3), for his time Stresemann had an extraordinarily broad vision of ornithology, embracing both systematics and evolution. Recognizing Lorenz's potential, Stresemann encouraged him to create his own colony of jackdaws. Lorenz's jackdaw paper, published in 1927, launched his career as a scientist.

In addition to developing Heinroth's idea that instinctive behaviors can inform phylogeny,[36] imprinting was Lorenz's main claim to fame. Lorenz demonstrated that young precocial birds will attach to, and follow, the first object that they see upon hatching—images of Lorenz walking or swimming, followed by a troupe of goslings, are now iconic. Lorenz also realized that imprinting

could be *sexual*—a recognition forced upon him when one of his male jackdaws tried to court him—the bird was said to be "overjoyed" when Lorenz opened his mouth and uttered a begging note.

> This must be considered as an act of self-sacrifice on my part, since even I cannot pretend to like the taste of finely minced worm, generously mixed with jackdaw saliva. You will understand that I found it difficult to cooperate with the bird in this manner every few minutes! But if I did not, I had to guard my ears against him, otherwise, before I knew what was happening, the passage of one of these organs would be filled right up to the drum with warm worm pulp.[37]

Lorenz was larger than life, clearly a genius in his understanding of animal behavior, but also bullish and dogmatic. Looking like a cross between a "Victorian God, an orchestra conductor and a husky dog,"[38] he was often teased by Tinbergen about his disregard for formal experimental design and statistics; Lorenz once declared, "If I have one good example, I don't give a fig for statistics."[39] Nonetheless, his demonstration of a "sensitive period" early in an animal's life, during which imprinting occurs, was groundbreaking. Lorenz assumed that such imprinting was irreversible—if a young bird imprinted on the wrong species, like his hand-reared geese and his jackdaws did on him, the result was permanent. This was not true, as subsequent research by Pat Bateson (1966) and Johan J. Bolhuis (1991), for example, has shown.

After graduating from medical school in 1928, Lorenz worked part time in Hochstetter's laboratory and enrolled to do his second doctorate there, submitting his thesis on bird flight in 1933 (Krebs and Sjölander 1992). Unlike Heinroth, whose focus was entirely ornithological, Lorenz was thinking on a far grander scale, keen to develop a new way of studying animal behavior. In 1935, when Hochstetter retired and his new boss at the Anatomical Institute in Vienna forbade him from continuing with his part-time behavioral studies, Lorenz sought advice from Stresemann, who simply said, "You must give up anatomy. Your talents in the field of animal psychology are so prominent that it would mean an autotomy (and in addition a biologically detrimental one!) if you would now become intimidated and would act 'rationally' instead of instinctively.... Don't worry and plunge into the water like a young guillemot, you will surely be able to swim."[40]

Lorenz took Stresemann's advice and resigned, something that he never regretted even though he had no other job to turn to. It may be no coincidence that Lorenz carried out and published some of his most influential works during this period of unemployment. In 1935 he produced a landmark study of the social relationships in jackdaws: "Der Kumpan in der Umwelt des Vogels" (translated in 1937b as "The Companion in the Bird's World"). That paper appeared in two parts in the *Journal für Ornithologie* and was praised by Margaret Morse Nice, who called it "a great paper ... of interest and great value,"[41] concluding her review by saying that "Dr. Lorenz gives us a solid foundation on which to build; with his illuminating viewpoint we can study bird behavior intelligently, understand phenomena that before were baffling, analyze our observations, and build up the large body of fact for which there is such a crying need."[42] That paper outlined all of Lorenz's ideas to date, providing a theoretical framework for the developing science of behavior. It was a strong message, and one that contributed hugely to the growth of Lorenz's international reputation.

A central concept underlying "Der Kumpan" was the *Umwelt*, an idea developed by the Estonian biologist Jakob von Uexküll. Von Uexküll proposed that each organism lives in a subjective world (an *Umwelt*), in which only certain things are important: the *Umwelt* of a kiwi, for example, being very different from that of a swift. Von Uexküll also used the term *Kumpan* to refer to the roles that conspecifics play in an individual's life. Lorenz elaborated, discussing five different types of *Kumpan* (or social roles) for the conspecifics that an animal encounters: offspring, sibling, parent, and both social and sexual interactants. Being fed by a male jackdaw was evidence to Lorenz that he was perceived as a sexual *Kumpan* by the bird, triggering the behavior of courtship feeding.

In 1936 two things happened that were to have permanent effects on Lorenz. The first occurred at a lecture he delivered in Berlin in February. Until then Lorenz had accepted that reflex theory accounted for the production of a behavioral act (i.e., that something in the environment is detected by receptor cells, a signal is sent to the central nervous system, and a motor response—a behavior—is produced). Any other interpretation was, to Lorenz, a concession to vitalism. During the lecture, Gretl was sitting behind Erich von Holst, a young behavioral physiologist who was nodding enthusiastically and muttering, "*Menschenskind*, that's right, that's right."[43] But when Lorenz concluded his lecture describing the reflexive nature of instincts, Von Holst put his head into his hands and moaned, "Idiot."[44] After the lecture it took Von Holst only a matter of minutes to convince Lorenz forever that reflex theory was inadequate.

The second event took place just a few months later, at Leiden University in the Netherlands, when Lorenz presented a paper at a symposium on instinct. It was here that he and Tinbergen met for the first time, though they had been corresponding for months. They hit it off immediately, and Lorenz later remembered thinking that Tinbergen "was my superior in regard to analytical thought as well as to the faculty of devising simple and telling experiments."[45] According to Tinbergen, he

> became Lorenz's second pupil.[46] . . . But from the start "pupil" and "master" influenced each other. Konrad's extraordinary vision and enthusiasm were supplemented and fertilised by my critical sense, my inclination to think his ideas through, and my irrepressible urge to check our "hunches" by experimentation—a gift for which he had an almost childish admiration. Throughout this we often burst into bouts of hilarious fun—in Konrad's words, in Lausbuberei.[47]

The following year (1937), Lorenz invited Tinbergen and his wife, Elizabeth, to visit the family home in Altenberg, Austria. This four-month stay forged their lifelong friendship and also marks the beginning of modern "ethology" through their first collaborative studies. The two men were very different in their approach: Lorenz liked rearing and studying animals in captivity, whereas Tinbergen was a field ornithologist, preferring to watch wild animals from a hide (blind). They likened each other to farmer and hunter, respectively. Yet their differences were complementary: it was the combination

of Lorenz's vision and Tinbergen's field experience that allowed them to forge the new approach of ethology so successfully.

Niko Tinbergen was born in 1907—the third of five children—and grew up in The Hague in the Netherlands. A keen naturalist from an early age, he kept sticklebacks as pets and loved watching gulls on the coast, much preferring them to songbirds: "I hate craning my neck all the time to follow them [songbirds] moving through the tree tops and to miss the most interesting things because they happen behind some thick branch."[48] As a teenager he joined the Dutch Youth Association for the Study of Nature and was a member of a group of amateur ornithologists led by ex-officer and teacher Gerard Tijmstra, whom Niko later called "the greatest influence on my biological thinking and its application to problems of animal behaviour."[49] He was hardly a natural scholar, being far more interested in sport—and hockey in particular, which he played at international level. Indeed, in the fall of 1925 family friends urged Niko's father to send him to Vogelwarte Rossitten (chapter 4), in the hope that he might be excited by biological research in action. It worked: Tinbergen was so inspired by his three-month stay at Rossitten he enrolled as a biology student at Leiden University as soon as he returned to the Netherlands for Christmas.

The university experience, however, was not what he expected:

> I realized that I simply had to . . . grind through the stuff that, until I knew better, I considered dull and boring . . . it was not until, as a young instructor, I had to teach vertebrate comparative anatomy and, later, the biology of certain taxonomic groups that I became

genuinely interested in this kind of subject, in which I learned to give attention to functional anatomy, adaptive radiation, and evolutionary aspects in general. That I had not widened my horizons earlier was partly due to my own intellectual limitations, partly to the fact that I started my studies in Leiden at the tail end of a period of the most narrow-minded, purely "homology-hunting" phase of comparative anatomy, taught by old professors just before they were succeeded by the younger generation.[50]

There were two people in Leiden who inspired him: Anton Portielje, who was in charge of animals at Amsterdam Zoo, and Jan Verwey, said to be Holland's most gifted naturalist (Röell 2000). Portielje was heavily influenced by Lloyd Morgan, and conducted studies on instinct that paved the way for Tinbergen's and Lorenz's concepts of "fixed action patterns" and "releasing stimuli." Morgan found, for example, that captive Eurasian Bitterns would show stereotyped defense behavior (pecking) even if they were simply presented with a piece of cardboard with two discs stuck to it, simulating eyes. But it was Portielje's recognition of "stereotypical" behavior in birds—the fact that a display is often performed in an identical manner every time—that really caught Tinbergen's eye.

Verwey instilled in Tinbergen an enthusiasm for the professional study of animal behavior, convincing him that it was scientifically respectable (Röell 2000), and encouraged him to develop an interest in the social behavior of birds. Exactly at the time Tinbergen started at Leiden, Verwey was undertaking a pioneering study of pair formation behavior in Grey Herons—conveniently

nesting in his back garden and watched from his bedroom (Verwey 1930). Verwey was responsible for the professionalization of ornithology in the Netherlands and was particularly irritated that others considered ornithology to be a "science for Sundays and holidays."[51]

One day in the summer of 1929, after he had completed his final examinations, Tinbergen was wandering over the dunes near his home thinking about his future. Serendipitously, he spotted a bright orange-yellow wasp busying itself on the bare sand. As he watched, Tinbergen realized that there was a whole "city" of digger wasps. "My worries were over; I knew what I wanted to do. This day, as it turned out, was a milestone in my life."[52] His research on this species of wasp resulted in an extremely concise PhD thesis that "the Leiden Faculty passed only after grave doubts; 32 pages of print were not impressive enough."[53] The thesis was short because it was a bit of a rush job—Tinbergen had been offered a place on the small Dutch meteorological expedition to East Greenland, for the International Polar Year of 1932–33.[54] Immediately after graduating he married Elizabeth Rutten, and together they left for Tasiilaq (now Ammassalik), where they lived with the Inuit for fourteen months, studying birds, huskies, and the native community itself. The Tinbergens conducted a detailed analysis of the breeding behavior of Snow Buntings but interrupted this for two weeks to take advantage of a rare opportunity to study the courtship behavior of Red-necked Phalaropes, one of the few bird species with reversed sexual dimorphism.

Back in the Netherlands in 1935, Tinbergen obtained what he called a "minor instructor's job" teaching animal behavior and comparative anatomy at Leiden University; it was here at the instinct symposium in 1936 that he met Lorenz for the first time. This was a propitious year for ethology. Germany started an animal behavior society, the Deutsche Gesellschaft fur Tierpsychologie (DGT, or Society for Animal Psychology) with Carl Kronacher (a reproductive biologist) as president and Lorenz as an active member. In Britain the Institute for the Study of Animal Behaviour (ISAB) was founded with Julian Huxley as president, together with Frederick Kirkman, several psychologists, and Solly (later Baron) Zuckerman.

During Tinbergen's stay at Altenberg in 1937, he and Lorenz conducted a variety of experiments, notably on the egg retrieval behavior of Greylag Geese and their now classic studies of the response of young birds to flying raptors, mentioned at the start of this chapter. The crux of all these experiments was that individuals responded in a particular way to a particular stimulus. To use their terminology, a "fixed action pattern" was elicited by a "sign stimulus," also referred to as a "releaser" if it occurred in a social situation. These concepts became the core of the new approach to ethology.

The begging behavior of young herring gulls is an iconic example of sign stimuli and releasers. In 1928 Oskar Heinroth noticed that European Herring Gull chicks were predisposed to peck at red objects, suggesting (erroneously) that this might be because they were fed on meat. Friedrich Goethe, who completed a PhD in 1936 at the University of Münster on herring gull behavior, also commented that gull chicks were attracted to red. In a simple (slightly macabre) experiment, he noted that when he presented a dead gull's head to a gull chick, it was highly motivated

Lesser Black-backed Gulls: the chicks peck at the spot on the adult's bill to get fed, just as in herring gulls.

to peck at the red spot on the lower mandible. When the red spot was painted over in yellow, the chicks pecked less at the beak. He also reported that chicks would peck at other red objects, including the red soles of bathing shoes. Picking up where Goethe left off, Tinbergen decided to test whether the red spot could be considered a releaser and the pecking a fixed action pattern. He noted that "the very fact that reactions to crude dummies were not rare, showed that the chick's sensory world must be very different from ours, for we would never expect a bathing shoe to regurgitate food."[55]

Tinbergen and his undergraduate students carried out the tests, taking herring gull chicks from the nest and presenting them with various cardboard models of gull

263

heads with different colored spots on the beak. The results confirmed Goethe's observations: the chicks preferentially pecked at the red spot on a yellow beak (Tinbergen and Perdeck 1950; Tinbergen 1953b). These striking findings have been reported in textbooks ever since as examples of the simplicity and power of the ethological approach.

Twenty-five years later Carel ten Cate, an ornithologist and professor of animal behavior at Leiden University, was browsing through Tinbergen's original data sheets archived at the university and "noticed that the data sheet and the 1949 publication did not quite seem to match, and when I consulted my copy of the *Study of Instinct* (Tinbergen 1951) it showed yet another picture of the experiments, one that no longer seemed congruent with the original data."[56] The data were a mess, and the accounts, published in three different books, were all slightly different.

Concerned, ten Cate (2009; ten Cate et al. 2009) repeated Tinbergen's experiments to check whether they merited their status as a classic study. They did, but it was also clear that the original procedures were sloppy, the peer review process inadequate, and the statistical tests less rigorous than commonly employed today (Kruuk 2003). A theme that emerged repeatedly as we conducted our research for this book is how much our approach to science has changed—improved—over time. This is partly due to new analytical techniques but also results in a clearer view of the problems to be resolved.

A discrepancy also exists with the famous hawk-goose model experiments mentioned at the beginning of the chapter. Wolfgang Schleidt and colleagues (2011) provided a detailed history of these studies and pointed out that, even in the 1930s, Tinbergen and Lorenz failed to agree on the interpretation.

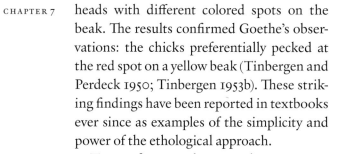

Artist's impression of the way Tinbergen and Lorenz conducted their hawk-goose model experiments with a pulley system that allowed them to fly hawk and other shapes over turkey poults. Inset (top) shows the range of shapes used in the experiments.

Tinbergen felt that the key stimulus eliciting a fear response in a range of bird species was the short neck, an idea he promoted strongly in *The Study of Instinct*. Lorenz on the other hand, felt that the key stimulus was the speed at which the model moved across the sky and that only young turkeys exhibited the fear response, not the other birds.

To test his ideas, in 1951 Lorenz took on Schleidt as an assistant to repeat the hawk-goose experiments. As well as using a hawk-goose dummy, Schleidt and his wife presented circular discs and rectangular shapes over the test birds. Young domestic fowl responded the same way to a plain black disc and the silhouette of a buzzard. Indeed, typical fear responses were elicited by any object, except that the birds habituated and stopped responding if the stimulus was presented repeatedly. Lorenz's idea that slow speed was the key stimulus was also confirmed. Schleidt made minute rectangular dummies—just 7.5 millimeters (0.30 inches) in length—and when he flew these over the birds at various speeds, he found that the fear response occurred only when the dummies moved slowly (i.e., resembling high-flying predators). When flown quickly (more like insects), there was little response. Using hand-reared turkeys, Schleidt found that object novelty was the important feature eliciting the fear response. Collectively, these experiments led to the "selective habituation hypothesis," which Lorenz welcomed. There was no support for Tinbergen's short-neck hypothesis—but for several decades this remained the most common interpretation in animal behavior textbooks (Schleidt et al. 2011).

When the Second World War started in 1939, dialogue between Tinbergen and Lorenz came to an abrupt halt. Tinbergen, like many of his fellow Dutchmen, spent part of the war in prison—in his case, as a hostage for protesting against the expulsion of Jewish professors from Leiden University. Tinbergen was at his home doing bird work when he was taken captive in September 1942 and sent to a prison camp in southern Holland.[57] Although relatively well treated, the constant threat of death in retaliation for Dutch resistance activities made incarceration stressful; on two occasions small groups of prisoners were taken out and shot. Despite his internment, Tinbergen was allowed to work as long as he did not use the English scientific literature. Even so, he wrote a draft of his now classic *Social Behaviour in Animals*, eventually published in 1953. He was released in September 1944.

In rather stark contrast, the jobless Lorenz opportunistically joined the Nazi Party in 1938[58] (Kalikow 1983). A public lecture on the behavior of domestic geese and crosses between domestic geese and Greylag Geese provided Lorenz an opportunity to show that his research had a bearing on the racial concerns of the Third Reich.[59] Lorenz argued that humans suffered the same fate, thereby providing "evidence" for superior and inferior sectors of society. In October 1941 Lorenz was drafted into the German army. Captured by the Russians in 1944, he was held prisoner until 1948, long after the war had ended. Like Tinbergen, Lorenz continued to be productive both during the war[60] and then in captivity, where he produced a manuscript on the comparative study of behavior, written, as he wrote to David Lack, "chiefly on cement-sack-paper with ink made of lamp soot and similar ingredients!"[61] Although Lorenz did not publish this manuscript when he was released from prison, it formed the basis for his book *Behind the Mirror: A Search for a Natural History of Human Knowledge* (1973).

Tinbergen and Lorenz were on different sides during the war, but their friendship resumed soon after Lorenz's release from prison, and their first postwar meeting, at Bill Thorpe's house in 1949, just before the joint Society for Experimental Biology / Institute for the Study of Animal Behaviour conference in Cambridge, was a highly emotional, joyous occasion. Unfortunately, Lorenz's association with Nazism tarnished not just his reputation but also that of ethology as a whole in Germany, as he discovered when he went there on a lecture tour early in 1950. Here's Lorenz writing to Thorpe shortly after:

> That poor country is at present visited by an epidemy of anti-Darwinian and existential philosophy and I had the most wonderful rows in some discussions. Once I found myself moved to say that it had been a complete mistake to lecture on comparative ethology at all to such an audience, the correct thing to do would have been to teach them the main facts about the origin of species. . . . After Cambridge it was as if I had stepped into H. G. Wells's time machine and travelled back a few hundred years! . . . even the young students are fanatically anti-Darwin, because they identify him with national socialism [Nazism]. One more cause for hating the latter, damn it![62]

The Second World War also bore witness to the collegiality among ornithologists of different nationalities. Erwin Stresemann, for example, was highly critical of Nazi politics from the outset and sent bird literature and color rings to British ornithologists in a prison camp near Eichstätt in southern Germany so they could study birds while incarcerated. John Buxton (studying Common Redstarts), Peter Conder (European Goldfinches), John Barrett (Eurasian Tree Sparrows), and George Waterston (Eurasian Wrynecks) were among those who benefited (Snow 1990). In America Margaret Morse Nice raised funds for European ornithologists and appealed to American ornithologists to send aid and ornithological literature to the European ornithologists who were hit hardest by the war. She also promoted ethology in the United States by making the new ideas coming out of Europe during the 1930s and 1940s known to English-speaking ornithologists by publishing translations in *The Auk* and *Bird-Banding*.

Ringing her first pair of Song Sparrows in 1929, Nice had a major impact on the study of life histories. Her aim was to build a complete picture of the lives of "her" birds. Inspired by the scientific approach of the European ethologists to study behavior, she even wrote a paper about the *Kumpan* of Song Sparrows (1939). The challenge she faced in trying to promote ethology in 1940s America was considerable, for by then the dominant approach to studying animal behavior was comparative psychology, in particular behaviorism.

ANIMAL BEHAVIOR IN AMERICA

At the beginning of the twentieth century, Charles Otis Whitman was "arguably the most influential biologist in America."[63] Interested in birds and taxidermy from an early age, Whitman studied for his doctorate under the great German zoologist Rudolf Leuckart in Leipzig. Working mainly on invertebrates, he became a leading reproductive and developmental biologist. Whitman's horizon was broad, however, and he spent several years doing experimental work on heredity and evolution of both pigeons and flickers. He felt strongly that field research should include studies of the entire life history, genetics, geographic distribution, habits, and intelligence of a particular species, or a group of closely related species. His views on animal behavior were set out in two lectures given at Woods Hole Marine Biological Laboratory in 1898, covering methodology, instinct, intelligence, evolution, and the importance of both ontogeny (development) and phylogeny for behavior. Whitman anticipated one of Heinroth's main ideas by a decade, ideas that Lorenz considered to be at the core of ethology: "Instincts and structure are to be studied from the common standpoint of phyletic descent."[64]

Whitman also believed that behavior should be studied under natural conditions and hence was an early advocate of field studies or—with respect to his beloved pigeons—free-living animals. In his fifties Whitman became obsessed with pigeons, and recognizing their enormous potential as

Charles Otis Whitman (photo in 1900 at age ca. 58).

study organisms, he used his own funds to establish an elaborate research facility at his home. Sadly, the more time he spent watching his birds the less he published, and the monograph he had planned on pigeon behavior never materialized. Worse, his pigeon obsession was eventually the end of him: on the first day of December 1910, during a sudden bitter spell, Whitman spent the afternoon moving his pigeons into their winter quarters. The next day he was in a coma, and five days later he was dead of pneumonia. One of his students, Oscar Riddle, spent several years assembling Whitman's work, but "the publication of these posthumous volumes [in 1919] was a nonevent for most of the scientific community."[65] Perhaps if he had been a better communicator, Whitman's innovative studies of pigeon behavior and phylogeny would have gained the same respect as Heinroth's waterfowl studies, conducted at around the same time.

Wallace Craig, another of Whitman's students, used the vocalizations of different pigeon species to trace their evolutionary history. Craig received his doctorate in 1908, but for a variety of reasons he struggled to succeed in American academia, despite his thesis work, which had the potential to make a significant contribution to the early study of behavior. Later, when Bill Thorpe was delivering a series of lectures on modern developments in animal behavior at Harvard in 1951, he referred to the contributions of Whitman and Craig. After the second lecture he was informed that Craig was in the audience—a complete surprise to Thorpe, who had assumed that Craig was long dead. On asking around, he discovered that only two people in the audience knew who Craig was and what he had done: "It was characteristic of his retiring nature that he did not make himself known to me."[66] Craig's work had a very important effect on Lorenz in the 1930s, after Nice put them in touch, and Craig in turn introduced Lorenz to Whitman's work. Lorenz considered Craig, along with Hochstetter and Heinroth, to have had major influences on his thinking and acknowledged this by saying that Craig had effectively written half of his paper "The Establishment of the Instinct Concept" (Lorenz 1937a; see Burkhardt 2005: 58).

John Watson was the third American figure to conduct important ethological work on birds. This is ironic, given that he later both rejected field biology for behaviorism and fostered the "science of rat learning"[67] that typified American comparative psychology for many decades. Although Watson is best remembered for his work after he had finished studying birds in the field, his early work on bird behavior was, as we have seen, exceptional (e.g., see Todd and Morris 1986).

Whitman, Craig, and Watson came within a whisker of founding the field of ethology, but there were huge differences between these early American pioneers and those in Europe. As Burkhardt has said, "For a whole complex of reasons—personal, professional, conceptual, methodological, and institutional, the American biologists who took an interest in animal behaviour in the first quarter of the twentieth century did not succeed in establishing animal behaviour studies.... The new discipline of ethology was accomplished not in America but . . . in Europe."[68] Perhaps if Watson had pursued his early interest in bird behavior with the same enthusiasm that he showed for behaviorism, progress would have been different. As it was, Whitman and Craig were eclipsed

by Watson's behavioristic approach, whereas European ethologists were able to capitalize on changes in attitude that were occurring in ornithology. What's more, the strong natural history ethos in European ornithology meant that those interested in natural behavior would not let a behaviorist approach take over. As the ethologist Donald Griffin later put it, "Behavior is much too important to be left to the psychologists."[69]

ETHOLOGY IN EUROPE AFTER WORLD WAR II

Soon after the end of the Second World War, Tinbergen started to look for a position outside the Netherlands, viewing himself as a "missionary" able to bridge the gap between central Europe and the Anglo-American world and help to spread ethology (Tinbergen 1985). Encouraged by Ernst Mayr, he undertook a three-month lecture tour in the United States during the winter of 1946–47. On his return he was offered a chair at Leiden, but appalled by the bureaucratic burden this would entail, he continued to look elsewhere. In 1948 David Lack invited Tinbergen to visit Oxford, and the following year the head of Oxford's zoology department, Alister Hardy, offered him a university lectureship, with the remit of establishing a center of research and teaching in animal behavior.

Tinbergen was excited by the recent developments in ecology and evolution—encapsulated by the Modern Synthesis (chapter 2). He was also excited by his move to Oxford, as he felt that the time was right to develop ethology in Britain: "I learned about the as-yet embryonic national society, the forerunner of the Association for the Study of Animal Behaviour, which published

(I believe) the 'Bulletin'; it seemed to me that the climate was favourable for our purpose and that after all the early work of Selous, Huxley, Howard had left 'Latent' effects."[70] Working with Bill Thorpe (of whom more later) in Cambridge, Tinbergen organized an international meeting there in July 1949 on the physiological mechanisms in behavior, bringing together ethologists of all stripes and generating new enthusiasm for behavioral research.

This was a landmark meeting, something that Lorenz realized as soon as he got his invitation: "It is my firm conviction that this congress is going to be of the utmost importance. Comparative Ethology . . . is quite doubtlessly developing into a school at least as original and important as Behaviorism and Pavlov's Reflexology and quite certainly a much nearer approach to an exact natural science than both. It is certainly high time to come together and give it a name!"[71] At that conference, Lorenz launched his famous—now infamous—psychohydraulic model of instinct and Tinbergen introduced many of the key ideas (based on his lectures in America in February 1947) that would appear later in his book *The Study of Instinct*. This meeting provided a much-needed opportunity to thrash out the concepts and terminology of ethology. It was also where Tinbergen and Lorenz were reunited after the war, and as Thorpe later said, "None present at this meeting will fail to remember it for the rest of his life."[72]

The Study of Instinct, published in 1951, consolidated ethology's position. As in other research domains, a key volume often provides both a foundation and a set of directions for further research. *Instinct* did both, and most reviews were enthusiastic, partly due to Tinbergen's lucid, enthusiastic style and partly because it was essentially a handbook for *doing* ethology. Not surprisingly, birds featured prominently in the book, though the behavior of many kinds of animal, both invertebrates and vertebrates, was discussed. Lorenz and his students at Altenberg were "appalled" by the title—they did not like the term "instinct"—and this marked the beginning of a disparity between Lorenz's and Tinbergen's thinking.

One of Tinbergen's main projects for his Research Unit of Animal Behaviour at Oxford (later, the Animal Behaviour Research Group, or ABRG) was the comparative study of gull behavior, beginning in 1950. He had been alerted to the intricacies of life in a gullery by Gerard Tijmstra in Holland, and Jan Verwey had told him that social interactions would be the most challenging aspect of animal behavior research (Tinbergen 1985: 438). Inspired by the comparative analyses of Heinroth (1911) and Lorenz (1941), Tinbergen (1959) decided that gulls would be a perfect system: "The aims of our studies were: first, a description of the behaviour of as many species as possible; second, as complete a coverage as possible of the entire behaviour pattern of each species; and third, analyses of the functions, the causation and the origin of the displays, with the ultimate aim of understanding how they could have originated and diverged in the course of speciation."[73]

Tinbergen first decided to work on Black-headed Gulls in England, setting up a study site at Ravenglass, Cumbria, but he later switched to herring gulls at Walney on the Lancashire coast. Tinbergen's students and colleagues studied a range of other species, including Sabine's Gulls in Alaska (Dick Brown, Nick Blurton Jones, and Dave

Hussell), Franklin's and Bonaparte's Gulls in Manitoba, Canada (Martin Moynihan), Lesser Black-backed Gulls in Britain (Mike Cullen), and Ivory Gulls in Spitzbergen (Svalbard) (Pat Bateson and Chris Plowright), constructing for each species an "ethogram"—a catalog of their displays and calls showing the situations in which they were employed (e.g., Tinbergen 1959). As with the earlier studies of ducks, a major finding was that the displays used by different gull species were very similar, with the Black-legged Kittiwake—studied by Esther Cullen on the Farne Islands off northeastern England—a remarkable exception.

Most gulls nest on the ground, but kittiwakes breed on tiny cliff ledges, usually high above the sea. As a result, pair members are forced into much closer proximity at their nest site than most other gull species. The effect of cliff nesting on the kittiwake's behavior is striking, and they differ from most other gulls in several respects, having a more specialized form of fighting and neither mobbing predators nor removing eggshells from their nests (Cullen 1957). Because their nest ledges are so inaccessible to predators, their chicks are not cryptically colored like those of other species.

While these differences suggested several behavioral adaptations to cliff nesting, the comparison was relatively weak, since it comprised only a single example: the kittiwake versus all other gulls. Recognizing that the Swallow-tailed Gull of the Galápagos also nested on tiny cliff ledges, Jack Hailman used that species as an independent test of Cullen's ideas, confirming that the two species share many behaviors not shown by ground-nesting gulls (Hailman 1965).

Public awareness of Tinbergen's studies of bird behavior increased dramatically in 1953 with the publication of *The Herring Gull's World* in the iconic New Naturalist series of monographs. Reviewers heaped praise on Tinbergen, not just for his innovative research but also for his ability to make his studies accessible. Here is Robert Storer, a major figure in American ornithology: ". . . [the book]should be studied by laboratory psychologists who too often forget the value of studying animals in their natural surroundings,"[74] and British ethologist Robert Hinde: "This monograph gives a deeper insight into the life of an animal than any other book on bird behaviour. . . . Scientifically it is a book of the utmost importance."[75]

Tinbergen must have been delighted: *The Study of Instinct* (1951), *The Herring Gull's World* (1953b), and *Social Behaviour in Animals* (1953a) put ethology in the academic spotlight. Success, however, was to be short lived. In 1953 a devastating critique of Lorenzian ideas was published, changing forever the course of ethology.

FLAWS IN THE ETHOLOGICAL APPROACH

As a teenager in New York City, Danny Lehrman was a keen birder, leading him, eventually, to volunteer in the Department of Experimental Biology (later the Department of Animal Behavior) at the AMNH, working under the curator, herpetologist G. Kingsley Noble.[76] With Noble, the young Lehrman went to nearby New Jersey to study the behavior of the Laughing Gull during incubation (Noble and Lehrman 1940). At twenty-two, Lehrman wrote a very favorable review of Lorenz's *Vergleichende Verhaltensforschung* [Comparative Behavior Studies]: "This is certainly one of the most important comprehensive papers on animal behavior that is known to the present reviewer.

Silver Gulls courting on a beach in Tasmania.

Whatever may be the eventual status of the individual aspects of Lorenz's theories, he has provided both a theoretical attitude and a methodological approach that bids fair to become essential for the investigation and understanding of behavior."[77] He was not particularly enamored of Lorenz's physiological explanation for instinctive behavior, but he concluded by saying that "knowledge of these analyses will enable the field observer of bird behavior to obtain an insight into the causes of the behavior he sees that will be, I think, superior and more fruitful than any that can be obtained otherwise."[78]

During the Second World War, Lehrman worked as a cryptanalyst in Italy, deciphering the communications of German airplane spotters to devise safe routes for American planes (Rosenblatt 1995). Becoming fluent in German, he was, after the war, able to translate all of Lorenz's early

scientific writings—thereby discovering Lorenz's Nazi sympathies, which must have been shocking for Lehrman, who was Jewish. As a teenager, Lehrman had met Theodore Schneirla at the AMNH, and inspired by Schneirla's approach to behavior, he returned after the war to do a PhD with him from 1948 to 1954. He was passionate about bird watching throughout his life, and this combination of interests in ornithology and rigorous psychology placed him in the almost unique position of being able to understand both the psychological and ethological approaches to animal behavior.

Encouraged by Schneirla, Lehrman published a detailed attack on Lorenzian ethology in 1953. First, he felt the distinction that Lorenz and Tinbergen made between innate and acquired (learned) behaviors was too rigid; "[they] consistently speak of behaviour as being 'innate' or 'inherited' as though

271

these words surely referred to a definable, definite, and delimited category of behavior."[79] He also did not like their tendency to attribute behavior to either "environmental" or "inherited" factors, feeling that this approach was too simplistic, neglecting both the wealth of complex interactions that occur between organisms and their environment and the different developmental stages at which these interactions take place: "... to say a behavior pattern is 'inherited' throws no light on its *development* except for the purely negative implication that certain types of learning are not directly involved."[80]

Lehrman was also unconvinced by Lorenz's idea that "isolation" experiments provided evidence for innate behavior. While the animal may be reared in isolation from other individuals, it "is not necessarily isolated from the effect of processes and events which contribute to the development of any particular behavior pattern."[81] More specifically, Lehrman felt that the question "Is the animal isolated?" was far less important than "From what is the animal isolated?"

While Lehrman praised Lorenz for his studies that used behaviors to infer phylogenetic relationships, he felt it was logically flawed to assume that all species-specific behavior patterns were "innate." Ignoring how the behavior developed (ontogeny), he said, could have drastic consequences for the estimation of taxonomic relationships.

Lorenz's hydraulic model of motivation also came in for criticism, as Lehrman linked it to the vitalistic theory of William MacDougall (which—ironically—Lorenz had previously criticized). Lehrman correctly pointed out that there was no neurophysiological evidence for any kind of hydraulic centers in the brain. Lorenz's hydraulic model, which seemed to many then—and

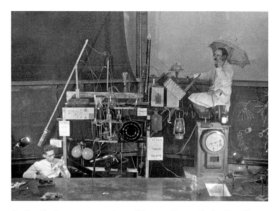

A physical representation of Lorenz's psychohydraulic model, constructed by Desmond Morris and Aubrey Manning for the 1953 International Ethological Conference, held in Oxford, and designed to demonstrate the implausibility of that model.

to us now—to have its roots in human male sexual motivation, proposed that "reaction specific energy" was stored up somewhere in the nervous system, and that this reservoir of energy was "dammed up" by a releasing mechanism until the organism was stimulated (by a "sign stimulus") to discharge the energy. The discharge then caused a motor response, a particular behavior. Lehrman was right, the hydraulic model *was* naive—just the previous year Aubrey Manning and Desmond Morris (both students with Tinbergen) had built an elaborate model of it—including a water-filled reservoir—for the International Ethological Conference in Oxford. In demonstrating its ludicrous underpinnings, they managed to soak the front few rows of the audience.[82]

Although Lehrman's critique of Lorenz was audaciously confrontational, in some respects it was very restrained. Following advice from several leading scientists, including Ernst Mayr,[83] Lehrman removed from his manuscript most of the material relating to Lorenz's Nazi ideologies in case that personal component detracted from the scientific arguments. Even so, his critique challenged not just Lorenz's life's work, but by hinting at his Nazi sympathies,[84] his credibility as a moral being as well.

Lorenz first met Lehrman in 1954 at a meeting in Paris, organized by the Singer-Polignac Foundation. Desmond Morris (who attended in Tinbergen's place) recalls that Lorenz clapped Lehrman on the back and roared, "Now that I know we are both fat men, it is impossible for us to be enemies!"[85] The pair seemed to get on well, and Lorenz wrote to Thorpe after the Paris meeting that "Lehrman . . . is a much more intelligent person than you would suppose from his paper."[86] Nonetheless, while openly jovial, Lorenz probably never forgave Lehrman (Marler 2004)—or at least never forgave his critique.

In contrast, Tinbergen was much more sanguine about the critique and invited Lehrman to Oxford to discuss their differences. That visit took place early in 1954, and Tinbergen's students were apprehensive:

> We were all on the defensive. . . . Lehrman certainly appeared as something of a threat. I am speaking of the graduate students here—with hindsight I can see Tinbergen was way ahead of us. Of course he was critical of some of Lehrman's ideas and certainly ready to defend ethology, but he had more wisdom than we did and saw at once that ethology would benefit from having to take good criticism on board. . . . Lehrman's impact was even greater than we expected because he was so entirely different from the image we had formed of him. We were quite unprepared for his enthusiasm for animals in all their diversity and for his knowledge of natural history—he too was a fanatical birdwatcher. It rapidly became

clear that he was in fact deeply sympathetic to the ethological approach and also hostile to much of white-rat experimental psychology. He wanted us to be more critical in our approach to behavioural development— that was really the crux of the matter— and who can dispute that he was absolutely correct.[87]

As much as anything, it was Lehrman's enthusiasm and friendly demeanor, as well as his expertise as an evolutionary biologist, naturalist, and ornithologist, that united rather than divided European and American approaches to animal behavior. Unlike most ethologists, though, he was also an expert in comparative and developmental psychology and endocrinology. The eventual meeting of the minds was a month-long workshop in 1957, organized by Frank Beach at the Center for the Advanced Study of Behavior in Palo Alto, California, and attended by the likes of Tinbergen, Robert Hinde, Donald Hebb, Eckhard Hess, Jay Rosenblatt, and Lehrman.

Lehrman was not the only one critical of ethology. Around the same time the evolutionary biologists J. B. S. Haldane and his wife, Helen Spurway, launched their own attack on Lorenz, pointing out that much of ethology in the mid-1950s was incompatible with evolutionary ideas (Haldane 1956; see also Griffiths 2004). In addition Haldane, with some justification, felt that Lorenz had not been all that original. In his opinion, Douglas Spalding rather than Lorenz should get the credit for being the founder of ethology (Haldane 1954). He could also see that there was an uncanny similarity between William MacDougall's earlier model of instinct and emotions and Lorenz's hydraulic model. But even Haldane and Spurway

Robert Hinde (at age ca. 42, left) with Danny Lehrman (at age ca. 46) in ca. 1965.

could not be totally objective, as they were both Marxists and "had no doubt that Konrad Lorenz was a thorough Nazi, and ethology a thoroughly Nazi antiscience."[88] Their critique of Lorenz was further complicated by the fact that Spurway and Lorenz had had an affair in 1949—albeit with Haldane's approval (Burkhardt 2005: 390)!

ETHOLOGY AFTER LEHRMAN

Lehrman's critique might have finished ethology had it not been for Tinbergen's willingness to embrace new ideas. The crux of the issue was the nature-nurture chestnut and the fact that Lorenz and Tinbergen were too easily convinced that particular behaviors were innate. A more rigorous, self-critical approach was essential, and that meant adopting some of the comparative psychologists' ideas. Change had to work both ways, though, and Lehrman's paper served to bring the two sides together—something that Lorenz grudgingly admitted in 1955:

My American trip was very tiring but altogether a success. The best of the learning theorists like Beach and others are just beginning to see for themselves that learning alone cannot explain "everything." Curiously enough,

Lehrman's objectionable paper seems to have helped considerably to the recognition of this fact. The best means to convince people that there is such a thing as instinctive movements is the film. I played duck films to Frank Beach until he nearly fainted, he got seriouser and seriouser and in the end he said in a small voice: "You know I did not believe a word of it and now I believe everything."[89]

The mid-1950s saw a new generation of ethologists starting to emerge, including William (Bill) Homan Thorpe, who was later famous for his studies of song acquisition in birds. Thorpe was brought up in a deeply religious family whose interests included both music and natural history. As he remarked later in life, "Looking back now I feel that my course in all essential respects was set before I was in my teens, and that everything of importance that has happened since has somehow been the natural, inevitable course of development from my earliest boyhood."[90]

Thorpe was not particularly academic and did not think he'd get into university, but after hearing that economic entomologists were in short supply, he managed—after much Latin cramming—to get into Jesus College, Cambridge, to study agriculture. A keen bird watcher, he became friends with fellow undergraduate Edward Armstrong—who was reading religion and would later become Reverend Armstrong—and together they helped found the Cambridge Bird Club in 1925, in memory of Alfred Newton. Armstrong went on to write several popular and influential books on bird behavior (e.g., Armstrong 1942).

After obtaining a second-class degree in 1924, Thorpe completed a diploma in plant pathology, then spent several years as an entomologist investigating how instinct and

Bill Thorpe (photo in 1936 at age ca. 34).

learning influenced the choice of food plants by insects. Then, in the late 1930s:

> ... after reading the five great papers of Konrad Lorenz [I] realized for the first time where my studies were leading me—straight into the ethological field. I concluded that what I was really primarily interested in as an entomologist and insect physiologist was the relationship between instinct in the ethologists' sense and learning in the sense of the American psychologists. ... I had been a keen field ornithologist for many years and it struck me very forcibly that for the particular work I hoped to do, birds would provide the most promising material.[91]

It was precisely because birds exhibited such an obvious mixture of instinctive and learned behaviors that Thorpe switched from entomology to ornithology. But there was no ethological research station in Britain at the time, so, after the end of the Second World War, Thorpe set out to establish one in Madingley, a village just outside Cambridge. He first asked Reg Moreau to be curator, but after he and his wife spent a couple of months living at Madingley,

Moreau declined because he was worried he would be "mis-cast" in the role (Moreau 1970)—and possibly also (we feel) because, like Lack, he did not consider himself to be an "ethologist." Thorpe also considered hiring Otto and Lilli Koenig—expert bird breeders—who had founded the Konrad Lorenz Institute at Wilhelminenberg, Vienna. Thorpe even considered Lorenz, who was then (1950) negotiating for a position at Bristol University that would allow him to conduct research on Peter Scott's magnificent collection of ducks and geese at Slimbridge. Had Lorenz moved to Britain, the course of ethology might have been different, but no sooner had he accepted Bristol's offer than the Max Planck Society offered him a salary to stay in Germany—a proposition that was just too good to turn down.

Thorpe's ethological field station at Madingley became a reality in 1950, with Thorpe as director, Robert Hinde as curator, and Gordon Dunnett as technician. Thorpe and Hinde first met in August 1946, as a result of Hinde—a keen bird watcher—finding a pair of Moustached Warblers nesting at the Cambridge sewage farm, the first breeding record for Britain.[92] This was a propitious event because among the hordes of ornithologists that descended on the birds was David Lack. On meeting Hinde—then aged twenty-three—Lack offered him a position at the EGI in Oxford, as a research assistant studying the feeding ecology of Rooks and jackdaws. Hinde found the corvid project boring and soon persuaded Lack to allow him to study the behavior of the Great Tit instead. Receiving little in the way of supervision from Lack, whose interests had by then shifted from behavior to ecology, Hinde sought advice from Tinbergen, who had

arrived in Oxford in 1949 while Hinde was midway through his studies. Completing his thesis in just two years, Hinde returned to Cambridge, to Thorpe's Madingley field station, in 1950.

Thorpe realized that the recently developed sound spectrograph (or "sonograph") offered a unique opportunity to understand how birds acquired their songs. It had been known for several centuries—through the deliberate cross-fostering practices of bird breeders—that birdsong had both a learned and an inherited component, and Thorpe's initial aim at Madingley was to disentangle the two. In so doing he was one of the first to bridge the deep chasm between the ethologists (studying instinct) and the comparative psychologists (studying learning). His review of the learning abilities of birds (Thorpe 1951a, b) provided ample evidence that birds were far more complex than the reflex automatons that previous scientists had assumed. With extraordinary foresight, Thorpe anticipated how the study of birdsong would successfully meld instinct and learning: "Thus, where the innate powers of recognition can only carry the animal a part of the way towards its goal, the process is completed and adjusted by a proclivity to attend to certain aspects of a situation and learn in certain restricted times and directions (as in the tendency of a bird to learn and copy the song of its own species in preference to the song of another) so that experience completes for the individual the process initiated by its inherited constitution."[93]

The sound spectrograph was developed for military use during the Second World War by Bell Telephone Laboratories in the United States, to break codes and to identify aircraft through the analysis of sound.

Bill Thorpe (at age ca. 60) at his ornithological field station at Madingley, near Cambridge, in 1962.

But military use was not the only application for the technology. At Bell Labs Ralph Potter was heading up a sound spectrograph project, and it was his "Visible Patterns of Sound" paper in *Science* (1945)—which included spectrographs of five bird species' songs—that pioneered the use of the sound spectrograph for research into both human speech and birdsong. Potter originally used the birds' songs because they showed a wide range of tone modulation, but he clearly recognized the enormous applications of this new technology: "With such patterns as these it will be possible to analyze, compare, and classify the songs of birds, and, of even more importance, it will be possible to write about such studies with meaningful sound pictures that should enable others to understand the results."[94] By 1947 acoustic researchers from all over the world were requesting sound spectrograph machines from Bell Laboratories. Kay Electric Co., an offshoot of Bell, developed such a machine exclusively for acoustic research, which they called the Sona-graph, first released in 1948.

By 1950 the technology was there—in the form of both reliable, portable tape recorders (also developed during the Second World War) and sonographs for the analysis of recorded songs. The scientific study of birdsong was born. The first scientific papers to show sonograms of birdsong started to appear (e.g., Bailey 1950; Borror and Reese 1953; Collias and Joos 1953), but Bill Thorpe (1954) was the first to use them to ask biological questions about song (as opposed to simply using the sonogram to describe the songs). Thorpe was the key player in promoting experimental studies of bird vocalizations, and his legacy lived on through the work of Peter Marler and subsequently in Marler's students, including those who developed a strong interest in the neurophysiology of birdsong, such as Fernando Nottebohm and Mark Konishi (chapter 6).

Initially, Thorpe used the only sonograph machine in Britain—located at the Admiralty Research Facility at Teddington—but soon a generous benefactor provided the funds for Thorpe to buy a sonograph for Madingley. He chose the Common Chaffinch as his main study species: "Its song is short, yet sufficiently complex to provide ample room for modification of detail without being so involved as to be unduly difficult of analysis. Besides the virtues of being common, hardy and sexually dimorphic, it has also been known since the late 17th or early 18th centuries that there were local song variants which were presumed to be learned, not inherited."[95]

Thorpe's first project was to determine how song developed in young chaffinches, and to achieve this he took chicks from the nest at various stages after hatching and reared them in groups under different conditions. The results of this classic study revealed strong differences in the songs produced by each group. Birds that had been reared normally by their parents—but were isolated later—produced normal chaffinch song the following spring. In contrast, the song of birds that had been taken from the nest at only a few days of age—and subsequently prevented from hearing any experienced bird's song—was quite abnormal. Chicks taken from the nest but housed together in a group went on to produce a song that was highly consistent within the group but highly abnormal for the species as a whole, implying that some aspects of their song were learned early in life. It was precisely because chaffinch song comprised both learned and innate components that Thorpe considered it an ideal model for testing ideas in ethology and psychology.

Thorpe's work on birdsong launched an exciting new subdiscipline within animal behavior. He had hoped that Hinde would also work on song, but taking his lead from Tinbergen, Hinde chose to conduct comparative studies on finches. But Madingley soon attracted others, including Peter Marler, who arrived in 1952. Marler had already completed a PhD in botany but was ready to start a second PhD research project in ethology, supervised by Thorpe and eventually examined, in 1954, by Tinbergen. Marler remained at Madingley as a postdoctoral researcher, but in 1957 he accepted a post at the University of California in Davis. His move to America (whose researchers were still in the grip of behaviorism) was of great importance for the spread of ethological ideas, and this, coupled with the steady stream of Americans who traveled to Europe to conduct periods of study in ethological laboratories (Dewsbury 1995), contributed to the advance of ethology in North America.

Peter Marler (photo in 1957 at age ca. 30).

Marler made birdsong his life's work. One of his studies at Madingley concerned predator alarm calls. Naturalists had known for a long time that small birds, like tits or finches foraging on the ground or in vegetation, utter distinctive "seep" alarm calls on seeing a flying predator. Tinbergen had recognized the evolutionary significance of such calls, reflecting that "conflict between individual security and security of the flock . . . [with] each utterance of an alarm call . . . is disadvantageous to the individual.[96] Lorenz believed that these alarm calls were arbitrary signals—one reason why they were thought to be useful for constructing phylogenies—but the more Marler studied alarm calls, the less convinced he was of this assumption.

Hinde had filled the Madingley aviaries with different species of European finches for his comparative study of courtship behavior, so Marler took advantage of this to carry out a comparative study of their vocalizations (Marler 1955). Marler also carried out some field studies, and while looking at the "hawk alarm call" of the chaffinch, he noticed something interesting: "I found that this 'hawk alarm call' has a curious ventriloquial quality. Sound spectrographic analyses revealed that these birds had, in fact, converged on a sound pattern that minimized the cues available for localization, disseminating alarm while reducing the risk of attracting attention."[97] This field study was published in 1957 and rapidly became a "citation classic,"[98] providing an incentive for others to examine the design of animal alarm calls. Importantly, that study explicitly linked signal structure to signal function, addressing Tinbergen's mechanistic and functional questions. Marler also discovered that the alarm calls of several unrelated species, including Great Tit, Common Blackbird, and chaffinch, were very similar, suggesting convergent evolution and refuting Lorenz's idea that these were arbitrary signals.

Thorpe's birdsong studies helped to bridge the gap between the ethologists and comparative psychologists, not least because "the bitterness of the controversy [resulting from Lehrman's critique] inhibited many of the next generation from investigating or even acknowledging the importance of the genetic side of the developmental equation. . . . Thorpe was one of the few who maintained a more balanced view point."[99] Indeed, the genetic basis of behavior was a research topic that many avoided in the decades following the Second World War, and Tinbergen's question about the development of behavior (ontogeny) became something of an "orphan" (Marler 2004). By providing unequivocal evidence for the existence of both genetic and learned aspects of song acquisition in birds, Thorpe and his colleagues—in particular Robert Hinde—helped to redress the balance.

Robert Hinde also recognized that ethologists had much to learn from the psychologists—especially a more quantitative approach, better controlled experiments, and more attention to ontogeny. This broader view epitomizes Hinde's classic textbook, *Animal Behaviour: A Synthesis of Ethology and Comparative Psychology* (1966).

Hinde arrived at Madingley in 1950, the same year that Lorenz published his psychohydraulic model of motivation, and immediately set out to examine the model's assumptions. Hinde theorized that if there was a reservoir of "action specific energy" damned up inside the nervous system, then it should be possible to measure the reservoir by studying how behavioral responses decline with time. He used the predator mobbing behavior of chaffinches as his model system, studying them in some of the sixty small aviaries available at Madingley. Hinde spent the winter crawling around on the cold aviary floors, counting the number of "chinks" given by the chaffinches in response to a model owl. The results were astonishing: Hinde realized that the mobbing response faded after just one presentation and never again reached its original intensity. He presented his findings at a meeting organized by Lorenz in 1952 at his new ethological research station (funded by the Max Planck Society) in Buldern—the First International Symposium of Comparative Behavior Researchers. Hinde's paper was a "bombshell," one of the most significant of the entire meeting (Burkhardt 2005: 376).

Remarkably, Lorenz was pleased with Hinde's work, because it was a direct experimental study of his principle of action-specific energy. He wrote to Thorpe afterward, congratulating him on his choice of Hinde as an assistant: "He really has become a first-rate ethologist and his paper was one of the most interesting—if most disturbing—that were [sic] read at our congress."[100] But he went on to say:

> What irks me, is that this fading cannot be natural! It would jeopardize the whole survival value of the mobbing, would it not, if the whole thing could function only just once with its full intensity!? It certainly is one of the most intriguing problems, and one that I subconsciously know for years, but did not realize before Hinde put it in clear words. Good for him!!!"[101]

At that point, he did not realize how devastating Hinde's findings would be for his model; this simple study had identified one of the most serious weaknesses of Lorenzian ethology.

Over the next decade other ethologists, including Hinde's student, Pat Bateson, began to reevaluate other tenets of ethology, including Lorenz's ideas on imprinting. Bateson, another keen ornithologist, studied zoology at Cambridge before conducting his doctorate on imprinting under Hinde's supervision. He showed that, under the correct environmental conditions, imprinting could be manipulated, suggesting that imprinting was not the unique behavioral phenomenon that Lorenz made it out to be. For Bateson, imprinting was more like other forms of perceptual learning, and he presented these findings in 1963 at the International Ethological Conference at The Hague. Lorenz, seated in the front row, was incensed. When Bateson had finished his talk, Lorenz turned to Hinde, berating him for encouraging his student to say such heretical things about imprinting.[102] Bateson recalls sitting down on the stage and smoking a cigarette while Lorenz and Hinde battled it out (Bateson 2009).

Bateson, with his Cambridge colleague Gabriel Horn, a neurophysiologist, went on to identify the location of the neural changes involved in imprinting. Lorenz, however, never accepted these findings and became increasingly frustrated at the way that ethology was changing, referring contemptuously to Tinbergen, Hinde, and their students as the "English-speaking ethologists."[103]

Tinbergen's contribution to the study of ethology and animal behavior as a whole is immeasurable. One of his most cited publications, *Aims and Methods in Ethology*, published in 1963, spells out exactly what it is that ethologists actually study. Also known as "Tinbergen's four questions," or four "whys," these are: (1) evolution, (2) function (adaptive significance), (3) development, and (4) causation. In other words, these are four different aspects of behavior that must be studied for a full understanding of behavior. For example, the bunching response shown by starlings in flight on seeing a peregrine, could be studied in terms of its (1) evolutionary history, by constructing a phylogeny and looking at the distribution of similar behaviors among the starling's closest and more distant relatives; (2) adaptive significance, by testing whether starlings that bunched together were less likely to be killed by a predator than those that didn't; (3) development, by determining whether totally naive (hand-reared) birds exhibited the behavior, or whether it was learned (Tinbergen's original hypothesis was that bunching together was an instinct, but he later concluded that it was learned; we feel the issue remains unresolved simply because the distinction between instinct and learning is now so much more difficult); and finally, (4) causation, by, for example, using model predators to ensure that potentially confounding variables were properly controlled.

Of these four questions, three had previously been identified by Julian Huxley (1942) as the main problems in biology—as Tinbergen fully acknowledged. In his modest way, Tinbergen says he merely added "development," but it was Tinbergen who brought these four questions into sharpest focus and as a result provided a clear set of directions for those interested in behavior that continue to be invaluable to this day.

ETHOLOGY'S LEGACY

In 1973 Lorenz and Tinbergen—with Karl von Frisch—were jointly awarded the Nobel Prize in Physiology or Medicine. Lorenz's initial response was insolent: "That's one in the eye of behaviorism,"[104] but Tinbergen was amazed: "I certainly did not dream of helping to get a new branch of science, 'ethology,' off the ground!"[105] With typical modesty, he felt that he and others had been attributed with too much forethought and that winning the prize did not reflect the "haphazard, kaleidoscopic attempts at understanding animal behavior done by the future ethologists."[106]

It is, at first sight, difficult to see how ethology could be eligible for a Nobel Prize in Physiology or Medicine, but the Karolinska Institute's press release of October 1973 cited the ethologists for their discoveries concerning "organization and elicitation of individual and social behaviour patterns."[107] Specifically, Lorenz was praised for both his work on imprinting in waterfowl, and his work on the fixed action pattern. Tinbergen was applauded for his "comprehensive, careful and quite often ingenious experiments,"[108] such as his work on the pecking response of young herring gulls to the red spot on their parents' beaks. The press release also linked the potential applications of ethology to human psychiatric and psychosomatic medicine, probably, as Dewsbury (2003) suggests, in an attempt to justify making the award to researchers of animal behavior.

The ultimate irony is that the scientific discipline of ethology, in its original form, more or less ceased to exist soon after the Nobel Prize was awarded. The disintegration had been going on for some time; Pat Bateson and Peter Klopfer (1989) felt "that ethology as a coherent body of theory ceased to exist in the 1950s,"[109] and indeed, throughout the 1960s "one by one the concepts and theories succumbed to critical analysis."[110]

Tinbergen died on 21 December 1988, following a long period of depression, sadly and incorrectly convinced that his contribution had been meaningless. Lorenz died a few months later, on 27 February 1989, defensive of his original ethological ideas to the end.

What did ethology accomplish? As Robert Hinde pointed out to us, ethology's achievements were considerable and include the development of ways of describing and measuring behavior; the recognition that behavior has evolved in response to the environment in which an animal lives; and of course Tinbergen's four questions. These achievements are far reaching and have been utilized and assimilated by other disciplines including physiology, psychology, psychiatry, animal husbandry, and behavioral ecology. We would add that the study of ethology helped expose the false dichotomy between instinct and learned behavior—best exemplified by bird studies of imprinting and the ongoing and remarkable developments in song acquisition (Marler and Slabbekoorn 2004; Catchpole and Slater 2008). Ethology has also told us a great deal about the behavior of birds, most notably resulting in a vast body of descriptive material on social signals and other behaviors. One has only to look at comprehensive handbooks, such as *Birds of the Western Palearctic* or *Birds of the World*, to see how those early descriptions of displays and postures have stood the test of time. It was precisely these descriptions that provided a solid foundation for research in behavioral ecology (chapters 8 and 9). Ethology also provided the methods for studying behavior in birds and other animals and formed the foundation for comparative studies, originally pioneered by Darwin; developed by Heinroth, Lorenz, and later Tinbergen and

Esther Cullen; and in recent years helped by detailed molecular phylogenies (chapter 3).

Once animal behavior researchers were exposed to individual selection in the late 1960s, they rapidly realized that Tinbergen's question about function (i.e., the adaptive significance of behavior), which had so far attracted only limited attention, offered extraordinary opportunities (Brown 1975). This shift in focus became behavioral ecology, a field of study that has been extraordinarily successful, especially with studies of birds. Interestingly, as behavioral ecology came of age in the late 1990s, there was an increasing tendency for researchers to return to Tinbergen's remaining three questions—not to concepts like fixed action patterns, but to a better integration of the other levels of analysis (chapters 8 and 9).

The year 2011 saw the 31st International Ethological Conference, but ethology's original aims, including its focus on causation and development (the false dichotomy between instinct and learned behavior), together with concepts of sign stimuli and fixed action patterns, are now seriously outdated. However, as Bill Thorpe pointed out in 1979, the fact that such concepts become outdated is a measure of how rapidly a particular field is developing. The ethological approach continues to be important, especially in terms of the behavior and welfare of farm animals, including poultry (Perry 2004).

Ethology might not exist as it was in the 1950s, but its principles have influenced numerous other areas of study, a view that Robert Hinde echoed when we asked his opinion on the present status of ethology in March 2010: "In my view, the basic issues of ethology (the four whys, observation in the natural environment as a baseline) are what matters. Most of the Lorenz/Tinbergen concepts were wrong or needed modification. The basic issues have been taken up by other behavioural sciences, and that is where ethology can best be seen."[III]

CODA

The story of ethology, which resides almost entirely within the twentieth century, beautifully encapsulates the process of science and the study of bird behavior in particular. Over time, ideas that once seemed useful for explaining behavior were reassessed and re-evaluated and, if necessary, replaced by better ideas and better explanations. Tinbergen and Lorenz were responsible for many of those ideas initially, and Tinbergen, especially, was open to what must have been a painful reassessment of these ideas by Lehrman. Partly because of this openness, Tinbergen has remained ethology's hero, despite the fact that some of his classic experiments— the bill pecking by young gulls and the hawk-goose model experiments—have since been shown to be poorly done. In the wrong hands, these less-than-perfect studies might be used to undermine the value of ethology, but viewed through a sophisticated lens they serve as exemplars of the ongoing nature of science. While the early experiments can be criticized for their design and execution, and their results open to various interpretations, the fact remains that the descriptions of behaviors provided by the ethologists will stand the test of time.

BOX 7.1 Robert Hinde

My earliest memory of being interested in the natural world was when I was ten, and became ill with German measles. Confined to my bedroom, I made a bird table outside the window and took photos of greenfinches feeding on it. I attended Oundle School, and was lucky in that both the headmaster and house-master were keen ornithologists. Peter Scott had attended this school, and his tradition lived on. Furthermore, James Fisher was the headmaster's son, and he became very important to the development of ornithology in this country, through his New Naturalist series.

After I left school, I went into the war for five years. I trained as a pilot in Southern Rhodesia and was subsequently posted to numerous places where there was excellent bird watching—South Africa, Ceylon, and the north of Scotland. When the war finished I left the air force as soon as I could and enrolled to study zoology at Cambridge. My father was a GP, and I had intended to be a doctor, but it seemed like an awfully long course to start when I was already incredibly old at twenty-three! I didn't realize that it actually takes longer to qualify as a zoologist when you add on the PhD and postdoc.

With a little help from my former headmaster (nepotism was rife in those days), I was awarded a closed exhibition at St. John's College, Cambridge, and was accepted. During my degree I spent a lot of time out at the sewage farm, which was a fantastic site for bird watching. One day I saw a bird that I didn't recognize, and I was convinced it was a Moustached Warbler. I went to Rev. Edward Armstrong, who was a local expert on birds, but he wouldn't have anything to do with it, and he passed me on to W. H. Thorpe, who was then a lecturer in zoology. The bird stuck around and started breeding, and I tried to catch the young in a butterfly net because I didn't want to shoot them. I couldn't, but the sighting proved to be tremendously important for me, because eminent ornithologists from all over the UK flocked to see this poor wretched bird. The sighting itself was a first breeding record of this species in the UK, but more importantly it allowed me to meet all these top ornithologists, including David Lack, James Fisher, and W. B. Alexander.

The Moustached Warbler episode prompted David Lack, who had just taken over as director of the Edward Grey Institute (EGI) in Oxford, to invite me to Oxford. He wanted me to work on the feeding ecology of Rooks and jackdaws, but I refused, wanting instead to do a behavioral study on the Great Tit. I owe Lack a great debt because he agreed and let me do as I wanted through my DPhil—I should have been devastated if I'd had to study the differences in feeding behavior between Rooks and jackdaws! So for two years (because I was an ex-serviceman), with a wage of £300 per year, I wandered around Wytham Woods with a

notebook and pencil and wrote down what I saw—a piece of cake, really.

At that time the EGI was housed in little prefabricated huts in the grounds of St. Hugh's College. These were shared with Charles Elton's Bureau of Animal Population, with the theory that the two groups would benefit from interacting, but in reality there was little integration, because Elton and Chitty and others had little interest in behavior. I also had little direct supervision from Lack, since I'd stuck my neck out to do a behavioral study. Fortunately, my time at Oxford was coincident with the arrival of Niko Tinbergen. Tinbergen was setting up his animal behavior group and had no pupils of his own to begin with, so he and I were able to wander around the parks and talk a great deal. He was a wonderful, charismatic man and a far greater influence on my life than Lack.

After finishing my DPhil I moved back to Cambridge. Thorpe was setting up his Ornithological Field Station at this point, and he tried to recruit Lorenz, and then Moreau, to come and look after it. Lorenz considered it but went elsewhere after receiving a big job offer in Germany. Moreau came and lived in Madingley for a while but decided against it, so I was appointed as curator. Initially it was only Thorpe, Gordon Dunnett (lab assistant), and me, and we spent the first year (from October 1950) constructing and painting aviaries. Then, because Tinbergen had been interested in understanding gull courtship in terms of conflicting drives, I started a similar study with finches. It was a comparative study, and not an outstandingly new idea, but it helped to show that Tinbergen's approach was worthwhile.

I also started to work on nest building in canaries, because Thorpe was interested in the relation between instinctive and intelligent behavior. This work led me to meet Danny Lehrman, who was working on a similar problem at Rutgers University, using ring doves. He became my closest friend—he was a wonderfully charismatic man and very important to the development of ornithology. He was also a passionate bird watcher: I remember once landing in New York at five a.m., to be met by Lehrman, who said, "Come on, Robert, we're going birding." By five past ten I'd seen fifty-one species of North American birds!

After being back in Cambridge for a few years, I was asked by Tinbergen to write a textbook with him, but he started to suffer from depression and eventually gave it up. So it should be acknowledged that my book *Animal Behaviour* owes a lot to Tinbergen. It was also Tinbergen's influence that got me started on imprinting work, and this led me to John Bowlby, a London psychoanalyst, who invited me to participate in his seminars on parent-offspring relationships. Bowlby had treated some juvenile delinquents, and many of these had had separation experiences as children; Bowlby felt that parent-offspring relationships were key to the development of behavior. I wanted to explore this area

further, and I also wanted to do something to make the world a better place. When I married for the first time, my then mother-in-law had thought it was ridiculous to study birds, because it wouldn't enhance the human condition. So when I got a chance to work with Bowlby on monkeys, I took it like a shot. And eventually I moved to work on children. It is no exaggeration to say that I was strongly involved in the development of Bowlby's attachment theory, which is very influential in child psychology today.

Things have certainly changed since the 1950s and '60s. For one, it's a lot harder to get funding now—I once had a grant application returned from an American foundation saying that I hadn't asked for enough money, and could I use twice as much! Another big difference is the increased sophistication of fieldwork today. In my day you simply walked around the woods with a notebook and binoculars. My daughter had 140 nest-boxes, cameras in the roof of many boxes, a loudspeaker so that she could play begging calls, microphones to record everything, and wires coming out in every direction! She also used far more sophisticated statistics. It is such a different world from the one in which I was starting out.

CHAPTER 8

Behavior as Adaptation

An explicit melding of natural history and biology.
—GEOFF PARKER (2006), DESCRIBING THE FIELD OF BEHAVIORAL ECOLOGY

BEGINNINGS OF A REVOLUTION

BY THE EARLY 1970S THE FIELD STUDY OF animal behavior was in need of new ideas and new approaches. Not surprisingly, then, the experimental, hypothesis testing, evolutionary-based advent of sociobiology and behavioral ecology was enthusiastically welcomed, especially by younger researchers. Selection thinking triggered a major change in attitude and ultimately generated a new corpus of knowledge (Birkhead and Monaghan 2010).

As with all revolutions, there was resistance, especially from the old guard. John Krebs and Nick Davies, editors of the textbook *Behavioural Ecology* (1978), bore the brunt of this criticism in Britain. Krebs[1] told us: "I recall going to give a seminar at Madingley in the 1970s on foraging theory and being relentlessly attacked by Robert Hinde,

and more substantively, John Kennedy, in his book *The New Anthropomorphism* (1992), was very critical of 'teleological' labels such as 'foraging'. He preferred to use the more neutral 'locomotor behaviour with a reduced threshold of response to food stimuli.'"[2] There was also a perception that behavioral ecology was simplistic—one review of Krebs and Davies's (1978) textbook, for example, suggested that all that was needed was a lively group of friends, a pub, and few beers.[3] Others of the old guard considered the behavioral ecologists' emphasis on hypothesis testing to be "pretentious,"[4] presumably because the usual approach to science up to that time was largely "find an effect and explain it."

Already well established as part of science, in theory if not always in practice, hypothesis

Painting from the 1970s by Raymond Harris Ching, one of several images for a fantasy field guide.

Nick Davies (left, at age 25) and John Krebs (right, at age 31) in 1977 at the International Ethology Conference in Bielefeld, Germany, where they began to plan the first (1978) edition of their edited volume *Behavioural Ecology: An Evolutionary Approach.*

testing was one of several core components of behavioral ecology. The philosopher of science Karl Popper was at the height of his popularity in the mid-1970s, and his idea that the falsifiability of hypotheses was the essence of science was extremely influential (Mayr 1982: 26). Another core component of behavioral ecology was its focus on quantitative analysis, with behavior being analyzed in a reductionist manner by evaluating costs and benefits. Behavioral ecology's blending of ecology, behavior, and evolution and its unambiguous focus on individual selection provided both a theory base and an array of exciting new hypotheses. In addition, several of the key architects of behavioral ecology were ornithologists, including Jerram "Jerry" Brown, Nick Davies, and John Krebs. The result? A transformation in our understanding of bird behavior.

The starting point was the publication of Vero Wynne-Edwards's overtly group selectionist *Animal Dispersion in Relation to Social Behaviour* in 1962 (chapter 10). John Maynard Smith (1964), Jerry Brown (1964), George Williams (1966a), and David Lack (1966) all responded sharply, criticizing the fallacy of group selection and reaffirming—with devastating lucidity—the logic of individual selection.

Here is the essence of the group selection approach to animal behavior and why it is wrong. Under group selection, animals behave for the benefit of the group (or population or species). For example, a bird that gives an alarm call on seeing a predator, allowing others to take cover, appears to be acting selflessly (altruistically) and hence for the good of the species. In exactly the same way, a bird that helps to feed the young of another breeding pair seems to be acting for the good of the species. The central issue is altruism, performing a behavior that is costly to one's self but beneficial to others, one that appears to be explicable only in terms of being "for the good of the species." As Wynne-Edwards's opponents pointed out, natural selection simply does not operate in this way. The clearest way to see this is to use one of Wynne-Edwards's own examples. He argued that animal populations are self-regulating and that when food is scarce some individuals refrain from breeding so the population does not overeat its food supply—that is, reproductive restraint occurs for the good of the species. The key point is how genes for reproductive restraint (or any other kind of altruistic behavior) could persist. Wynne-Edwards's opponents illustrated the fallacy of group section thinking by saying something like this: If you imagine a population with two mutants—one a Wynne-Edwards mutant that refrained from breeding when food was scarce, the other a selfish mutant that reproduced as fast as it could when food was scarce—which would leave the most descendants? The answer is obvious: the selfish mutant. Natural selection operates on individuals, not on groups, populations, or species.

The problem was that group selection has an intuitive appeal, elegantly but erroneously accounting for selfless behaviors. While Wynne-Edwards's opponents knew that group selection was wrong, it was not clear how one explained selfless behaviors. Darwin (1859) had been plagued by the same problem when he tried to explain the apparently altruistic behavior of social insects, recognizing that without an explanation firmly based on individual selection, the existence of altruism was fatal to his theory of natural selection.

A plausible solution did not emerge until the early 1960s, when Bill Hamilton started thinking about the genetics of apparently altruistic individuals, showing that their genes gained a fitness advantage when they helped their close relatives to reproduce. His idea became known as "kin selection," referring to the fact that an apparently altruistic behavior actually benefited the individual performing it, through its relatives or kin.

From this time on the case for individual selection started to gain momentum and by the mid-1970s was compelling, exemplified by Richard Dawkins's *The Selfish Gene* (1976). A year earlier E. O. Wilson had published his blockbuster, *Sociobiology*—the name he coined for the integration of evolution and natural history in the study of social systems and social behaviors. In his final chapter Wilson discussed the evolution of human behavior, and in doing so he created a firestorm of controversy, especially in the United States, where the idea of genetic determinism in humans was anathema. It was because of this vitriolic opposition to "sociobiology" that researchers interested in birds and other nonhumans adopted the term "behavioral ecology" (Segerstråle 2000).

Wilson's and Dawkins's two books are often grouped together on the assumption that they presented an identical new paradigm. They didn't. Wilson failed to define exactly what he meant by terms like "group selection," "kin selection" (see below), and "individual selection." Dawkins, on the other hand, was much clearer, and although he was concerned with selection operating at the level of the gene, his central theme was about the functioning groups of genes in *individuals* (Segerstråle 2000: 71; Birkhead and Monaghan 2010).

Prior to the 1970s most field ornithology—with a few notable exceptions—was descriptive natural history. Behavioral ecology, by focusing on whether behaviors or other traits, such as elaborate plumage or songs, were adaptive, allowed ornithologists to generate hypotheses and test specific predictions within an individual selection framework (Brown 1964; Krebs and Davies 1978). In so doing, behavioral ecology changed the very nature of ornithology, elevating its scientific status and influencing where ornithologists published their results. Prior to the mid-1970s, ornithologists aspired to publish their results in journals like *Ibis* or *The Auk*, but once behavioral ecology took off these journals were passed over in favor of new, concept-based publications, such as *Behavioral Ecology* and *Behavioural Ecology and Sociobiology*. The introduction of journal impact factors and citation indices in the 1980s and 1990s further damaged bird journals, and indeed, after the mid-1970s many researchers who worked on birds ceased calling themselves ornithologists at all, because that now seemed too narrow. Instead they often referred to themselves as behavioral ecologists or, sometimes, evolutionary biologists—who happened to study birds.

Hamilton's ingenious solution to the paradox of altruism prompted new interest in cooperative breeding in birds. We therefore start with this topic and then consider three other central areas of behavioral ecology where the study of birds, informed by individual selection thinking, changed or is changing our understanding: foraging, brood parasitism, and cognition.

COOPERATIVE BREEDING

It is hard to imagine what Costa Rica must have been like in the 1930s, when the biologist Alexander Skutch decided to live there. American by birth, Dutch by descent, Skutch trained first as a botanist, studying bananas for his doctoral research. On seeing a Rufous-tailed Hummingbird building its nest just outside his lab window in Panama in 1928, he switched from "foliage to feathers" and never looked back:

> I found this [the hummingbird], and the many other birds that nested in the garden, so fascinating that I decided to learn more about tropical American birds. After my return to the United States, I delved into the literature and found that nearly every species had been collected, named, and minutely described, but that very little was known about their habits. I concluded that I could do nothing more important and satisfying than to learn the intimate details of their lives.[5]

After more than a decade traveling about Central America, Skutch bought a farm (Los Cosingos) in Costa Rica, where he remained, with his wife, Pamela, but without electricity or a telephone line,[6] for the rest of his long life. He continued to study and write about birds, eventually becoming the premier Neotropical ornithologist. When asked, at the age of ninety-six, what he considered his most important ornithological contribution, he said, "Writing about the life histories of many tropical birds, whose nesting and other habits were previously little known or

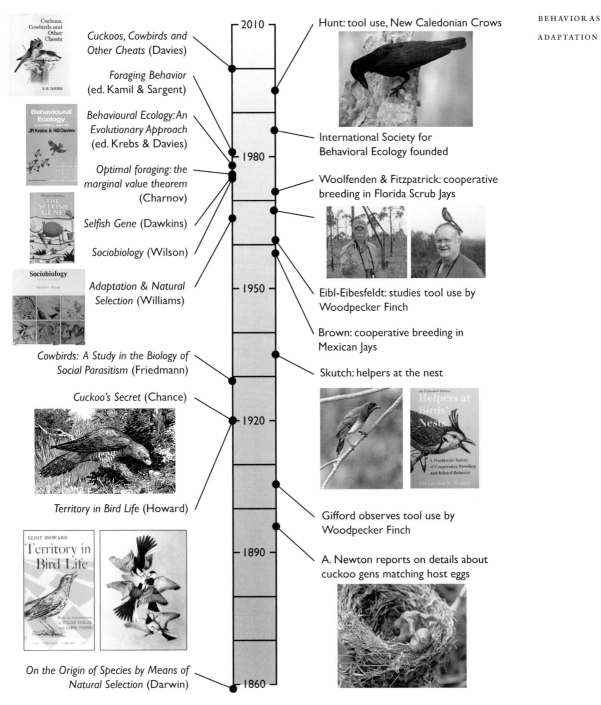

2010 —

— Hunt: tool use, New Caledonian Crows

*Cuckoos, Cowbirds and
Other Cheats* (Davies)

Foraging Behavior
(ed. Kamil & Sargent)

— International Society for
Behavioral Ecology founded

*Behavioural Ecology: An
Evolutionary Approach*
(ed. Krebs & Davies)

1980 —

— Woolfenden & Fitzpatrick: cooperative
breeding in Florida Scrub Jays

*Optimal foraging: the
marginal value theorem*
(Charnov)

Selfish Gene (Dawkins)

Sociobiology (Wilson)

— Eibl-Eibesfeldt: studies tool use by
Woodpecker Finch

*Adaptation & Natural
Selection* (Williams)

1950 —

— Brown: cooperative breeding in
Mexican Jays

*Cowbirds: A Study in the Biology of
Social Parasitism* (Friedmann)

— Skutch: helpers at the nest

Cuckoo's Secret (Chance)

1920 —

Territory in Bird Life (Howard)

— Gifford observes tool use by
Woodpecker Finch

1890 —

— A. Newton reports on details about
cuckoo gens matching host eggs

*On the Origin of Species by Means of
Natural Selection* (Darwin)

1860 —

TIMELINE for BEHAVIORAL ECOLOGY. Left: Covers of Davies (2000), Krebs and Davies (1978), Dawkins (1976) and Wilson (1975); Common Cuckoo visiting a host's nest; cover of Howard (1920); two pairs of Pied Wagtails fighting over territory depicted in Howard's *Territory in Bird Life*. Right: New Caledonian Crow using a stick to extract a grub from a crevice; John Fitzpatrick (left) and Glen Woolfenden (right), each with a Florida Scrub Jay at Archbold Biological Station; Brown Jay; cover of Skutch (1987); Common Cuckoo chick ejecting a host egg from a nest.

completely unknown—and, in particular, calling attention to the prevalence of cooperative breeding[7] (helpers at the nest)."[8]

Costa Rica, along with the rest of Central America, is extremely rich in birds, giving Skutch extraordinary scope for making discoveries. His philosophy was to observe but not disturb; he disapproved of catching and banding birds, and almost uniquely among American ornithologists of his day, he never collected (killed) a bird. As his colleague Gary Stiles said, "His interpretations of his observations were often permeated by his philosophical leanings: he viewed nature as a harmonious association of species living together."[9]

Living in the bananas at his farm were Brown Jays. Intrigued by their noisy behavior and the fact that variation in bill color allowed him to recognize individual birds, Skutch soon realized that the incubating female, and later the young, were fed by more than just the female's partner. He called these additional birds "helpers at the nest."

The Brown Jays weren't unique; Skutch also found helpers in American Bushtits and Band-backed Wrens (Skutch 1935).

His interpretation of cooperative breeding now seems quaint: "And so among birds help is given and received entirely in a spirit of good fellowship, neither those who give nor those who receive compromise their independence nor lose their self sufficiency."[10] In an update published in 1961, he presented an overview of different forms of helping behavior, and only in a short section toward the end of the paper does he discuss the evolution of helping, saying:

> If an accelerated rate of reproduction were highly advantageous to any species whose young mature slowly, it appears that this could be more efficiently achieved by hastening the advent of the adult state than by making helpers of the innubiles [immatures] . . . Why birds which have long been fully grown should pass from one to six breeding seasons without themselves reproducing, certainly poses a problem . . . The existence of this large nonbreeding class is impressive testimony that its members are not needed as breeders; that, in fact it would be disadvantageous to the species to have them engage in reproduction.[11]

He finishes by saying, "The presence of helpers, and especially innubile helpers, is, then, one more link in the lengthening chain of evidence pointing to the conclusion that birds have a considerable store of unused reproductive potential and that their reproductive rate has, in many cases, been delicately adjusted to the conditions of their lives rather than pushed to the limit of their power to rear offspring, as some have contended. . . ."[12]

Alexander Skutch.

Skutch was fiercely opposed to individual selection thinking, commenting that

> some biologists seem to delight in detecting repressed hostility in acts that appear friendly, and they use all their ingenuity to disclose selfish deceit or cheating in ostensibly altruistic behavior. They appear to take perverse satisfaction in exposing the nastiness and harshness of the living world. Harsh it undeniably is, but not so unmitigatedly as it is often painted. Nature has a gentler side that is too frequently overlooked. Ranking high among the more amiable aspects of the natural world is the helpfulness of birds.[13]

While Skutch is generally acknowledged as the discoverer of cooperative breeding in birds, his philosophy and training prevented him from using a behavioral ecology approach to interpret his findings.

Helping behavior in birds flew in the face of individual selection thinking because, at first sight, it appeared to be altruistic. However, Hamilton's (1963; 1964a, b) papers on kin selection, which made little immediate impact because they were difficult to understand, now came into their own, providing an explanation for apparently altruistic behavior that was entirely consistent with individual selection thinking. Kin selection provided behavioral ecologists with a potential explanation for the evolution of helping behavior in birds.

Jerry Brown, one of the founders of the modern study of cooperative breeding, and of behavioral ecology itself, read Hamilton's first paper on the evolution of altruistic behavior soon after its publication in 1963 and immediately recognized its significance. He assumed that others would do the same and that the field of cooperative breeding would

be revolutionized (Brown 1994). They didn't and it wasn't. Brown says that it was not until he presented data on helping behavior in Mexican Jays, in 1970, that "widespread interest in the relevance of Hamilton's rule to helping was aroused."[14] Brown attributes the subsequent increase in studies of cooperative breeding to the subject being "theory driven," specifically by Bill Hamilton's (1963, 1964a, b) concept of "inclusive fitness," which became a cornerstone of behavioral ecology: the idea that one's genetic legacy can come from one's self or via one's relatives (or kin). As Brown said, "The first reaction by many in the early 1970s to my position that it was useful to view helping in the context of inclusive fitness theory was a mixture of disbelief of my observations, scepticism of any theoretical interpretation, and hostility based mainly, in my opinion, on misunderstanding and conservatism."[15] This is not an unusual response to new ideas in science, so in some ways Brown should not have been surprised, but the concept of inclusive fitness was, and still is, a thorny issue—a difficulty captured by Steve Stearns, who said, "Fitness is something everyone understands but no one can define precisely."[16] J. B. S. Haldane was blunter, once saying to John Maynard Smith, "Fitness is a bugger!"[17]

On the face of it, kin selection seemed to provide a straightforward explanation for helping behavior. If helpers were close relatives of the breeding pair, as observations suggested, then they obtained an indirect (genetic) benefit from helping. A key question was whether helpers actually helped, and to test this Brown conducted a pioneering study of Grey-crowned Babblers in Australia that involved comparing the breeding success of intact groups with groups from which helpers had been experimentally removed. The effect of helpers was clear cut: they increased the parents' breeding success (Brown et al. 1978).

The idea that kin selection explained helping behavior was attractive, but some were not convinced by its apparent all-explanatory power. In their monograph *The Florida Scrub-jay: Demography of a Cooperatively Breeding Bird* (1984), Glen Woolfenden and John Fitzpatrick said, "We are aware that kin selection could lead to helping behavior, but it remains to be shown in even one natural system that kin selection was necessary for helping to evolve. In short, we emphasize that the *relative* importance of the direct versus the indirect components to inclusive fitness still is unclear."[18]

Woolfenden started to observe Florida Scrub Jay nests at Archbold Biological Station, Florida, in the spring of 1969. After banding the parents, he was intrigued by the fact that more than just the breeding pair attended six of the eight nests he was watching. Two years later John Fitzpatrick, then a student, met Woolfenden and on seeing the banded population recognized its potential. The extraordinary tameness of the jays, combined with their very limited dispersal from the nest meant that individuals could be followed throughout their whole lives—from being banded at eleven days old to fifteen or more years later, when some are still breeding.[19] As we write in 2013, the population is still being studied.

Woolfenden and Fitzpatrick were less enthusiastic than Brown about indirect, kin-based benefits of helping, as Fitzpatrick

294   Florida Scrub Jay.

explained to us when we asked whether they felt that Jerry Brown and others had seized upon Bill Hamilton's ideas too eagerly: "Yes, absolutely, it was an alluring idea to explain helping behavior, and frankly became something of an intellectual fad. Glen and I definitely were regarded as old-school for arguing that selfish interests and direct selection could select for the same thing, and that the indirect component of selection for cooperation was often very tiny and eventually quantified for our chapter in Ian Newton's *Lifetime Reproduction in Birds*."[20]

A key question for the study of cooperative breeding was why young birds of some species hang around and help their parents, why they don't simply go off and breed on their own like most other birds. In 1982 Steve Emlen provided an answer in two landmark papers, published back to back in *American Naturalist*, introducing the idea that "ecological constraints" forced families to remain together. The central idea was that if opportunities for independent breeding were limited, or constrained in some way because the habitat was "full" or because of climatic conditions, it would pay offspring to remain at home. Research in the 1960s and 1970s had laid the groundwork for this theory: Robert Selander's study of cooperatively breeding *Campylorhynchus* wrens, for example, led him to propose that it was a lack of available breeding territories that prevented offspring from dispersing: the habitat saturation hypothesis (Selander 1964). A similar situation existed in the Florida Scrub Jays (Fitzpatrick and Woolfenden 1989; for a different viewpoint, see Brown 1989[21]).

Researchers began testing the ecological constraint idea in the 1990s. Potential constraints included a shortage of territories,

high dispersal costs, a shortage of breeding opportunities, and a low probability of successfully breeding even if a territory is established (Hatchwell and Komdeur 2000). Experimental studies supported the idea of constraints. For example, if a resident male Superb Fairywren is experimentally removed, helpers readily move in, but only if the resident female is still present (Pruett-Jones and Lewis 1990), suggesting that the availability of both a territory and a breeding partner are important ecological constraints in this species. Similarly, when Seychelles Warblers were translocated to previously unoccupied islands, individuals chose to breed independently until all the available territories began to fill up. When the only available territories were of low quality, individuals chose to remain at home with their parents—if their parents held high-quality territories. If their parents held low-quality territories, the young birds left home to breed independently (Komdeur 1992).

While there is strong support for ecological constraints *within* a species, it isn't clear whether, across species, ecological constraints help to explain the occurrence of cooperative breeding. There have been numerous attempts to identify ecological correlates of cooperative breeding, but with limited success. One of the most recent shows that variability in environmental factors such as rainfall is important (Jetz and Rubenstein 2011), but even this explains only a small amount of the variation in cooperative breeding (Cockburn and Russell 2011). Moreover, as Ben Hatchwell and Jan Komdeur (2000) point out, the central problem is this: all species face some kind of ecological constraint on their breeding, yet only 9 percent (Cockburn 2006) of birds breed cooperatively.[22] As they say, what really needs to be explained is the absence, rather than the presence, of cooperative breeding.

Whether, and to what extent, helpers increase reproductive output is probably best determined by experiment. Throughout the 1970s and early 1980s, researchers were naturally reluctant to disrupt their ongoing, long-term studies by performing disruptive removal experiments. The few studies that were undertaken gave mixed results. For example, in the Seychelles Warblers helpers can either help or hinder, depending upon territory quality (Richardson et al. 2002). In the Florida Scrub Jays, removal of helpers resulted in a decrease in productivity, through an increase in nest predation (Mumme 1992), while in the Common Moorhen the removal of helpers had no effect on productivity (Leonard et al. 1989).

These disparate results eventually led to a reassessment of the role of kinship in cooperative breeders, capitalizing on the recently developed molecular methods for establishing parentage (chapter 9). As the results of parentage studies began to emerge, it became increasingly obvious that helpers were not always relatives, and therefore that kinship was not always the basis for helping. Reassessment of the relative importance of direct and indirect benefits highlighted the tendency of researchers to assume that if helpers were found to be helping nonrelatives, helping must have evolved through direct benefits, suggesting that kin selection was unimportant (Cockburn 1998; Clutton-Brock 2002).

Hatchwell has argued that if the costs of helping are relatively low, as they are in the Long-tailed Tits that he studies, then it is unlikely to be important if birds make the occasional mistake of helping a nonrelative. The current consensus is that some cooperative

breeding systems (like Long-tailed Tits) are entirely kin based, whereas others—like the secondary helpers in Pied Kingfishers (Reyer 1980)—are driven entirely by direct benefits. Overall, about 85 percent of all cooperatively breeding birds involve kin (Hatchwell 2009). The main debate now is what phylogenetic, life history, or ecological factors favor cooperative breeding, and the key to this is understanding family structures rather than cooperation per se.[23]

Recent research suggests that a crucial factor in understanding the evolution of cooperative breeding is sexual monogamy. Using theory derived from social insects (Boomsma 2009), it has been suggested that cooperation depends on relatedness, and relatedness between offspring depends on reproductive females being sexually monogamous. A promiscuous female is equally related to all her offspring, but those offspring are less related to one other if the mother mated with more than one male. Charlie Cornwallis and colleagues (2010) recently tested this monogamy hypothesis in birds and found strong support. Using a sophisticated phylogenetic analysis, they found that bird families that were ancestrally monogamous were more likely to include cooperative breeders than were those whose ancestors tended to be promiscuous. Despite some convincing results overall, a number of striking exceptions remain; notably two Australian species: the Superb Fairywren and Australian Magpie, both of which are cooperative and both highly promiscuous (chapter 9)—and therefore still require an explanation.

FOOD, FORAGING, AND OPTIMALITY

In an attempt to find out what birds eat, tens of thousands were killed and their stomach contents examined during the first few decades of the twentieth century. Some ornithologists made such studies their life's work; Foster Beal, for example, employed by the US Biological Survey between 1892 and 1916, examined the stomach contents of 37,825 birds (Taylor 1931). He was by no means unique. On the basis of this kind of research, species were categorized as useful (and preserved) or harmful (and persecuted). Writing in 1913, ornithologist Walter Collinge said, "The nature of the food of many of our wild birds has hitherto been largely guesswork, for, with the exception of a very few species, no detailed investigations have been carried out, and without these details . . . it is impossible to arrive at any sound conclusions respecting their economic status."[24] "Economic ornithology" provided the basis for much of what we know about the diets of birds, including regional, seasonal, and annual differences (Taylor 1931; see also Wenny et al. 2011). Also of interest was how much food birds consumed—because of its economic implications—and such studies subsequently led to a better understanding of avian energetics.

Economic ornithology declined through the 1930s and was replaced by more academic studies of food and feeding. The ornithological journals between the 1930s and 1960s contain numerous descriptive papers with titles like "The Feeding Habits of . . . ," mostly from a species point of view. During the latter part of this period, and probably inspired by Lack's (1945) work on Darwin's finches, some of these studies were concerned with the avoidance of interspecific competition for food (e.g., Kendeigh 1945; Lack 1971).

Starting in the 1960s, ecologists and evolutionary biologists began to think about the foraging behavior of birds from an evolutionary point of view. Two papers published in 1966—one by Robert MacArthur and Eric Pianka and the other by John Merritt Emlen—introduced the idea that the *way* birds foraged was shaped by natural selection. A key assumption of this work was that food is rarely superabundant, and even when it is, it will still pay individuals to forage efficiently because this will free up time for other activities—like looking out for predators, preening, defending a territory, or trying to attract a mate. Despite the wide range of food types consumed by birds—plankton, seeds, fruits, nectar, invertebrates, fish, or other birds—they all face similar choices of where to forage, how much time to spend foraging, and what items to select. The approach is one of identifying the adaptive significance of foraging in a particular way, viewing foraging in terms of its benefits and costs, and identifying the optimal trade-off between those costs and benefits that maximizes fitness. MacArthur and Pianka (1966) thought about the efficiency with which individuals searched for patchily distributed food; Emlen (1966) thought about the selection of food items in terms of its calorific intake and handling time, recognizing that intake could be optimized by judicious choice of food types.

The paper that perhaps best captured the spirit of this new, concept-driven approach to foraging was Gordon Tullock's (1971) "The Coal Tit as a Careful Shopper," published in *American Naturalist*. Tullock was an eminent American economist, who, while reading Lack's *Population Studies of Birds* (1966), discovered John Gibb's work on the feeding ecology of Coal Tits. In winter, Coal Tits forage for the larvae of the moth *Enarmoria conicolana*, which lie dormant in small crevices on the surface of pine cones and are located by the bird tapping the cone. Gibb's graph, showing the consumption of larva by Coal Tits, looked to Tullock like a standard economic supply and demand graph, suggesting that the behavior might be interpreted using economic principles. Gibb's own interpretation was that the birds were "hunting by expectation" that is, they had formed an expectation of the density of prey items based upon their sampling and then remained in a patch according to this expectation. Tullock's explanation was much simpler: "The birds are behaving much like a careful housewife . . . shopping in the cheapest market . . . any animal, in order to remain alive, must be economical in its use of energy to obtain food. An inherited behavior pattern which led to such economy would have survival value, not only in consumption of *E. conicolana* but in many other types of food as well."[25] The basic approach exemplified by Tullock was to use (micro)economic principles to better understand foraging (Ydenberg et al. 2007). "This motivation fused with developing notions about natural selection (Williams 1966) and the importance of energy in ecological systems to give birth to 'optimal foraging theory' (OFT)."[26]

Here we provide four examples of the types of question that optimal foraging theory (OFT) aims to address:

(a) **Where to forage?** Food is often patchily distributed in space: imagine dungflies—a favorite food of wagtails—on cowpats in a field. If all the wagtails converged on just one cowpat, all else being equal it would pay some birds to move to another site where there would be less competition. The end point is an "ideal free distribution" (Fretwell and Lucas 1969), analogous to the way that humans respond to queues at supermarket checkouts. Empirical tests of the ideal free distribution confirm that under some circumstances at least some birds do forage in this way (e.g., Harper's (1982) study of Mallard ducks).

(b) **What to eat?** The idea is that the choice of prey items is determined by the trade-off between calorific value of the prey and the time taken to handle or process that prey: optimal diet or prey choice (MacArthur and Pianka 1966; Houston et al. 1980). Empirical tests of this model were made on Great Tits in captivity by John Krebs and colleagues (1977), who presented subjects with a conveyor belt on which were two sizes of mealworm segment and for whom the "search time" was manipulated by varying the distance between successive prey items on the belt. Theory predicts that when the search time is reduced birds should switch from generalizing and eating both large and small prey in similar proportions, to eating only large prey—which is exactly what the birds did. John Goss-Custard (1977) conducted a similar experiment with Common Redshanks feeding on ragworm—the birds preferring to take large worms, varying their response to small worms as a function of the number of large worms eaten and thereby maximizing their intake rate—as predicted by theory.

(c) **When to move from one food patch to another?** Here the idea is that the longer a bird forages in a certain patch, the lower its rate of gain, because as it forages resources are used up—that is, gains are subject to diminishing returns. Gibb's "hunting by expectation" hypothesis was compared with predictions from optimal

A Great Tit engaged in an optimal foraging task, choosing between different-size pieces of mealworm moving past on a conveyor belt in an experimental study by John Krebs.

foraging theory by Krebs and colleagues (1974), who found that foraging Black-capped Chickadees did not learn to expect a fixed number of prey but instead showed a "giving up time" inversely proportional to the average capture rate for an environment. Foraging animals, under the "marginal value theorem" formalized by Eric Charnov (1976), are predicted to leave a patch when the rate of gain declines to the marginal rate of gain in the habitat. Craig Benkman (1987) tested this in the field with Two-barred Crossbills feeding on pine cones. Crossbills extract seeds from under the scales of the cones, but the seed mass decreases along the length of the cone, and crossbills typically stop before they reach the tip of the cone, leaving

some seeds untouched. It seemed odd that birds would leave perfectly edible seeds in a cone, but this is exactly what the marginal value theorem predicts. Benkman's analysis based on that theorem predicted that crossbills would maximize their rate of gain by moving onto a different cone after scale 7 or 8: that is exactly what happens with crossbills, most often moving on after scale 8.

(d) **Central place foraging.** When birds (or other animals) forage around a fixed location—such as a nest where parents bring food for their nestlings—individuals have to consider the energetic cost of the round trip, which might influence the optimal number of individual items to collect in a distant patch of food. The precise

299

relationship between optimal load size and travel time can be determined using a graphical model based on the marginal value theorem, as shown initially by Orians and Pearson (1979). Alex Kacelnik (1984) conducted what is now considered to be a classic study of central place foraging in Common Starlings, looking at how long parent birds foraged in different "patches," recording the number of prey swallowed, number of prey brought back to the nest, and the total time spent in the patch. He manipulated patch quality (determined by the rate of encounter with food) and patch distance from the nest, and he tested different models about optimal patch time. As predicted by theory, starlings stayed longer to forage in patches that were farther from the nest.

While many regarded optimal foraging as an exciting approach to the study of foraging behavior, it was not without its critics. As one commentator put it, optimal foraging "sparked some of the most acrimonious debate ever seen in a scientific discipline."[27] Part of the criticism was philosophical (or political) and focused on the issue of "optimality" and adaptation in general (see also Gould and Lewontin 1979), while another concentrated specifically on optimal foraging theory. A particularly vocal critic was John Ollason, who at a symposium at Brown University in 1984 said that "a labyrinthine tautology has been constructed that is based on assumption piled on assumption."[28] A few years later he published, with Graham Pierce, an article titled "Eight Reasons Why OFT Is a Complete Waste of Time" (Pierce and Ollason 1987), in which he gave the details of his arguments. In that paper the

authors asserted that optimization was an inappropriate way to investigate the products of evolution, that animals should not be expected to be optimal, and that it was impossible to actually test whether they are optimal. This was a major misunderstanding. Optimal foraging researchers did not mean to imply that organisms were optimal, only that "optimality" refers to the investigative, mathematical technique.

By the mid-1990s interest in optimal foraging theory was waning, partly because of its success in addressing most of the major issues, but also because criticisms encouraged researchers to modify their approach, dropping the term "optimal" and placing less emphasis on "theory." If one plots the number of publications in which the term "optimal foraging" appears in the title, the boom and bust of this area of behavioral ecology is clear, starting in the mid-1970s, peaking in the early 1980s, and declining thereafter. Most areas of science show a similar pattern, but different subdisciplines endure for different periods of time. The study of optimal foraging, based around the simple models of the ideal free distribution, marginal value theorem, and others, evolved into the study of more general issues of decision making and rationality (Cuthill and Houston 1997). The book by David Stephens, Joel Brown, and Ronald Ydenberg (2007), titled simply *Foraging: Behavior and Ecology*, illustrates how the discipline has changed, compared with its predecessors, *Foraging Theory* (Stephens and Krebs 1986) and *Foraging Behaviour: Ecological, Ethological and Psychological Approaches* (Kamil and Sargent 1981).

The study of how birds forage continues, but in a different form: simplicity and coherence have been left behind, but diversity, richness of texture, and understanding have been gained. The tentacles of foraging theory, in its broadest sense, have extended to form links with neuroethology, behavioral economics, life histories, animal learning, game theory, and conservation biology.[29]

BROOD PARASITES

That the Common Cuckoo parasitizes the parental care of other species has been a source of fascination since its discovery in the fourth century BC (Schulze-Hagen et al. 2009). It subsequently became clear that the Common Cuckoo is one of around a hundred interspecific brood parasites, including other cuckoo species, honeyguides, cowbirds, finches, and even a duck (Davies 2000).

Edward Jenner (1788) was the first to provide a detailed description of the way the recently hatched Common Cuckoo chick ejects host young from the nest. Prior to this the cause of the host chicks' disappearance was both unknown and subject to much speculation. The similarity in appearance of the cuckoos' eggs and their hosts' eggs was not noticed until the mid-1700s and triggered an interest in egg collecting that among a minority of fanatical individuals continues to the present day.

It was Darwin (1859) who recognized the adaptive significance of brood parasitism, and as Nick Davies says, "Darwin packs more good ideas into these four sentences [in chapter 8 of *The Origin*] than all previous commentators on the Cuckoo since Aristotle."[30] However, not everyone agreed with Darwin's interpretation, and the idea that brood parasitism was somehow "degenerative" seemed a persuasive explanation for the lack of nest building and brood parasitic

behavior (see Schulze-Hagen et al. 2009). A century after Darwin, however, William J. Hamilton III and Gordon Orians (1965), and David Lack (1968)—all of whom adopted an evolutionary perspective—felt that the most plausible starting point for the study of brood parasitism was to assume that it had evolved as a response to nest loss during laying, leading to the opportunistic takeover of other species' nests.

Brood parasitism raised many questions. One was how the Common Cuckoo placed its eggs in host nests—especially domed nests and those with a tiny entrance hole. Many thought that the female cuckoo picked up its own egg and deposited it in the host's nest with its bill. It was not until Edgar Chance—a wealthy businessman and egg collector—started to observe cuckoos in detail in the early 1900s that the truth was revealed. By removing all but one potential host nest from his study area, Chance was able to predict where, and when, a cuckoo would lay, providing an excellent opportunity to witness cuckoos (1) laying directly in the host's nest; (2) removing one or more of the host's eggs before they laid; and (3) depositing their egg very rapidly—typically in just a few seconds. Chance employed film maker Oliver Pike to make this new knowledge available to the public, but old beliefs die hard, and Pike's film *In Birdland* (1907), while making Chance a minor celebrity, failed to convince everyone about the cuckoo's habits.

Chance later summarized his discoveries in *The Cuckoo's Secret* (1922), a book that failed to create the impact he hoped for. Ornithologists were reticent in their praise, on the one hand commending him for showing that an egg collector could work in a scientific

A Common Cuckoo chick being fed by its much smaller Eurasian Reed Warbler foster parent.

manner, but on the other criticizing him for overstating some of his results. Damned by faint praise, the book was considered an "interesting addition to our knowledge [even] if the author's new facts were not all quite so new to others as they were to himself."[31] Chance's second book, *The Truth about the Cuckoo* (1940), also received mixed reviews; one, titled "Myths Die Hard," alluded to the continuing controversy over placement of the cuckoo's egg in host nests. In another review, David Davis suggested that Chance write a monograph on the cuckoo's life history, urging him to "avoid the lapses into the anthropomorphic and mystical viewpoints."[32]

Chance was despised by the cognoscenti, a view reinforced when in 1926 he was convicted of stealing Red Crossbill eggs in Norfolk. In his defense, Chance claimed that he had collected the eggs for the Reading Museum, but the curator, H. M. Wallis, denied this (Wallis 1926). To avoid embarrassing the BOU, Chance offered to resign, but not without a parting shot, stating in an open letter to all BOU members: "It may not be generally known, that some of the Committee who have adjudicated on my case are themselves ardent collectors and have taken quite a number of Crossbills' eggs. But they do not happen to have been summoned for so doing!"[33]

At a special meeting on 8 December 1926, at the Zoological Society in London, a motion was proposed condemning the BOU for its handling of the Chance case and presumably voting for Chance to be reelected. But after some discussion the motion was rejected, since only 12 of 103 members voted in favor of it.[34] Chance was not reinstated, and when he died in 1955 the BOU made its feelings known by not publishing an obituary.

Despite Darwin's (1859) perceptive interpretation of brood parasitism, and the recognition that hosts are duped into accepting a parasite's eggs, much confusion and misunderstanding regarding brood parasites persisted until the mid-twentieth century. Behavioral ecology, with its focus on individual selection thinking, gave the study of brood parasitism an enormous boost in the 1970s and 1980s, making it a model system for understanding coevolution—in this case between parasite and host. Although the Common Cuckoo in Europe and the Brown-headed Cowbird in North America became the "model" avian brood parasites, it is now clear that there are numerous ways of being a brood parasite. Since the 1980s the main questions asked by ornithologists have been concerned with identifying adaptations and counteradaptations in brood parasites and their hosts.

The idea that cuckoos might mimic host eggs in color and pattern to avoid detection by the host was first suggested by an egg collector, Eduard Baldamus (1853), who also recognized that individual females lay eggs of a consistent color, and that females lay in the "wrong" nest only if no appropriate host nest is available. He also proposed the existence of what we now refer to as "gentes" (see below). Baldamus rejected the popular idea

that cuckoos could adjust the color of their eggs to match those of a host (see Schulze-Hagen et al. 2009).

Stuart Baker (1913)—a brood parasite enthusiast—seems to have been the first to recognize that it was rejection by hosts that drove the evolution of egg mimicry in Common Cuckoos: "It is not the cuckoo that evolves a perfectly adapted egg, but the foster parents, which gradually eliminate the types of cuckoos' eggs that are so ill-adapted as to attract their notice."[35] The Reverend Francis Jourdain perceptively added that "it is not the closeness of the mimicry or the reverse, as Baker suggests, which gives us the key to the age of the practice, but the activity of selection by the fosterer or its indifference."[36]

Remarkably, the first experimental tests of whether hosts would eject "foreign" eggs were conducted as early as the 1770s (Schulze-Hagen et al. 2009), but it was the work of Steve Rothstein on Brown-headed Cowbirds in the mid-1960s that made researchers realize how powerful these sorts of tests could be. Rothstein made dummy cowbird eggs from plaster of Paris, painted them with acrylic paints and shellac, placed them in the nests of potential host species, and recorded whether the eggs were accepted or rejected. Eight species were "rejecters" and twenty-three were "accepters," with very few being intermediate (Rothstein 1975). Rothstein concluded that "accepters and rejecters present a confusing array of species. . . . The puzzle . . . is why rejection behavior has not evolved in all species in which it would be adaptive."[37] He suggested that many accepters would do better if they rejected the cowbird egg and that their "acceptance" was a case of an evolutionary lag in response to a relatively new selection pressure.

Wealthy businessman, amateur ornithologist, and enthusiastic oologist Edgar Chance (photo date unknown).

We now know that an important host adaptation is to minimize variation in egg color within clutches laid by the same female and maximize egg color variation between clutches laid by different females. Charles Swynnerton (1918) referred to the latter as a polymorphism, recognizing that it "may be of use against cuckoos,"[38] that is, that markings on eggs are "signatures." He was right, both with regard to the Common Cuckoo and also, as elegantly demonstrated recently in a system with even more extreme egg polymorphism, in the Cuckoo-finch and its prinia hosts (Spottiswoode and Stevens 2010).

Collections of the eggs of brood parasites and their hosts, in museums or in private hands, have been extremely useful in understanding the evolution of egg mimicry and other adaptations (Moksnes and Røskaft 1995). Some collectors, for example, noticed that when they removed the contents from their eggs, cuckoo embryos were often at a more advanced stage of development than host embryos. David Lack, in *Ecological Adaptations* (1968), alluded to this more advanced stage of development resulting from egg retention by the cuckoo: "I have known for many years that this was said of *C. canorus* [the Common Cuckoo], but have failed to find a reference; it is confirmed by a single recent record by C. M. Perrins, pers. comm."[39] In fact it was George Montagu

(1802), in his *Ornithological Dictionary*, who suggested that fully formed eggs are retained in the oviduct to give the cuckoo chick a head start. Although Montagu apparently did not know it, the eggs of Common Cuckoo (and Greater Honeyguide) are laid at forty-eight-hour intervals, whereas their hosts, and indeed most other small birds, lay at twenty-four-hour intervals. In all species embryo development starts a few hours after fertilization, and because cuckoos (and Greater Honeyguides) retain their eggs for an additional twenty-four hours, their embryos are at a relatively advanced stage when their egg is laid. A comparison with other species showed that, by retaining the egg for an additional twenty-four hours at body temperature (40°C), the young brood parasite gains a thirty-one-hour hatching advantage over host young (Birkhead et al. 2011). Further adaptations for early hatching include rapid embryo development and small egg size—smaller eggs have faster incubation times but may also be better mimics of host eggs (Moksnes and Røskaft 1995).

In the 1880s egg collectors also noted that brood parasites often produce relatively small, thick-shelled eggs (Schulze-Hagen et al. 2009), and it was Swynnerton (1918) who proposed what is now known as the "puncture resistance hypothesis"—that a thick-shelled egg is an adaptation to avoid host ejection—although this is still unconfirmed (see Antonov et al. 2006). Reduced egg size is also thought to be an adaptation to engaging in brood parasitism, an idea confirmed by a comparison of parasitic and nonparasitic cuckoos (Payne 1974; see also Krüger and Davies 2002, 2004).

Alfred Newton (1896) introduced the term "gens" (plural, gentes) to describe the different color forms of Common Cuckoo eggs, recognizing the adaptive significance of host specificity and parasite-host egg matching:

> Hence the supposition may be fairly regarded that the habit of laying a particular style of egg is also likely to become hereditary. Combining this supposition with that as to the Cuckow's [sic] habit of using the nest of the same species becoming hereditary, it will be seen that it requires but an application of the principle of "Natural Selection" to shew the probability of this principle operating in the course of time to produce the facts asserted.[40]

Heinroth and Heinroth (1924–34) appear to have come up with the same idea independently, suggesting in addition that the female cuckoo inherits her egg color pattern from her mother rather than her father (see also Southern 1954). An extensive survey of cuckoo eggs from museum collections shows that the Common Cuckoo comprises at least fifteen gentes (Moksnes and Røskaft 1995), some of which have been confirmed as being genetically distinct by molecular studies (Gibbs et al. 2000). Other brood parasites, including the Cuckoo-finch, also have well-defined gentes (Spottiswoode and Stevens 2012).

In North America the most common brood parasite is the Brown-headed Cowbird, so named because it is often found foraging among domestic livestock and must once have followed the huge herds of bison that roamed the Midwest. Despite the bison's decline, changes in the landscape driven by agriculture have resulted in a massive eastern range expansion of the cowbird, such that it is now considered a threat to many of its host species (e.g., Brittingham and Temple 1983). As the cowbird spread across eastern North

America, those conducting studies of other species, such as Kirtland's Warbler (Mayfield 1961) and the Prairie Warbler (Nolan 1978) in the 1960s and 1970s, could not fail to see the deleterious consequences of brood parasitism by cowbirds. Conservation was therefore an important motivation for cowbird studies, accelerated by the interest in evolutionary questions arising from behavioral ecology in the 1970s.

Herbert Friedmann—one of the fathers of avian brood-parasitism research—was among the first to study cowbirds and provided a solid account of their natural history (Friedmann 1929). A gentle, enthusiastic scholar, Friedmann—who also studied other cowbird species in South America and honeyguides and cuckoos in Africa—recognized the enormous potential of brood parasites to inform us about coevolution, social behavior, and the reproductive strategies of birds. A PhD student of the great ornithologist

Herbert Friedmann (right, at age ca. 53), here with Alexander Wetmore (at age ca. 61), looking at specimens at the US National Museum, Washington, DC, in 1951.

Arthur Allen at Cornell University, Friedmann's doctoral thesis became his classic book *The Cowbirds: A Study in the Biology of Social Parasitism* (1929). He continued to study cowbirds and other brood parasites throughout his life, publishing numerous papers and several monographs—many of which are still cited today. Friedmann was president of the AOU between 1937 and 1939, elected to the National Academy of Sciences in 1962, and awarded the Brewster Medal from the AOU in 1964.[41]

The Brown-headed Cowbird is an extreme generalist, its eggs having been found in the nests of 220 other species and successfully raised in the nests of 144 species (Friedmann and Kiff 1985)—far more than the Common Cuckoo (Davies 2000). The cowbird does not mimic the eggs of its hosts, and individual females often parasitize a range of species. However, the Brown-headed Cowbird does share many features in common with the cuckoo, including flexibility in its mating system and rapid laying (which occurs early in the morning while the host parent is absent). It also pays for the cowbird eggs to hatch first, but unlike the cuckoo, this is achieved by rapid embryo growth rather than by internal incubation (Kattan 1995).

Inspired by Rothstein's experimental studies of cowbirds, Nick Davies and Mike Brooke in the late 1980s used model eggs, matched for color and mass to resemble those of the Common Cuckoo, to investigate acceptance and rejection among potential hosts. They were especially pleased when the ornithologist Bruce Campbell, unaware of their work, mistook one of their model eggs for a genuine cuckoo egg. Davies's and Brooke's studies (e.g. Davies and Brooke 1989a, b) marked a new era in brood parasitism research, testing a suite of ideas suggested by a range of earlier workers.

They tested, for example, Alfred Russel Wallace's (1889) idea that cuckoos produced mimetic eggs as camouflage against predators, but they found no evidence for this. They also tested some of Chance's (1922, 1940) ideas, asking why Common Cuckoos lay in the afternoon and finding that, as predicted, hosts were less attentive at this time. Davies and Brooke also showed that by laying just after the host had started to lay, cuckoos were more likely to have their eggs accepted than if they laid in the host's nest before the host had started laying eggs.

It had been a long-standing question why cuckoo hosts often react so strongly to a "foreign" egg, but so readily accepted a foreign chick—even though a cuckoo chick is usually very different from host chicks. Arnon Lotem (1993) addressed this question using a theoretical model, pointing out that it would be maladaptive for hosts to learn by imprinting on their first brood, because, if their initial brood is parasitized, they would then discriminate against their own offspring in subsequent broods. In other words,

"misimprinting" (on the cuckoo chick) might result in hosts failing to raise all subsequent broods.

Langmore et al. (2003) later showed that Superb Fairywrens in Australia, which are parasitized by both the Shining Bronze Cuckoo and Horsfield's Bronze Cuckoo, *do* discriminate between their own chicks and those of cuckoos and abandon parasitized nests. The reason that chick discrimination occurs in the fairywren, but not in Eurasian Reed Warblers, is probably because the higher incidence of parasitism among wrens outweighs any costs of misidentification and because the fairywren's longer breeding season allows them to compensate for losses due to brood parasitism. Studies of other brood parasites subsequently demonstrated chick recognition and ejection behavior by hosts (e.g., Grim 2007; Sato et al. 2010).

New technologies will undoubtedly further our understanding of brood parasitism. For example, in Claire Spottiswoode's study in Zambia, 20 to 30 percent of all prinia nests are parasitized by Cuckoo-finches, yet she rarely sees or hears female Cuckoo-finches. Different types of tracking devices, now tiny

The Cuckoo-finch (male left, female second from left) is a brood parasite of the Tawny-flanked Prinia (right), producing remarkably variable, mimetic eggs (host eggs in the outer ring; Cuckoo-finch eggs in the inner ring).

enough to fit on small birds, will help resolve this conundrum (chapter 4). More generally, molecular techniques will both help resolve why egg rejection behavior varies so much within and between host populations and provide a better understanding of the evolution of brood parasites.

COGNITION

For an entire year between 1905 and 1906 the Galápagos archipelago was invaded by the California Academy of Sciences (chapter 3). It was a collecting expedition, and among their most avid members were Edward Winslow Gifford and Rollo Beck, expedition leader. In addition to killing and skinning birds, they also made casual behavioral observations, and on the island of Albermarle (now Isabela), Gifford "watched a bird feeding in a leafless, dead tree. It was apparently searching for insects, for it inspected every hole carefully. Finally, it found one too deep for its bill. It then flew to a neighbouring tree and broke off a small twig, about half an inch [1.25 cm] in length. Returning to the hole, the bird inserted the little stick as a probe, holding it lengthwise in its bill. . . . Mr [Rollo] Beck and Mr King said they had noted similar instances elsewhere."[42] This is the first published observation of tool-use by the Woodpecker Finch—and indeed by any bird.

Before this, John Gilbert, who was working for John Gould in Western Australia in the 1830s and 1840s, was told by aborigines about tool use in Black-breasted Buzzards. Allegedly, the buzzards drove Emus from their nests and then dropped rocks onto the eggs to break them. It sounded like folklore, but the behavior was eventually verified (Leitch 1953).

Inspired by Gifford's tantalizing observations, and encouraged by Konrad Lorenz, who "thought he might learn something,"[43] Irenäus Eibl-Eibesfeldt traveled from the University of Vienna to the Galápagos to make a more detailed study of tool use by the Woodpecker Finch in 1954. His observations showed that the birds tap branches with their bill, bringing their head close to the branch, and appearing to listen for concealed insect larvae. If the larva is inaccessible, the birds use their bill to modify a twig or cactus spine to probe for it.

In the 1960s tool use by nonhumans was widely considered to be a form of "insight learning" and was therefore of special interest to psychologists and, to a lesser extent, to ethologists. Following Eibl-Eibesfeldt, George Millikan and Robert Bowman (1967) conducted experiments on captive Woodpecker Finches and found that the birds manufactured tools by breaking off twigs and shortening them when they were too long to be easily used.

Prior to Darwin the dividing line between human and nonhuman behavior was considered clear-cut: animals operated largely via instinct, humans via intelligence. Darwin (1871) blurred this stark distinction between animal and human minds, seeing only evolutionary continuity: "The difference in mind between man and the higher animals, great as it is, is certainly one of degree and not of kind."[44] Darwin's ideas generated a torrent of anecdotes about animal "intelligence,"[45] resulting in books like *Animal Intelligence*, by his friend George Romanes (1881). Just a few years later the American psychologist Edward Thorndike produced his own book, titled *Animal Intelligence*, and said that "in the first place, most of the books do not give us a psychology, but rather a *eulogy*, of animals. They have all been about animal *intelligence*, never about animal *stupidity*. . . . The history of books on animals' minds thus furnishes an illustration of the well-nigh universal tendency in human nature to find the marvelous wherever it can."[46]

Ethologists like Tinbergen and Lorenz steered clear of considering "intelligence," concentrating instead on "instinct" and the ways in which instinctive behavior could provide information about phylogeny (chapter 7). Ethologists focused on genetically hardwired behavior, leaving animal learning to the comparative psychologists of the day. Studies of intelligence—mainly tool use, spatial orientation, reasoning, and abstraction—were carried out in primates and rats but also in pigeons, quail, and the chicks of domestic fowl, and the conclusion was that birds were generally less intelligent than mammals. This was consistent with the commonly held view at the beginning of the twentieth century that evolution was progressive: "simple" animals were capable only of simple behaviors, and humans sat at the top of the evolutionary ladder. Here is Thorndike: "Sooner or later clear learning appears, and then, from crabs to fish and turtle, from these to various birds and mammals, from these to monkeys, and from these to man, a fairly certain increase in sheer ability to learn . . . can be assumed."[47]

In the late nineteenth century, the neuroanatomist Ludwig Edinger pioneered research on the structure of the vertebrate brain. Using new staining methods to identify specific brain regions, Edinger and colleagues noticed that bird brains were very different from those of mammals; most notably, they lacked the "layered" outer cortex.

Edinger could not see a layered cortex in birds' brains; instead, their brains looked similar to the part that was *underneath* the cortex in mammals where it is responsible for controlling the most basic instincts. These differences made sense to Edinger, who considered birds inferior to mammals, and he labeled the avian brain accordingly: regions were suffixed with *-striatum*, rather than *-cortex*, and as a result birds were assumed to be capable only of species-specific, instinctive behavior. Pigeons became a model species for basic research on learning, but few researchers subscribed to the idea that birds might be useful for the study of intelligence. The mood of the time was summed up by the comparative anatomist Charles Herrick: "It is everywhere recognized that birds possess highly complex instinctive endowments and that their intelligence is very limited."[48]

Even so, there were a few attempts to study avian intelligence. At the turn of the twentieth century James Porter, an instructor in psychology at Indiana University, made a study of the psychology of the House Sparrow. Porter's study was designed to establish whether the sparrow's invasion success in America was the result of higher intelligence compared to the native birds: "No one has made a study of the psychical life of mature birds which shows the full range of their ability to profit by experience, to discriminate, to inhibit useless actions and reinforce useful ones."[49] He investigated a number of aspects of the sparrow's cognitive ability, including its response to novel food, its problem-solving ability and subsequent learning speed, its performance on various discrimination tasks, and even its ability to solve a maze (Porter 1904). He found no evidence for reasoning or forward planning, but sparrows were rapid learners.

Other researchers adopted a more naturalistic approach. In 1920 Robert Miller studied American Bushtits in the field for his master's degree under Joseph Grinnell at the University of California–Berkeley. Dissatisfied with the way birds were being studied, Miller wrote:

> The study of birds from a behavioristic standpoint has been relatively neglected, and those investigators who have given the matter some attention have usually gone to one of two extremes: the field observers, being better naturalists than psychologists, have interpreted the behavior of birds in an extravagantly anthropomorphic fashion; and the experimentalists, being better psychologists than naturalists, have with amusing seriousness taken caged birds into the laboratory and assumed that they would there behave in normal fashion (cf. Porter, 1904 and 1906). What we need would seem to be a new science of "field psychology" which should combine in due proportions the observational and experimental methods.[50]

He tried to do just that with his studies of the flocking behavior of bushtits, but his research had no impact, and the little subsequent research on bird cognition continued to be influenced by psychological methods, with captive subjects confronted with artificial and arbitrary experimental procedures, such as maze learning (e.g., Sadovinkova 1923; Allee and Masure 1936).

By 1951, with his facility at Madingley established (chapter 7), Bill Thorpe published a lengthy review of avian cognition in two papers in *Ibis*, titled "Learning Abilities of Birds." Thorpe synthesized all that was

known about habituation, conditioning, trial-and-error learning, insight learning, imitation, imprinting, and memory, also discussing anatomical and endocrinological factors relevant to avian learning. The field was crying out for such a synthesis, and Thorpe's two papers pointed the way forward. Reviewing those papers in *The Auk*, Margaret Morse Nice said that "Dr Thorpe has presented a most valuable review article on this important subject, organized and presented in masterly fashion. It is an important contribution to the study of bird behavior and one which will be particularly enlightening to us on this side of the Atlantic."[51] Curiously, Thorpe's papers have been cited just twenty-five times[52]—and may be worth another look.

"Insight" was one of the topics that particularly interested Thorpe, which he defined as "the sudden adaptive reorganization of experience or the sudden production of a new adaptive response not preceded by . . . random trial behaviour. . . . Thus insight learning itself seems to provide evidence of action by hypothesis and has often been regarded as showing that ideational processes are going on in the animal's mind."[53] Up to that time, few considered that birds could show insight. Yet there were some indications, for example from studies of "string pulling." Reviewing all the available evidence, including work by his student Margaret Vince, Thorpe (1951a, b) concluded that in most cases trial-and-error learning was an insufficient explanation and therefore that species such as Eurasian Blue Tits, Great Tits, and European Goldfinches, which all excelled at the string-pulling task, might be demonstrating insightful learning.

Forty years later Bernd Heinrich, at the University of Vermont, revisited string pulling, this time using Northern Ravens. He provided his birds with meat tied to a string suspended from a perch and found that some individuals spontaneously pulled the string up on their first trial. This strongly suggested that the birds planned their behavior in advance, rather than learning by trial and error, which would take a number of trials (Heinrich 1995). As we'll see below, corvids (including ravens) excel at many types of problem-solving, forcing a rethink about the evolution of intelligence.

The comparative psychologists' approach to avian cognition was typically anthropocentric: behaviors like tool use were taken as indicators of intelligence because they are behaviors at which humans excel. In contrast, behavioral ecologists preferred an alternative, "adaptive specialization" approach, focusing on the socioecological factors that promote the evolution of particular behaviors. In this approach, cognitive ability is largely determined by the types of ecological problems that particular species face. Emery (2006) proposed six socioecological variables that correlate strongly with advanced cognitive ability in animals in general: diet, social structure, capacity to innovate, brain size, life history, and habitat. Typically "intelligent" animals tend to be omnivorous, that is, generalist foragers that are highly social and innovative, with large relative brain size, a long developmental period, and extended longevity, and are living in a variable habitat. Two groups of birds in particular meet these criteria: parrots and corvids.

In the 1940s and 1950s the cognitive abilities of birds were investigated by several German researchers, notably Otto Koehler, who was the first to convincingly show numerical competence in animals (e.g., Koehler 1949).

Koehler's work received more attention than most because the paper he presented at the Institute for the Study of Animal Behaviour (ISAB) conference in Cambridge (chapter 7) was included in the conference proceedings and published in English (Koehler 1950). Koehler's studies were ahead of the times—he recognized the problems associated with the study of intelligence and carefully controlled for extraneous factors in his experiments (such as the "Clever Hans" effect[54]). Koehler tested a range of bird species, including parrots and corvids, in a task requiring them to match a "key" stimulus, which had a particular number of dots on it, to a lid with the same number of dots, stones, or mealworms. His belief was that counting in animals might be informative with respect to the origins of human language (Koehler 1950).

In 1977, soon after finishing her doctoral research in theoretical chemistry at Harvard, Irene Pepperberg began working on vocal learning in parrots. She was inspired by a television program called "Why Do Birds Sing?" in which Peter Marler (chapter 7) described his work on vocal dialects, as well as his own switch from chemistry to ornithology. Pepperberg realized that no one was studying birds in the same way as they were studying primates. She got her first subject, a Grey Parrot she named Alex,[55] from a pet shop, and over the next thirty years he (and some other male parrots) revealed some previously unimagined behaviors of birds. Pepperberg asked whether Alex could go beyond what Koehler and others had shown by using numbers symbolically. Providing Alex with large sets of objects that differed in color, shape, and texture, Pepperberg asked him various types of questions.

Alex correctly answered questions about the number and color of objects, and he also appeared to have learned about concepts such as shape. If asked a specific question, such as "How many blue blocks?," he could answer correctly. Perhaps more strikingly, he often combined words into sentences that he had not been taught but that seemed appropriate to the situation; for example, he was trained to request objects by saying "I want X," but sometimes he was heard to mutter "Wanna go back" when bored with an experiment. Pepperberg concluded that "the data suggest that a non-human, non-primate, non-mammalian animal has a level of competence that, in a chimpanzee, would be taken to indicate a human level of cognitive processing."[56]

Pepperberg's research has been the topic of much debate, and she has described her struggles with funding and institutional support for her work (Pepperberg 2008). Reviews of her academic book *The Alex Studies* (1999) were mixed, with reviewers appreciating the apparent complexity of Alex's behavior but being less sure of the theoretical rationale behind the research: "Pepperberg's book . . . illustrates the unsettled state of current animal cognition research . . . researchers are left to select for themselves problems to be emphasized, training and test methods, and data analysis strategies. This situation is not entirely satisfactory."[57]

Members of the crow family (corvids) also have a long-standing reputation for being clever, and among the early efforts to investigate corvid intelligence was Charles Coburn's (1914) study of two American Crows that he tried to train to learn to discriminate between two options. The birds failed, and Coburn concluded that they were "temperamentally ill-suited" to that kind of task.[58] Research in the last twenty years provides better evidence that corvids are clever, and in some cases provides evidence for specific myths, such as Aesop's fable of the thirsty crow dropping pebbles into a container to raise the water level so that he could drink. When Christopher Bird and Nathan Emery (2009) presented naive captive Rooks with this problem, the birds solved it immediately. The range of tasks that corvids excel on has led to them being called "feathered apes" (Emery 2004).

Lying 1,600 kilometers (1,000 miles) off the east coast of Australia, New Caledonia is home to the new pinup of comparative cognition. In 1972 Ronald Orenstein reported an instance of tool use by New Caledonian Crows, but it was not until 1996 that Gavin Hunt, of Auckland University, described how wild New Caledonian Crows routinely make and use tools to probe rotting timber and leaf litter for invertebrates, such as the enormous larvae of wood-boring beetles. Since then Hunt and his colleagues have collected a wealth of information on the natural tool-oriented behavior of these crows, showing that their use of tools is comparable to that in chimpanzees—traditionally regarded as both the most sophisticated nonhuman tool users and the most intelligent nonhuman animals. The finding intrigued Oxford University's behavioral ecologists Alex Kacelnik and Jackie Chappell, who decided to examine the New Caledonian Crows' tool-oriented behavior under controlled conditions in captivity.

Studies of hand-reared birds in captivity show that juveniles do not need to see tool use in order to acquire the behavior (Kenward et al. 2005), although the fact that they remain with their parents for up to a year after fledging suggests that social learning may be important (Holzhaider et al. 2011). Laboratory studies have produced some exciting results: one female—named Betty—was particularly innovative in her construction of tools, spontaneously bending wire into hooks to secure an otherwise inaccessible bucket of food (Weir et al. 2002). More recent work shows that New Caledonian Crows can use sequences of up to three tools to secure food, implying a capacity for forward planning (Wimpenny et al. 2009; Taylor et al. 2007). Intriguingly, these crows have also been observed using tools in situations when they are not looking for food (Wimpenny et al. 2011). The success of the New Caledonian Crow research rests on the fact that it is based largely upon natural behavior and is guided by principles from psychology, evolution, and ecology.

The advanced cognitive abilities of other corvids is elegantly demonstrated by Nicky Clayton's work on the food hoarding and recovery behavior of California Scrub Jays. Clayton was intrigued by birds as a child, and by corvids in particular: "I was fascinated by the quizzical look that some birds give you and wondered what was behind that beady eye."[59] Unable to read for a joint degree in zoology and psychology at Oxford, she took the advice of her interviewer, John Krebs, and did zoology instead, recognizing that she could learn about bird behavior and still think about psychology on the side. After a PhD on birdsong learning at St. Andrews University, Clayton moved back to Oxford to work on the neurobiology of food caching and spatial learning in birds. During this period—in the late 1980s—she struggled to convince other neuroscientists of

the worth of studying the avian hippocampus: "As somebody studying the bird brain I was a second class citizen . . . those studying human brains were number 1, those studying monkey brains were number 2, beta class was non-primate mammals, rodents in particular, and then the real second class citizens, the poor old gammas were those doing birds."[60] In 1995 Clayton took up an assistant professorship at the University of California–Davis, and, while watching scrub jays in the parks on her lunch breaks, she became sufficiently interested in their behavior to do some experiments with them. When she moved to Cambridge University, back in England, in September 2000, the scrub jays came too.

As well as remembering where they have cached food, the jays seem to remember what they have cached and when—a phenomenon called an "episodic-like" memory[61] (Clayton and Dickinson 1998). Scrub jays also appear to be able to plan future events (Raby et al. 2007), a finding that has generated much discussion because many believe that only humans are capable of planning for events not related to their current motivational state (e.g., Suddendorf et al. 2009). The jays also have to deal with the possibility that their caches might be raided by other jays, and Clayton's research has revealed some remarkably sophisticated antithievery tactics; when jays are watched by other jays as they cache, they re-cache food items more often than if they are alone. Remarkably, re-caching behavior also seems to depend upon the cacher's own previous experience as a thief: individuals that had previously stolen other birds' caches re-cached more often, suggesting that they may use their own experience to decide how to behave (Clayton et

313

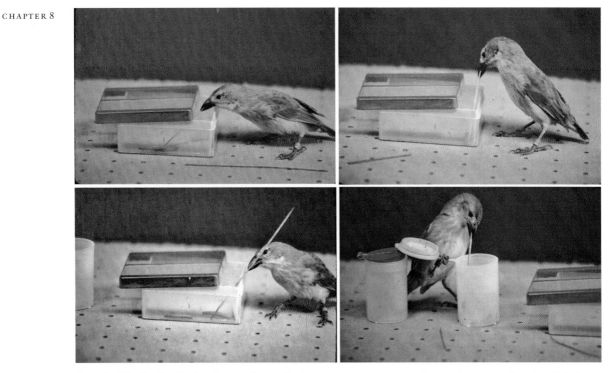

A captive Woodpecker Finch using a tool to get food from covered containers, in experiments conducted by Bob Bowman in 1965 at the California Academy of Sciences.

al. 2007). As with the studies of New Caledonian Crows, the strength of Clayton's research is that it is examining a natural behavior, allowing questions to be asked about both evolution and mechanism.

We will finish by returning to the Woodpecker Finch and more recent research by Sabine Tebbich of the University of Vienna. Thorpe (1951a) singled out the Woodpecker Finch as the very best avian tool-user (as it was, at that time). Yet he also recognized that by itself tool use reveals nothing about the cognitive underpinnings of the behavior, and he suggested that "to study carefully the individual development of the tool-using behaviour of a . . . Galápagos Woodpecker-finch reared apart from its congeners would be a most valuable contribution to the study of avian psychology."[62]

Fifty years later Tebbich did just that, asking whether juvenile birds needed to watch their parents using tools in order to develop the behavior. When young Woodpecker Finches in captivity were allowed to play with twigs and cactus spines, they all became tool users, but non-tool-using adults, even after watching a "tutor" bird using tools, never acquired the behavior. This suggests that Woodpecker Finches acquire their tool-using skills through trial-and-error learning, during a specific phase of development (Tebbich et al. 2001). On the Galápagos, the incidence of tool use is highest in populations from the dry, arid zone where about half of the birds' prey is obtained through tool use; in contrast, Woodpecker Finches occupying the humid zone, where food is much more abundant, rarely use tools (Tebbich et al.

2002). Young birds are therefore prepared for the ecological pressures they will encounter as adults, learning tool use only when it will be of considerable benefit. These results show how important ecological factors are in determining the incidence of tool use and how essential it is to consider these factors in order to understand the evolution of the behavior.

In July 2002 Edinger's outdated terminology was finally overhauled at a meeting of the Avian Brain Nomenclature Forum at Duke University (Reiner et al. 2004). The twenty-eight members of this forum argued that avian brains were far less primitive than Edinger and others had assumed. Even without a layered cortex, birds still have the capacity for intelligence; they have simply achieved it in a different way. Rather than just governing instinctive behavior—as Edinger assumed—birds have a cortex that allows for complex "higher" behaviors, enabling them to solve social, physical, and spatial problems at levels previously only thought possible in primates.

The study of animal cognition has expanded considerably since its beginnings in the early years of the twentieth century, when it was pursued almost entirely by psychologists. Much of this expansion in the past thirty years has come from the increased interest of bird behavioral ecologists who recognized that animal intelligence could be studied from a life history perspective (Ricklefs 2004). Unlike other aspects of behavioral ecology, animal cognition does not yet have a strong theoretical foundation and thus runs the risk of becoming top heavy, a mass of primary research overpowering the underlying theory. Scientific disciplines develop differently but generally begin with an accumulation of observational information that provides the basis for theoretical foundations, and this theory then generates predictions that researchers test.

The study of animal cognition grew out of a mixture of disciplines, including comparative psychology, anthropology, ethology, and behavioral ecology, and it may be that this mix of "parents"—while providing important interdisciplinary perspectives—has also made it difficult for the field to progress along a clear path. As Alex Kacelnik told us, "Many people, and I include myself, don't know yet even how to ask the questions properly. . . . So this field is still waiting for a major intellectual advance, and that probably should happen soon, because there is now an accumulation of facts about things that organisms do, which is not explained either by conventional associative learning, or by anthropomorphic reference to a mental life." When asked what was necessary for the field to progress, Kacelnik answered simply that, "we need a Bill Hamilton for cognition."[63]

CODA

Of all the directions taken by ornithology since Darwin, behavioral ecology has probably been the most successful in terms of both the breadth and depth of new bird research it has generated. Tinbergen's ethological approach, so successful initially, suffered eventually from the lack of a theoretical foundation (chapter 7). Individual selection, coupled with quantitative cost-benefit models inspired by economics, provided a solid theoretical foundation for field studies of behavior. The ability to formulate and test predictions from theory made behavioral ecology more "scientific," more

315

rigorous, and ultimately more successful than ethology. Behavioral ecology also provided excellent, clear-cut projects that could be completed within a typical three-year doctorate. Research students also found the absence of a substantial past literature appealing. One of the most exciting aspects of the behavioral ecology approach was that it addressed "general" problems. "Altruism," for example, was not simply of interest to ornithologists studying cooperative breeding; it was equally relevant to those studying social insects and primates. Similarly, the questions that female promiscuity raised (chapter 9) were as relevant to entomologists as they were to ornithologists.

Behavioral ecology itself has evolved. In the initial gush of enthusiasm some early studies were poorly done, but harsh criticism—especially from outside the field (e.g., Gould and Lewontin 1979)—helped to raise standards. New paradigms typically go through a bandwagon period as researchers strive to become part of the new field; referees are uncertain about what's good and what's bad, and journal editors consciously or unconsciously sacrifice quality to secure exciting new work for their journals. Over time, however, self-policing helps to impose a sense of what constitutes good science (e.g., Montgomerie and Birkhead 2005).

Subdisciplines within behavioral ecology have had different life spans, some short, some more enduring. Optimal foraging research, for example, enjoyed a relatively brief, successful life while others, like the study of sexual selection, discussed in the next chapter, have continued to flourish. The history of science makes it clear that boom and bust is exactly what we should expect: in some areas questions are rapidly answered,

leaving little new to be discovered; in others—such as cooperative breeding, cognition, and reproduction—the diversity and complexity of these subjects across the animal kingdom means that each new finding reveals yet more to be discovered.

An important factor affecting the continued development of particular subdisciplines in behavioral ecology has been the result of practitioners extending their reach beyond function to embrace "causation" and the underlying anatomy and physiology of behavior. This extended approach—a return to Tinbergen's other questions—has proved immensely rewarding. Precisely because of the boom-and-bust nature of science (see Horgan 1996), some scientists, writing in the early twenty-first century, have suggested that behavioral ecology is in decline or even moribund (e.g., Caro and Sherman 2011). While it is obvious that the initial excitement generated by new theories is over, the phase of what Kuhn (1962) called "normal science"—the ongoing, day-to-day testing of hypotheses in behavioral ecology—shows little sign of decline. The reasons are obvious. Natural history is of fundamental interest, fueling our understanding of ever-increasing conservation issues; and the diversity of the natural world is so extraordinary that new discoveries—even new species—continue to emerge.

The success of behavioral ecology rests largely on its "scientific" approach of testing theoretical ideas. Understanding theory and formulating hypotheses isn't always straightforward, and testing of predictions in the most elegant, efficient manner possible requires imagination and creativity. The ability to execute high-quality research in this area is both demanding and intellectually satisfying, but it has made behavioral ecologists eminently employable—accounting for the creation of numerous new university posts since the 1970s.

Many behavioral ecologists who study birds started out as bird watchers, and many of them—even those involved in behavioral ecology from the beginning and now reaching the end of their careers—are still excited by studying how birds behave. The constant stimulation provided by the extraordinary variation of the natural world is what drives most behavioral ecologists and has motivated some, like Bernd Heinrich, John Alcock, John Marzluff, and others, to write personal accounts of their research.[64]

Because birds have been the behavioral ecologists' favorite taxonomic group, the increase in our understanding—over the last forty years—of topics as diverse as foraging, reproduction, cooperative breeding, parental care, brood parasitism, and cognition in birds is truly extraordinary.

BOX 8.1 Andrew Cockburn

My early academic experience was shaped by the pervasive intellectual tradition in Australian vertebrate biology: marsupials define our fauna, and hence are the most exciting research subjects. I took this attitude with me when I gained tenure in 1984 at the Australian National University. I had been profoundly influenced by reading *Red Deer: Behaviour and Ecology of Two Sexes* (Clutton-Brock et al. 1982), which set out an agenda for measuring reproductive success across a whole lifetime, and I decided to commit to long-term study of antechinuses. They were ideal for lifetime fitness measures, as the defining feature of their life history is a brief annual rut that culminates in the death of all males before the annual cohort of young are born. This simplicity throws otherwise intractable problems into sharp relief.

However, in between regular trips to the rain forests of the New South Wales escarpment to trap these small mammals, I was slowly seduced by the rather more in-your-face antics of the local avifauna and became fascinated by White-winged Choughs and a little later by Superb Fairywrens. Pioneering work by Ian Rowley provided excellent background on the cooperative breeding exhibited by both species.

This ornithological foray proved timely. First, as I read the literature on cooperative breeding, it became clear that here was a research culture curiously focused around the long-term studies of individual species that I had admired in mammals. Indeed, it seemed the discipline at that time had a fairly simple recipe: find a cooperatively breeding species, fall in love with the soap opera of its social life, and study it obsessively for the remainder of your career. By 1990 several studies had exceeded two decades of intensive study, and the practitioners tended to dismiss upstarts who did not have at least a decade under their belt. Second, a chance conversation at a mammal conference alerted me to Alec Jeffreys's discovery of DNA fingerprinting. Third, Charles Sibley and Jon Ahlquist had produced the first compelling evidence that the Australian passerine birds were as phylogenetically distinctive as the marsupials but, unlike them, had gone on to take over the world in one of the great (and still unexplained) examples of ecological replacement. All in all, cooperative breeding seemed to draw together many of my academic influences and was easier to study within minutes of my office door than anywhere else in the world. Choosing to focus on this problem was an easy choice and one I have never regretted.

Intricate long-term studies allow investigation of many questions that cannot be tackled in the three-year time frame of a doctoral program or a research grant, but this disciplinary research style can actually be an impediment to sorting out problems of prevalence. Devoting twenty or thirty years of life to a single species encourages the belief that you are studying the general phenomenon rather than the species, encouraging extrapolation. Where genuine comparison occurs, it has often involved contrasts with the first cooperatively breeding species whose natural history and demography was the subject of a major synthesis: the Florida Scrub Jay. These birds are confined to a dwindling habitat, which is full of jays, consistent with early suppositions that young birds might stay at home with their parents because there was no unoccupied space in which to breed. Validation of this idea seemed to come from comparison with the closely related California Scrub Jay. This species lives on the other side of the USA, is more generalist in its habitat use, and does not breed cooperatively. This comparison encouraged almost universal acceptance that cooperative breeding was a consequence of a saturated habitat, or in later parlance, ecological constraints on dispersal. The helping behavior of the philopatric young could then be explained by a variety of advantages, of which helping their parents (kin) raise offspring seemed most general.

The trouble with this worldview is revealed by peering over the Mexican border, where most of the New World jay radiation is found. Virtually all the species, and all the southern congeners of the Florida and California Scrub Jays are cooperative breeders, regardless of their degree of habitat specialization. More interesting, their mating systems are incredibly diverse. Hence a broader comparative palette suggests that the peculiar habitat descriptions of the Florida Scrub Jay can be generalized. Indeed, it remains very difficult to reconcile philopatry by just one sex with ecological constraint ideas. The weakness of this original model became a key theme of my own work.

My interest in a broader synthesis was helped through a little good luck. While I am typical in having become obsessed with a single species, in my case the Superb Fairywren, I had the good fortune to have started work on White-winged Choughs at the same time, and I soon realized that these two species had very little in common. Fairywrens also drove home a point that had largely been ignored by my predecessors: sex matters. DNA fingerprinting revealed that the brilliant male plumage was directed at an unusual audience, females from nearby territories. Most fertilizations are won by foreign males, who contribute nothing to the rearing of the young they sire, throwing the problem of altruism into even sharper relief and providing an unusual opportunity to study a different theoretical conundrum, the genetic benefits females gain from careful choice among males.

Recognition that cooperative societies occur in diverse social and ecological settings, and evolve along diverse trajectories, provides us with an exciting new perspective to seek common patterns. Knowing that the Florida Scrub Jay may be exceptional rather than a standard model will allow us to seek new empirical models and hopefully achieve the holy grail of prediction rather than prescription. Along the way we can continue to enjoy these most fascinating and charismatic of birds.

BOX 8.2 Nick Davies

One of my earliest memories is making a hide [blind] out of deck chairs and using opera glasses to watch chaffinches. We lived thirteen miles north of Liverpool and were surrounded by wildlife. Pink-footed Geese used to fly low over our house, and I have worshiped them ever since. You could also hear nightjars churring in the dune slacks and natterjacks calling in old bomb craters. I was captivated by

wildlife from a very young age. The rest of my family had no interest in birds, so I must have had some other, genetic predisposition to enjoy bird watching. As a little boy I was quite obsessive about it—I liked to make lists, and I filled many notebooks, which I still have.

I attended the Merchant Taylor School in Crosby. One very formative thing is that we went out on field trips. We'd go out at dusk in June to hear and see nightjars. We also used to go to Hilbre Island in the Dee estuary, which I loved. We'd walk over at low tide and stay on the island when the tide came in. We'd get fantastic views of Red Knot, Purple Sandpipers, and Dunlin. While at school we also took three trips to Bardsey Island. George Evans was the warden, and I was astonished by his ability to identify all the tiny birds that we could see in the distance. My ambition then was to become the warden of a nature reserve.

My parents secretly hoped that I'd become a doctor, because most other people who did biology did this. But I realized it might be possible to study birds as a career when Rhys Green and I carried out a study of warblers and presented the results at an EGI conference. It was very flattering to be asked questions and talk with people like Arthur Cain. At this first EGI conference I saw Tim [Birkhead] give a talk on herons, and the next year he got into the EGI to do a PhD. I wanted to do the same; unfortunately, I wasn't great at exams, but I think our project caught Chris Perrins's eye.

I was lucky to have been around at a time where so many new ideas were being born, such as Parker's and Trivers's. They really changed the way that we see birds and behavior. When I was an undergraduate I had bought Lack's *Ecological Adaptations* (it cost £4.50 from Heffers, which was more than my week's allowance at the time. I didn't dare tell my mum!). The main message was that to understand mating systems you must know about ecology. Trivers and Parker came along and changed our perceptions—they said that we must think about an individual's actions and uncover conflicts of interest. These theories were inspirational. If you had just a pair of binoculars and a spare afternoon you could go out and see things in a new light; in those days just going out and watching what animals did could lead to new research and papers. New techniques do determine progress, but mainly it's new ideas that are important. I know that I was enormously lucky to be working when I did.

The idea of breaking behavior into decision making with costs and benefits was also a very big influence. When I started my DPhil I went up to Port Meadow and sat on the bank watching pied wagtails from dawn to dusk. I didn't really know why I was doing it, but I thought it was important to sit there and feel what it was like to be a wagtail. So I filled my notebook and took it to my supervisor, Euan Dunn, and proudly showed him this mass of data on feeding rates. I said I'd discovered that

a wagtail has to eat one insect every three seconds to meet its energy budget, and he said: "That's very interesting—and what hypothesis are you testing?" It stopped me dead in my tracks because I didn't have a hypothesis; I was just collecting data to see what it showed, and I realized then that the art of doing good science is not so much working hard to collect good data but spending more time thinking about questions.

It was a talk by John Krebs that really showed me the importance of asking questions and thinking critically about theory. In those days we were all very species oriented—my thesis was going to be a descriptive study of pied wagtail territoriality—and the title of John's talk, and this was the one that really changed my thinking, was "Optimal Foraging in Herons and Chickadees." I thought he was nuts; it just seemed outrageous that you would have, in the same talk, two so very different birds. John had just come back from Vancouver, and he showed that, although herons and chickadees are very different birds, they face the same fundamental decisions when they are foraging, namely, how to give up optimally from a patch with diminishing returns and go on to the next foraging patch. This was an absolute revelation to me, and I immediately saw how my study of pied wagtails could be improved by thinking more clearly about costs and benefits. It revolutionized my approach to animal behavior; this and Trivers's and Parker's ideas on social behavior really have been the theories that have guided my work.

Oxford was a very exciting place to be in those days; I think never since has there been the same buzz. I remember going to talks by John Maynard Smith and talking to Trivers about parent-offspring conflict. These were really new ideas, and hard to grasp. You couldn't just go to Trivers's talk and understand his ideas; they had to gradually sink in. At this time Dawkins was writing *The Selfish Gene*, and he would try his chapters out in lectures. Dawkins was so clear thinking and way ahead of his time in understanding Hamilton's ideas and implications. Having him asking Trivers and Maynard Smith about their seminars was really amazing, and made the department a very exciting place to be.

There are still lots of new unanswered questions to address. There are also big unsolved problems, such as linking microevolutionary questions (function of behavior) through to consequences for populations and evolution of species (macroevolutionary patterns). This link has not really been firmed up at all. We also need to go downward to look at the genetic mechanisms that govern behavior.

There's one other important development. When I began as a lecturer in Cambridge in 1979, we thought very little about conservation. Now it's a major concern and features in all our courses. Many of the birds I've studied have declined even during my lifetime (Common Cuckoos and Spotted Flycatchers, for example). Looking back, we had the luxury of being intellectually free to do what we wanted and enjoy nature without having to worry much about what was happening to it.

PARADISORNIS RUDOLPHI, Finsch.

CHAPTER 9

Selection in Relation to Sex

The external beauty of form and colour which birds present, has so far proved
a serious distraction, so that ornithologists, captivated thereby, have paid
but little heed to the possible factors to which these features are due.

—WILLIAM PYCRAFT (1910: VII), IN HIS *A HISTORY OF BIRDS*,
POINTING OUT HOW LITTLE HAD BEEN DONE TO FOLLOW UP
ON DARWIN'S INSIGHTS INTO SEXUAL SELECTION

A SWEDE IN AFRICA

IN KENYA'S FERTILE GREEN HIGHLANDS, thirty-three-year-old Swedish ornithologist Malte Andersson is on vacation with his wife, escaping the worst of a Scandinavian winter. Every few minutes a male Long-tailed Widowbird emerges from the long grass in a magical floating display flight. Like a black sparrow sporting a tail 50 centimeters (20 inches) long, the male widowbird is distinctly unbirdlike. The female, on the other hand, is dull, brown, and all but tailless. Seeing the male display for the first time, Andersson is captivated.

A bird watcher since childhood but with an interest in physics, Andersson trained first as an engineer. At university he switched to biology and animal behavior after reading books by Eric Fabricius and Niko Tinbergen.

As well as liking birds, Andersson was fascinated by the explanation of evolution by George Williams, whose book *Adaptation and Natural Selection* (1966) was one of the foundations of behavioral ecology.

After his first degree, Andersson went on to do a PhD on the behavior of skuas, inspired by Tinbergen's (1959) comparative studies of gull displays. Andersson's research compared the social signals of the skuas and jaegers and demonstrated unexpected behavioral similarities between Great Skuas and one of the smaller species, the Pomarine Skua. It was in 1975, after completing his PhD, with its public examination as the finale, that Andersson and his wife headed for Africa.

Fascinated by the huge difference in appearance between male and female

Male (right) and female (left) Blue Bird-of-paradise illustrated by William Matthew Hart. Hart did his best to imagine the display, not knowing that the male actually displays hanging upside down.

widowbirds, by the males' elaborate displays, and by their polygamous mating system, Andersson came home and started thinking, not just about how he could study these wonderful, long-tailed birds, but about how he could test one of the really big ideas in evolutionary biology: sexual selection. He had watched his Long-tailed Jaegers point their elongated central tail feathers skyward during their own courtship displays; he had measured them and noticed how the tails of individual males got longer year after year. Sexual selection needed to be tested and tested by experiment. Andersson's training in physics had instilled in him an experimental approach—an approach reinforced by reading about Tinbergen's elegant field experiments.

Why males and females of the same species should look and behave so differently has fascinated and confused biologists for centuries. Darwin's solution was sexual selection and comprised two interlinked ideas: aggressive competition between males and female choice. The spurs on the legs of cockerels and many other game birds could be crucial in male fights. On the other hand, the elaborate plumage of male cotingas, peafowl, and widowbirds could be crucial in making a male attractive to a female. Male competition was so obvious that few biologists doubted it, but Darwin's idea of female choice was a trickier notion from the start and remained controversial for a full century after he suggested it. Many biologists doubted whether female choice existed at all, and specifically, whether female choice was responsible for the elaborate and costly ornaments that males flaunted.

Intrigued by the strength of the theory, Andersson also recognized that empirical

Malte Andersson on the Kinangop plateau, central Kenya, in 1981, holding one of his experimental male Long-tailed Widowbirds with an artificially elongated tail (photo in 1981 at age 39).

evidence was scarce. What was needed—as Darwin knew[1]—was an experiment in which male ornaments were altered and the preference of females tested. The long tails of widowbirds were ideal for such experiments because they could be experimentally lengthened or shortened. If female choice influenced the evolution of male tail length, those with longer tails should be more attractive and those with experimentally shortened tails less so. Andersson thought that the Indian Peafowl would be a good study species, but a reconnaissance trip to Sri Lanka soon quashed that idea—abundant elephants and boisterous buffaloes made fieldwork a potential nightmare! So it was back to Kenya in 1979 to conduct a pilot study on widowbirds, cutting some birds' tails to make them shorter and making others longer by inserting the pieces removed from the cut tails. "Superglue"—a recently developed, exceptionally strong and fast-drying adhesive used in veterinary surgery—made all this possible. Andersson initially used a brand called Hot Stuff. Hot stuff indeed! Little could the inventors of superglues, who were trying to develop a clear plastic lens for gun sights,[2] have known that they would change the study of sexual selection.

The fieldwork was tough. In temperatures of around 30°C (86°F), Andersson and his field assistant worked hard to catch thirty-six males and manipulate their tails. The trickiest part—even with the assistance of a local Kikuyu teenager—was finding nests in the tall, dense grass. Like males in many polygynous species, the male Long-tailed Widowbird plays little or no part in nesting: females make the nest, incubate the eggs, and rear the chicks alone. Finding nests was the only sure way for Andersson to know how many females each male had attracted to his territory, and this, after all, was the information needed to test his ideas. Only as the breeding season drew to a close did Andersson have enough time to check his results, and, sure enough, the predicted pattern was emerging. The males whose tails he had lengthened had more females than the "control" males, which in turn had more females than males whose tails he had shortened.

Andersson's results were published in *Nature* in 1982, and the paper transformed the study of mate choice and sexual selection. A decade later he wrote *Sexual Selection*, the definitive book covering evidence and theory across the entire animal kingdom (Andersson 1994).[3]

DARWIN DEVELOPS THE THEORY

On his return from the *Beagle* voyage in October 1836, Darwin began developing his ideas about natural selection. He also started thinking about traits that could not be explained by natural selection. Colorful plumage, spectacular song, or the elaborate antics of male birds demanded energy and

hence were "costly" and did nothing to enhance survival. Indeed, those traits looked as though they would do just the opposite, diverting energy from daily maintenance and making their owners more conspicuous or vulnerable to predators. On the face of it these were traits that natural selection should have eliminated. The fact that animals retained these traits indicated that they must be beneficial in some way. Darwin's genius was to recognize that even though conspicuous male traits may reduce their owner's survival, this could be more than compensated for by greater reproductive success.

It was a clever idea, and Darwin envisioned increases in reproductive success occurring through aggressive competition between members of the same sex, and by mate choice. As he and Wallace said, in addition to natural selection,

> . . . there is a second agency at work in most unisexual animals, tending to produce the same effect, namely, the struggle of the males for the females. These struggles are generally decided by the law of battle, but in the case of birds, apparently, by the charms of their song, by their beauty or their power of courtship, as in the dancing rock-thrush of Guiana [Guianan Cock-of-the-rock]. The most vigorous and healthy males, implying perfect adaptation, must generally gain the victory in their contests. This kind of selection, however, is less rigorous than the other; it does not require the death of the less successful, but gives to them fewer descendants.[4]

Darwin introduced sexual selection briefly in *Origin*, but developed it fully in *The Descent of Man and Selection in Relation to Sex*, published in 1871. In typical fashion, it took Darwin twelve more years after publication

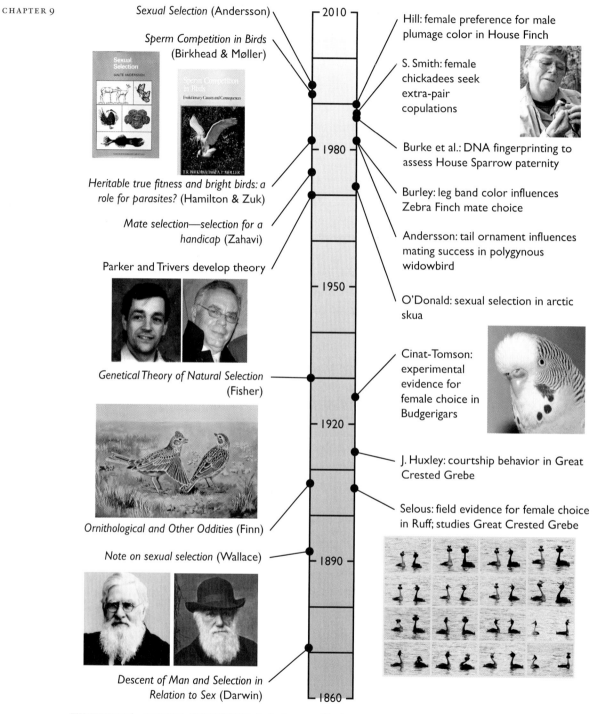

Sexual Selection (Andersson)

Sperm Competition in Birds
(Birkhead & Møller)

Heritable true fitness and bright birds: a
role for parasites? (Hamilton & Zuk)

Mate selection—selection for a
handicap (Zahavi)

Parker and Trivers develop theory

Genetical Theory of Natural Selection
(Fisher)

Ornithological and Other Oddities (Finn)

Note on sexual selection (Wallace)

Descent of Man and Selection in
Relation to Sex (Darwin)

2010

1980

1950

1920

1890

1860

Hill: female preference for male
plumage color in House Finch

S. Smith: female
chickadees seek
extra-pair
copulations

Burke et al.: DNA fingerprinting to
assess House Sparrow paternity

Burley: leg band color influences
Zebra Finch mate choice

Andersson: tail ornament influences
mating success in polygynous
widowbird

O'Donald: sexual selection in arctic
skua

Cinat-Tomson:
experimental
evidence for
female choice in
Budgerigars

J. Huxley: courtship behavior in Great
Crested Grebe

Selous: field evidence for female choice
in Ruff; studies Great Crested Grebe

TIMELINE for SEXUAL SELECTION. Left: Covers of Andersson (1994), Birkhead and Møller (1992); Geoff Parker
(left) and Bob Trivers (right); Eurasian Skylarks courting as depicted in Finn's *Ornithological and Other Oddities*; Alfred
Russel Wallace (left) and Charles Darwin (right). Right: Susan Smith with a chickadee; head of Budgerigar showing
spots manipulated in Cinat-Tomson's (1926) experiments; Great-crested Grebe courtship sequence.

of *Origin* to accumulate sufficient material to make what he felt would be a convincing case for sexual selection, and he acquired this material through a vigorous correspondence, pumping naturalists across the world for information. Birds were crucial to the development of Darwin's ideas about sexual selection, and he dedicated four chapters of *Descent* to birds, more than to any other animal: "Secondary sexual characters are more diversified and conspicuous in birds, though not perhaps entailing more important changes of structure, than in any other class of animals."[5]

Darwin was convinced that in all animals there was a struggle between males for access to females: "This fact is so notorious that it would be superfluous to give instances."[6] Interestingly, Darwin also identified cases where competition for mates was reversed; female Barred Buttonquails (now known to be a role-reversed polyandrous species) were recorded fighting against each other for access to males.[7]

He also believed that in certain species females actively chose males—an idea he obtained from animal breeders such as the cage-bird aficionado Johann Matthäus Bechstein (1795; see Birkhead 2008). In contrast, most naturalists assumed females to be entirely passive, waiting to be chosen by males. Here is Darwin again: "All those who have attended to the subject, believe that there is the severest rivalry between the males of many species to attract by singing the females. The rock-thrush of Guiana, birds of Paradise, and some others, congregate; and successive males display their gorgeous plumage and perform strange antics before the females, which standing by as spectators, at last choose the most attractive partner."[8]

Darwin was not the first to suggest an active role for females in mate acquisition. The naturalist and artist John James Audubon, who spent a lot of time watching birds in the field, wrote extensively of courtship routines that ended when the female bird had made her choice of mate. Although Darwin did not spend much time watching wild birds, he did study his domestic pigeons, read widely, and gleaned information from an army of informants.

To many of Darwin's contemporaries, female choice seemed downright implausible. Men dominated Victorian society, and women were dominated by men; the idea that females might be responsible for male form or behavior was completely unacceptable and challenged male superiority. Moreover, many commentators on Darwin's *Descent* naively assumed that choice required conscious thought and felt that female animals simply did not have the cognitive capacity to make such decisions.

Alfred Russel Wallace—codiscoverer of natural selection—also disagreed with Darwin about sexual selection, although not for naive social reasons. "How can we imagine that an inch in the tail of the peacock, or ¼-inch in that of the Bird of Paradise, would be noticed and preferred by the female?"[9] Indeed, as Mayr (1982) has said: "Nothing demonstrates Darwin's intense interest in the subject [sexual selection] better than his extended correspondence with Wallace on the causation of sexual dimorphism."[10]

Wallace's view was that bright coloration was the default state, reflecting life's "vital energy" (and for him at least, also the vivid hues of internal organs). Natural selection, he felt, favored cryptic plumage in females to protect them from predation while incubating. In a letter to Darwin, he wrote, "I impute the absence of brilliant or conspicuous tints in the female of Birds (when it exists in the male) almost entirely to this *protective* adaptation because in Birds, the female while *sitting* is much more exposed to attack than the male."[11]

Wallace had three objections to female choice: (1) it requires females to have a humanlike aesthetic sense; (2) even if females do prefer certain males (as Darwin's evidence suggested), their secondary sexual traits had no bearing on this; and (3) even if they do choose males on aesthetic grounds, female choice is too fickle and inconsistent to give rise to male ornaments as perfect as the exquisite patterns on the Great Argus's tail.[12]

Wallace also pointed out what he saw as a shortcoming in Darwin's view of sexual selection. Darwin presumed that females chose males on the basis of "good taste," simply preferring males with certain attributes to others. Wallace, however, felt that if females were going to discriminate between males, they should do so on the basis of "good sense" and choose males for their resources, but also for their vigor, health, and stamina. In other words, females needed to benefit from their choice. A century later, this idea was elaborated and formalized with the development of "good genes" theories.

Although most researchers later agreed that Wallace was wrong to have objected to sexual selection (e.g., Cronin 1991), in the decades after the publication of *Descent* Darwin's ideas on female choice found little favor. Worse, Darwin's attempts to keep the two processes of sexual selection (male-male competition and female choice) separate failed, and most people eventually lost interest. Natural selection remained the focus,

and prior to the rediscovery of Mendel's work in the early 1900s, those who thought about it at all assumed that bright coloration could be explained through the inheritance of acquired characteristics. The zoologist J. T. Cunningham (1900), for example, felt the repeated erection of feathers in male display encouraged feather growth and possibly the deposition of color in feathers.

Once Mendel's ideas became known at the turn of the twentieth century, there was renewed interest in understanding the inheritance of traits, but sexual selection continued to be ignored. The geneticists, however, were not interested in distinguishing natural from sexual selection. Instead, they were primarily concerned with gene frequencies, defining fitness simply "as the contribution of a gene to the gene pool of the next generation."[13] Those who did think about sexual selection largely opposed it; for example, Thomas Hunt Morgan, the geneticist and embryologist whose pioneering studies of inheritance in fruit flies earned him a Nobel Prize, published a list of twenty problems with sexual selection theory and concluded that "the theory meets with fatal objections at every turn."[14]

PUTTING DARWIN TO THE TEST

Early in the 1900s two English pioneers of field ornithology—Edmund Selous and Eliot Howard (chapter 7)—made independent, and groundbreaking, contributions to Darwin's ideas about sexual selection. Selous was an excellent field worker, and his account of the lekking behavior of the Ruff in the Netherlands provided the first field evidence for female choice. Inspired by a newspaper article on Ruffs, Selous set off for Texel in the spring of 1906 and after just a week or so was able to conclude that "selection on the part of the reeves [female ruffs] is most evident. They take the initiative throughout and are true masters of the situation."[15] The following year Selous conducted a similar study of lekking Black Grouse in Sweden, recognizing that females appeared to choose to copulate with the most dominant male on the lek (Selous 1927).

Despite these extraordinarily perceptive observations, Selous was rarely acknowledged as a key player in scientific ornithology during his lifetime,[16] and his contributions to the study of sexual selection have been virtually ignored. For example, Joel A. Allen's review of Selous's *Bird Watching* (1901b)

asserted that Selous's discussions of sexual selection "had better have been wholly omitted from an otherwise excellent book."[17] Selous himself was frustrated by the lack of recognition for his work on sexual selection: "I did everything in my power, to further scientific truth, and indeed have produced immensely strong evidence in favour of the Darwinian theory of sexual selection. It would seem, however, that, since the theory is (officially) out of favour, such evidence is not wanted."[18]

Although Lack (1959) later applauded Selous's observational skills, he also had little to say regarding his contribution to sexual selection (Lack 1959, 1968). In striking contrast, Margaret Morse Nice (1935a) identified

Male Ruffs at a lek, fighting.

329

sexual selection as the most important part of Selous's (1927) book *Realities of Bird Life*.

Howard was also aware of Darwin's ideas on sexual selection, but he did not set out to explicitly test them as Selous did. Rather, his initial goal was to learn more about the minds of birds. Seeing male warblers fighting among themselves in the absence of females—before the females had returned from migration—he reasoned that Darwin's idea that males competed for females must be wrong. Unlike Selous, Howard saw no evidence for female choice—after the females arrived back they simply turned up in a male's territory without obviously observing and choosing that male (Howard 1907–14; 1920). Howard concluded that "sexual selection as a rational explanation of the phenomena is impossible."[19]

The editor of the *British Bird Book* (1910–13), Frederick Kirkman, was also interested in sexual selection, and in the preface to the first volume he acknowledged the great potential that ornithologists had to provide information on this topic:

> To take one example only, it is not possible at present to give, in the case even of many of our commoner birds, a detailed reliable description of the differences in the nuptial displays that occur at the beginning of the breeding season; yet one has only to turn over the pages of Darwin, Wallace, and their successors to realise how important is the evidence that the ornithologists might bring to the solution of the vexed question of sexual selection.[20]

Kirkman tried to include in the *British Bird Book* all the information that he could about sexual displays, and in this he was aided by his contributors. One of them, William

Pycraft, was also enthusiastic about Darwin and sexual selection, though he found fault with certain aspects of the theory. Pycraft was a professional zoologist at the British Museum[21] who, had he not suffered poor health, would have made a first-class field naturalist.[22] He was a great admirer of Howard, and he reiterated the latter's concerns about the details of Darwin's theory: "There is evidently here no sexual selection in Darwin's sense: no choice from among a number of males of the individual which most excites desire within her; but the mating of the most mettlesome, most virile males has been determined before her arrival and by a double sieve."[23] Pycraft believed that the elaborate structures and colors found in many birds were simply "expression points" of individual variation and had nothing to do with sexual selection: "Finally, it is contended, the facts garnered during recent years show that the theory of Sexual Selection, as Darwin propounded it, especially in so far as birds are concerned, is no longer tenable: but it is not an exploded theory, it has only undergone modification."[24]

More optimistically, in a paper titled "Courtship in Birds," published in *The Auk* in 1920, the American physician and amateur ornithologist Charles W. Townsend[25] wrote this about sexual selection:

> After all is said this theory, if not taken too literally, explains the facts better than any other. It is not necessary to assume that the female critically examines the display of color, dance or song of the rivals and balances them in her mind, but if we admit, as Pycraft is willing to do, that she is attracted and influenced by these, even if only in a reflex or sub-conscious way, we have practically admitted the truth of

Darwin's theory. The fittest male in any or all of these respects will be more likely to perpetuate the race.[26]

Several other ornithologists, now largely forgotten, were also committed to sexual selection. John Winterbottom (1929), for example, felt that the presumptive evidence for sexual selection was great, even though direct evidence was lacking. He cited the work of Frank Finn, whose book *Ornithological and Other Oddities* (1907) described how, on placing a female and two male Red Avadavats in a three-compartment cage, the female chose the male with the reddest plumage and the rejected male died soon after.[27]

Darwin (1871) had made several suggestions for experimental tests of female choice, including dyeing the bright pink breast of a male Eurasian Bullfinch dull brown to see how that would affect its attractiveness. He also wondered about clipping the tails of peacocks, but recognized that this might be frowned upon: "It would be a fairer trial to cut off the eyes of the tail feathers of male peacocks; but who would sacrifice the beauty of their bird for a whole season to please a mere naturalist?"[28] Darwin went as far as asking some of his correspondents to carry out experimental manipulations: first, he suggested to William Tegetmeier the possibility of trimming the tail feathers of an already mated male cockerel to see whether this would reduce his attractiveness.[29] This experiment was never done, but Tegetmeier did volunteer to stain a white male pigeon magenta and see how his mate would respond. The results yielded nothing of interest,[30] and Darwin—a year after first suggesting these manipulations but no more enlightened—had to resign himself to

the fact that his theories would not be experimentally supported: "I see I shall have to trust to mere inference from the males displaying their plumage, and other analogous facts. I shall get no direct evidence of the preference of the hens."[31]

Darwin's ideas about experiments may even have been inspired by information on the behavior of Long-tailed Widowbirds in Africa. Writing to William Tegetmeier in 1867, he said that "it has been stated that if two long feathers in the tail of the male Widow-Bird at the Cape of Good Hope are pulled out, no female will pair with him."[32] The following year, when corresponding with John Jenner Weir, he referred again to widowbirds: "Barrow asserts that a male *Emberiza* (?) at the Cape has immensely long tail-feathers during the breeding season; and that if these are cut off, he has no chance of getting a wife."[33] Remarkably, this is the same species whose tails Malte Andersson cut nearly a century later to test the idea of female choice, although he was unaware of the coincidence until after his own work was complete.[34]

Finn reasserted Darwin's point that experimental manipulation was essential if ornithologists were to explore female choice further: "Until females of their respective species are introduced to couples of males, one of which has had his characteristic adornments more or less shorn, and rejection of the disfigured suitors is noted, we are not justified in saying positively that the *raison d'être* of these decorations is the attraction of a wife, though à priori reasoning certainly leads to this conclusion."[35]

Somewhat later, Hilda Cinat-Tomson of Riga University, Latvia, conducted an ingenious experimental study of mate choice in

Budgerigars that provides the first conclusive experimental evidence of female choice. She started by making an observational study of pair formation using natural plumage variation and found that males with a greater number of black throat spots were the most attractive to females. She then tested this apparent mate choice in an elegant pairwise experiment in which the attractiveness of males was modified by either reducing or increasing the number of throat spots. The results were striking: for males that were initially attractive to females, removing spots rendered them less attractive, while for males that were initially unattractive, adding more black spots (pieces of feather) with glue increased their attractiveness. Published in 1926 in *Biologisches Zentralblatt*,[36] a fairly obscure German journal, Cinat-Tomson's study never received the attention it deserved.

In addition to these early empirical studies of sexual selection there were also important advances in theoretical work. Sir Ronald Fisher developed mathematical models showing that female choice could evolve (Fisher 1915, 1930), based on the newly revealed features of Mendelian genetics. The key to Fisher's contribution was that for female choice to evolve, females must gain some benefit from choosing particular males—something that Darwin had brushed under the carpet with his weak explanation of a female aesthetic sense. The benefit for females, Fisher said, was in producing attractive sons. Males with a particular attribute—such as beautiful plumage or song—are more attractive to females and as a result leave more offspring. In Fisher's novel synthesis, male attractiveness and female preference evolve together: "The two characteristics affected by such a process, namely plumage development in the male,

and sexual preference for such developments in the female, must thus advance together, and so long as the process is unchecked by severe counterselection, will advance with ever-increasing speed."[37] This "runaway" theory of sexual selection would eventually revolutionize the field but at the time made little impact. Wrapped up in his book *The Genetical Theory of Natural Selection*, the theory was obscure, even though Fisher purposefully left out the mathematics of the models so that people weren't put off.[38] Because the concept of female choice was still unpopular, Fisher's ideas were ignored at the time. It should also be remembered that Fisher's active interest in eugenics means that his motivation for developing the model may have been stimulated more by his desire to improve human society than to further the cause of sexual selection per se.[39]

HUXLEY SPELLS TROUBLE

Selous's early book *Bird Watching* (1901b) provided inspiration for a young Julian Huxley,[40] grandson of Darwin's bulldog, Thomas Henry Huxley. In turn, Julian Huxley supported[41] and promoted the work of both Selous and Howard. Growing up in privileged circumstances, Huxley was interested in biology from an early age, prompting his grandfather to remark that "Julian evidently inclines to biology—How I should like to train him!"[42] His lifelong interest in bird watching was ignited at the age of thirteen by seeing a European Green Woodpecker at a pond near Tring; thereafter he tried to take note of not just the appearance but also the habits and language of birds (Huxley 1970). Given his grandfather's connections, it isn't surprising that Huxley became interested in both natural and sexual selection. With a scholarship to Eton in 1900, followed by another to Balliol College, Oxford, Huxley secured a first-class degree in zoology. In 1910 he was appointed demonstrator in the zoology department in Oxford, where he taught what was then classical zoology. However, he found bird behavior much more rewarding and in his spare time conducted a detailed study of the courtship behavior of Great Crested Grebes at Tring reservoir, with his brother Trevenen, in just two weeks during the Easter vacation of 1912 (Huxley 1914). This is truly a landmark study, describing in exquisite detail—and with evocative terms such as the "penguin dance" and "ghost dive"—the courtship behavior of a single species in terms of sexual selection.[43] Huxley was a professional zoologist, and he went further than simply recording what he saw—he sought to understand the mechanisms underpinning the behavior and the way that behaviors had been shaped by evolution. Huxley's grebe study became iconic for the entire field of animal behavior—as he himself noted in his autobiography; the paper "proved to be a turning point in the scientific study of bird courtship, and indeed of vertebrate ethology in general."[44]

The "courtship" displays of Great Crested Grebes comprise an elaborate and beautiful series of stereotyped movements, performed on the water, in which the birds show off their elaborate head plumage, mirroring each other's movements. Huxley marveled at the serenity of the courtship routine, the strong bond that apparently existed between pair members, and the equal vigor of male and female in displaying.

The grebe study may have marked a turning point in Huxley's belief in female choice, for he was unable to see how female choice could account for displays occurring

after copulation. Nor could he reconcile the concept of sexual selection with the facts that both sexes performed the same courtship routine and both possessed the same secondary sexual characteristics. Darwin had anticipated this, suggesting that in monogamous species ornamental traits were the result of choice in both sexes, calling it "mutual sexual selection." Over a century later Ian Jones and Fiona Hunter (1993) confirmed the existence of this phenomenon in the monogamous Crested Auklet—a species in which both sexes have elaborate facial ornaments. Huxley accepted that a form of choice had to occur in the breeding of birds, but he proposed that females were simply excited to a greater or lesser extent by male displays, without any involvement of an aesthetic sense. For Huxley, this aphrodisiac

Crested Auklet.

effect reached its highest levels in humans, when men and women fell in love.[45]

Thus, after initially accepting female choice,[46] Huxley proceeded to dismiss it as an agent of selection. Instead he favored a group selection explanation for male-male competition, proposing, first, that male combat purged weakness from the population, and thus benefited the species, and, second, that displays in which males simply had to meet some minimum standard to stimulate females had evolved to maximize the *efficiency* of reproduction—again, for the benefit of the species. Third, his interest in the relative size of organs and secondary sexual traits such as plumes or antlers deluded Huxley into seeing elaborate male traits as mere consequences of allometry with little or no adaptive significance.

By the mid-1920s Huxley was calling for a new "synthesis" of the study of sexual behavior in birds: "The time has come when the field-observation of the behaviour of birds, especially their sexual behaviour, should be fully and properly investigated along carefully-thought-out lines, and, when sufficient facts have been elicited, that the new synthesis demanded by the partial overthrow of the Darwinian theory of sexual selection and the discovery of new facts and the rise of new disciplines should be attempted."[47]

Huxley believed that the secondary sexual characters used by Darwin as evidence for sexual selection were in the same general category as "accessory male characters such as copulatory organs, or primary male characters such as those of sperms, which were not regarded by Darwin as having been evolved by sexual selection."[48] He highlighted a need to distinguish between the "*component of general advantage*" and the "*component of (purely)*

individual advantage."[49] He proposed that the phrase "intrasexual selection" be used when referring to selection involving competition between members of one sex (usually males) for reproduction and to account for the evolution of male weapons. He suggested that the term "epigamic selection" be used for female choice (today called "intersexual selection," to include those rare cases—like buttonquails—where males choose females).

With respect to natural selection, at least, Huxley *was* a committed Darwinian, although he seems to have had an anthropomorphic view of evolution itself, believing in "progress," and that the most harmonious societies had made the greatest progress—a view consistent with his idea that selection operated at the level of the group. He disliked Darwin's version of sexual selection because it seemed to favor individual traits, some of which he felt disrupted the "harmony" of the species. For example, in a study of the reproductive behavior of Mallards, Huxley (1912a) noted how, despite their apparently monogamous mating system, groups of males sometimes attempted to forcibly copulate with females, some of whom were drowned as a result. For Huxley this "disharmony" demonstrated that ducks had not "progressed" as far as his beloved grebes.[50]

As we now know, in the absence of individual selection thinking, extra-pair copulations defy convincing explanation (see later). Huxley believed that monogamy, exemplified by his "seraphic symbol" the grebe (Juliette Huxley 1987), was the most harmonious system and that humans should model themselves upon it (Bartley 1995). There's a deep irony here, since Huxley's own marriage was distinctly disharmonious. On his honeymoon in 1919 he built himself a one-man hide (blind) in which he sat alone each morning watching grebes while his wife, Juliette, bored, cold, and hurt, waited for him outside (Juliette Huxley 1987).[51] Then, in 1931, Huxley began an affair with an American girl less than half his age that he met while traveling to Africa. During a subsequent lecture tour of America, he wrote to Juliette of his wish to continue seeing the girl: "I must confess that the combination of adventureness and efficiency with femininity is extremely stimulating to me—and, my dear one, I can't help it and you must make the best of it!"[52]

His own extra-pair copulations aside, Huxley sounded the death knell for studies of sexual selection—for the time being at least. He was so authoritative and influential that others readily followed his lead, and it was decades before sexual selection became a serious focus of bird research. Here, for example, is James Fisher—a colleague of Huxley's with whom he had written the introduction to a new edition of Howard's *Territory in Bird Life* (1948)—in 1953: "Practically all Darwin's important conclusions have been proved to be correct, but not this one [sexual selection]. It is highly probable that a great element of sexual selection is found in bird (as in most animal) evolution; but it is equally probable that the bright colours and adornments of certain birds have as their primary biological purpose intimidation and threat rather than attraction."[53]

A quirky addition to the sexual selection literature was the publication of Major Richard W. G. Hingston's book *The Meaning of Animal Colour and Adornment* in 1933. Like most of his predecessors, Hingston rejected female choice but instead assumed that the conspicuous colors and appendages, crests, manes, and tufts of feathers in males have all evolved

Julian Huxley and his long-suffering wife, Juliette, seen here on a Christmas card they sent to David and Elizabeth Lack with the message "From Dr and Mrs Julian Huxley, Paris 1948–1949."

as a result of anger and rivalry—a view no doubt influenced by his military training. Superficially Hingston's ideas are consistent with male-male competition, but his book is really an attack on sexual selection as a whole. The book was not well received by reviewers. As one said: "What in birds we have been accustomed to call courtship-display, antics, song—is all [in Hingston's view] directed, not towards the female, but against a rival male, and even in the final act of mating, the emotion of the male is that of triumph over his rivals rather than that of any purely sexual satisfaction, while a similar triumphal anger against all other females animates the female in this final act."[54]

Despite the studies by Selous and Cinat-Tomson, and Ronald Fisher's elegant (if somewhat obtuse) theory, by the 1930s sexual selection was all but dead, laid to rest almost

335

single-handedly by Huxley, who substituted Darwin's prescient insights with his own muddled ideas of group, natural, and sexual selection (Andersson 1994: 18). Huxley's social and scientific status, together with his ability to popularize science, enabled him to impose his views on a wide audience over several decades (Erlingsson 2009). Huxley was even able to impose his views on David Lack, ornithological champion of individual selection thinking but—bizarrely—blind to female choice.

DAVID LACK DOUBTS FEMALE CHOICE

During the 1940s, while studying the behavior of the European Robin, David Lack seemed to accept that some aspects of sexual display were compatible with Darwin's idea of sexual selection. He considered both male-male competition and even female choice to be plausible mechanisms, and in *The Life of the Robin* (1943a) he wrote: "Often a hen does not take the first unmated male which she encounters. . . . Since in an average year about one-fifth of the cock robins fail to get mates, there is ample scope for the sort of selection which Darwin supposed to exist."[55] However, Lack felt that displays that occurred between pair members (Huxley's "epigamic" displays), such as female copulation solicitation, could not be sexually selected because most birds were monogamous, and therefore such displays would be seen only by the female's partner. Lack, of course, did not know that extra-pair paternity was widespread in socially monogamous birds and concluded: "Hence epigamic display and associated colour patterns have survival value, though not at all on the same grounds as those postulated by Darwin."[56]

Later, in his *Ecological Adaptations for Breeding in Birds* (1968), Lack analyzed the links between avian mating systems—monogamy, polygyny, polyandry, and promiscuity—and the dispersion of food. Although he identified some associations—for example, that "unusual pairing habits seem linked indirectly with diet, though the link is less certain than in the case of nesting dispersion"[57]—it is hardly surprising that he draws few conclusions, as so little information was available at that time. He wrote that monogamy seemed to be associated with a carnivorous diet, whereas polygyny tended to be associated with seed or plant diets and promiscuity with fruit diets, but he was forced to conclude that there was nothing special about the ecology of certain species to account for their mating systems. Lack also explained the association between bright male coloration and polygyny through strong male-male competition rather than female choice. Although he presented the views of Darwin and Wallace, Lack then dismissed Darwin's idea[58] that the most attractive males pair and breed earlier in the season and hence leave more offspring: "The date of breeding in birds has been evolved primarily in relation to . . . the food supply for the young and the capacity of the female to form eggs, and hence Darwin's and Fisher's suggestions are not very cogent."[59] On top of all this, Lack describes sexual selection as "merely a special form of natural selection."[60]

While the Darwin-Fisher theory may not have seemed cogent to Lack, it makes perfect sense today. We know, for example, that breeding success often declines with laying date, so it now seems obvious that not only are early breeding birds likely to be of better quality, they are also likely to be more successful. Indeed, several recent studies have used date of breeding as an index of fitness in birds (chapter 5). Lack also dismissed conclusions from the studies of Ruff and Black Grouse conducted by Selous, who, he says, "tried rather ineffectively to demonstrate the validity of female choice."[61] Clearly, Lack aligned his views very closely with those of Huxley,[62] acknowledging that male-male competition is sexual selection but rejecting female choice, and believing that sexual differences in plumage serve as interspecific rather than intersexual signals.

Although Lack was unconvinced by female choice, presumably because his interests were ecological rather than behavioral, his chapter "The Significance of the Pairbond, and Sexual Selection" in *Ecological Adaptations* (1968) was extremely influential, appearing to provide a much-needed synthesis of mating systems, ecology, and (some) evolution.

INDIVIDUAL SELECTION THINKING

As Lack points out, a large part of his inspiration for *Ecological Adaptations* was John Hurrell Crook's work on weaverbirds. Crook had a long association and friendship with Lack, having attended the Edward Grey Institute annual conference during his last year at school. During that year, he had studied the ways that time and tide influence the roosting behavior of gulls, divining the first glimmering of links between foraging and social behavior. After National Service in Hong Kong—where he developed an interest in social anthropology and Buddhism—Crook started a PhD in 1954 with Bill Thorpe at Cambridge University, on the economics of bird-rice interactions, but soon realized that this was not what he

wanted to do. Encouraged by Peter Ward, then employed by the British Colonial Office on "the Quelea problem," Crook switched to investigating why some weaver species were social and others solitary, the outcome of which marked the beginnings of modern behavioral ecology. He soon discovered that weaver species that bred in colonies tended to feed in flocks on seeds in grassland, whereas the solitary species tended to breed in woodland and feed on insects. The grassland species were mostly polygynous, while the forest species, monogamous (Crook 1964). Crook later showed the same pattern in primates (e.g., Crook and Gartlan 1966). David Lack was sufficiently intrigued that he referred[63] to his *Ecological Adaptations* as "Crook's book." As Crook said:

> In his kindly way, he had asked me if I was preparing a book before he got on with his own. But I knew I did not have anything like his grasp of population biology. Needless to say, I was close to David at that time and we had numerous intense discussions. His knowledge of birds was formidable. One afternoon we spent two hours wandering around the meadows behind Magdalen discussing these issues, bird family by family—totally exhausting me before I was to give a talk at one of his mini-conferences that evening. Great times.[64]

With a faculty position at Bristol University, Crook continued to develop his ideas (see Hall and Crook 1970), and he started to write an account of the social behavior of mammals. The death of Crook's colleague Ronnie Hall, however, "unhinged" the research at Bristol for some time and delayed the work: "Eventually, I realised that Ed Wilson's new book, the massive *Sociobiology*, by including genetics—especially Hamilton's

John Hurrell Crook (photo in ca. 1957 at age ca. 27).

theories of inclusive fitness, was reaching to a wider scope of thought than my own. It still amazes me how Tinbergen failed to realise the significance of Hamilton, thus allowing the impetus of our work to pass to Harvard."[65]

Crook was as interested in humans as he was in birds and primates, and in the late 1960s he started to develop his work on Buddhism:[66]

> My work then became somewhat schizoid—struggling to maintain my purely zoological interests while developing these new and challenging psychological ideas that often involved me personally. Peter Marler, in a discussion with him at Rockefeller, N.Y., kindly asked me whether I was going to the dogs! Actually I then produced *The Evolution of Human Consciousness* (1980) which made its mark for a time—which was however not yet ripe for a three-way interpretation of the mind from evolutionary biology, psychology and Asian thought![67]

Crook was later able to apply his ideas about mating systems and ecology to provide a compelling explanation for human polyandry in Ladakh, a mountainous region in northern India.[68]

One of the illustrations in Lack's *Ecological Adaptations* depicts the extraordinary convergent evolution between the weaverbirds of Africa (subfamily *Ploceinae*) and the New World blackbirds (family *Icteridae*). The information on the icterids came largely from Gordon Orians, another of Lack's colleagues, who lived in the United States. A bird watcher since early boyhood, Orians discovered at the age of thirteen that it was possible to have a career in ornithology, and he never looked back. As an undergraduate he was taught by John T. Emlen, who had worked on Tricolored Blackbirds with David Lack in the 1930s[69] and who encouraged Orians to approach Lack. On graduating from the University of Wisconsin, Orians was awarded a Fullbright Fellowship to spend several months during 1954–55 with Lack's group at Oxford. Orians had already collected three years of data on competitive interactions between Red-tailed Hawks and Great Horned Owls in Wisconsin (see Orians and Kuhlman 1956) and was keen to discuss the manuscript describing his findings with Lack. That first meeting with Lack changed Orians's life forever. His manuscript had been well received by his mentors in the United States, and he was quite proud of it, but Lack

> returned my manuscript, which was immersed in a sea of red ink . . . he had not bothered to comment on minor matters of grammar and sentence structure, but had confined his remarks to major matters of logic, data presentation and the like. After an hour of "Gestalt therapy" I staggered back to my office and slumped down into my chair. . . . That hour permanently affected how I approached science and how I taught courses throughout my career.[70]

Orians enjoyed the intellectual buzz of Oxford, attending Tinbergen's Friday evening discussions, but returned to the United States to undertake a PhD at the University of California–Berkeley with Frank Pitelka, comparing the ecology and behavior of Red-winged and Yellow-headed Blackbirds. Inspired by Lack's interest in ecology and individual selection, Orians started to think about the evolution of mating systems. His polygyny threshold model, which elegantly explained how it could pay a female to pair with an already-paired male—rather than pairing with an unpaired male (and hence being monogamous)—was a major advance in the understanding of mating systems (Orians 1969). Orians later consolidated all of his ideas into a book in the Princeton series Monographs in Population Biology titled *Some Adaptations of Marsh-Nesting Blackbirds* (1980). Like Lack, Orians created a dynasty of outstanding biologists during his academic career at the University of Washington—including Mary Willson, Jared Verner, Henry Horn, Jim Wittenberger, and Eric Charnov—modestly denying any part in their success.

David Lack's synthesis in *Ecological Adaptations* rather eclipsed both Crook and Orians and inspired a new generation of ornithologists. Lack died in 1973 at about the same time that the Oxford zoology department was beginning to buzz with the beginnings of behavioral ecology, as Richard Dawkins prepared *The Selfish Gene*.

BEHAVIORAL ECOLOGY
CHANGES EVERYTHING

The spotlight on individual selection thinking in the 1960s (chapter 8) kick-started a renewed interest in sexual selection. By focusing on individuals, rather than groups or species, researchers began to consider how a female might gain a fitness advantage by discriminating between different males. Two publications, George Williams's (1966) book *Adaptation and Natural Selection* and Robert Trivers's (1972) paper "Parental Investment and Sexual Selection"—published as a chapter in a book celebrating the centenary of Darwin's *Descent of Man*—propelled both sexual selection and parental care into the forefront of behavioral ecology research. Williams discussed parental care in terms of costs and benefits. Trivers then expanded on this approach in his chapter, creating a cornucopia of ideas that opened up a new vista for the study of sexual selection by focusing on the adaptive significance of behavior and other traits, such as plumage. Trivers's paper was inspired by a much earlier, but largely forgotten, study of fruit flies that Ernst Mayr suggested he read, wherein Angus Bateman[71] (1948) had pointed out the different costs and benefits of reproduction for males and females (Trivers 2002).

Both Williams (1966) and Trivers (1972) promoted the idea that sex was cheap for males—who could walk, swim, or fly away after copulating with no further involvement—but expensive for females, who had to produce eggs and then nurture subsequent offspring. The amount of effort (investment) that each sex made in reproduction, they argued, diminished the parents' future reproductive success and thereby dictated the strength of sexual selection. Because females generally invest more than males, they become a limiting resource for males, resulting in male-male competition. As a result, females can benefit from being coy, and Trivers proposed a suite of possible benefits females might accrue from being choosy, opening up a host of exciting research possibilities. Trivers also acknowledged the possibility that sexual selection might continue after pair formation and copulation, citing Geoff Parker's (1970) groundbreaking studies[72] of sperm competition (in dung flies, *Scatophaga*), which later launched a new field of ornithological research (see below).

Trivers pointed out that in birds like grouse, birds of paradise, and bowerbirds, females undertake all parental duties, whereas the males—and in particular the most striking male birds—compete vigorously at communal display sites for their favors and provide no parental care. These species thus exhibit potentially the most intense form of sexual selection. Among socially monogamous species, like the Great Crested Grebes that Huxley studied, parental duties are often shared, implying similar intersexual selection on each sex and resulting in near-identical plumages in both males and females. The avian example that clinched Williams's and Trivers's argument was the phalaropes—sex role–reversed species in which males undertake the major share of parental care—where, they argued, females should be fighting over males, and males choosing between females, as proved to be abundantly true (Reynolds and Cooke 1988; Colwell and Oring 1988).

Stephen Emlen and Lewis Oring (1977) applied this reasoning—building upon the work of Crook, Orians, Verner, and Pitelka—to develop a much-needed ecological classification of avian mating systems. The most important feature of Emlen and Oring's approach was that the evolution of mating systems and the intensity of sexual selection were driven by the ability of certain individuals to control

A pair of Great Crested Grebes displaying.

access to mates. In turn, the potential for individuals to monopolize mates was—they suggested—determined by environmental factors, specifically the spatial distribution of resources and the temporal pattern of mate receptivity (which encompassed their concept of the operational sex ratio[73]). Resources or mates evenly distributed in space and time minimized the likelihood that they could be monopolized, resulting in monogamy, while the opposite encouraged monopolization and polygamous mating systems—and as a result increased the intensity of sexual selection.

Thinking about female choosiness in terms of individual selection led to the development of new theories concerned with "good genes." Wallace (1889, 1892) had touched upon the basic idea by proposing that female choice could only evolve if the males that were chosen were the most vigorous and healthy, and therefore the most likely to produce surviving offspring. However, Wallace had interpreted this as showing that natural, rather than sexual, selection was at work, making female choice irrelevant. In 1975 Israeli biologist Amotz Zahavi—also building on Williams's ideas—proposed the first good genes hypothesis, called the Handicap Principle.[74] Instead of continuing to ask how elaborate or costly male traits could evolve (the status quo approach at the time), Zahavi turned the question around and proposed that it was precisely *because* of its cost that an elaborate ornament, such as a long tail, was a good signal. He suggested that elaborate secondary sexual characters "tested" the male, and females could accurately assess male quality by their performance on this "test": only males with superior genetic makeup would be able to overcome the added energetic and predation costs associated with displaying the longest tail, brightest colors, or most complex song.

Zahavi's theory, which he presented as a verbal model, was not immediately applauded by theoreticians, appearing as it did in the heyday of mathematical modeling: "The simple argument of the handicap principle was considered by theoreticians to be 'intuitive'; they insisted on having mathematical models to show its operation in evolution. For some reason that I cannot understand, logical models expressed verbally are often rejected as being 'intuitive.'"[75] It was another fifteen years before the Handicap Principle was taken seriously. Theoretician Alan Grafen redeemed Zahavi's theory through a mathematical analysis that demonstrated that female choice could evolve under certain conditions (Grafen 1990; Iwasa et al. 1991). The theory's most outspoken critics had to reconsider:

I was cynical about the idea when I first heard it, essentially because it was expressed in words rather than in a mathematical model. This may seem an odd reason, but I remain convinced that formal models are better than verbal ones, because they force the theorist to say precisely what he means. However, in this case my cynicism was unjustified. It has proved possible to formulate mathematical models showing that what Zahavi called the "handicap principle" can lead to the evolution of honest signals.[76]

Armed with the ideas of Williams, Trivers, Parker, and Zahavi, ornithologists turned their attention to sexual selection in the 1970s and 1980s with a fervor that continues to the present day, exploring its influence on plumage color, songs, ornaments, mating systems, speciation, and reproductive anatomy. We cannot do justice to all of this research so we focus on a few examples that we consider to be interesting and representative.

In the late 1950s Peter O'Donald, then a PhD student at Cambridge University, started to collect data on breeding success in Parasitic Jaegers (arctic skuas) on Fair Isle, Shetland (chapter 2). He tested the logic of Darwin's and Fisher's idea that sexual selection *could* operate within socially monogamous birds, acting on differences in mate quality to result in earlier and more successful breeding. Fisher had previously presented verbal models, but O'Donald (1972) developed rigorous mathematical models based on his field data to test his ideas, confirming that female preferences were important. Parasitic Jaegers maintain a stable polymorphism in their plumage color—individuals fall into either pale, intermediate, or dark forms—and O'Donald was able to show that males of the dark form paired earlier and therefore had higher breeding success. By running simulations of mate preferences along with fitness and breeding components, he proposed that sexual selection was important in maintaining the polymorphism: females preferred dark males, but males paired with any female, and a constant influx of pale males into the population prevented assortative mating resulting in speciation.

In the 1980s Nancy Burley demonstrated experimentally that sexual selection was important in another monogamous species, the Zebra Finch. Using captive birds, she found that the addition of different colored leg rings (bands) influenced the attractiveness of males to females. Red rings on males enhanced their attractiveness, while green rings reduced it. Attractive birds had twice the reproductive success of unattractive birds, for two reasons: (1) assortative pairing—attractive birds paired with high-quality partners—and (2) the partners of attractive birds were prepared to work harder and invested more in their offspring (Burley et al. 1982; see also Price 2008).

Perhaps the most striking experimental demonstration of the existence of sexual selection in a socially monogamous bird species in the wild was Anders Pape Møller's work on the Barn Swallow in Denmark.[77] Using a more tractable study species, Møller essentially duplicated the tail manipulation experiment Andersson had conducted on widowbirds (see above). Møller found that male swallows with elongated tails obtained partners sooner than control males, which in turn did so sooner than males with experimentally reduced tails. Early breeding

increased reproductive success by allowing pairs to produce a second brood. In addition, however, he also showed that longer tails decreased male survival, a key result since it explains why males do not grow even longer tails. Published in *Nature* (Møller 1988), this remarkable result was soon followed by another exciting result: male swallows whose tails were asymmetrical (i.e., one tail streamer longer than the other) were less attractive to females (Møller 1992).

These papers marked the beginning of a highly productive but controversial career for Møller.[78] Typically, Møller combined novel insights with exceptionally clear evidence from field studies. His remarkable productivity stemmed partly from an excellent grasp of theory and knowledge of the relevant literature combined with an extraordinary efficiency in the collection, analysis, and publication of studies on a wide variety of subjects.

The explosion of sexual selection studies eventually led to the study of color and its role in social and sexual interactions. Wanting to follow the experimental approach taken by Andersson and Møller, Geoff Hill began his PhD studies at the University of Michigan in the late 1990s by looking for a suitably colorful study species. He had a young family then, and wanting to work close to home, he eventually settled on the House Finch, a species that had only recently moved into eastern North America and was already a suitably common and accessible breeder on the Ann Arbor campus. Using a technique pioneered by Sievert Rohwer (1977) in an experimental study of plumage signaling in winter flocks of Harris's Sparrows, Hill decided to see how changing the colors of a male House Finch's plumage would affect

Two male House Finches display simultaneously toward a female (center). Female House Finches prefer red males over yellow (Hill 1990).

female choice. He could see that the red plumage of males varied considerably, even to his eye. Bringing the birds into captivity, he made some males more and some males less red. In a series of carefully controlled experiments, the females were clear on their choices—redder males were preferred (Hill 1990). Moreover, redder males were preferred because red was produced by carotenoids, and carotenoids turned out to be an indicator of male quality and condition (Hill 1991; Hill and Montgomerie 1994). The serious study of mate choice for plumage coloration had begun, 140 years after Darwin (1871) had suggested such experiments and 75 years after Hilda Cinat-Tomson (1926) had first implemented them in a study of mate choice in birds (see above). Since the early 1990s the study of plumage coloration has

dominated the literature on sexual signaling in birds (Hill and McGraw 2006).

Red colors produced by carotenoids were the first to be studied, but attention soon turned to the blue colors produced by nano-structural variation in feathers (Hill and McGraw 2006). Although Burkhardt (1989) provided evidence for strong UV reflectance in bird plumage, the evidence for UV signals emerged only in the early 1990s. Maier and Bowmaker (1993) investigated whether UV influenced female mate choice in the Red-billed Leiothrix (Pekin robin), an Old World babbler in which males and females are morphologically indistinguishable, and found that male birds viewed by females through UV-transmitting glass were significantly preferred to males viewed through UV-absorbing glass.

343

A few years later Innes Cuthill and Andy Bennett provided convincing evidence for the role of UV reflective plumage in mate choice, in a series of elegant experiments with Zebra Finches, in which female birds could assess four males in a four-arm cage. A filter in front of each male either allowed UV light through, blocked UV light, or was a neutral-density control. Females showed a clear preference by perching in the arm in which they could see the males' ultraviolet plumage (Bennett et al. 1996). Subsequent research has shown that many birds have areas of plumage that reflect UV particularly strongly, and that this is often important in their mate choice (Cuthill et al. 2000).

In addition to UV sensitivity, recent studies have shown the importance of considering the fine details of the avian visual system in interpreting sexual displays. Evolutionary ecologist John Endler started studying animal coloration and crypsis in the 1990s, examining the influence of light on the displays of lekking birds. Endler recognized that an organism's coloration is only one aspect of its total display and should be considered together with the ambient light and characteristics of its habitat. With Marc Théry, Endler studied three lekking bird species in French Guiana (Endler and Théry 1996), showing that each species displays in only a small subset of the available light environments and, more remarkably, that a striking interplay exists between the light characteristics of the chosen spot and the birds' own coloration. At their display sites on the lek, displaying individuals show off their plumage with maximum contrast, but when off the lek (or not displaying) they are much less conspicuous. Color patterns and display behavior may have therefore coevolved to maximize the effectiveness of the signal—but at minimal predation risk.

Endler, together with his postdoc, Lainy Day, then decided to tackle the question of what birds actually see, using the Great Bowerbird as their study species. Great Bowerbirds tend to prefer white and gray objects—including man-made objects such as golf balls and pieces of metal—to ornament their bower (part of their "extended phenotype," according to Dawkins in 1982[79]). It had previously been proposed that because white objects are relatively scarce, males that accumulated more of these showed that they were higher quality (Borgia et al. 1987). Endler and Day criticized this idea on the basis that birds and humans have very different visual systems, so it may not be white per se that the birds prefer, and, even if so, this may not have anything to do with displaying genetic quality. By providing bowerbirds with a series of different-colored sections of chalk, whose colors were controlled for brightness, and using a physiological model of the avian eye (see Endler and Mielke 2005) that enabled them to visualize what the bird was seeing, they found that bowerbirds chose objects that contrasted most with their plumage, the colors of the bower, and the background environment (Endler and Day 2006).

With convincing evidence that females were discriminating between males on the basis of the size or quality of their ornaments, Darwin's idea was vindicated. But explaining why females choose males on the basis of apparently deleterious traits was still contentious. In the early 1980s Bill Hamilton and Marlene Zuk published a new hypothesis that proved to be very influential. They proposed that the quality of male secondary sexual characteristics provided information on parasite resistance; in other words, females gained information about a potential mate's *genetic* quality simply by assessing the quality of his ornament. Under this hypothesis, longer tails or more vibrant color effectively indicate a stronger immune system, and females should therefore base their choice upon these traits. Hamilton and Zuk (1982) supported their hypothesis with evidence from a comparative study of North American birds, observing that higher infestation of chronic blood parasites was positively correlated with plumage brightness and extravagant displays. Although a similar study of European birds (Read 1987) corroborated this result, the parasite theory—in common with all other theories of female choice—has suffered strong criticism, and little interspecific evidence for it exists. Intraspecific studies show more promise, but in many cases it may be more likely that females are simply avoiding parasitized males for direct fitness benefits. Furthermore, the difficulties involved in objectively assessing plumage coloration, as well as parasite prevalence, make it very difficult to experimentally assess the theory (Møller 1990).

The current (2013) approach to the study of mate choice is an integrated one—worlds away from the simplicity of Cinat-Tomson or even Andersson's pioneering studies. Today researchers examine how multiple processes (including direct benefits, costs, selection on the sensory system in other situations, and indirect selection) influence the evolution of mate choice, capitalizing on developments in the fields of genomics, quantitative genetics, bioinformatics, and population biology to revolutionize our understanding of mate choice and mating systems (Brooks and Griffith 2010).

The shift to individual selection thinking that epitomized behavioral ecology not only revolutionized the study of mate acquisition, it also caused a fundamental change in the way biologists thought about events *after* mate selection and copulation. Geoff Parker's (1970) studies of promiscuity in dungflies provided the springboard for this new focus, showing that males compete with each other not only before copulation but at a postcopulatory level too. Parker called this phenomenon "sperm competition."

The earliest studies of sperm competition in birds highlighted how little field ornithologists knew about the reproduction of birds. When exactly are female birds fertile? No one seemed to know. Poultry biologists knew, but such was the gulf between field ornithology and poultry biology that the transfer of knowledge between the two was very limited. The fact that hens could store sperm and produce fertile eggs for several weeks after removal of the cockerel was well known (and had been elegantly summarized by Alexis Romanoff in 1960): the difficulty was in translating these results into something relevant to the study of sperm competition. A key paper, titled "Gamete Production and the Fertile Period with Particular Reference to Domesticated Birds" by Peter Lake (1975), a poultry biologist, introduced the concept of the fertile period—the time when a female's ova can be fertilized. Remarkably, the basis for sperm storage in birds—the existence of sperm storage tubules—had been discovered only ten years previously (Bobr et al. 1964; Birkhead 2008: 84–85).

Birds store viable sperm for days or weeks, depending on the species, prior to using it for fertilization. The fertile period therefore starts days or weeks before the onset of laying and ends when the last ovum is fertilized, which in most species is twenty-four hours before the last egg of a clutch is laid. In the majority of birds, including the domestic fowl and virtually all passerines, ova are fertilized twenty-four hours before the egg is laid. The fertile period in birds therefore spans several days or weeks, during which time any copulation (within-pair or extra-pair) has some chance of fertilizing eggs. Compared with mammals the fertile period of birds is relatively long. Early observations of extra-pair copulations confirmed that most occurred shortly before the onset of egg laying and hence within the female's fertile period (Birkhead and Møller 1992).

While this confirmed that extra-pair copulations occurred at "the right time," and hence were adaptive, what was really needed to demonstrate the adaptive significance of extra-pair copulations was that they resulted in extra-pair offspring. Fortuitously, a major breakthrough occurred at this time when Alec Jeffreys at Leicester University developed DNA fingerprinting for genetic studies of humans. Terry Burke, who was also at Leicester at that time, had recently completed a PhD on House Sparrows and attempted to infer paternity using allozyme markers, the only technique then available. When Burke heard about Jeffreys's discovery, it seemed too good to be true, and he recognized that if it also worked for birds it would revolutionize the study of avian reproduction. In contrast to the allozyme method that Burke, Dave Westneat (1987), and others had been using, DNA fingerprinting

provided unequivocal evidence of paternity. The technique was tricky, but Jeffreys had the molecular equivalent of green fingers, and Burke was able to learn from Jeffreys how to apply the method to birds. As it happened, Burke's ex-PhD supervisor, David Parkin, was also interested in the new technique and with his student Jon Wetton had submitted a paper to *Nature*, which Burke was asked to review, prompting Burke to finish his own paper within three to four days.[80] After that things continued to move quickly, and within a couple of weeks the two papers appeared in the same issue of *Nature* (Burke and Bruford 1987; Wetton et al. 1987).

Terry Burke had previously highlighted the enormous potential of DNA fingerprinting at the International Ornithological Congress in Ottawa 1986, generating a great deal of interest. One of those inspired by the news of DNA fingerprinting was Nick Davies, who realized what accurate paternity assessment could do for his study of Dunnock mating

The significance of the Dunnock's bizarre precopulatory display, first observed by Selous in 1902, was recognized by Nick Davies in 1983 as an adaptation to sperm competition. The male pecks at the female's cloaca, inducing her to eject sperm from previous copulations, including those of other males.

systems—which it did (Burke et al. 1989); the same was true for others who quickly followed suit. Since then Burke has collaborated with numerous ornithologists and provided a service analyzing genetic material and training new researchers in these techniques. DNA fingerprinting, and subsequent molecular refinements—minisatellites and microsatellites—in combination with field studies of sperm competition in birds, transformed our understanding of avian mating systems (Avise 2004). Among other things, the term "monogamy" had to be redefined as either social monogamy or sexual (genetic) monogamy—the latter proving to be the exception rather than the rule.

A turning point in the behavioral aspect of sperm competition occurred with the publication of Susan Smith's fourteen-year study of color-marked Black-capped Chickadees. Starting during her doctoral studies, Smith routinely collected data on which individuals copulated with which. Then, as the general awareness of extra-pair copulation and sperm competition increased, she

> simply got tired of reading paper after paper addressing the topic of cuckoldry, which either stated or implied that if only a male could prevent any other male from entering his breeding territory, he would thereby avoid any danger of being cuckolded. The unstated, unexamined assumption in these papers was that females don't behave—or, at the very least, don't move. But I had seen a number of extra-pair copulations in my colour-marked chickadees and knew this typically occurred in the territory of the "other" male. After reading one too many of the fatuous papers assuming females don't move, I sat down with my field notes and pulled it all together.[81]

The results confirmed that females actively sought extra-pair partners—and did so selectively, targeting males that were of higher social rank than their current partner, when they were all together in winter flocks (Smith 1988). Subsequently Bart Kempenaers and colleagues (1992) showed that female Eurasian Blue Tits behave in exactly the same way, although for some reason they failed to cite Smith's prescient paper. Because it was published in a high-profile journal, the blue tit study launched a bandwagon such that within a few years it was generally (and erroneously) assumed that all or most extra-pair copulations were female initiated. As Dave Westneat and Ian Stewart (2003) later pointed out, this was not true. Nonetheless, Susan Smith deserves to be recognized for being the first to identify the active role of females in extra-pair copulations.

Prior to Smith's study, the focus of many researchers studying sperm competition in birds had been primarily on males. Part of the reason for this was that ornithologists followed the lead of theoreticians: since the reproductive potential of males was so much greater than that of females, sexual selection was assumed to operate more intensively on males (Parker 1984; Trivers 1972). Male traits were also much easier to study than the more subtle female traits, and it is possible that there was also some unconscious bias in concentrating on males.[82] Nonetheless, the initial focus on males was fortuitous because it allowed researchers to "deal with" the male aspects of sperm competition before tackling the more difficult female aspects and the even more difficult interactions between males and females (Birkhead 1998a, b).

An important question was what determines whether an extra-pair copulation results in fertilization. Some early observations of poultry showed that if females were inseminated by two males in succession, the second fertilized the majority of eggs. Studies by Birkhead and colleagues (1995) showed that the mechanism responsible for this "last male sperm precedence" was the passive loss of sperm from the sperm storage tubules, but that the likelihood of an extra-pair copulation resulting in fertilization was determined by several factors operating simultaneously, including sperm numbers, sperm quality (velocity), and the timing of insemination relative to egg laying.

An additional factor influencing the likelihood of a copulation resulting in fertilization is the female herself. The ability of females to "choose" between the sperm of different males was first suggested by Randy Thornhill (1983), who referred to it as cryptic female choice, an idea later elaborated by Bill Eberhard (1996). In practical terms, distinguishing convincingly between male effects and cryptic female choice is difficult (Birkhead 1998a). Tom Pizzari and Tim Birkhead (2000) showed that female fowl ejected the sperm of subordinate (i.e., unattractive) males immediately after insemination. They also showed that even without any information about male quality, female fowl could preferentially use sperm from particular males, probably through their immune response (Birkhead and Brillard 2007). The most plausible explanation for the evolution of cryptic female choice is that it is a way to avoid incompatible gametes, for example between related individuals.

The most likely situations in which cryptic female choice might occur is when females are subject to forced extra-pair copulations by males, as in fowl (above), and waterfowl (McKinney et al. 1983). In waterfowl, cryptic female choice seems to be mediated through anatomical adaptations of the oviduct, rather than through sperm ejection. In contrast to all other birds examined so far, the oviduct of certain ducks is distinguished by a spiral device at the junction of the vagina and uterus (shell gland) just before (going up the tract) the sperm storage tubules. In addition, and again uniquely, the waterfowl vagina itself has several side branches. In a comparative study Patricia Brennan and colleagues (2007) showed that in those species of waterfowl in which forced extra-pair copulation was common, the male had a particularly well-developed penis. The spiral device in the female's vagina may—during forced extra-pair copulations—deflect the penis down one of the side branches, thereby preventing males from depositing semen in the optimal site near the sperm storage tubules. Brennan et al. (2010) predicted that during ordinary pair copulations the spiral is relaxed, permitting successful insemination, but as yet this hypothesis remains unconfirmed.

The study of sperm competition in birds continues to be of interest (e.g., Kempenaers and Schlicht 2010), but the benefit of extra-pair copulations to females remains unclear. Some studies have identified particular benefits (e.g., enhanced immune function of extra-pair offspring; Sheldon et al. 1997), but there is no consensus. It has even been suggested that low levels of extra-pair paternity may simply reflect nonadaptive consequences of copulation behavior (Reyer et al. 1997). One of the most detailed and comprehensive tests of the hypothesis that females gain indirect (genetic) benefits from extra-pair copulations came from the long-term study of Song Sparrows on Mandarte Island, Canada. Jane

Reid and colleagues at Aberdeen University used eighteen years of data from more than eight hundred maternal half-sib broods and showed that contrary to much theory, extra-pair offspring were *less* likely to survive than within-pair young. They also detected sex-specific differences in fitness, with female extra-pair young (EPY) significantly less likely to survive than within-pair young (WPY), while male EPY had slightly greater survival than WPY. This suggests that females engaging in extra-pair copulations may suffer an indirect fitness cost through their female EPY offspring but a small benefit via their sons. Sardell et al. (2011) were also able to compare the lifetime reproductive success of EPY and WPY, again showing the EPY had lower fitness that WPY. They also found no evidence that male extra-pair paternity was heritable. Overall, the Song Sparrow study suggests that females do not benefit from EPC.

CODA

Although precopulatory female choice now seems to be well established in birds (Andersson 1994), we feel that it would still be worthwhile to undertake a critical reappraisal of this topic. As this chapter shows with respect to Julian Huxley's influence, it is all too easy for an entire generation of ornithologists, including some senior figures, to become trapped within a particular paradigm. To make sure that we are not now trapped uncritically within a female choice paradigm, some reappraisal is necessary.

Why should there be any cause for concern? The main reasons are that so few studies are replicated and there is immense pressure on researchers to obtain results that confirm the current paradigm. Behavioral ecologists are often reluctant to replicate studies because environmental conditions can differ between locations and between years, reducing the chances of obtaining exactly the same results. In the few instances where studies have been repeated, the results sometimes confirm the previous study, but others do not. The effects of tail-length manipulation on reproductive success in Barn Swallows reported by Møller were repeated by Henrik Smith and colleagues (1991), confirming some, but not all, of the previous findings. In the Zebra Finch there have been numerous tests of female choice in captivity (when it should be easier to control for confounding factors), but the results are remarkably mixed (Forstmeier and Birkhead 2004). A further difficulty is that the environmental conditions that can make replication difficult may also affect females directly—as shown by Alexis Chaine and Bruce Lyon (2008) in a study of Lark Buntings.

The other major ongoing challenge is postcopulatory female choice: why do females of socially monogamous species engage in extra-pair copulations? Despite a considerable amount of work (summarized by Kempenaers and Schlicht 2010), there is still no consensus. The recent results from the Song Sparrow study on Mandarte Island provide convincing evidence that females might not benefit from extra-pair dalliances.

Finally, although Huxley was responsible for the demise of sexual selection from the 1930s to the 1960s, one of the issues he raised remains important. That is, why do socially monogamous species—like Great Crested Grebes—continue to perform elaborate "courtship" displays well after pair formation has occurred? We still don't know (see Price 2008).

BOX 9.1 Amotz Zahavi

I have been watching birds since I was a small boy. As no one around me in my small native town of Petach Tiqva knew anything about them, I had to invent my own names for many of the birds I encountered. I learned their proper names only when, at the age of twelve, I met H. Mendelssohn, who was the director of a small zoo and bird collection at the Pedagogical Institute in Tel-Aviv, which years later became the Department of Zoology at Tel-Aviv University. He introduced me to systematic bird watching, convinced me to study zoology at the Hebrew University, and supervised my study (an MSc project) of the birds of the Huleh swamp and lake. The time I spent at this beautiful site impressed me about the importance of conservation.

Niko Tinbergen's book *The Study of Instinct* influenced my decision to spend a year with him at Oxford. With a recommendation from Colonel Meinertzhagen, whom I accompanied on his last tour of Israel, I received a British Council scholarship to go to Oxford. Tinbergen and his student group meetings

introduced me to the study of behavior, and I spent the spring of 1955 at Ravenglass with Uli and Rita Weidman. While watching the incubation of the Black-headed Gulls from a hide (blind), I wondered whether the birds were pressured to leave the nest by the presence of the mate that was eager to replace them. I tested this by building an additional, adjacent nest in their territory with an opaque partition between the two nests. Both birds incubated simultaneously more than twice their usual shifts. The findings from that experiment (which I have never published) prepared me, years later, to gather data on the competition displayed among babblers to serve their group.

In 1953 I was among the small group of naturalists who established the SPNI (Society for the Protection of Nature in Israel). Returning from Oxford at the end of 1955, I opted for the position of secretary of the society rather than that of a demonstrator at the Hebrew University. I was its secretary-general from 1955 to 1969, during which the SPNI succeeded in promoting conservation legislature and building a system of field study centers that serve thousands of youth and tourists annually. Like in other countries, here too bird watchers were the pioneers and more active members of the SPNI.

In the late 1960s I studied the wintering White Wagtails around Tel-Aviv. M. Cullen, a close friend from our time as Tinbergen's students, suggested that a modern PhD project demands some experimentation. Hence, by changing the dispersion of the wagtails' food, I manipulated them into altering their dispersal from territorial behavior into that of flocking.

I returned to Oxford to the Edward Grey Institute in 1970, to write up the wagtail study, suggesting that the wagtail roosts function as information centers. That year I met with P.

Ward and together we pooled our field experience, with various bird species, to write a paper on bird gatherings as information centers.

At Oxford I was exposed to the dispute between Wynne-Edwards and Lack on the level of selection. It is interesting to note that this dispute, which is still going on, started between two groups of ornithologists.

I visited Wynne-Edwards's research group working on the red grouse [Willow Ptarmigan] and was impressed by their study and the conclusions they derived from the fact that the red grouse defended territories much larger than required for their food. I was even more impressed by Lack's logic, in that he insisted on searching for explanations based on individual selection.

I arrived at the idea of the Handicap Principle in 1972, as a consequence of a remark by one of my students, who questioned the logic of Fisher's model of mate choice.

That principle revolutionized the way I understand signaling and social behavior. In 1971 I started a study of the behavior of the Arabian Babbler, a cooperative bird species, resident in the hot deserts of Israel. This study is still going on. We tamed about twenty groups around Hazeva and were able to observe them while walking among them. Arabian Babblers collaborate to defend their common territory.

I divide my time, at present, between watching the babblers at Hazeva and exploring with my students, at Tel-Aviv University, the way by which the Handicap Principle can help us to understand messages encoded in hormones and neurotransmitters in relation to their chemical patterns. We use the principles developed from explaining signaling among babblers, by means of the Handicap Principle, as a basis for our studies of these hormones and neurotransmitters.

Avishag, my wife, has been my partner throughout all these years in the long-term study of the babblers and the development of the Handicap Principle, a principle that has changed the way we see the world and understand social behavior and signaling.

BOX 9.2 Peter O'Donald

I went to Enfield Grammar School, finishing in 1953, then spent two years in the air force. When it came to going to university, it seemed that Trinity was the best college in Cambridge, so I applied. While at school I had read E. B. Ford's *Mendelism and Evolution*. I was struck by what Ford wrote on how the actual phenotypic effects of genes could be altered by the selection of modifiers acting upon them, based upon Fisher's theory of the evolution of dominance. I was very interested by this, so I looked up a bit about Fisher and decided that since Fisher was at Cambridge, this was where I wanted to be!

Right at the beginning of my first term Fisher gave a lecture course called "The Basis of Genetics," which was held on Monday, Wednesday, and Friday nights from five to six p.m. Quite a lot of people went to begin with, but Fisher was so incomprehensible to most that the majority abandoned it. But if you stuck with it after the end of the lecture, you might be taken for a drink at the Bun Shop, an old pub where many Cambridge scientists went after work. So I was lucky in that I got to know Fisher from my first year. He was very nice to students but very critical of staff he disagreed with—he thought that students shouldn't necessarily know better but the staff should!

In 1957 I went with a group of friends to Fair Isle. Peter Davis had just become warden of the Bird Observatory. As far as I can remember, Ken Williamson, the former warden, was also there, showing Peter Davis his study of the arctic skua [Parasitic Jaeger] population on Fair Isle. The arctic skua is polymorphic for pale, intermediate, and dark phases in its populations. When I became a research student at Cambridge in the following year, I saw a paper on the arctic skua by H. N. Southern showing a distinct cline of pale to dark as you went from high latitudes southward. Since a stable cline can only be maintained by a balance between selection and diffusion, this was an obvious opportunity to analyze the selection, since demographic data was being collected on the three color morphs. I was also influenced by a remark in Fisher's *Genetical Theory of Natural Selection*. Darwin's theory of sexual selection in monogamous birds depends on the assumption that there is a relationship between breeding date and breeding success, earlier birds being more successful. Darwin assumed that the males would come back to the breeding grounds before the females, and the earlier females to arrive would

have the first choice of males. The males that bred with the earlier females would thus produce more offspring, and sexual selection could operate even in monogamous birds on this basis. Fisher remarked: "Whether or not there is such a correlation [between breeding date and breeding success], it would seem no easy matter to demonstrate." Fisher's remark convinced me that the arctic skua would be a good research project: the correlation could be tested, since data on breeding date and breeding success was obtainable.

My DPhil was examined by David Lack, and while I can't recall his having much of a reaction to my work, his opinions of sexual selection rather followed some earlier papers by Julian Huxley published in 1938. In my book on sexual selection (O'Donald 1980), I was highly critical of Huxley's papers; I thought they were hopelessly confused—sexual selection confused with natural selection, natural selection confused with group selection. Yet Huxley's papers were strangely influential, particularly his denial of the effect of female choice. As late as 1968, David Lack in his *Ecological Adaptations for Breeding in Birds* followed Huxley in his criticism of the theory of sexual selection. No one seemed to have taken any notice of Fisher's theory of the evolution of mate choice and the "runaway process of sexual selection." Fisher's discussion of the runaway process was very brief and lacking in detailed explanation. He'd actually worked out all the maths on sexual selection but never put it in his book, not wanting biologists to be put off by maths. So he tried to describe his mathematical ideas in ordinary words, not always successfully. Saying that, the brief paragraphs in the Genetical Theory are, I think, pretty clear. And they solved Darwin's main problem with sexual selection: how female preference evolved.

After my DPhil I moved to Bangor as a lecturer. I then went to Brown University, but in 1968 I got an unexpected offer of a job in the genetics department back in Cambridge, and I came back in 1970 after a year in Liverpool. That's when I took up the arctic skua study again. I went back to Fair Isle in 1973 with John Davis, a very competent ornithologist in every way, as my research associate. He devised very efficient methods of catching the adults, far better than the clap nets that had previously been used. We were able to ring all the adults with uniquely coded color rings so that we could identify them with binoculars. By this time the colony had got up to about 130 pairs, so we had quite a lot of demographic data and thus could calculate out the selection acting on the population.

I published most of my results in my arctic skua book (O'Donald 1983), which I now think was wrongly titled. It should have been called *Darwinian Selection in a Seabird: A Study of the Arctic Skua*, because *The Arctic Skua: A Study of the Ecology and Evolution of a Seabird* doesn't give the right emphasis. That book received very favorable reviews in France and the United States, much less so in Britain. I think partly this was because it appeared to be a book about a particular species of bird, but actually was full of maths and demographic statistics that had never been put into bird books before; it would have been better to focus on the evolutionary theory aspects of the subject. How important it is to get the title of a book right!

The big change for sexual selection came when a number of people who, at that time, would have called themselves sociobiologists got into the subject in the mid '70s. In particular Russell Lande, who worked out a theory for quantitative traits that was very similar to the one Fisher had already worked out but never published in 1930. (It also took about forty years before Fisher's Fundamental Theorem of Natural Selection was finally understood!) Sexual selection in birds quickly became a hot topic, brilliant work of direct observation and experiment revealing female choice in many species.

Quite an interesting development has taken place quite recently. There is evidence for a conserved genetic basis of melanism related to mate choice in a number of different species of birds (Mundy et al. 2004). And that opens up quite an interesting prospect for studying the biochemical relationship between the genes and the behavior. I have been thinking of the possibility of producing a second edition of my book, taking in some of these more recent developments, trying to see how they fit in with the sort of theory I had devised before.

Population Studies of Birds

It was hard for contemporary ecologists to understand the importance of population regulation then [1960s] and consequently the major importance of behavioral regulation of population density by territorial behavior.

—JERRY BROWN (2011), IN HIS UNPUBLISHED AUTOBIOGRAPHY,
DESCRIBING THE ORIGINS OF BEHAVIORAL ECOLOGY

DON'T DO IT

IN OCTOBER 1962 ROBERT MACARTHUR, A talented, thirty-two-year-old bird ecologist at the University of Pennsylvania wrote to David Lack in Oxford:

My main motive for writing is the understanding derived from Gordon Orians and Evelyn Hutchinson that you are contemplating a book whose purpose would be in part to refute Wynne-Edwards. May I take the liberty to urge you not to? . . . First, I think the quickest and surest fate of incorrect science is oblivion. . . . Second, and more important, I am sure that an analysis of group selection would be premature.[1]

Four months later, in February 1963, Charles Sibley, later a prominent avian systematist (chapter 3), wrote to Lack on the same topic:

I am pleased to see that you plan to answer Wynne-Edwards quickly. I truly believe that his book will be quietly dropped without causing much of a stir but nevertheless it would be well to have the nonsense pointed out. . . . As always, I am fascinated to see how rapidly a person goes under and drowns as soon as he lets go of the firm rock of natural selection. In Wynne's case, the paradox is that he doesn't realise that he has let go of the rock but continues to believe that he has a firm grip even while he is resting under thirty fathoms of pointless examples.[2]

David Lack and Vero Wynne-Edwards were the key protagonists at the heart of one of the greatest twentieth-century debates in ornithology and ecology. For several centuries it had been recognized—through simple

Red grouse (Willow Ptarmigan) males fighting. Painting by Rodger McPhail in 2009.

arithmetic—that the reproductive potential of all animals would result in larger populations than are ever observed and that, as a result, something must regulate their numbers. The question was what.

Almost from the moment Wynne-Edwards started to promote his ideas about population regulation, they were considered "nonsense." Even so, they proved surprisingly difficult to dislodge, possibly because Lack took Sibley's advice over MacArthur's. However, as Albert Einstein famously once said, "A crisis can be a real blessing. . . . For all crises bring progress."[3] Wynne-Edwards's "nonsense" did precipitate a crisis, and it did indeed result in progress.

A MOST UNUSUAL MAN

Vero Copner Wynne-Edwards[4] was born in 1906 in Leeds, England. His father was a canon of the Church of England and headmaster of Leeds Grammar School, which Vero attended. Interested in natural history from an early age, Wynne, as he was known, watched birds in the rural outskirts of Leeds. At thirteen he was sent to Rugby School, where, among other things, he enjoyed listening to visiting speakers such as Ernest Shackleton and Julian Huxley, then a lecturer at Oxford, and whom Wynne-Edwards thought lectured "awfully well" (Newton 1998b). After wondering whether he should train as a doctor, in part because he fancied being an expedition medic, Wynne-Edwards applied to read zoology at New College, Oxford, where he hoped Huxley might be his tutor.

Ironically, a year after Wynne-Edwards arrived at New College in October 1924, Huxley left to take the chair of zoology at

Vero Copner Wynne-Edwards (self portrait in pencil in 1926 at age 19).

Kings College, London. This was not quite the disappointment it might have been, since Huxley's student, Charles Elton, took over as Wynne's tutor, igniting his interest in population biology. With such zoological giants as Gavin de Beer, John Baker, and E. B. (Henry) Ford on the staff, and ornithologists Bernard Tucker and Max Nicholson nearby, Oxford was a stimulating place. Part of the excitement was that Wynne-Edwards's time in Oxford coincided with the beginnings of the Modern Synthesis, the melding of natural selection and genetics, in which Julian Huxley played a prominent role (chapter 2).

Charles Elton had been a zoology undergraduate at Oxford, where he served as Julian Huxley's assistant on a 1921 expedition to Spitzbergen. That expedition provided an excellent opportunity to see at first hand adaptations to the Arctic environment.

Elton graduated the following year and remained in Oxford for the rest of his career, constructing the foundations on which the discipline of ecology was built. Elton's ideas relating to animal populations were inspired by Alexander Carr-Saunders's *The Population Problem: A Study of Human Evolution* (1922). In 1932 Elton established the Bureau of Animal Population[5] at Oxford and in the same year became first editor of the *Journal of Animal Ecology*. Among his many achievements, Elton is often credited with our current concept of the ecological niche.

The idea of the ecological niche, however, originated in 1917 with Joseph Grinnell in a paper on the niche relationships of the California Thrasher. Grinnell's pioneering contributions include a system of recording field observations, an accomplishment that today seems trivial but was groundbreaking at the time.[6] Elected to the AOU in 1901, their youngest member ever at twenty-four, he subsequently served as editor of *The Condor* from 1906 until his death in 1939. In 1907 he met the wealthy naturalist Annie M. Alexander, who asked him to establish and direct a natural history museum, the Museum of Vertebrate Zoology at Berkeley in California, which he did.

Through extensive observation and the collection of over 20,000 bird specimens, Grinnell recognized that each species has an ecological niche: "a different and very special combination of requirements."[7] These were "separate cubby-holes or dwelling places or habitats (in the narrowest sense), which differ in essential respects from one another."[8] Said another way:

> Each niche is separately occupied by a particular kind of bird, and the locality supports just

as many species of birds as there are niches; furthermore the numbers of individuals of each bird are correlated directly with the degree of prevalence or dominance of the niche to which that particular bird is adapted. . . . In other words,—and here is the crux of the idea—both the number of the species and the number of the individuals of each species, in a locality, are directly dependent upon the resources of the environment.[9]

Elton, like Grinnell, understood that bird numbers are limited by resources and that population sizes and bird communities are determined by the type and number of ecological niches. However, while Grinnell thought of niches in terms of habitats, Elton focused on a species's *role* in a community—a definition still in use today, although the concept of the niche underwent further refinements through the 1960s and 1970s (for an overview see Schoener 1989).

Elton's book *Animal Ecology* (1927) was a landmark in the study of animal populations and may have been the source of Wynne-Edwards's ideas. However, with the benefit of hindsight, Elton's use of language seems almost designed to mislead. Part of the problem was that evolutionary ideas in the 1920s were still muddled (chapter 2). Lamarck's notion of the inheritance of acquired characteristics seemed as plausible as Darwin's natural selection, and few biologists distinguished explicitly between natural selection operating on individuals or on groups and populations. This muddled thinking is apparent in Elton's writings: when he talks about the enormous potential for increase in animal populations, he refers to the "desirable density of numbers."[10] Desirable by whom? By the animals themselves? It isn't

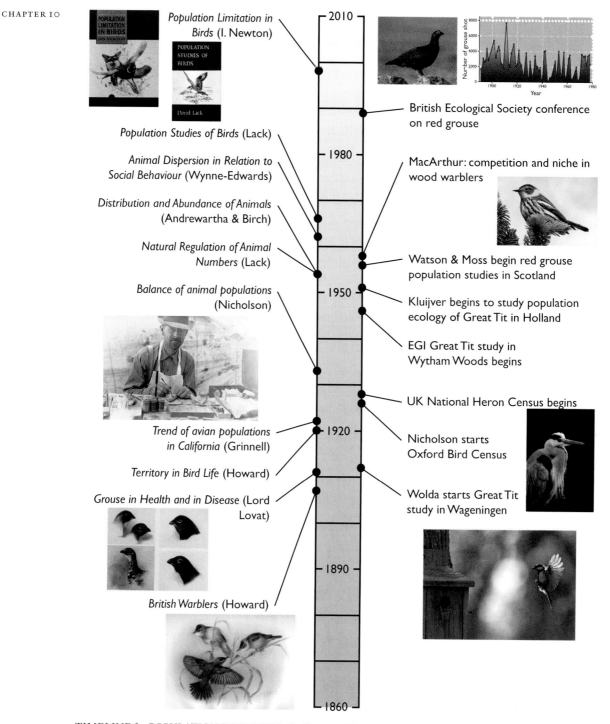

Population Limitation in Birds (I. Newton)

Population Studies of Birds (Lack)

Animal Dispersion in Relation to Social Behaviour (Wynne-Edwards)

Distribution and Abundance of Animals (Andrewartha & Birch)

Natural Regulation of Animal Numbers (Lack)

Balance of animal populations (Nicholson)

Trend of avian populations in California (Grinnell)

Territory in Bird Life (Howard)

Grouse in Health and in Disease (Lord Lovat)

British Warblers (Howard)

British Ecological Society conference on red grouse

MacArthur: competition and niche in wood warblers

Watson & Moss begin red grouse population studies in Scotland

Kluijver begins to study population ecology of Great Tit in Holland

EGI Great Tit study in Wytham Woods begins

UK National Heron Census begins

Nicholson starts Oxford Bird Census

Wolda starts Great Tit study in Wageningen

2010

1980

1950

1920

1890

1860

TIMELINE for POPULATION ECOLOGY. Left: Covers of Newton (1998), Lack (1966); Joseph Grinnell preparing a specimen in a field tent (photo ca. 1922 at age ca. 45); red grouse as depicted in Lord Lovat's *Grouse in Health and in Disease*; painting of immature Sedge Warblers at play, from Howard's *British Warblers*. Right: Red grouse and graph showing population cycles based on number of grouse shot each year from 1890 to 1980; Cape May Warbler; Grey Heron; Great Tit.

clear. He also talks about an "optimum density"[11]—again, optimal for whom? Elton asks, "How do animals regulate their numbers so as to avoid over-increase on the one hand and extinction on the other?"[12] To us, these statements read like the seeds of the group selectionist ideas Wynne-Edwards later developed. However, Elton also recognized that while "the amount of food available sets an upper limit on the increase of any animal," predators and parasites are important too, "keeping numbers down well below the point which would bring the population in sight of starvation."[13]

Elton seems to have planted two ideas in Wynne-Edwards's mind. The first was that animals did not sit around waiting to be acted upon by natural selection; they can move in search of food and other resources. The second, derived from Carr-Saunders's book, was the novel suggestion that "contrary to popular Malthusian concepts, the primitive tribes that had survived into modern times had all, virtually without exception, practiced population control in one form or another."[14]

Wynne-Edwards graduated in 1927 with a first-class degree. He took a position as "senior scholar" at New College, Oxford, where he wrote papers on the function of communal roosting in Common Starlings—papers that anticipate his links between social behavior and population regulation (Wynne-Edwards 1929, 1931). In 1930 he was offered, and accepted, an associate professorship at McGill University in Montreal, traveling to Canada by ship later that year. During that North Atlantic crossing, Wynne-Edwards made detailed observations of the distribution of seabirds, recognizing that different species had different niches and were associated with particular marine zones. He also noted the low density of birds at sea, commenting on the lack of any overt intraspecific struggle for existence. The struggle for these birds, he deduced, was not with conspecifics but with the environment.

Keen to make further observations, he persuaded Cunard to allow him to travel, free of charge, on four more round-trip transatlantic crossings.[15] The result was a pioneering study of the distribution of birds at sea, showing clearly that there were distinct inshore, offshore, and pelagic communities. His paper on that survey (Wynne-Edwards 1935) was submitted to the Boston Natural History Society for publication and won their Walker Prize for science in 1934, as well as a check for $100. (The normal prize was $60, but the committee felt that his paper was of exceptional merit.[16])

In 1937 Wynne-Edwards set off on an expedition to Frobisher Bay at the southern end of Baffin Island, traveling via Labrador and collecting plants and birds as he went. The Arctic environment played a crucial role in shaping Wynne-Edwards's thinking because it "lacked the 'Darwinian' ferocity of struggle and competition."[17] It was during this trip that he formulated his hypothesis about intermittent breeding in Northern Fulmars, drawing together several diverse strands of evidence. First, many fulmars that he observed in the summer during his previous transatlantic crossings were molting, suggesting that they were nonbreeding individuals—since breeders do not molt until September. Moreover, sixteen of the birds he shot at Cape Chidley, Labrador—650 kilometers (404 miles) south of the southernmost known breeding colony—"had their gonads and genital ducts reduced and inactive,"[18] further supporting the idea that these were not breeding that year.

Second, Wynne-Edwards noticed that some of the fulmars he collected had needle-sharp claws, indicating that they hadn't spent much time ashore, whereas the claws of others were blunt, suggesting that they were breeders: "A visit to the breeding colonies leaves a small but permanent mark on the bird. It seems very reasonable to suppose that an individual with sharp claws . . . is a young bird which has never, or only once or twice, bred, whereas a bird whose claws are broadly rounded . . . is very old, having bred many times."[19] Thus Wynne-Edwards was sure that some of those nonbreeders were adults who had bred in earlier years.

Third, his examination of the ovaries of four females revealed a relationship between the numbers of resorbed follicles—an index of the number of previous ovulations—and the sharpness of the claws, "confirming" that birds with sharp claws had bred less often than birds with worn claws. For Wynne-Edwards, the presence of molting fulmars with both worn claws and many resorbed ovarian follicles during the normal breeding season was evidence of intermittent breeding. The idea that seabirds sometimes take a year off from breeding was not new—it had been suggested previously by others (Bertram et al. 1934; Murphy 1936)—but nonbreeding was a central component of Wynne-Edwards's big idea about group selection.

When Canada entered the Second World War in 1939, Wynne-Edwards retained his position at McGill, where he trained radar mechanics in electronic physics, navigation, and the history of naval warfare—as well as undertaking fishery surveys in the Yukon and Mackenzie Rivers. But a brief visit to Britain at the very end of the war left Wynne-Edwards "overwhelmed with nostalgia,"

and he decided that he needed to return to England. When his children complained that there'd be no skiing, he replied, "We'll have to go to Scotland instead."[20] By coincidence, shortly after returning to Canada from that brief trip, his wife, Jeannie, whom he left back in Britain, wrote to tell him that the Regius Chair of Natural History at the University of Aberdeen had just been advertised. His application was successful, and he started there in 1946, eventually creating one of the largest departments of ecology anywhere in the world. Wynne-Edwards encouraged colleagues to begin what would become a long-term population study of red grouse [Willow Ptarmigan] in the Cairngorms of Scotland, headed by David Jenkins and Adam Watson, and of fulmars on the island of Eynhallow in Orkney, headed by George Dunnet. In 1958 Wynne-Edwards helped to establish the Culterty Field Station on the Ythan estuary some ten miles north of Aberdeen, with Dunnet as director.

Wynne-Edwards was described by Adam Watson as "a naturalist with a great intellect and prodigious mental and physical energy."[21] A tall man, Wynne-Edwards was also extremely competitive, being a rapid hill walker and champion cross-country skier, and thought nothing of breaking the ice on a highland burn to bathe before breakfast. He was still hill walking in his eighties. Adam Watson described Wynne-Edwards as "a good enabler, but not a very good leader."[22] Like many academics of his day, Wynne-Edwards provided little or no supervision for his thesis students. As Watson later recalled,

It was sink or swim. . . . [Wynne-Edwards] said the best research was done by picking the right people and letting them get on with it the way they wanted to do it. With hindsight, I believe this may be appropriate for later work, but every student benefits from critical discussion of thinking, fieldwork and writing. Certainly he would have had no time for today's smothering of scientific imagination by pedestrian funding agents with more money than sense, who long ago lost the ability to think critically if they ever even had it in the first place.[23]

1950S: WYNNE-EDWARDS AND LACK

The year 1954 witnessed several key events in bird population biology, including the publication of two important books: *The Distribution and Abundance of Animals*, by Herbert Andrewartha and Charles Birch, and *The Natural Regulation of Animal Numbers*, by David Lack. The two could not have been more different, and this was largely a consequence of the organisms and environments in which these respective authors worked (Orians 1962).[24] Andrewartha and Birch were entomologists based in environmentally unpredictable regions of Australia, whereas Lack studied birds in temperate English woodlands. Andrewartha and Birch felt there was no population regulation as such—they said that environmental conditions alone determined numbers. Lack, on the other hand, considered density-dependent mortality—mediated by food, predators, or disease—to be the key factor regulating animal numbers.

Lack's view, that bird populations were regulated, came from various sources that showed bird numbers varying from year to year, but within certain limits. The Grey Heron was one of his premier examples, and Lack made use of a run of data going back to 1928, when Max Nicholson initiated an annual census of heronries in England and Wales, using amateur ornithologists. Although heron numbers fluctuated from year to year, they never exceeded a certain level. There was also clear evidence that numbers declined sharply following a hard winter but then bounced back to previous levels over the following two or three years.

The heron example illustrated several key principles. First, there was a ceiling on numbers, which suggested that numbers were indeed regulated, with that ceiling reflecting the carrying capacity of the environment. Second, the potential for a rapid increase in numbers was evident from what happened after a crash induced by a hard winter. Finally, once numbers had returned to the ceiling, a balance between birth and death rates—it was presumed—kept the population size relatively stable, implying the existence of density-dependent processes.

Another source of circumstantial evidence that bird numbers are regulated in a density-dependent manner came from studies in North America. Robert Stewart and John Aldrich (1951), for example, removed (shot) all the birds from an area of forest, to assess the impact of those birds on an insect pest. In the following year an almost identical number of birds had recolonized the wood—providing clear evidence for stability of numbers.

In a review of Lack's and Andrewartha and Birch's books, Dennis Chitty—a population ecologist who worked for twenty-six years with Elton at the Bureau of Animal Population—wrote that they "demonstrated how hard it was to create a science out of data which are still preponderantly descriptive."[25] What was lacking was "verification through theoretical and experimental science."[26] J. B. S. Haldane praised Lack's book, saying that it was "the first to deal with one of the central problems of biology, a problem which is becoming increasingly relevant to our own species."[27] But Haldane would have preferred more quantification, chastising Lack for being "a little too frightened of mathematics,"[28] even though Lack (1954: 3) had explained how he had deliberately avoided statistics to make the book more accessible to naturalists. The ornithologist Bill Vogt wrote to Lack saying, "What a superb job it is! And how improbably pleasing to find such a piece of scholarship so well written—I suspect that, at least in this country [United States], it will do more to stimulate the study of ornithology than any book yet published."[29]

The second major event of 1954 occurred in early March, when David Lack visited four Scottish universities to speak on population regulation. At Aberdeen Adam Watson considered him a good speaker, but he later remembered Wynne-Edwards whispering to him during the talk: "He's ignoring slow breeding seabirds and elephants . . . but I daren't ask a question in case he steals my ideas. I'll have to publish mine first."[30]

It is interesting that Wynne-Edwards saw Lack as a potential competitor, when in reality Lack was already fed up with the whole question of population regulation. In 1955 Lack wrote to Wynne: "I will be interested to see your paper, when it appears. . . . Actually, I have had to close my mind completely to population questions as I was so stale on it all—that book was very exhausting. I am just completing one in lighter vein on swifts."[31] A little later, after Wynne-Edwards had asked Lack to comment on his paper:

"Reg [Moreau] has not shown me your paper on populations. As I think I said to you earlier, I feel completely stale on the subject at present, and incapable of criticising anything on it."[32] Lack's apathy may have allowed Wynne-Edwards to feel he had gained the upper hand.

The 11th International Ornithological Congress (IOC) in Basel, Switzerland, took place in late May and early June 1954. It was here that Wynne-Edwards presented a paper on low reproductive rates in birds, tentatively introducing the idea that population regulation "could be largely intrinsic . . . depending for its operation on the behaviour-responses on the members of the population themselves."[33] According to Borrello (2010), this marked the beginning of Wynne-Edwards's commitment to the theory of group selection. Another, related event stimulated Wynne-Edwards to present his ideas as a book. After returning from the IOC, Wynne-Edwards submitted a paper on slow-breeding seabirds to *Ibis*, a manuscript that, to his surprise and intense disappointment, was rejected. The editor of *Ibis* was Reg Moreau, a close friend of Lack's, which suggests that Adam Watson's idea that the paper was bounced by the "Oxford establishment" was probably correct.[34]

Around this time, in 1954, Wynne-Edwards was asked by the editor of *Discovery* magazine to review Lack's book. He did so—albeit after a delay of several months—praising its breadth but criticizing Lack's idea that mortality (through density dependence) adjusted itself to natality (birth rate). He also complained that Lack had mistaken fecundity for fitness (which, as we'll see, was correct, but for a different reason). Wynne-Edwards then went on to present some of

his own ideas, including the notion that selection could favor low reproductive rates,
"differentially permitting the survival of
populations that live in harmony with their
environments."[35]

While writing his review of Lack's book,
Wynne-Edwards had the feeling that, contrary to Lack's idea that population regulation
is a passive process, a much more plausible
way of keeping numbers in check seemed to
be active self-regulation. "It was very exciting
and yet seemed so simple that others would
surely be drawing the same conclusions any
day. My life was already busy and if the hypothesis was to be adequately supported there
was a huge literature to survey. In fact it took
me seven years to finish the book!"[36] During
that time, Wynne-Edwards did little other
than marshal evidence for his hypothesis.
Outside Aberdeen he kept a low profile, with
one exception—albeit by proxy. In 1959, the
British Ornithologists' Union celebrated its
centenary with a meeting in Cambridge that
included a symposium on population ecology.[37] Wynne-Edwards was concerned that
because he had accepted a previous invitation to lecture in the United States he would
not be able to attend the Cambridge meeting.
Seeking Bill Thorpe's advice, he wrote, "I am
working on what I may be forgiven for thinking is a tremendous theory . . . [it] might be
difficult for someone else to present in my
stead, but I do not say it is impossible."[38]

As it turned out, Wynne-Edwards's colleague George Dunnet gave the paper for
him, reiterating Wynne-Edwards's view
that (1) because they can move, most animals are able to regulate their own populations, and (2) while food is indeed critical,
birds maintain their populations below a
level when starvation might occur through

"social conventions"—notably, through social gatherings (such as the huge pre-roost
flocks of Common Starlings), behavior that
evolved so that animals can assess their own
population density and respond accordingly.
Wynne-Edwards also proposed that the territorial behavior of birds had evolved so that
they could avoid starvation by limiting their
numbers. In other words, like human hunters, he suggested, animals avoid overex-
ploiting their prey through a consensus of
restraint.

Max Nicholson opened the symposium and was followed by Dunnet reading
Wynne-Edwards's paper. The next speaker
was the Oxford-based ornithologist Mick
Southern, who devoted the first half of
his talk to an attack on Wynne-Edwards's
paper, which he had almost certainly seen
beforehand—Wynne-Edwards had sent a
copy to Lack prior to the meeting, and Lack
and Southern were colleagues. Southern's
attack was premeditated.[39] Dunnet wrote
to Wynne-Edwards: "It was not a nice or a
good attack . . . but . . . mainly by the use of
slight sarcasm and his manner, he carried
the audience."[40]

Lack, who did not give a paper at that
symposium, stood up and briefly summarized his objections to Wynne-Edwards's
ideas: mainly that natural selection operates
on individuals rather than groups and that
territoriality cannot have evolved to regulate
populations—largely because birds don't
always feed in their territories. Lack also
pointed out that he had already responded
to all of Wynne-Edwards's points (in Lack
1954), but that Wynne-Edwards "nowhere
refers to these views, either when they agree
with his own, or, more often, when they
differ."[41]

Dunnet reported back to Wynne-Edwards in a letter two weeks after the event, telling him that the population session had been "a flop" and that "all the Oxford people [i.e., Lack, Southern] came prepared to object strongly to your paper."[42] He described how, over lunch, James Fisher had told him that he was "going to town" with criticisms of Wynne-Edwards's paper, "saying that if we were to believe this, then there was no longer any possibility in believing in Darwinism and natural selection!"[43]

Lack also wrote to Wynne-Edwards after the meeting, thanking him for sending the copy of his paper, but saying, "As I expect you will realise, I am in considerable disagreement with part, though not all, of it."[44] As Borrello (2010: 72) later pointed out, there is no copy of any reply to Lack's letter. Our examination of the archives confirms that Wynne-Edwards rarely bothered to reply to Lack and his criticisms.

Dunnet concluded his letter by saying, "I can't help feeling that you are going to find it very difficult to put across these ideas to these people who are so obviously set against them."[45] Unfortunately for Wynne-Edwards, this would prove to be something of an understatement.

1960S: WYNNE'S MAGNUM OPUS

Wynne-Edwards's *Animal Dispersion in Relation to Social Behaviour* appeared in 1962. Just before publication, Wynne-Edwards told the ornithologist David Snow that he considered it to be an evolutionary advance as important as Darwin's.[46] Adam Watson was disappointed that Wynne-Edwards did not give the manuscript to anyone prior to publication, and although he showed Watson the

proofs, he did so at such a late stage it was impossible to make any substantive changes. Wynne-Edwards was unshakable in his belief that group selection was the basis for population self-regulation, but he remained remarkably defensive. After a seminar in his own department in Aberdeen, in which he described the enormous pre-roost gatherings of starlings as "epideictic" displays, his long-time research technician, Sandy Anderson, commented that it sounded doubtful and asked whether he had any evidence to support his ideas. Wynne-Edwards was furious and responded, "Sandy, I know you mean well, but it is not for you to question the ideas of the Regius Professor of Zoology at Aberdeen."[47] He later told Watson and Dunnet that he hoped his ideas would never again be criticized in front of his students.

In his own words, Wynne-Edwards's 653-page *Dispersion* postulated: "(1) that animals collaborate socially for the benefit of the group, because in no other way can their numbers be matched to the carrying capacity of their habitats; (2) that they compete conventionally for property and status, rather than for the real necessities of life, and the losers patiently accept their lot; (3) that animals are not, as Darwin supposed, always striving to increase their numbers; instead they are programmed to regulate them."[48]

In the sense that he accepted that animal populations were regulated, and that food played a crucial role, Wynne-Edwards agreed with Lack. But that's where any similarity ended. For Wynne-Edwards regulation occurred through social behavior—which enabled animals to assess their own population density—and through voluntary reproductive restraint, which in turn enhanced the survival of the group (or population).

Northern Fulmars.

In other words, Wynne-Edwards's vision of population regulation was unashamedly group selectionist: groups or populations made up of unselfish individuals outsurvived groups comprising selfish individuals.

Wynne-Edwards's zeal was evangelical: every facet of natural history was interpreted as evidence for his theory, including the "nonbreeding" fulmars he had seen on his various oceanic voyages. That he continued to use fulmars to support his ideas is surprising, given the results emerging from his student George Dunnet's study of this species in Orkney. Dunnet's data provided pretty good circumstantial evidence that Wynne-Edwards's "nonbreeding" fulmars were failed breeders—birds that had attempted to breed but had lost their egg or chick—rather than birds refraining from breeding (Carrick

and Dunnet 1954). Wynne-Edwards ignored this straightforward explanation, claiming instead that what happened in Orkney was unlikely to be the same as what happened in the Canadian Arctic, sticking to his belief that the birds he saw there were taking a year off from breeding. He said that in Britain the fulmar population was rapidly increasing but in Canada it was static or even declining, and therefore "the demand for breeding stock [in Britain] may be much less and the surplus of non-breeders on a far larger scale."[49] It has since become clear that the "nonbreeding" birds observed by Wynne-Edwards were likely to be a mixture of failed breeders, pre-breeders, and some birds genuinely taking a year off from breeding—although, for reasons other than that proposed by Wynne-Edwards.[50]

365

Many of Wynne-Edwards's colleagues greeted *Dispersion* with polite if cautious enthusiasm—at least to the author's ear. Julian Huxley wrote to say how interesting he was finding it. Charles Elton wrote that "your book does look very interesting and packed with fascinating ecology and natural history."[51] But later, in a review in *Nature*, Elton (1963) said, "The theory is set forth with enthusiasm, often pontifically (if a bishop can wear blinkers), sometimes in a sort of Messianic exaltation which admits of no other important processes affecting population-levels."

Lack refrained from public comment and instead invited Wynne-Edwards to give the main talk at the EGI conference, in Oxford, in January 1963. Keen to promote his ideas, Wynne-Edwards couldn't resist, even though it meant walking into the lion's den. The winter of 1962–63 was a hard one, and with snow piling up outside, the atmosphere in St. Hugh's College, where the conference took place, was sharp with anticipation. Inevitably, Wynne-Edwards's talk elicited much critical comment. Both Niko Tinbergen and David Lack—who were sitting in the front row—appeared to be affronted and were "didactic and deprecatory and not a little arrogant."[52] Arthur Cain and his colleague Philip Sheppard, evolutionary biologists formerly in Oxford, both gave caustic—but from the audience's point of view, constructive—comments, explaining why Wynne-Edwards was wrong. For his part, Wynne-Edwards "withstood the barrage of criticism with amazing patience and charm,"[53] but refused to rise to the challenge. His response to every comment was simply to say, "Read my book . . . it's all in the book."[54] The students at the conference were disappointed—the clash

of titans they were hoping for simply did not materialize because Wynne-Edwards was so unresponsive.[55] One slightly bizarre aspect of the meeting was that Wynne-Edwards's grouse colleagues, Adam Watson and David Jenkins, also attended, and although Wynne-Edwards had used data from their red grouse study to support his theory, Watson's presentation interpreted the data in a way that barely supported Wynne-Edwards.[56] This was to be a familiar theme, and one gets the impression that Jenkins and his colleagues were embarrassed by Wynne-Edwards.

Despite MacArthur's plea that Lack ignore Wynne-Edwards's book, Lack immediately started to prepare a new book whose aim was to demolish Wynne-Edwards's group selection ideas. Before looking at that, we need to backtrack, to consider the origin of Lack's ideas about population regulation.

1960S: DAVID LACK

David Lack's early career is described in chapter 5 (see also Anderson 2013). His thinking about population regulation probably started as an undergraduate, as his tutor Bill Thorpe later recalled: "I have a recollection of his displaying considerable interest in the biology of insect parasitoids and in the population problems raised by the then rapidly expanding field of the biological control of insect pests. . . . Perhaps this was the very beginning of his fascination with population ecology."[57] The onset of the Second World War provided an unexpected opportunity for Lack to develop his ornithological ideas. After submitting his name to the Central Register of Scientific Workers, he was given the task of visiting the radar stations along the British coast. This allowed him to see how migrating birds were detected by radar (chapter 4), as well as providing an opportunity to meet many other biologists, including George Varley. It was Varley who encouraged Lack to take another look at A. J. Nicholson's *Balance of Animal Populations* (1933) and to reconsider Gause's principle of competitive exclusion, both of which Lack had glanced at but had not yet fully appreciated. Lack was then persuaded by Nicholson's argument that animal populations were regulated in a density-dependent manner through competition.

After moving to Oxford to take up the directorship of the Edward Grey Institute in 1945 (chapter 5), Lack had to decide what the institute should focus on. He identified two fields in which studies of birds had already led to major advances: (1) the origin of species from subspecies (chapter 2) and (2) instinctive behavior (chapter 7). "In both these fields, principles originally developed for birds are now being widely applied to other animals."[58] A third field, not yet so well established, concerned population regulation.

David Lack (photo in 1965 at age ca. 55).

"With a small research staff, the Institute had to concentrate on only one of these three problems, and I decided on the third.... Briefly then, the main object of study at the Institute is to find out why birds are as numerous as they are."[59]

Initially, Lack hoped that his beloved European Robins might be suitable for a long-term population study, but he decided that their nests were too difficult to find. After visiting the Dutch ornithologist Huijbert Kluijver in 1945, he realized that a nest-box population of Great Tits had many advantages for the kind of work he envisaged.

Kluijver's study was built on work done by Gerrit Wolda, a forester who had erected nest-boxes in woodland near Wageningen in 1912 and monitored their contents and fate each year. The original aim of Wolda's project was to encourage tits that—by consuming caterpillars—would improve the growth of the trees (Brouwer 1954; Savill et al. 2010). During the 1920s, Kluijver worked under Wolda's supervision, assessing the damage that birds inflicted on agricultural and forestry interests: at that time there was considerable concern on the Continent, as well as in both Britain and North America, about whether certain bird species were harmful or helpful to man, for example, by controlling insect pests (Ritchie 1931). Kluijver took over and developed Wolda's study, and his subsequent paper on the ecology of the Great Tit (Kluijver 1951) was the first to demonstrate a density-dependent effect on reproductive success, showing that both clutch size and the percentage of second broods declined as population density increased. Reviewing Kluijver's paper, Lack (1952) wrote: "This is far the most comprehensive study yet made of a bird population, and probably of any animal population in the wild."[60] Lack concluded by saying, "Ornithology is advancing from the descriptive to the quantitative stage, a sign that the science is maturing, and Kluijver's study is an important milestone on the road."[61] Chris Perrins said in Kluijver's obituary that the study "must surely rank among the greatest of the pioneer studies in the ecology of bird populations."[62]

Back in Oxford, Lack employed John Gibb as a field assistant to establish the Oxford Great Tit nest-box population. Gibb had originally studied law at Oxford, but after the war he returned as EGI field assistant. During the vicious winter of 1946–47, he erected about one hundred boxes in Wytham Woods on the outskirts of the city. Subsequently, as assistant but also later for his DPhil studies, Gibb monitored the birds' population size and their timing of breeding, clutch size, breeding success, chick diet, and survival (of both adult and immature birds). Robert Hinde—who was in Oxford at this time—described Gibb as Lack's "general factotum" (presumably because Lack did no fieldwork on tits himself[63]), emphasizing Gibb's essential role in initiating the long-term tit study.[64]

Also working in the EGI at this time was Peter Hartley, who made a detailed study of the feeding habits of six species of tit in various locations across Oxfordshire. David Snow (1949, cited in Hartley 1953) had previously described the height distribution of feeding tits in woods near Uppsala, Sweden, finding consistent differences between species. In Oxfordshire Hartley (1953) obtained similar results but was also able to show the importance of food abundance in determining this ecological separation: there was more overlap in feeding habits between species when food was superabundant.

Gibb (1954) then showed that each of the six species of tit breeding in Wytham had a distinct feeding ecology—as predicted by Gause's competitive exclusion hypothesis—thereby allowing them to coexist. He also demonstrated that both inter- and intraspecific competition were at play and that competition was most intense in the winter when food was scarcest. Gibb went on to suggest that "mortality from food shortage in winter is thus likely to be strictly density-dependent,"[65] adding that mortality among juvenile birds in late summer is probably due more to inexperience than to food shortage and so may or may not be density dependent. After Gibb left Oxford, his place was taken first by Dennis Owen (1952–57) and then in 1957 by Chris Perrins, who extended and continued the Great Tit population study.

Lack's *Population Studies of Birds* (1966), written partly in response to Wynne-Edwards's book, was also a way of pulling together the available evidence for the role of density dependence in the regulation of bird numbers. The book comprised an overview of thirteen major studies and eleven minor ones, across a wide range of species, including thirteen passerines, predominantly from temperate regions.

Lack's commitment to density dependence—expressed originally in his 1954 book—was largely intuitive. Even by 1966, the evidence was still sparse: "When I wrote my earlier book of 1954, the existence of density-dependent mortality still rested largely on theoretical considerations, supplemented by data from laboratory populations of various insects. . . . The evidence from natural populations is not much stronger now."[66]

Although he addressed his hypothesis on a wide front, simultaneously seeking evidence for density-dependent effects on clutch size, adult mortality, immigration, and emigration—and for evidence that food was the limiting factor—the end result was disappointing. Lack applauded the increase in studies of reproductive rates of birds but bemoaned the fact that there had been so little progress in understanding the regulation of bird numbers. He reiterated his view that reproductive rates in birds have evolved through natural selection and are as rapid as the environment permits; that mortality rates balance reproductive rates through density-dependent mortality; that this is mediated mainly by food, via starvation outside the breeding season; and that birds are "dispersed broadly in relation to food supplies, through various types of behaviour which are as yet little understood, but which are to be explained through natural selection."[67]

The latter point was another swipe at Wynne-Edwards, who had suggested that territoriality in birds had evolved as a mechanism to distribute individuals across the environment in such a way that they did not overeat their food supplies. Lack's point was that this was not consistent with the way natural selection operated; territoriality had evolved for other purposes, and the fact that it distributed birds across the environment was a *consequence* of its main function.

The truth was that evidence for density dependence was difficult to obtain. Several of the studies Lack reviewed in *Population Studies of Birds* confirmed that the mortality of young birds was often very high—an idea that initially elicited ridicule—and high enough to operate in a density-dependent manner. Hardly evidence, but these observations were at least consistent with Lack's ideas.

These studies had started mainly in the 1940s—some inspired by Lack but most initiated independently. They consisted of measuring a suite of population parameters, including the survival rate of adult birds, the age at which birds first started to breed, clutch size, and breeding success. Lack's study of European Robins (chapter 5) was among the first to use ringed birds to obtain information about mortality. He calculated that three-fifths of adult robins die each year, a figure other ornithologists considered ridiculously high; indeed, Lack was laughed at when he presented this at a wartime meeting of the British Ornithologists Club. Only Arthur Landsborough Thomson (chapter 4) believed Lack, probably because his interest in bird ringing had allowed him to reach similar conclusions. Only when ringing became an integral part of ornithological studies did researchers begin to realize how many birds actually die each year.

WYNNE-EDWARDS, LACK, AND RECONCILIATION

Lack's studies of the survival rates of several species of British birds, conducted during the 1940s, came to the startling conclusion that, after the first year of life, the annual survival rate remained more or less constant. In America Margaret Morse Nice (1937) had reported a similar pattern in her studies of Song Sparrows, as did Don Farner (1945), in his study of American Robins. This pattern, which was found to be generally true of birds (e.g., review by Deevey 1947), contrasted with the survival patterns seen in most other animals. In externally fertilizing fish, for example, mortality is extremely high among very young animals, and with long-lived mammal species like ourselves, annual survival after a period of high infant mortality remains similar until old age, when it decreases rapidly. Although Lack was broadly correct about survival rate being more or less constant with age, this was partly an artifact of the small numbers of relatively old individuals available in such studies. Mortality *had* to increase in old age, otherwise, as Daniel Botkin and Richard Miller (1974) pointed out, some individuals would be immortal. The problem was that the required age-dependent rate of annual mortality was so small—for example, just 0.0028 for the Northern Royal Albatross (Botkin and Miller 1974)—that demonstrating a decrease in survival in old age required very large sample sizes. It was several years before sufficient data had been accumulated to confirm this effect (e.g., McDonald et al. 1996; Newton and Rothery 1997).

Lack's conclusions included a reaffirmation of his ideas on clutch size (chapter 5), and a powerful reminder of his commitment to density dependence. Adult mortality rates are higher in species in which the reproductive rate is high, which Lack considered "a simple consequence of population balance through density dependent regulation of the mortality."[68] There was no evidence for "prudential restraint" as proposed by Wynne-Edwards, since there was little evidence that the number of young produced in any year affected the density of breeding pairs.

Another point was the age of first breeding. Most passerines begin to breed at one year of age, but storks and seabirds (and a few passerines, even) at two to seven years. Lack claimed:

> There is no evidence for the view of Wynne-Edwards (1955, 1962) that such deferred maturity has been evolved through group selection in long-lived species to reduce the number of young and so prevent over-population. There

is equally no proof of my alternative view that, in such species, breeding is difficult and individuals which try to breed when younger than the normal age leave, on the average, fewer not more surviving young than those that start later. My view depends on two conditions: first, that young parents are less efficient at raising young than older parents, of which there is suggestive evidence in several species . . . and secondly, that breeding somewhat lowers the chances of survival of young adults, of which there is as yet no evidence, though it seems not unlikely on general grounds.[69]

The idea of a possible trade-off between survival and reproduction later became a key feature of life history studies (chapter 5). Lack also felt that food shortage was probably the main density-dependent mortality factor, although he acknowledged that certain groups of organisms—gallinaceous birds, deer, and phytophagous insects—were largely limited by predators and parasites (Lack 1966: 287).

Lack's criticism of Wynne-Edwards's book meant that their relationship became somewhat frosty. However, early in 1968 Lack asked Adam Watson if he might help to break the ice. The Lack family was interested in trying to find a rare alpine plant, the brook saxifrage (*Saxifraga rivularis*), in the Cairngorms, and asked Watson if he could arrange for Wynne—who was an excellent field botanist—to accompany them. In July 1968 the Lacks traveled north, where they met up with Watson and Roger Bray (another game bird researcher). As Watson drove them up Glen Quoich, Lack impressed everyone by spotting a Eurasian Wryneck—a rare bird in Scotland. On meeting up with Wynne, they set out in search of the saxifrage. As Bray later recounted, "We all knew that it was an historic occasion and I well remember that the two protagonists enjoyed a relaxing day."[70] As the group finished their picnic lunch, Lack tentatively alluded to "competition for limited food resources," but all mention of the controversy evaporated almost immediately when Lack spotted a group of nine Eurasian Dotterel. The search for the saxifrage continued, and prophetically, perhaps, it was the Lacks and Watson who found it rather than Wynne-Edwards. Even so, the meeting was a success, and Lack returned to Oxford singing Wynne's praises.[71]

In February 1968 the British Ecological Society had accepted Wynne-Edwards's suggestion that his department hold a symposium to be titled "Food Resources and Animal Numbers." David Jenkins drew up the program but found it difficult to decide on the order of speakers. With such a high level of perceived controversy between Lack and Wynne-Edwards, should he explicitly set it up so that they were "against" each other, or try to be more tactful? The following letter[72] from Lack to Wynne-Edwards—reproduced in full—provides an interesting angle on the perceived controversy between the two men:

> Dear Wynne,
> The programme drawn up by David Jenkins for the B.E.S. Symposium next March keeps changing the role that I am supposed to play. Originally I was asked to open, and later to provide a concluding summary, which is more to my liking as I have nothing new to say. I now find they have put you and me together for the last afternoon and anyone reading this programme will conclude that we are to take part in a public controversy. I do not know what you would think about this, but I would hate to do so. I do not think that it would be appropriate at this particular meeting. I have said what I wanted to say on this subject (I mean, your views when they differ from mine) and I have thought of nothing fresh to say, so think it much better from my viewpoint that I should now leave it alone. I am not, of course, suggesting in the very least that you should not discuss this topic, but only that I, myself, would rather not do so, particularly if it meant a public dispute. I have, therefore, written to David Jenkins, asking him to think again. This new proposal was not put to me at all beforehand, and if I had been asked to do this in the first place, I should have refused.
>
> My difficulty now is that, even if we avoid controversy completely, other people will think, from the way the programme has been framed, that this is what is going to happen and human nature being what it is, will be disappointed if we don't argue.
> Yours ever,
> David

Clearly, Lack preferred to avoid rather than encourage public dispute. He did not want an argument, and, with no new data, was reluctant to rake over old ground. The symposium took place in March 1969 and Wynne-Edwards gave the final talk. Continuing to defend his theory, Wynne-Edwards ended on an exhortation, and Adam Watson remembers Mick Southern, who was sitting next to him, shaking his head and groaning, "Oh Wynne," thinking he had gone too far.[73]

GROUSING: WYNNE-EDWARDS AND BEYOND

The red grouse, central to Wynne-Edwards's ideas, has since Victorian times been among Britain's most economically important game birds. Grouse shooting along with deer stalking are the British equivalent of big game hunting, conferring status on the hunters and generating important income for landowners. Grouse numbers in Britain started to decline in about 1880, and such was the concern that in the early 1900s Lord Lovat established the Committee of Inquiry on Grouse Disease, whose findings became one of the first population studies of birds: *The Grouse in Health and in Disease* (1911).

The decline continued and was particularly noticeable from around 1930, when numbers were so low that shooting became almost uneconomic. Worried, in 1956 the Scottish Landowners Federation asked the Nature Conservancy (a government body) to investigate. They declined, but Wynne-Edwards volunteered, and the following year he appointed David Jenkins as a senior research fellow to undertake the study. Jenkins sought out Adam Watson, who was then a schoolteacher, and persuaded him join the team. Although Wynne-Edwards left the design and running of the study to Jenkins and Watson, his motivation for initiating the project was that it allowed him to test his idea that grouse regulated their own numbers through the social conventions of dominance and territoriality.

Jenkins and Watson confirmed that male red grouse compete vigorously for territories in the autumn and those failing to acquire a territory usually died—either from disease or predation—before the next breeding season. Population density in the grouse appeared to be determined by the aggression of the territorial males: the greater the aggression the larger the territories, and the lower the breeding density. Experiments, in which territorial males were removed, showed that nonbreeders rapidly took over empty territories, demonstrating that they were capable of breeding and hence had been prevented from doing so by the territorial behavior of established males.

To Wynne-Edwards, these results were entirely consistent with his ideas: "No one can doubt what is controlling the grouse numbers. They are doing it themselves. The cocks decide, by their individual contests in October and their subsequent acceptance of females, who are to live and who are to die."[74]

However—as Wynne-Edwards knew—for his hypothesis to be correct, food had to be the limiting factor, and initially at least this seemed unlikely. Red grouse feed on heather, and in their moorland habitat they appear to be surrounded by food, making it difficult to imagine how it could be limiting. Jenkins and Watson's study provided a clue: in the two years when grouse territories were noticeably larger, the heather was in particularly poor condition. In 1963 Robert Moss, a biochemist, joined the team and found that despite an apparent superabundance of food, grouse eat only the nutritious tips of the heather, and because the quality of the heather can vary both between years and between areas, food could indeed be limiting. Experiments in which moors were carefully fertilized provided convincing evidence that, as heather quality improved, territory size decreased and population density increased. Wynne-Edwards had a vivid demonstration of the efficacy of this experiment as he flew south from Aberdeen one summer's day in 1965, where from the aircraft window the fertilized side of the moor blazed with blooming pink shoots, whereas the other side was dull brown.

By the late 1960s the views of Wynne-Edwards and Lack were both divergent and convergent. Both agreed that grouse numbers were regulated, that regulation was by food, and that it occurred in a density-dependent manner. Where they differed was in the mechanism by which the regulation occurred. For Wynne-Edwards it was through the social convention of territory. By suggesting this, he implied that territoriality had evolved specifically to regulate populations, distributing individuals across suitable habitat in such a way that they did not overeat their food supply. In other words, to him territorial behavior had evolved for the good of the species. An almost identical view had been expressed many years earlier by the Irish ornithologist Charles Moffat (1903), who Wynne-Edwards cites extensively in his book. Moffat's idea stemmed from his reluctance to accept that for a population to remain stable as many as 90 percent of young birds must die: "I am altogether unable to find grounds for believing in so great a death-rate," he said. "Territory resulted in such a parcelling out of the land as must limit the number of breeding pairs to a fairly constant figure, and prevent indefinite increase . . . at the same time condemning the less powerful individuals to unproductiveness rather than death."[75] Moffat, like Wynne-Edwards, simply could not accept the fact that nature was red in tooth and claw. In contrast, Lack's view of territoriality was that it had evolved because selfish individuals should acquire as much food as possible in order to reproduce as rapidly as possible.

Despite two decades of research by Watson and Jenkins, by the early 1980s Scottish landowners were increasingly unconvinced that grouse numbers were regulated entirely by territorial behavior. If it was true, the killing of raptors, corvids, and foxes was a waste

Adam Watson (photo in 2009 at age 79).

of time. Fearing for their livelihoods, the gamekeepers couldn't accept that predators had no role in limiting numbers. Nor could landowners. The fact that predation by raptors had been virtually ignored up until this point was due to the numbers of birds of prey in Britain being at a very low level when the grouse project started, having been virtually eliminated over much of the country by persecution and the use of organochlorine pesticides (chapter 11). What about sheep ticks, they asked? It was well known that ticks acted as vectors for the louping-ill virus, which causes serious illness in sheep and high mortality in grouse. It was also known, since the work of Thomas Spencer Cobbold (1873) and Lord Lovat (1911), that grouse suffered from another parasite, a nematode that causes the disease strongylosis.

Dick Potts, who had worked extensively on the Grey Partridge, had noted the occurrence of strongylosis at high partridge population densities and wondered whether something similar occurred in red grouse when their populations became artificially high as a result of low predator numbers. Potts became research director of the Game Conservancy in 1977, which initially (at least) worked amicably with the Scottish red grouse team. But in February 1978

373

Lord "Willie" Peel, who had a grouse moor on Gunnerside in northern England, asked Potts to initiate an independent project, the North of England Grouse Project. Potts, together with Peter Hudson, started an investigation the following year, but they found increasing evidence for the role of strongylosis and little evidence for the role of territorial behavior. As a result, relations with the Scottish researchers deteriorated. Relationships declined further when the Natural Environment Research Council (NERC), the main government funding body, openly questioned why it was necessary to have two teams studying the same problem. As Dick Potts put it, "At its most basic our team were finding inimical effects on grouse arising from their environment and relevant to management. Their team was invoking cycles, social consequences of kinship, effects of hormones and suchlike all a bit reminiscent of Wynne-Edwards although they hotly denied this."[76]

In the 1980s, several years after Lack's death (but with Wynne-Edwards still going strong), the grouse wars entered a new phase. With Hudson as their main grouse researcher, the Game Conservancy's explicit focus was on the idea that strongylosis controlled grouse numbers. Conversely, the Institute for Terrestrial Ecology at Banchory, Scotland, continued to focus on territorial behavior and the role of aggressiveness in determining population density.

An important and unusual feature of red grouse populations, first recognized by Lovat (1911), was that numbers fluctuate in a cyclical manner, peaking every four to ten years. Any explanation for the natural regulation of grouse numbers had therefore to take these oscillations into account. The crux of the issue was this: to explain population cycles, density-dependent effects had to operate with a delay, giving the populations time to oscillate widely, rather than being damped on a year-to-year basis. Whatever was causing the cycles had to include this delayed density dependence. Hudson, inspired by the influential theoretical models of Anderson and May (1978), recognized that parasites could create the necessary delayed effect and drive population cycles.[77]

In a series of ambitious field experiments, in which grouse were dosed with an antihelminthic drug, Hudson showed that treated female grouse produced more young, and that with a reduction in the parasite burden, population declines were reduced as well (Hudson et al. 1992, 1998). These results flew in the face of the Scottish researcher's ideas that cycles were driven by between-year variations in male aggressiveness and therefore in territorial spacing behavior—for them, parasites were a consequence of population cycles, rather than their cause. Intriguingly, Steve Redpath and colleagues (2006) replicated Hudson's population-level antihelminth treatments in both Scottish and English populations of grouse and found that the effect of parasite removal was stronger in England than Scotland. This suggested that parasites may play a greater role on English moors than Scottish moors, which may in turn explain some of the differences in interpretation between the two studies.

These opposing viewpoints generated strong emotions, and so acrimonious were the relations between the two research groups that in December 1989 the British Ecological Society decided to get them together to thrash out their differences. The meeting took place at Silwood Park near Windsor, with the leading British ecologists—Bob May, John Lawton, Dick Potts, Ian Newton, John Krebs, and Mick Crawley—in attendance. It was an intense couple of days, and the overall consensus was that Hudson had "won" the debate, or at least his arguments were more convincing (Krebs and May 1990). But, as Ian Newton pointed out,[78] Hudson had prepared his material better and was more open to criticism than were Watson and Moss. Moreover, Hudson had framed his ideas around the new and influential disease theories of Roy Anderson and Robert May.

The grouse wars entered yet another phase in the 1990s, when Stuart Piertney at Aberdeen University was able to test Moss and Watson's (1991) "kinship hypothesis." This hypothesis proposed that changes in the levels of aggression through a population cycle—driven by changes in kin structure—accounted for grouse population cycles. There was already good evidence that more-aggressive males (supplemented with testosterone) defended larger territories and that aggressiveness could therefore play a role in determining population density (Moss et al. 1994). But how could this cause cyclical fluctuations? At low population density, Moss and colleagues suggested, territorial males are less aggressive to related males than to unrelated males, facilitating the recruitment of kin into the population and the creation of clusters of related males. As population density increases, the benefits of helping kin in this way are offset by the costs of increased competition; aggression increases and the population starts to decline. Using molecular techniques to ascertain kin structure and using comb size as an index of aggressiveness, Piertney et al.

(2008) confirmed that both varied through a population cycle in exactly the way predicted by the hypothesis.

The red grouse story is a long and convoluted one; many individuals' careers have hung on it, and the debates between the different factions have been as aggressive as any in ornithology. What is remarkable is that the "answer," the truth of what regulates certain grouse populations, lies partway between Lack and Wynne-Edwards's disparate views. Group selection has no role, but kin selection does appear to be involved. The story is not yet over.

AFTERMATH

The Lack and Wynne-Edwards controversy dominated British ornithology throughout the 1960s. It was more than a simple disagreement over how bird numbers are regulated; the debate struck at the very heart of biological principles, begging the question of the level at which selection operates: groups or individuals. Individual selection won, giving rise on the one hand to behavioral ecology (chapters 8 and 9)[79] and on the other to a new style of ecology. Yet while Lack and Wynne-Edwards were battling it out, the rest of avian ecology did not stand still, and in particular there were great advances in North America, as we discuss below. From the 1960s, the study of avian ecology expanded enormously—much of it motivated and informed by the slowly unfolding revolution in evolutionary thinking. This ornithological expansion is nicely captured by a comment by John Coulson in the introduction to his monograph on the Black-legged Kittiwake: "In the early 1950s the prospects of being employed as professional ornithologists were poor, with perhaps no more than a handful of people so employed at that time in Britain. How things have changed in 60 years!"[80]

The 1960s saw a new era in population studies of birds, driven largely by the expansion of European and North American university systems. Student numbers more than doubled during the 1960s (Perkin 1972), and with more people at university, and more studying subjects like zoology, it was almost inevitable that there would be more research on birds. And there was.

To see how this change came about we need to go back to the early years of the twentieth century. The Laboratory of Ornithology at Cornell University was founded by Arthur A. Allen in 1916 and was the only institution in the United States to have ever offered a PhD in ornithology.[81] Students flocked to Cornell, and by the time Allen retired from teaching in 1953,[82] more than ten thousand students had taken his undergraduate and summer school courses, and more than a hundred had been awarded advanced degrees (Pettingill 1968). Allen's popular films, books, and lectures made him one of the most significant figures in the expansion of American ornithology, and, indeed, many of his students went on to establish their own ornithological research groups. A little later—during the 1940s, 1950s, and 1960s—David Lack did much the same in Britain. Through his scientific papers and highly readable books, and particularly through his annual student conference on bird biology, Lack inspired generations of ornithologists. Visitors were drawn from across the world to the Edward Grey Institute to meet Lack and his students, many of whom later became academics and researchers at other institutions, fueling a massive increase in the output of ornithological science.

From the late 1960s, research on the population ecology of birds started to expand on three broad fronts: (1) life history studies, (2) long-term studies, and (3) what we loosely refer to as "new" areas of ecology. We concentrate on life history studies in chapter 5, so we focus here on the other two topics.

LONG-TERM STUDIES

At around the same time that Lack was watching robins in southern England, Margaret Morse Nice was undertaking a similar study of Song Sparrows in Ohio. Interested in birds since childhood, Nice conducted a detailed investigation of the behavior, breeding biology, and survival of this species during the 1920s. Her results were so novel that she had difficulty getting them published in the United States, but her mentor, Erwin Stresemann, helped her to publish them in *Journal für Ornithologie*, which he edited, in 1934. Nice later summarized her findings in *Studies in the Life History of the Song Sparrow* in 1937.

Further studies of individually marked birds followed, including Lack's studies of Great Tits; George Dunnet's study of Northern Fulmars in Orkney (starting in 1950); John Coulson's study of Black-legged Kittiwakes (starting in 1954); Ron Murton's study of Common Wood Pigeons (started in 1955); Jamie Smith's Song Sparrows on Mandarte Island, Canada (started by Frank Tompa in 1959); and Fred Cooke's study of the Snow Goose at La Pérouse Bay, Manitoba, Canada (started in 1968; chapter 2). Abundant research funding during the 1960s allowed these studies to flourish, and at that time there was also a boom in fundamental questions about population ecology and the evolution of social behavior. These topics were rapidly diversifying, and it became clear that answering them was possible only with long-term studies of marked birds. This encouraged more researchers to start their own long-term investigations.

Perhaps the single most important generalization that arose from these population studies was the confirmation of density dependence. Across a range of life history traits—including clutch size, the incidence of second broods, breeding success, fledging mass, recruitment, and juvenile and adult survival—density-dependent effects were apparent (reviewed by Newton 1998a).

Ian Newton's thirty-year study of the Eurasian Sparrowhawk provides an elegant example. The study started because birds of prey were badly affected by pesticides (chapter 11), yet little was known about the population biology of raptors. Lack had even advised Newton against the study, saying that birds of prey would be too difficult, and indeed—compared with Great Tits breeding in nest-boxes—they were. Newton's investigation, however, proved to be a model population study, confirming that sparrowhawk numbers were limited in a density-dependent manner by both food and nest sites and that density-independent factors also played a role in limiting numbers. The population fluctuated within narrow limits and was relatively stable, with increases in numbers in one year tending to be followed by a decrease in the following year, as one would expect with density-dependent regulation. However, the extent to which numbers fluctuated from year to year was determined by weather, and rainfall in particular. In dry years, breeding success

and adult survival were relatively high, and numbers tended to increase, whereas in wet years—which made hunting more difficult for the birds—both survival and breeding success were lower, and numbers declined (Newton 1986).

An innovative aspect of Newton's sparrowhawk study was his estimate of lifetime reproductive success (LRS)—the first for any bird. By the late 1980s there were sufficient long-term studies for Tim Clutton-Brock (1988) and Ian Newton (1989) to summarize data for about twenty bird species. Intriguingly, most of these studies revealed similar patterns to those seen in the sparrowhawk— the longer an individual lived, the greater its LRS, and only a small proportion of individuals in a population contributed to the next generation. In the Eurasian Blue Tit, for example, a mere 3 percent of those that survive to fledge produce 50 percent of the next generation's offspring (Newton 1995).

This result raised the question of what determines how long an individual survives: is it due to stochastic (random) processes, or are there inherent differences between individuals? These inherent differences could be genetic, or they could be nongenetic differences—what are now called "maternal effects," in which, for example, females may provision eggs with more or less nutrients and thereby influence the eventual success of their offspring (Schwabl 1993; Groothuis and Schwabl 2008). The notion that there are individual differences in quality and/ or attractiveness was pursued by behavioral ecologists—with mixed success (chapter 9).

Newton's study confirmed for the sparrowhawk what Lack believed to be true for most birds—that food was the single most important factor limiting populations. For

Lack, predators and disease were relatively unimportant, although his views were based on logic rather than hard data. Newton was later able to test the predation idea empirically, by comparing songbird numbers in southern Britain in three periods: (1) before the pesticide-induced decline in sparrowhawk numbers (in the late 1940s to 1960), (2) when sparrowhawks were absent (1960 to the early 1970s), and (3) then again when sparrowhawks were present (early 1970s to the 1990s). He found no evidence that sparrowhawks had any effect on the breeding densities of small birds (Newton and Rothery 1997).

On the other hand, in a landmark review of the factors limiting bird numbers, Newton found that the breeding densities of ground-nesting birds, such as grouse, partridges, and ducks, *are* affected by predators. In the majority of studies in which predators were removed, breeding success increased, although this was not always followed by an increase in either post-breeding numbers or subsequent breeding population sizes. Overall, he concluded that about half of all predator-removal studies resulted in an increase in breeding density, suggesting that, for ground nesting birds, predation had an important role in limiting numbers (Newton 1998a).

There is also little direct evidence that parasites and pathogens limit bird numbers. In part this is because parasites and pathogens are difficult to study, and in part because they usually operate in conjunction with starvation and other stress factors such as toxic chemicals to cause mortality and poor breeding performance. However, as we have seen, red grouse populations are affected by both louping-ill disease and strongylosis,

and there is evidence that epizootics may be the main limiting factor for some waterfowl populations in North America (Newton 1998a). Although ectoparasites are sometimes common and appear to have a negative effect on chick growth, as in the American Cliff Swallow (Brown and Brown 1986), there is no evidence that this limits subsequent breeding. Brood parasitism, however, can sometimes limit host numbers, as in the case of the Brown-headed Cowbird in North America. This species has spread dramatically in the last century, presumably as a result of changes in agriculture, and levels of brood parasitism have increased—for example, in the White-crowned Sparrow from 0 to 50 percent between the 1930s and 1990s, reducing breeding success in that species to a level that is likely to limit its numbers (see Newton 1998a: 281–82).

From the early 1970s on, there was a long, slow decline in research funding, and those committed to long-term population studies have had to fight to keep them going. This happened despite the fact that long-term population studies of birds have been disproportionately successful in terms of the quality of science they generate and hence in our understanding of avian ecology. They also provide data that can be used to address questions that no one had thought of when the study began, such as the effect of climate change (e.g., Charmantier et al. 2008).

A NEW ECOLOGY

When Robert MacArthur wrote to David Lack in 1962 urging him to ignore Wynne-Edwards, he did so in a spirit of friendship. MacArthur had spent a year at the Edward Grey Institute in Oxford on completing his

PhD at Yale in 1957. With a master's degree in mathematics from Brown University and a PhD supervised by the great founder of American ecology, G. Evelyn Hutchinson, MacArthur combined a love of quantification with natural history to create a new approach to avian ecology. His scientific genius was helped by an ability to form fast friendships with his collaborators. One of those, E. O. Wilson, described him thus: "He was medium tall and thin, with a handsomely angular face. He met you with a level gaze supported by an ironic smile and widening of eyes. He spoke with a thin baritone voice in complete sentences and paragraphs, signaling his more important utterances by tilting his face slightly upward and swallowing."[83]

MacArthur's main interest was in community ecology—the factors determining what combination of species coexist together in the same habitat, exemplified by his thesis work on five species of North American boreal forest warblers: Cape May, Myrtle, Black-throated Green, Blackburnian, and Bay-breasted. Most previous researchers had assumed that these warblers had very similar ecological requirements, and "thus it appeared that these species might provide an interesting exception to the general rule that species either are limited by different factors or differ in habitat or range (Lack 1954)."[84] In setting out the logic for his investigation, MacArthur acknowledges Joseph Grinnell (1922), for introducing the notion of the ecological niche, and cites Gause's (1934) more formal, theoretical approach. He also cites Charles Kendeigh's (1945) work on wood warblers in western New York State but does not seem to have noticed that Kendeigh said that "the warblers are so prominent in the

Robert MacArthur (photo in 1958 at age ca. 28).

mixed evergreen-deciduous forest because they fill so many of these diverse niches."[85] Nonetheless, MacArthur's way of quantifying niche use by the warblers—all feeding in coniferous trees—was highly innovative. He divided the trees into sixteen different zones, based largely on height above the ground and distance from the trunk, and showed that—contrary to the earlier assumption by some authors that they had identical roles in the community—each warbler species exploited trees in a different way, providing support for Lack's original idea that competition—both between and within species—was important in population regulation and community structure (MacArthur 1958).

The notion that competition between species is an important agent in natural selection was absent from early ecological texts (e.g., Hesse et al. 1937) and was actively disputed in others (Andrewartha and Birch 1954). Yet Darwin was clearly convinced of its importance: "We have reason to believe that species in a state of nature are limited

379

in their ranges by the competition of other organic beings quite as much as, or more than, by adaptation to particular climates."[86] The work of Lack and his students, based around Gause's theory, helped to bring that idea back into focus. MacArthur's subsequent studies as well as those of his mentor, Hutchinson, provide further evidence for the role of competition such that it became accepted as conventional wisdom. However, in the 1970s competition's central role was challenged and an active debate ensued (Schoener 1982). In this particular controversy, the unit of selection was irrelevant—those arguments had moved on, focusing instead on the relative importance of different factors in determining the composition of ecological communities.

Of those ecologists who opposed the idea of competition, one school argued against the importance of any kind of biological interaction and felt that communities were structured by physical factors; the other school advocated predation (rather than competition over food) as being important in determining community structure. The debate—which had broad implications for community ecology as a whole—was generated mainly by those studying birds. John Wiens (1977), for example, argued that interspecific competition is too intermittent or too rare to be important in natural selection. In contrast, Jared Diamond (1978) argued that competition was the driving force in natural selection. As Thomas Schoener (1982) commented, "Here is an example of the kind of controversy about how the world works that makes contemporary ecology simultaneously so exciting and so frustrating."[87] Schoener favored a compromise, emphasizing the error of focusing on single-factor

explanations: "Certainly we were never justified in thinking that the ecological world was so simple as to be largely explainable on the basis of a single interaction. . . . We are well on our way to developing a multifaceted theory to match what is clearly a highly diverse natural world."[88]

During the late 1950s and 1960s, Robert MacArthur—based first at the University of Pennsylvania and later at Princeton—revolutionized community ecology and the evolutionary ecology of birds. His new evolutionary approach to avian ecology was theory driven, quantitative, innovative, and immensely exciting. As Stephen Fretwell (1975) pointed out, before MacArthur's 1957 paper on the relative abundance of bird species, only about 5 percent of papers published in the journal *Ecology* tested predictions. By 1975 about half the papers contained tests of theoretical predictions.

MacArthur was happy to test his ideas with limited data, and his students soon followed suit. Thus, one of his major contributions was to provide a role model for the next generation of ecologists. Stephen Fretwell, one of MacArthur's students in the 1960s, said that "his contribution to the people who let themselves be positively affected by him was both to point the way for a radically different perspective on how ecologists should proceed and to get famous enough so that we could follow this direction without undue harassment."[89]

The result was an explosion of avian ecology studies that went far beyond the idea of population regulation while still retaining links with it. This "new ecology" showed some continuity with what Lack had started, but population studies became increasingly refined, both in execution but especially in analysis (e.g., Lebreton et al. 1992) and population modeling (e.g., Levins 1966; Caswell 1989). MacArthur's death from cancer at the age of forty-two in 1972 was a major loss not just to ornithology but to the rapidly expanding fields of population biology, life history theory, and theoretical ecology, all of which had benefited enormously from his genius.

Just as Lack's influence lives on in the Edward Grey Institute in Oxford and through the subsequent research of his many students, so MacArthur's continues at Princeton, and especially through the landmark series of Princeton Monographs in Population Biology. The series started with MacArthur and Wilson's enormously influential *Theory of Island Biogeography* (1967) and included Martin Cody's (1974) *Competition and the Structure of Bird Communities*; Glen Woolfenden and John Fitzpatrick's (1984) *The Florida Scrub Jay: Demography of a Cooperatively Breeding Bird*; and Walt Koenig and Ronald Mumme's (1987) *Population Ecology of the Cooperatively Breeding Acorn Woodpecker*.

That the major advances in this new ecology occurred in North America is perhaps not surprising, given that ornithologists there were not as preoccupied by the population regulation debate as were British researchers. In truth, there was a sense of bewilderment about the intensity of feeling surrounding population regulation, as Gordon Orians (2009) makes clear:

> Although I became an ardent proponent of density-dependence, not too many years elapsed before I came to regard the dispute as a tempest in a teapot. Too much of the argument centred on the definition of the term "regulation" that one adopted. Also,

I realized that density-dependence and density-independence were not mutually exclusive concepts. Fortunately, that debate has subsided, not because one side won, but because the debate came to seem less and less interesting to most ecologists. Nevertheless, thinking about that debate focused my attention on the importance of asking the right questions.[90]

Orians (1962) was the first to use the term "evolutionary ecology" in the sense that it is used today, and he strove to better integrate the fields of behavior, ecology, and evolution:

> During much of the past century many students of animal behavior paid little attention to the ecological theatre, even though an ecological perspective was central to Niko Tinbergen's approach to animal behavior. Ecologists, in turn, have had an uncertain relationship with the field of animal behavior. Behavioral ecology has been an important component of ecology, but population ecologists have little incorporated the rich results of behavioural ecological research in their models.[91]

CODA

The original motivation for studying populations was to understand what causes bird numbers in a particular area to increase, decrease, or remain stable. The question was addressed at several levels. The first concerned population parameters: birth rates, death rates, emigration, and immigration all influence numbers, and ornithologists had to devise methods for censusing birds and measuring these parameters (average values derived from individually marked birds). Armed with these, they could then perform the necessary calculations to predict population trajectories. Information on numbers and population parameters is now routinely collected—often by "citizen scientists"—by organizations such as the Cornell Laboratory of Ornithology in the United States and the British Trust for Ornithology in Britain. Censuses are also an important aspect of most long-term population studies described in this chapter. Long-term studies have been disproportionately successful because they allow researchers to really get to know their birds, to appreciate year-to-year variation, and to witness long-term changes in their biology. For these reasons they are also much more likely to provide the "right" answer to biological questions than most short-term projects favored by government research grants.

The second level focuses on the *way* numbers are controlled or regulated—what, for example, keeps a population stable? The term "control" refers to any factors that affect numbers, either upward or downward. Regulation, on the other hand, explicitly assumes that there is a ceiling or equilibrium level on numbers, imposed by competition for resources via density-dependent processes. Do populations fluctuate widely, or are numbers regulated and kept within certain limits? The stability of some bird populations implied that they were regulated.

There is a third level of explanation, epitomized by the debate that dominated ornithology in the 1960s: both sides agreed that bird numbers are regulated, but while Wynne-Edwards favored intrinsic processes,

Lack favored extrinsic processes. Notwithstanding Orians's comment (above), Lack eventually won and emerged as a hero of twentieth-century ornithology, not just because he resolved the way populations were regulated, but because he made others aware of the level—the individual—at which selection operated.

A fourth level concerned the extrinsic factors that regulated numbers: was it food, predators, or disease? Lack, for all his extraordinary biological intuition, always favored single-factor explanations and decided that competition for food was what regulated bird populations. Research since the 1960s—especially from the relatively small number of long-term studies—has made it abundantly clear that ecology is rarely if ever "simple," and that single-factor explanations are likely to be naive.

Controversy is part of the scientific process. That between Wynne-Edwards and Lack was conspicuous, although it was not as acrimonious as the gossip suggested: both parties behaved in a civilized way toward each other. Controversy usually results in the polarization of views, with the true answer later discovered to lie somewhere in the middle. What made population regulation especially difficult was that there was no middle ground and the two viewpoints could not be resolved by an experiment. Distinguishing group selection from individual selection relied partly on logic and partly on evidence from other areas of biology. Lack's defeat of Wynne-Edwards and group selection heralded a new era of avian population ecology, including behavioral ecology (chapters 8 and 9).

BOX 10.1 Ian Newton

I first became interested in animals as a boy, and as I grew this interest became increasingly focused on birds. No one in my family had any academic leaning, but my father had a mild interest in birds, and both my parents encouraged me. They saw my interest merely as a youthful passion, something I would grow out of, eventually becoming a medical doctor. We lived in the large industrialized village of New Whittington, just north of Chesterfield, but our house bordered on open countryside which I could explore. Beyond the age of twelve none of my school friends shared my interest, so much of my birding was done alone, without binoculars. As a teenager in the 1950s, I was dismayed that many bird-rich localities were being cleared or developed in some way and that birds were disappearing. This realization may have subconsciously influenced two of my main interests in later life, namely in the factors that

affect bird numbers and in their conservation. In my early teens I developed a special interest in Eurasian Sparrowhawks, and within a few years I knew of about ten nests within easy cycling distance. I used to check the nests to see how they were doing and record prey remains, for no other reason than curiosity. A few years later my father started to keep canaries, allowing me space in his self-built shed to keep a few finches. This stimulated an interest in these birds, especially in their feeding habits, which I could study in the wild by watching which plants they fed from, and also in captivity by bringing in various plants, to see how different species extracted the seeds. By the time I was eighteen I probably had enough data for a PhD on the subject, but I had no clue on how to analyze or present the data or even if anyone would be remotely interested. In any case, I had never heard of a PhD.

I was educated at Chesterfield Boys' Grammar School and then read zoology at Bristol University. In my first term a notice advertised a student conference at the Edward Grey Institute in Oxford, inviting students to attend and speak. I offered a talk on Eurasian Bullfinch feeding habits, which seemed to go down well. The director of the Institute, David Lack, told me I should try writing a paper on the subject. I did this, and he helped me. The paper was published in *Bird Study*, but I cannot look at it now without feeling ashamed at my efforts. Everything is accurate, but it is entirely nonquantitative. I went to later EGI conferences and spoke on other finches, and when the time came, David Lack agreed to take me on as a DPhil student to study finches. This led on to a postdoc on bullfinches, on the factors that influenced their numbers, and the damage to fruit buds they were causing at the time. The EGI student conferences were immensely important to me, meeting people like Arthur Cain and Mike Cullen, and hearing the way in which ideas were discussed. I continued to attend, and speak at, most of these conferences for another forty years. They continue to this day, more than forty years after Lack's death.

Of all the books I read as a student, the one that most impressed me was Lack's *The Natural Regulation of Animal Numbers* (1954). To me this book was a revelation, simply and succinctly written, and I still refer to it today. Another person who influenced me greatly at the EGI was Reg Moreau, the amusing and lovable doyen of African ornithology, whose editorial skills were legendary. The way I write now was influenced by the lessons I learned from having Reg go through my drafts with his blue pencil.

The six years I spent at the EGI were crucial in shaping my interests in bird populations. It was a time when Wynne-Edwards's book *Animal Dispersion in Relation to Social Behaviour* was published, stimulating much discussion and controversy, and Lack wrote his book on *Population Studies of Birds* (1966), the chapters from which he showed me as they were typed.

On leaving Oxford I took up a post in Edinburgh with the Nature Conservancy to work on waterfowl, a group totally new to me. In winter I studied geese on farmland, partly to assess crop damage, and in spring-summer I went to the Outer Hebrides to study the relict population of native Greylag Geese resident there. I was fortunate in having David Jenkins as my boss, because he gave me freedom to follow my own interests. He was also, and still is, a great critic of papers, having read

much of what I have written in the last forty years. While working in Edinburgh, I met Hugh Boyd, who arranged for me to have a year in Canada, learning about waterfowl. I spent a fascinating summer in the Canadian Arctic, working with Fred Cooke and Charlie MacInnes on geese, and later on prairie ducks with Alex Dzubin and others in Saskatoon.

On returning to Britain, I continued with waterfowl but had the chance to work on birds of prey, which were declining at the time, owing to organochlorine pesticide poisoning. Sparrowhawks were one of the most affected species, yet in western Scotland they were still numerous enough to provide large samples. Drawing again on my boyhood experience, I was thrilled at the prospect of studying them. This work became a long-term study in which I could show that breeding densities in woodland varied systematically between regions, in association with soil fertility and the densities of songbird prey species. It also revealed a relationship between territory occupancy and nest success, and gave new information on age-related trends in breeding and survival, senescence, and lifetime reproductive success. During this time, Derek Ratcliffe, who had a passionate interest in Peregrine Falcons and was a key figure in the organochlorine battles, became the chief scientist of the Nature Conservancy Council. He became a close friend, who helped and influenced me in many ways.

I enjoyed my years on sparrowhawks, not just because of the excitement of discovering new things, but because of the field skills involved in finding the birds and the sheer physical challenge, involving up to twenty tree-climbs per day at the peak. Because of the problems caused by DDT and other organochlorines, research on birds of prey blossomed during the 1970s and 1980s, not so much in Britain, but certainly in continental Europe and North America. On the back of this burgeoning interest, my book *Population Ecology of Raptors* (1979) achieved a status it might never otherwise have reached.

Over the years, interest in birds became less species oriented and more ideas oriented, and the need for computing and statistical skills gradually took precedence over the need for field skills. The days of long-term population studies, of the type I was privileged to undertake, may be coming to an end, at least for professionals, partly because of the difficulties of getting such work funded consistently. But a new era is already upon us, based on the large-scale data sets assembled over the years by organizations such as the British Trust for Ornithology, based on the collective efforts of skilled volunteers. This has enabled questions to be addressed on much larger spatial scales, and the emphasis has shifted more to issues of conservation concern, such as the effects of modern land-use practices and climate change on bird populations.

BOX 10.2 Peter Hudson

For my fifth birthday, my parents gave me the *Collins Guide to the Birds of Britain* and my future career path was sealed; from that point on I wanted to spend my life understanding animals. I became fascinated by reptiles and at the age of twelve had the audacity to write to the British Herpetological Society requesting membership. The society dropped their minimum age by nine years, and I attended meetings every month for many years. I gave my first presentation on tortoise hibernation at the age of thirteen in the room of the Linnaean Society where Darwin and Wallace's paper on natural selection was read.

While the society meetings were wonderful, they were only once a month, and I craved the company of other biologists. I went on to read zoology at the University of Leeds, where I took every course I could on ecology and made a point of avoiding parasitology. I became interested in ecological entomology and made friends with Stephen Sutton, a lecturer in zoology and keen entomologist. Steve was invited on the Zaire River Expedition in 1974, and he took me along as his research assistant. We designed aerial moth traps and specialized sticky traps and published some of the first data on the vertical distribution of insects in rain forests. Ornithology was my hobby, and when I returned I joined Peter Gladstone and Sir Peter Scott to help establish a new wildfowl and wetland reserve at Martin Mere in northwest England. Digging ditches and watching birds in the snow and rain was good, but I wanted to do ornithological research, and so I applied to do my doctorate at the Edward Grey Institute in Oxford and moved there in the summer of 1975.

At long last I had everything I ever wanted: a group of comrades interested in talking about birds and biology and all working on interesting things. My supervisor was Chris Perrins, and I undertook fieldwork on the island of Skomer (west Wales) where I examined the population biology and behavior of auks. On Skomer—a wonderful secluded location, surrounded by birds—I was totally in my element. Among other things, I undertook experiments on parent-offspring conflict and showed that the parents didn't force Atlantic Puffin chicks to fledge as previously supposed: fledging date was determined by the chicks, but the provisioning rate was shaped by the adults.

After finishing my DPhil, I went to Panama to extend the studies I had started in Zaire looking at inflorescence size in hummingbird-pollinated plants. By this time my passion was for simple and elegant experiments that tested an important aspect of theory. While in Panama I heard about a position to examine grouse population dynamics, and within a couple of months I had got married, joined the Game Conservancy, and established a research base in North Yorkshire on Lord Peel's Gunnerside Estate. I spent many evenings talking with keepers and Lord Peel, and poring over the extensive grouse literature, and my days watching the birds. I became convinced that spacing behavior was not driving the population cycles in grouse through the processes the literature recorded; natural enemies and particularly parasites were having a more obvious impact. In Oxford I had

shared an office with Andy Dobson, and he pointed out a number of key papers on ecological parasitology; then Roy Anderson and Bob May published a nice modeling paper that identified the conditions when parasites could cause population cycles. The timing of the paper was perfect and provided me with a framework to test, so I started a series of individual level experiments and showed that parasites (1) reduced host fecundity, (2) did not have a big impact on grouse mortality, (3) had a distribution in the host population that was close to random—and that these features together could lead to instability. Andy and I expanded the Anderson and May models and showed that the experimental data and observations could account for the cycles. We could explain what we saw by integrating individual responses, but now we needed to test this at the population level and see if we could stop the cycles. Just as we began this experiment, the grouse debate started.

I had met both Adam Watson and Robert Moss on several occasions, and they had heard me talk about my ideas and experiments. We all accepted that ecological factors would act differently in different parts of a bird's distribution, but when the disparity between our studies was highlighted in the Scottish Sunday newspapers, there were questions in Parliament about government funding of research on grouse and concern that if natural enemies were important to grouse this may encourage the persecution of raptors. I was little more than an upstart of a postdoc, and these guys had been undertaking research on grouse since Wynne-Edwards had established the research in 1953—the year I was born. My feeling was that spacing behavior could operate through kin selection, but since fecundity was determined by parasites the two would interact and spacing behavior would accentuate the role of parasites.

In an attempt to resolve the differences between the two grouse teams, a scientific meeting was convened. It took place in December 1989 behind locked doors at Silwood Park with an audience of senior scientists—mainly fellows of the Royal Society. The Game Conservancy told me minutes before the meeting started that if the meeting went badly, I would lose my job! The meeting generated some good discussions, but there didn't seem to be any firm conclusions. It was clear to me that all that was needed was good science.

Soon after, I undertook a population scale experiment that effectively stopped the population cycles in four different populations by simply removing the parasites. This was the first clear experimental evidence to show the causes of population cycles. I developed a management tool sold as "medicated grit" to control worms in grouse, which is now used extensively across grouse populations and has effectively stopped the large grouse population crashes observed throughout the Pennines.

I moved the focus of my research base to Scotland in 1985 and showed that our fundamental understanding of the intimate relationship between parasite and host can provide insights into the interaction with predators, other infections, and food quality and could explain the spatial variation in grouse dynamics across its distribution. In 1995 I moved to the University of Stirling and expanded my research on parasites and birds. I worked on shared parasites between hosts and showed that the release of pheasants could cause the local extinction of partridges through parasite-mediated competition. I then moved to an endowed chair at Penn State University in the USA in 2002 and established the Center for Infectious Disease Dynamics. Subsequently I was asked to run life sciences at the university. While my day-to-day task is to provide science leadership, I still work on parasite-host systems. We are examining the invasion of mange into the newly established wolf population in Yellowstone National Park, pneumonia in bighorn sheep, and a similar disease in desert tortoises. I spend my spare time looking after my ninety-acre nature reserve and photographing birds.

THE EXTINCT GUADALUPE CARACARA   Ralph STEADman 2011

CHAPTER 11

Tomorrow's Birds

Birds are among the best indicators of a healthy environment.
—ERNST MAYR, IN STRESEMANN'S (1975: 396) *ORNITHOLOGY: FROM
ARISTOTLE TO THE PRESENT*, POINTING OUT WHY ORNITHOLOGISTS
WERE AMONG THE EARLIEST CONSERVATIONISTS

THE TRAGEDY OF SHIFTING BASELINES

SPIX'S MACAW IS ONE OF THE LARGEST AND most spectacular parrots in the world. It is also balancing on the brink of extinction, and in the wild it is almost certainly extinct. A victim of habitat loss in its native Brazil and a corrupt global cage-bird trade, Spix's Macaw currently (2013) consists of around just eighty-five captive individuals held at five locations around the world. Most are derived from a single breeding pair and are so inbred they are effectively clones of each other: clones with an unlikely future.

Discovered by Johann Baptist von Spix in 1817, the species was already in decline—the result of habitat destruction and exploitation along the Rio São Francisco corridor. Yet as it became rarer, Spix's Macaw became more attractive for bird dealers—rarity has always

bred demand—and it was this secondary cause that effectively sealed its fate. In 1978 Spix's Macaw was listed as "vulnerable" in the IUCN's Red List,[1] and just ten years later only three or four individuals were thought to be left in the wild. The rapid decline in the 1970s and 1980s was engineered by just two bird dealers known to have taken twenty-three individuals from the wild, which they sold for about $10,000 apiece. By 1990 only a single bird, a male, remained in the forests of Brazil. In a desperate attempt to save the species, a captive female was released nearby in 1995. Ironically, but fortuitously, this female was thought to have been the male's partner prior to her capture six years earlier. Just six weeks after the pair was reunited, the female disappeared and was later found dead

In 1900 Rollo Beck (chapters 2–3) collected nine individuals of the already rare Guadalupe Caracara from a flock of eleven he encountered on Guadalupe Island, Baja California. There was one further, unconfirmed sighting of this species in 1903, and the species was certainly extinct by 1906. Painting by Ralph Steadman.

beneath power lines. The male was last seen on 5 October 2000, and today, Spix's Macaw is listed as Critically Endangered (Possibly Extinct in the Wild).

Spix's Macaw epitomizes many of the problems of bird conservation, from the causes of its decline to the valiant efforts of those committed to saving it through captive breeding and ultimately, it is hoped, through the release of captive-bred birds. But Spix's Macaw is far from alone in being endangered: bird populations across the world are in decline. In 2004 it was estimated that one-fifth of all bird species are "extinction-prone" and 6.5 percent are functionally extinct. Even though only 153 species (1.5 percent) of all birds are known to have gone extinct since the Age of Exploration began in the fifteenth century, the total number of *individual birds* worldwide has declined by 20 to 25 percent since that time (Şekercioğlu et al. 2004; BirdLife 2008).

It is almost inevitable that the inexorable rise in the human population will result in a decrease in the numbers of birds and other wildlife. As a result, the human population as we write has topped 7 billion and is predicted to be 10 billion by 2050—the demand on space for food production is destroying natural habitat at an ever-increasing rate. The rate of extinction is also increasing—currently one in eight (12.5 percent) bird species is at risk from global extinction (BirdLife 2008). Çağan Hakkı Şekercioğlu and colleagues (2004) predict that by 2100 "6–14% of all bird species will be extinct, and 7–25% (28–56% on oceanic islands) will be functionally extinct"[2] It is a chilling thought that their predictions may actually be a "best case scenario."

Even seemingly limitless populations have declined to extinction since the early 1800s. The huge potential for animals to multiply was recognized centuries ago (chapter 10) and coincided with a belief that populations of birds and other edible species, notably fish, were inexhaustible. This was particularly true for species that congregated in large numbers, such as seabirds and the Passenger Pigeon once did. Except to a very few enlightened individuals—and those seeing massive destruction firsthand—the idea that man might ever reduce bird numbers to extinction was unthinkable. One of those enlightened individuals was the naturalist-cum-adventurer George Cartwright, who, on witnessing the boatloads of dead Great Auks being brought ashore from Funk Island, Newfoundland, in the 1780s, said, "If a stop is not soon put to that practice, the whole breed will be diminished to almost nothing."[3] Just sixty years later, in 1844, the last Great Auks were killed on the island of Eldey, 15 kilometers (9 miles) off southwestern Iceland.

The Passenger Pigeon was once thought to account for 25 to 40 percent of all wild birds living in North America and to be indestructible. It seems to have been a nomadic species, moving around en masse according to the availability of food—John J. Audubon reported seeing one flock in 1813 that may have comprised hundreds of millions of birds. For both native people and recent colonists, the Passenger Pigeon was an important source of food and was killed in huge numbers. By 1874 it was still common, but numbers were beginning to dwindle (Baird et al. 1874), and by 1900 there were only occasional sightings of wild birds and just a few birds held in captivity. Between 1909 and 1912 the AOU offered $1,500 to anyone who could find a Passenger Pigeon's nest or a nesting colony. A few captive flocks were established, with the intention of reestablishing the species, but lacking the social stimulation of a huge colony, captive breeding was a failure. The last record of a living bird was of Martha (named after Martha Washington), a captive bird in the Cincinnati Zoo who died on 1 September 1914, and whose corpse was autopsied by Robert Shufeldt, frozen into a block of ice, and donated to the Smithsonian Institution, where she was stuffed and honored with the following plaque:

MARTHA

Last of her species, died at 1 p.m.,
1 September 1914, age 29, in the
Cincinnati Zoological Garden.

EXTINCT

As these examples make clear, one of the most disturbing aspects of bird conservation is that it is such a recent phenomenon. The scientific study of birds is about five hundred years old, but, by contrast, the idea that we might need to conserve birds—other than on a local scale—is little more than one hundred years old. Certainly there were earlier concerns about particular species in particular locations, but it is only during the last century that there has been concern about bird numbers on a global scale. Conservation is a consequence of problems created by man and is concerned with the protection of species from man-made threats. These threats are both direct, such as the killing or taking of birds, their young, or their eggs, and indirect through habitat loss, the introduction of other species, the use of toxic chemicals such as pesticides and herbicides, fishing, and most recently through climate change (BirdLife 2008). We touch briefly on most of these topics in this chapter.

Why should we care if bird populations are decreasing?

For one thing, there is increasing evidence for the cultural value of birds (e.g., Collar 2007; Cocker 2013), but the so-called shifting baseline means that each generation adjusts to the present abundance of birds and assumes that what *they* experience is the norm. "Your grandfather saw thousands of skylarks, your father saw hundreds, you have seen dozens and your children will see the odd one, and each of you thinks that that is the way things normally are; that's what the baseline is. But it isn't, of course. The real baseline is . . . as Yeats put it; living things living everywhere in profusion."[4] Bird watchers who move from Europe or North America to New Zealand say that what they miss most is the sound and sight of birds. Devastated by introduced predators, endemic birds are almost entirely absent from much of the New Zealand mainland, rendering many wild areas eerily silent. Yet people born in New Zealand accept the absence of birds with barely a second thought.

It is precisely because we value birds that there is such huge conservation concern over their declining numbers by bodies such as the Audubon Society in North America, the Royal Society for the Protection of Birds (RSPB) in Britain, and globally with BirdLife International (see BirdLife 2008).

In their efforts to slow or halt the decline of birds, conservation bodies draw on every available piece of knowledge regarding the biology of particular species. It is sad that the exponential growth in our understanding of the biology of birds during the last century—described in the previous chapters—has been accompanied by an exponential decline in numbers. It is only through our knowledge of bird biology and

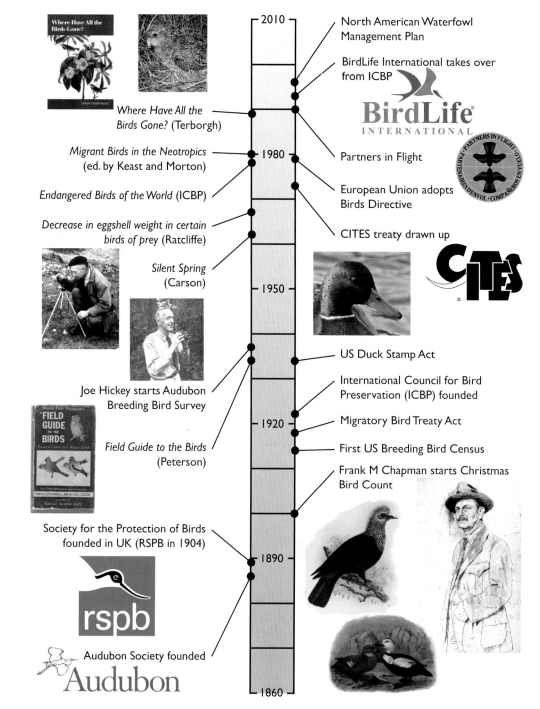

Where Have All the
Birds Gone? (Terborgh)

Migrant Birds in the Neotropics
(ed. by Keast and Morton)

Endangered Birds of the World (ICBP)

Decrease in eggshell weight in certain
birds of prey (Ratcliffe)

Silent Spring
(Carson)

Joe Hickey starts Audubon
Breeding Bird Survey

Field Guide to the Birds
(Peterson)

Society for the Protection of Birds
founded in UK (RSPB in 1904)

Audubon Society founded

2010

1980

1950

1920

1890

1860

North American Waterfowl
Management Plan

BirdLife International takes over
from ICBP

Partners in Flight

European Union adopts
Birds Directive

CITES treaty drawn up

US Duck Stamp Act

International Council for Bird
Preservation (ICBP) founded

Migratory Bird Treaty Act

First US Breeding Bird Census

Frank M Chapman starts Christmas
Bird Count

TIMELINE for CONSERVATION. Left: Cover of Terborgh (1989); Kakapo (endangered species); Derek Ratcliffe (left) and Joe Hickey (right); cover of Peterson Field Guide; RSPB logo; Audubon Society logo. Right: Logos of BirdLife International and Partners in Flight; Mallard, one of the species that benefited immensely from wetland conservation in North America; CITES logo; Mauritius Blue Pigeon (extinct in ca. 1830s); Frank Chapman; Labrador Duck (extinct since ca. 1878).

bird numbers, and in particular trends in numbers, that we have any chance of saving what we have left. As we will see, conservation is an interdisciplinary activity drawing on all aspects of bird biology—from population studies, systematics and molecular biology, migration, ecological adaptations, physiology, and behavior—in its efforts to save birds. Perhaps most important of all, it is now recognized that effective conservation depends upon a good understanding of the process of science—collecting evidence about bird population changes and their potential causes, and testing the predictions of several hypotheses. Nonetheless, as all conservationists know, most issues boil down to human behavior and politics: even the most carefully researched conservation plan has little guarantee of being implemented unless the public, the politicians, and the economists are all onside:

> We can no longer simply do the science and hope that someone else uses the information to make good laws that protect species and their ecosystems ... the major advances in conservation action will take place not in scientific laboratories or field research sites, but in the political and economic arenas, because present limitations in conserving biodiversity do not typically occur through lack of knowledge, but rather poor implementation—the policy arena.[5]

The general public interest in birds was given an enormous boost by the publication of Roger Tory Peterson's *A Field Guide to the Birds* in 1934, covering only the birds of the eastern half of the continent. An accomplished artist with a background in education, Peterson—who was just twenty-five at the time—recognized the value of simple,

Roger Tory Peterson (photo in 1966 at age 57).

aesthetic images that captured a bird's essence to aid identification.[6] His western bird guide was published in 1941, but the European "Peterson" did not appear for another thirteen years (Peterson et al. 1954).

It is no understatement to say that Roger Tory Peterson changed bird watching forever. His skills as a communicator, his celebrity status, and his commitment to conservation made him a powerful champion for birds. Officially involved in checking the effects of pesticides on birds almost from the start, he was among the first to anticipate their deadly effects. One of his biographers, Elizabeth Rosenthal (2008), paints an intriguing image of the man: a wonderful guy, but also inefficient, inconsiderate, ambitious, workaholic, self-centered, monomaniac, tight with money, a terrible driver, a savvy showperson, and a one-man conservation dynamo (see also Drennan 1998).

In this chapter we demonstrate how some of the advances described earlier in the book have informed—or have the potential to inform—conservation biology, a field that came into being only in the 1970s. We provide examples that serve to highlight both the staggeringly destructive abilities of mankind and the difficulties faced by conservation efforts.

FASHIONABLE DEATH

The economic boom that followed the industrial revolution in the middle of the nineteenth century created, among much else, a fashion industry. From the late 1800s women's hats decorated with birds' feathers became extremely popular. In addition to plumes, the wings, bodies, and even heads of birds became fashion accessories. As *Harper's Bazaar* magazine noted in 1897, "That there should be an owl or ostrich left with a single feather apiece hardly seems possible."[7]

The example set by European fashion houses was repeated in America, Australia, and the Far East, and feather companies sprang up all over the world. The plume trade was big business, and those running millinery companies cared little about the destruction caused by the brutal harvest of birds. Paris was the manufacturing center, but London, the main market, was described as "the head of the giant octopus of the 'feather trade,' that has reached out its giant tentacles into the most remote wildernesses of the earth, and steadily is drawing in the 'skins' and 'plumes' and 'quills' of the most beautiful and most interesting unprotected birds of the world."[8]

The birds most affected were herons, and the Great and Snowy Egret in particular: their nuptial plumes, euphemistically called "aigrettes" being among the most popular. Sold by weight, one ounce of aigrettes required the death of six birds. In a typical nine-month period during 1911, one investigator reported that four London feather firms sold 21,528 ounces of aigrettes—the equivalent of 129,168 egrets (Hornaday 1913: 121). Customers were duped into thinking that the feathers were naturally molted and simply picked up from the ground beneath the nesting trees. Reassuring, perhaps, but biologically implausible—birds do not typically molt while breeding, and, anyway, molted feathers are old and rarely in perfect condition.

The truth, of course, was very different, as Arthur Mattingley—a founding member of the Royal Australasian Ornithologists Union in 1901—discovered on visiting a heronry at Riverina, New South Wales. Approaching by boat, he found the carcasses of at least fifty Great and Intermediate Egrets shot off their nests, and the bodies of at least seventy nestlings that had fallen from those nests and drowned, weak from starvation. "What a sickening sight! How my heart ached for them. How could anyone but a cold-blooded, callous monster destroy in this manner such beautiful birds, the embodiment of all that is pure, graceful, and good."[9]

Mattingley's report alerted Australian ornithologists, who were outraged. It also reached Britain, and in 1909 the RSPB published his photographs in their *Bird Notes and News*.[10] Copies of his photos, some of which illustrated the methods of the plume hunters, were exhibited in shop windows. The previous year Lord Avebury had brought the first plumage bill into the British House of

Egg collecting on Laysan Island, North Pacific, in the early decades of the twentieth century was conducted on an industrial scale.

Lords, but it did not get through the House of Commons. In 1910 a government committee was formed to address bird plumage, but it was not until 1921 that the Importation of Plumage (Prohibition) Bill was passed (Doughty 1975).

There are numerous examples of birds being slaughtered for their feathers, but one of the most shocking took place in 1909–10 on the small island of Laysan—about 1,600 x 2,400 meters (1 x 1.5 miles) in extent—one of the northwestern Hawaiian Islands. Prior to the early 1900s, Laysan was an important breeding site for huge numbers of seabirds, including Laysan and Black-footed Albatrosses, Sooty Terns, Brown Noddies, shearwaters, and boobies. Those fortunate

enough to visit the island regarded the bird life there as one of the wonders of the natural world (Hornaday 1913: 137).

Laysan was inevitably a rich source of guano, and in March 1890 the island was leased to the North Pacific Phosphate and Fertilizer Company, which began digging two years later. In 1904 the manager, German-born Max Schlemmer, bought the mining rights from the company and ruled Laysan as if he owned it; throughout Hawaii he was known as the King of Laysan.[11] As the guano ran out, Schlemmer's thoughts turned to plumage, and after failing to persuade the US Territorial Government to allow him to start a feather business, on 22 December 1908, he signed a contract with

395

a Japanese firm, allowing them to remove and sell "products of whatever nature in and from the islands of Laysan and Lisianski."[12] In return, Schlemmer received $150 in gold each month, in a contract running for fifteen years.

Just six weeks after setting up the Japanese deal, he finally received a lease from the US Territorial Government for Laysan and the neighboring island of Lisianski—but a lease that would not allow the capture and destruction of birds. Almost at the same time, President Theodore Roosevelt issued an executive order, declaring the Hawaiian archipelago to be the Hawaiian Islands Bird Reservation.

Undeterred, Schemmer allowed the Japanese to continue. But late in 1909 reports of "piracy" reached officials in Honolulu, and the US revenue cutter *Thetis* was sent to investigate. Arriving at Laysan on 16 January 1910, they found fifteen Japanese workers with sacks of breast feathers and thousands of bird wings in various stages of curing. There were corpses everywhere, and the sight and smell was sickening.

The Japanese workmen were arrested and the plumage seized. It was later discovered that between April 1909 and January 1910 these workers had collected a total of 2.25 metric tons (4,960 pounds) of feathers and 310,000 bird wings. Schlemmer was removed from the island in 1911 but—incredibly— was not prosecuted. Even more improbably, he and his family were allowed to return in 1915, after Schlemmer pleaded that he was unable to live away from Laysan. They did not last long: one of Schlemmer's other contributions to the island had been to introduce rabbits and guinea pigs, and these multiplied and ate practically every green

shoot on the island. With no food left, the Schlemmers almost starved and had to be rescued.

Immediately following Schlemmer's removal from the island in 1911, Iowa State University organized a scientific expedition to Laysan led by Homer Dill, assistant professor of zoology. Dill and his colleagues had been planning—before the Japanese massacre—to produce an exhibition of Laysan's abundant birdlife.[13] What Dill discovered was an island stripped of life: there were bleached bones everywhere and evidence of much cruelty, including the remains of birds with broken legs and deformed beaks. In an old cistern hundreds of living birds had been left to starve to death, intentionally, for when the fatty tissue next to the skin was used up the skins were less greasy and required less cleaning. Dill was sickened: "[This] surpasses anything else done by these heartless, sanguinary pirates, not excepting the practice of cutting wings from living birds and leaving them to die of hemorrhage."[14]

The combined effects of the rabbits and the plumage harvesting were devastating, and Laysan never fully recovered. By the 1920s the endemic Laysan Rail and the Laysan race of the Millerbird (*Acrocephalus f. familiaris*), as well as the Laysan fan palm, were all extinct. The US Biological Survey attempted to exterminate the rabbits in 1913 but, after killing five thousand, ran out of ammunition, leaving a substantial number still alive.[15] Finally, in 1923, the USS *Tanager* expedition completed the elimination of rabbits on Laysan. Among the members of this ship's crew was Schlemmer's son, Eric.

Opposition to the feather trade resulted in the formation of the two best-known conservation bodies in the world: the Audubon

A page from Eaton's catalog of 1913 showing the various plumes or entire birds it was possible to purchase as fashion items.

Society in America and the Royal Society for the Protection of Birds (RSPB) in Britain. Ironically, perhaps, given that the feather trade was primarily the result of women's fashions, those responsible for establishing these organizations, and who campaigned for the cessation of feather millinery, were women.

The driving forces behind the early Audubon Society were Harriet Hemenway and her cousin Minna Hall, following an initial effort by the pioneering ecologist George Grinnell. Grinnell knew John Audubon and had been tutored by his widow, Lucy Audubon. From 1876 until 1911 he was editor of the magazine *Forest and Stream*, which he used as a vehicle to campaign against the killing of birds and taking of eggs. Grinnell established the Audubon Society in 1886, urging his readers to join. And they did, in extraordinary numbers: thirty thousand in the first three months alone. The workload, however, was too much for Grinnell and two years later he abandoned it.

The society was reconstituted in 1896—through the efforts of Hemenway and Hall—as the Massachusetts Audubon Society, whose main aim was to encourage women to boycott the wearing of feathers. Similar societies sprang up across the United States, and by 1906 they had united to form the National Association of Audubon Societies for the Protection of Wild Birds and Animals.

In Britain Emily Williamson established the Plumage League in Didsbury, Manchester, in 1889, while in Croydon, near London, Margaretta Smith and her neighbor, Eliza Phillips, arranged meetings of the "fur, fin and feather folk." The two groups merged in 1891 to become the Society for the Protection of Birds, with the Duchess of Portland as its president—a post she retained until her

death in 1954, aged ninety-one. The society attracted members of the aristocracy and by 1904 had obtained a royal charter. As with the Audubon Society, the RSPB's aim was to discourage the wearing of feathers and the killing of birds (except by sport hunting).

In 1892 Margaretta Smith married Frank Lemon, a barrister who drew up the RSPB's constitution, and she remained the society's driving force until her death in 1953. She was renowned, during church services, for writing down who was wearing feathered hats and then sending them a polite note, drawing their attention to the horrific way in which plumes were obtained (Hammond 2004). She was not an ornithologist (and was said to have distrusted ornithologists), but she was a formidable woman and commanded respect. Indeed, as Horace G. Alexander wrote, "There was no point in fighting Mrs Lemon. She would defeat you sooner or later. I decided quite early on that the best plan was to go along with her, and try to make sure that the birds did not suffer."[16]

Both the Audubon Society and the RSPB were successful, and the wearing of feathers went out of fashion almost as rapidly as it had emerged. Although there were to be small resurgences in feather fashion, by the Second World War no one was wearing feathers taken from wild birds. The societies could turn their attention to other issues.

COUNTING BIRDS

To conserve any bird species, you need to know how many there are. Censuses of people have taken place for the last two millennia, but the idea of censusing birds is more recent. The first efforts occurred in the nineteenth century, when, in 1811, Alexander Wilson counted birds in a 3-hectare (8 acre) botanic garden in Pennsylvania, providing the first (and only) density estimates of native birds prior to the arrival of the House Sparrow and Common Starling in North America. In 1861 Alfred Newton proposed a national census of British birds (Newton 1861), which was undertaken by More (1865).[17]

In America, the Christmas Bird Count was started in 1900. The brainchild of Frank Chapman, this annual event was set up as an alternative to the then popular Christmas Side Hunt, in which teams of hunters competed to see who could kill the most animals. Christmas Bird Counts continue to this day, and " . . . in combination with other avian monitoring techniques, they certainly appear to be surprisingly good indicators of spatial and temporal patterns in avian geographical ecology."[18]

In 1914 the Bureau of Biological Survey in the United States initiated its Breeding Bird Census (BBC), which involved counting all breeding birds—based on singing males—in sample plots, usually of forest habitat. The project fizzled out after ten years, but in 1937 the charismatic and dynamic Joseph Hickey reinvented it under the auspices of the Audubon Society.

Hickey was a salesman—albeit with a degree in history and a passion for birds—when he met Ernst Mayr (chapters 2 and 3) during the Great Depression. Mayr encouraged him to take a biology degree, which he did at night school. Mayr also fostered Hickey's interest in bird populations, which would eventually result in his monumental census of Peregrine Falcons (Cade and Burnham 2003). Hickey's master's degree—supervised by ornithologist and environmentalist Aldo Leopold and motivated by Max Nicholson's *The Art of Birdwatching* (1931)—was published as the influential *A Guide to Birdwatching*

(Hickey 1943). Hickey went on to complete a PhD with Josselyn Van Tyne at the University of Michigan, constructing life tables from banding data to obtain survival estimates for birds.

Before Hickey had even completed his thesis, Leopold invited him to become second professor in his thriving Department of Wildlife Management at Wisconsin. Just four months after Hickey's arrival, however, Leopold died and Hickey became head of the department, inheriting all of Leopold's teaching and administrative responsibilities. At the same time he arranged for the publication of Leopold's (1949) *A Sand County Almanac* (Rosenthal 2008). His interest in populations, and Peregrine Falcons in particular, meant that Hickey was to play a leading role in the pesticide battles to come. The Audubon Society's Breeding Bird Census—a plot-based spot-mapping census—continued for several decades, reaching its peak in the 1980s. Thereafter, the survey declined in popularity, but it still continues to this day, albeit at a low level.[19]

As the BBC declined, American bird watchers embraced the Breeding Bird Survey—a roadside transect comprising fifty three-minute counts at intervals of 0.8 kilometers (0.5 miles)—which started in 1966. Highly standardized, this is the monitoring program that now provides the best index of population status and geographic distribution across a wide range of habitats (Ralph and Scott 1981). Remarkably, the American BBS was designed to be done by car, in order to allow much greater distances to be covered, but also perhaps to appeal to the American way of life—the only nation to have a designated "motor nature trail."[20] The British Breeding Bird Survey, initiated in 1994[21] by the BTO, took a different approach, requiring volunteers on foot to count the birds they saw by walking two transects 1 kilometer (0.6 miles) each long for two and a half hours in a randomly allocated 1 kilometer (0.6 miles) square.

Once bird banding started in the 1920s (chapter 4), the numbers of birds captured at banding stations in successive years also provided an index of numbers (Kendeigh and Baldwin 1937). In Britain in 1926, Max Nicholson started an organization called the Oxford Bird Census—the forerunner of the British Trust for Ornithology—which began a national census of heronries in 1928 (chapter 10). The Oxford Bird Census was part of Nicholson's vision for a new ornithology, breaking away from what he called "the Victorian leprosy of collecting."[22] Nicholson was a mover and shaker, an innovator, and a conservationist. His heron survey involved four hundred amateurs and was made practicable by the availability of relatively cheap motoring. The choice of the Grey Heron was not motivated by conservation concerns but simply because it was a relatively easy species to count. Nicholson also introduced approaches for censusing other birds, including counts of singing males, which became part of a standard methodology (Nicholson 1931; Bibby et al. 2000).

There were similar initiatives elsewhere; for example, Ernst Schüz (1933) counted White Storks in East Prussia. In North America the idea of managing game birds for hunting—"wildlife management"—emerged rather later than it did in Britain, but starting around 1930 it had a profound effect on census thinking and resulted in the improvement of counting methods. One of the most notable techniques was Frederick C. Lincoln's capture-mark-recapture[23]

method, devised initially for estimating wildfowl numbers (Lincoln 1930).

DOOMED SURPLUS

One motivation for understanding the population dynamics of any species, including birds such as grouse or wildfowl, is the concept of "doomed surplus"—the birds that die naturally each year, mainly young individuals (Errington 1946). This is based on the recognition that most young individuals fail to survive to breeding age and that populations are maintained near their carrying capacity through density-dependent processes (chapter 10). In theory, killing or capturing the doomed surplus makes no difference to the size of the adult population, and this idea has been incorporated into conservation in two main ways. First, gathering up the doomed surplus can be used to start new populations, as with the New Zealand Robin, Stitchbird (often referred to as "hihi," its Maori name), and North Island Saddleback (Armstrong et al. 2002). Researchers have successfully used population models to estimate the size of the surplus and, hence, the number of birds that can be safely taken and translocated, as in the case of the North Island race of the New Zealand Robin (Dimond and Armstrong 2007).

A similar situation occurs in those species where females lay two or more eggs but only one chick ever survives, as in certain eagles, hornbills, and the Whooping Crane (Sutherland et al. 2004). In such species, the second chick has sometimes been taken to seed a new population, as in the case of the Southern Ground Hornbill in South Africa.[24] Similarly, by removing an entire clutch, females can be encouraged to produce a replacement clutch (known as "double clutching"), and this can provide a supply of eggs for hand rearing or fostering. In 1974 the Mauritius Kestrel was reduced to just four individuals in the wild, initially because of habitat destruction but compounded in the mid-twentieth century by widespread pesticide use (of which more later). Today, BirdLife estimates there to be 800 to 1,000 mature individuals, making this one of the most spectacular conservation success stories. The credit for this goes largely to Carl Jones. Rather than bringing all four birds into captivity for breeding, Jones followed the advice of Peregrine Falcon experts Tom Cade and Willard Heck, who advocated an intensive management program, of which double clutching was a key component.

The second way that harvesting birds has been incorporated into conservation is through recreational hunting. While this might seem paradoxical, hunting can be sustainable. On many private estates in Britain, for example, the harvest of nonmigratory game birds such as red grouse [Willow Ptarmigan], Black Grouse, and Western Capercaillie is effectively managed (albeit through the destruction of predatory birds). In North America the management of migratory ducks has also been successful. In 1934 the Migratory Bird Hunting Stamp Act was passed in the United States, meaning that federal licenses were required for hunting migratory waterfowl. All waterfowl hunters over sixteen must annually purchase and carry a Federal Migratory Bird Hunting and Conservation Stamp (or "duck stamp"). As well as allowing hunters to continue their pursuits legally, the scheme has had significant conservation benefits: 98 cents from every dollar goes toward the purchase

or lease of wetland habitat in the National Wildlife Refuge Program, and since 1934 more than $750,000,000 has gone into the fund, enabling the purchase of more than 5.3 million acres of breeding, migration, and wintering habitat.

Ducks Unlimited was also created by hunters. Founded in 1937 during an intense drought that caused many waterfowl populations to nose-dive, its aim is to preserve habitat and ensure an abundant supply of waterfowl for the future. Ducks Unlimited is the world's largest private waterfowl and wetlands conservation organization and has conserved more than 12.5 million acres of habitat across North America.

MIGRATION AND CONSERVATION

Migratory species are of special conservation concern because so many have undergone dramatic declines over the past fifty years, due largely to habitat loss on breeding and wintering grounds, as well as stopover points in between. In contrast, a small number of migrants have experienced a massive increase in numbers, and some, like the Barnacle Goose in the Netherlands, have altered their migratory behavior and, as a result, achieved pest status.

Winter goose hunting was once popular and widespread across much of northern Europe. For the Barnacle Geese from Russia that wintered in the Netherlands, the cessation of hunting in the 1970s resulted in a spectacular rise in numbers. Detailed monitoring of the behavior and ecology of their populations over several decades, by Bart Ebbinge and colleagues, has revealed an extraordinary situation. By the 1970s, instead of returning to their breeding sites on Novaya Zemlya, some geese started to breed at what had previously been stopover sites in the Baltic. As numbers continued to increase, some geese ceased to migrate at all, and by 1980 they were breeding in their Dutch wintering areas (Eichhorn et al. 2009). The advantages of not migrating were dramatic: nonmigrant birds remaining all year in the Netherlands had a 97 percent annual adult survival rate, whereas the survival rate of migrants returning to northern Russia was 55 percent.

Those geese that continued to migrate did so one month later than they had in the 1970s, and this was unlikely to be related to anthropogenic climate change, which predicted earlier migration. Instead, it is possible that either reduced food intake (as a result of greater competition) or increased predation risk on the Baltic stopover sites is driving the later departures (Jonker et al. 2010). One consequence of the delayed onset of spring migration is that there is now a gap between the end of parental care and the start of migration. It has been known for many decades that parental care in geese is protracted. In the 1940s Ernst Mayr had commented that parents and offspring did not separate until they were back on their Arctic breeding grounds the following summer and hence that migration behavior seems to be culturally transmitted (Mayr 1942: 242). In the 1990s it became apparent that Barnacle Goose spring migration was starting later and that breaking the tie between parents and offspring was probably the cause of the change in migratory behavior, with the "abandoned" young birds remaining to breed in their wintering area (Jonker et al. 2011).

The Barnacle Goose situation is unusual. Across much of the world, migratory birds are in decline. Between 1970 and 2000, 40 percent (48/119) of long-distance Eurasian migrant species showed a significant reduction in numbers (Sanderson et al. 2006). A similar pattern has occurred in North America (Holmes and Sherry 2001): the Audubon Society's *State of the Birds* report for 2010 showed that about 30 percent of bird species are in decline, with Hawaiian and oceanic birds the most seriously threatened. Initially it was assumed that these declines were the result of events—such as intensification of agriculture and habitat loss—in the *breeding* ranges, but starting in the 1960s and 1970s in Europe, and in the 1980s in North America, it became increasingly apparent that events in the *wintering* areas and on migration were also critical (Keast and Morton 1980).

The Palearctic-African migration system (covering Europe and Africa) is better known than the Nearctic-Neotropical system (covering the Americas). This is mainly due to European colonialism in Africa; many of those men sent to Africa from Europe in the 1800s were keen birders, and it was their records that enabled pioneers like Eliot Howard to produce winter distribution maps of warblers in Africa as early as the 1900s (Howard 1907–14). After Howard's pioneering, and essentially unnoticed, efforts, there was little further interest in African migrants until the 1960s, when Jean Dorst (1956, 1962) as well as François Bourlière and Gérard Morel (Morel and Bourlière 1962) considered the ecological impact of migrants on resident birds in tropical Africa. Building on these studies, Reg Moreau (1972) provided a more detailed summary of the Palearctic-African system, calculating that about one-quarter of the five hundred or so bird species that breed in Europe—some 5 billion individual birds—spend the winter in sub-Saharan Africa. The paradox for Moreau—as for Morel and Bourlière—was how all these additional birds could survive the winter in a place as arid as Africa.[25]

In the late 1960s the European population of Common Whitethroats crashed—the first indication that all was not well with the Palearctic-African system. Three-quarters of all whitethroats—several million birds—simply disappeared between the breeding seasons of 1968 and 1969 (Glue 1972). The crash was linked to a severe drought in the Sahel, where the rainfall in 1969 was 25 percent below normal, causing a massive reduction in the abundance of insects on which the whitethroat, and many other wintering warblers, rely. In the following years there were some signs of recovery, but numbers crashed again in the 1980s and again in the 1990s, as further droughts struck. Whitethroat numbers have remained depressed ever since.

As time went on, it became clear that other migrants were also decreasing. The cause was subsequently discovered to be a combination of events, including the intensification of agriculture and degradation of breeding habitat in Europe; habitat degradation, drought, and desertification in Africa; and hunting in southern Europe and Africa while the birds were on migration. The lack of water in Africa means that migrants have to fly longer distances before stopping; it also means less food, which in turn means that they cannot always time their migration optimally. In addition, the reduction in suitable habitat at stopover sites means that birds are more concentrated in fewer locales, making them more vulnerable to local hunting pressure. On top of all this, the widespread use of pesticides in Africa—especially those now banned in Europe and North America—may be accelerating bird population declines (Zwarts et al. 2009).

Summing up the Palearctic-African situation, Fiona Sanderson and colleagues (2006) said that "conservation action to address these declines is required under the Convention on Migratory Species and the Pan-European Biological and Landscape Diversity Strategy, to which most European countries are signatories and which aim, respectively, to conserve migratory species and to halt the loss of biodiversity by 2010. Our results indicate that more conservation action may be required outside Europe to achieve these targets."[26] Finally, the realization had dawned that it is not enough to work only on the European side of the problem.

A similar realization gradually emerged for many species that breed in North America and winter in the Caribbean, Central America, and tropical regions of South America. Using data from the Breeding Bird Survey for sixty-two migrant species, Chandler Robbins and colleagues (1989) showed that between 1966 and 1978 24 percent of species declined in numbers, and 76 percent increased. Between 1978 and 1987, however, the situation reversed with 71 percent declining and just 18 percent increasing. Initially it was thought that these declines were a consequence of deforestation in the breeding areas and the resulting "edge effects"—the increase in forest edge habitats as a consequence of forest fragmentation. The idea was that birds avoided forest edges, effectively further reducing the total amount of forest habitat available, but studies provided no evidence for this. It was then speculated that predation was higher on forest edges than elsewhere, and while there was some evidence for this, the effect was insufficient to account for the declines, at least in some species.

Until the 1970s the decline in migrant numbers was assumed to be a temperate zone

North American problem, rather than one that involved both breeding and wintering areas. This was partly because—in contrast to the colonial history of many European countries—the North Americans lacked information from their birds' wintering areas; the only ornithological information about Neotropical migrants[27] came from a handful of museum expeditions and the specimens they collected and the work of a few tropics-based stalwarts like Alexander Skutch (chapter 8). Central and South America have some 3,400 bird species, but until the 1950s there were no Neotropical field guides or handbooks. The turning point with the Nearctic–Neotropical system came in the mid-1970s, when Allen Keast and Eugene Morton organized a symposium on western hemisphere migrants, bringing together forty experts to discuss the birds' biologies. The proceedings were published in 1980 as *Migrant Birds in the Neotropics*—in many ways the counterpart of Moreau's pioneering *The Palaearctic-African Bird Migration Systems* (1972), although with a more explicit focus on what birds were actually doing in their winter quarters.[28]

One of those speaking at the Keast-Morton symposium in 1977 was John Terborgh, but it was not until he published his book *Where Have All the Birds Gone?* in 1989 that the scale of the problem became more widely known. Reviewing it, Russ Greenberg said that Terborgh's book "will likely become one of the most important books on bird conservation."[29]

Remarkably, when Terborgh wrote his book there were almost no data on migrant bird numbers. In 1989 Chandler Robbins used data from the Breeding Bird Survey to look for the effects of pesticides (see below), and the results were dramatic: long-distance migrants had undergone precipitous declines in numbers. These results were independently confirmed by Sidney Gauthreaux, who—using radar to monitor the movement of migrants—found a 50 percent reduction in the number of migrating birds between the 1960s and 1980s (Gauthreaux 1992). Detailed fieldwork dating back to 1969 by Dick Holmes—based at Dartmouth College, New Hampshire—and his colleagues, has revealed the subtle interactions between events in the wintering areas and breeding areas and how winter conditions can have carry-over effects into the breeding season, and vice versa (Holmes 2007).

Songbird declines triggered several conservation initiatives, including further analyses of population trends (RSPB/BTO: Ockenden et al. 2012; BirdLife: Burfield et al. 2004); meetings, including one at Vogelwarte Radolfzell in May 2008, with nineteen people from eleven countries whose aim was to establish a network within Africa to identify changes—such as deforestation[30]—affecting bird populations; and the Flyways Campaign initiated by BirdLife.[31] The aim of these initiatives is to slow the decline in bird populations, in the hope that they will stabilize and eventually grow. Currently, the so-called Natura 2000 sites[32] protect both birds and habitats across a network of locations that encompass summer and winter ranges, in which all parties have ownership. In Europe, for example, one hope is that, despite the ongoing intensification of agriculture, eastern regions, such as Belarus and Kazakhstan—areas that have so far avoided agricultural intensification—will retain their pools of potentially recolonizing species. Another possibility is that post-oil economies will return to low-intensity farming, enabling bird populations to recover.[33]

BEHAVIOR AND CONSERVATION

One of the main ways that an understanding of behavior contributes to conservation is through captive breeding programs. In some cases, captive breeding is the final resort, as in the case of Spix's Macaw or the California Condor. Ultimately, however, the aim of captive breeding is to reestablish a wild population. The alternative is to either move the endangered individuals away from their main threat—often predators—or to remove the threat, both of which have been implemented with many New Zealand birds. Captive breeding sounds straightforward but rarely is, because it often results in the production of inferior offspring whose chances of reestablishing a wild population are low. Studies of behavior have had an important role in resolving these problems.

The concept of misimprinting has been known at least since the 1500s, when it was recognized that, on hatching, the young of precocial birds, like ducks and fowl, would follow their human owners. Oskar Heinroth and Konrad Lorenz studied this phenomenon in detail in the early part of the twentieth century (chapter 7), but it was not until the 1970s that it was appreciated that interspecific cross-fostering or hand rearing—as part of a bird conservation program—could result in inappropriate imprinting. In an effort to save the endangered North American Whooping Crane in the mid-1970s, eggs were transferred into the nests of the much more common Sandhill Crane, to try to found a new breeding population of Whooping Cranes in Idaho. The Whooping Crane chicks were reared successfully and migrated to New Mexico with their foster parents. But when they became sexually mature, they failed to mate with other Whooping Cranes—because they were sexually imprinted on Sandhill Cranes. The cross-fostering project was discontinued in 1989 (Lewis 1995).

Exactly the same phenomenon occurred in the Chatham Islands, New Zealand, with the Black Robin, a species whose entire wild population in 1979 consisted of only a handful of individuals. In that year the world population of Black Robins comprised three males and two females (Old Blue, eight years old, and Old Green, a one-year-old). Both females produced clutches in 1979 and in the following seasons, but only Old Blue's offspring subsequently bred successfully. It was her sons and daughters, produced between 1979–80 and 1982–83—together with a spectacular conservation effort coordinated by Don Merton—that saved the species. Old Blue's eggs were cross fostered to Chatham Gerygones (a failure because the adult gerygones did not feed the foster chicks properly) and later to Tomtits, which was successful. Cross-fostering ceased in 1988–89 as numbers increased naturally, reaching 197 individuals at the start of the 1998 breeding season. But it was not plain sailing—some of the cross-fostered robins became sexually imprinted on the Tomtits and then, because the two species are closely related (in the same genus), subsequently mated with Tomtits to produce hybrid "tobins" or "robotits." The problem was identified early on and measures were taken to prevent further misimprinting, but also—after much debate—the existing hybrids and their parents were destroyed (Butler and Merton 1992). Nevertheless, a small amount of

CHAPTER II

Whooping Cranes.

genetic introgression still occurred through the 1990s (Kennedy 2009), and while numbers continue to slowly rise, the population is still critically endangered.[34]

Once the misimprinting problem was recognized, offspring were hand reared using glove puppets resembling the correct parental species, an approach pioneered with Peregrine Falcons (Cade and Fyfe 1977) and later used on other species, including Sandhill Cranes (Horwich 1989), Takahe (Maxwell and Jamieson 1997), and California Condors (Snyder and Snyder 2000). Whether hand-reared birds misimprint seems to depend on whether they are reared alone or with conspecifics of the same age: misimprinting is less of a problem if conspecifics are reared together (Curio 1996).

The phenomenon known as "predator blindness" was first noticed among hand-reared African Collared Doves (Klinghammer 1967). In virtually all subsequent studies, hand-reared birds proved to be less fearful of predators than parent-reared offspring. The mechanism for this is unknown, but it must involve cultural learning (Curio 1998: 174-5). Vulnerable species can be conditioned to avoid predators—as Curio (1969) did with Darwin's finches harassed by a cat, and as has been shown in Takahe chicks trained to avoid stoats (Hölzer et al. 1995; McLean et al. 1999). But as Curio (1998) pointed out, for this conditioning to have a lasting effect in the wild, these behaviors would need to be culturally transmitted across generations. So far at least, there's no evidence that this occurs.

Once thought extinct, New Zealand's South Island Takahe was rediscovered in 1948; conservation measures including captive breeding and translocations have resulted in their numbers increasing to 260 by 2012.

In an overview of the role of behavioral studies in conservation, Caro (1998) recognized both the advantages and disadvantages of employing behavioral knowledge. The main limitation is that behavioral studies are labor intensive and often take time, whereas conservation problems are often urgent and require a rapid response. Second, with a handful of notable exceptions—such as the Spotted Owl (Forsman et al. 1984), Seychelles Warbler (Komdeur et al. 1997), and Stitchbird (Ewen et al. 1999)—there have been relatively few behavioral studies of endangered species, so it is difficult to make appropriate generalizations.

DDT

The acronym DDT[35] is almost synonymous with environmental contamination. DDT is a particularly persistent pesticide that, once ingested, accumulates in fat stores and then builds up in the food chain with devastating effects on top predators. The removal of DDT from western agriculture was one of the major success stories in environmental protection during the twentieth century, and it set precedents for the removal and control of other toxic pesticides. Many people worked to achieve this goal, but the contribution of Rachel Carson is particularly significant.

407

Carson was brought up in rural Pennsylvania, where her mother instilled in her a deep love of nature. Graduating from Pennsylvania College for Women in 1929 with a degree in zoology, she completed a master's in marine zoology at Johns Hopkins University in 1932 and remained there as teacher for the next few years. When her father died in 1935, she was forced to find a better position, and being a keen writer, she became a radio scriptwriter with the US Bureau of Fisheries (renamed the US Fish and Wildlife Service in 1940). She rose through the ranks to become editor-in-chief of the Wildlife Service's publications, until she retired in 1952 to become a full-time writer. Her time at the Wildlife Service immersed Carson in science and nature conservation. As well as publishing technical articles, she wrote popular pieces, and in 1951 she published the award-winning book *The Sea Around Us*. Ten years later she published a book that sent shockwaves around the world. In the meantime, the use of toxic chemicals, including DDT, continued to increase.

In the mid-1940s the US Army was using DDT to kill mosquitoes in Florida. Rich Pough, of the Audubon Society, was one of the first to warn of the dangers. Pough was a force for conservation in America, one of whose first achievements was calling attention to the ruthless slaying of Northern Goshawks and other birds of prey migrating at Hawk Mountain in the Appalachians. In 1934 he was able to persuade socialite and amateur bird watcher Rosalie Edge to buy 1,400 acres of Hawk Mountain, establishing the world's first refuge for birds of prey. In 1938 he joined the Audubon Society, and in 1945 he published an article describing the effects of DDT spraying on birds in Pennsylvania (Rosenthal 2008: 175). The same year,

Pough also publicized his concern in *The New Yorker*, saying, "If DDT should ever be used widely and without care, we would have a country without fresh-water fish, serpents, frogs, and most of the birds we have now. Mind you, we don't object to its use to save lives now [during the Second World War]. What we're afraid of is what might happen when peace comes."[36]

Pough's comment had little impact and was contradicted by others, including Frank Kozlik, who stated, "Properly applied, light concentrations of DDT should give good insect kills without harmful effect to birds."[37] The only problem identified by Kozlik was that a reduction in insect numbers might deplete the birds' food supplies. In another test—conducted by Chandler Robbins of the US Fish and Wildlife Service—it was again found that woodland birds seemed to be unaffected by spraying, possibly because the actual amount of DDT reaching the woodland floor was low (Stewart et al. 1946). However, when that study was repeated in scrubland (where a higher proportion of DDT reached the ground), the effects were more serious: birds had tremors and were unable to fly, and the numbers of House Wrens and Prairie Warblers fell by 80 percent (Robbins and Stewart 1949).

Although there were numerous incidents involving the direct death of birds as a result of spraying or the use of toxic chemical as a seed dressing, it was the devastating nonlethal effects on the top predators—the ones in whom toxins had accumulated through the food chain—that really attracted attention and on which most of the influential research was carried out.[38]

Derek Ratcliffe—one of several ornithological champions in the pesticide story—began working for Britain's Nature

Conservancy in 1956. His first project was to conduct a survey of Peregrine Falcons, because the racing pigeon fraternity claimed the falcons had so increased in numbers that they were disrupting their hobby. Peregrines posed a threat to pigeons carrying messages during the Second World War, so huge numbers had been shot in Britain. After the war Peregrine Falcon numbers started to recover—possibly triggering the pigeon racers' concerns. Ratcliffe was seconded to the British Trust for Ornithology to conduct the survey that took place in 1961 and 1962. Instead of finding an increase in peregrine numbers, Ratcliffe discovered that, following their initial postwar recovery, their numbers had declined. There were reports of people finding dead and dying peregrines, as well as reports of female peregrines failing to lay—or if they did lay, of their eggs failing to hatch or breaking underneath the incubating bird. As awareness of the plight of the Peregrine Falcon became public, other countries—including the United States, France, Germany Sweden, and Finland—also reported similar declines.

Ratcliffe considered the various factors that might be responsible for breeding failure, feeling that pesticides were the most likely cause, but at that time, in the early 1960s, the evidence was circumstantial. On 1 July 1961, Ratcliffe climbed to a Peregrine Falcon eyrie in Scotland, where he found the female incubating two addled eggs. He took the eggs and sent one to Harold Egan (in the Laboratory of the Government Chemist), who, using gas chromatography, discovered a cocktail of toxic chemicals[39] in the egg, at concentrations that in laboratory studies of domestic poultry caused a significant reduction in hatching success.

A good friend of Ratcliffe's was Desmond Nethersole Thompson.[40] A superb naturalist, writer, and enthusiastic rattler of establishment cages, Nethersole Thompson's reputation as an egg collector, especially of rare birds' eggs, meant that most of the ornithological community treated him with suspicion. However, when Ratcliffe mentioned to him the problems of addled and crushed eggs that the peregrines were experiencing, it was Nethersole Thompson who suggested he check whether the shells of the unhatched eggs were unusually thin. Edward Blezard, curator of natural history at Carlisle Museum, allowed Ratcliffe to weigh the peregrine eggs in his collection to obtain an index of the shell thickness.[41] While there was a hint that the shells were thinner than in the pre-pesticide era, the number of eggs available was too few to be certain. Ratcliffe went back to Nethersole Thompson, who offered to "open the doors of the underworld,"[42] that is, to provide access to individuals with private (and illegal) collections. The collectors that Ratcliffe contacted all allowed him to measure their eggs, thereby providing sufficient data to show a very clear link between the introduction of DDT in the late 1940s and the drop in eggshell thickness.[43]

Avian physiologists initially puzzled over the cause of shell thinning, and no less than fifteen different hypotheses were proposed, eight of them predicting a decrease in blood calcium levels, meaning less calcium in the shell. Luckily, largely due to a burgeoning poultry industry and an economic interest in egg production during the 1960s, much of the basic science of egg and shell formation was well established. It was also known that the mechanism of shell formation was probably highly conserved across bird species (Gilbert 1979). However, while the domestic

Rachel Carson, with wildlife artist Bob Hines. In the early 1950s they visited wildlife refuges along the US Atlantic coast gathering material for US Fish and Wildlife Service pamphlets and technical publications (photo 1952 at ages ca. 45 and ca. 40).

fowl was extremely well known, it was not the ideal study species for this particular problem because it is relatively resistant to pesticide-induced eggshell thinning (Cooke 1973). The mechanism was finally worked out by David Peakall in the United States, showing that DDE (a metabolite of DDT with a half-life of ten years) inhibited the action of carbonic anhydrase—the enzyme responsible for supplying the carbonate ions used in eggshell formation—in the shell gland (the avian uterus) (Peakall et al. 1973).

As the evidence against DDT accumulated, the chemical industry started to fight back. Their target in North America was Rachel Carson, whose *Silent Spring* (1962) made the dangers of DDT and other pesticides clear to a wide public. Carson had begun work on the book in 1957, stimulated by a number of events. She was horrified by the US Department of Agriculture's

campaigns to eradicate both the gypsy moth and the fire ant—neither of which was actually causing the amount of damage that the USDA claimed. In 1956 one million acres across New York, Michigan, and Pennsylvania were sprayed with DDT in the moth campaign, and in 1958 some 20 million acres of the southern United States were sprayed with the newly developed dieldrin and heptachlor (neither of which had been extensively tested) in an effort to remove the ants. Bird numbers crashed in both cases, and worse, there was no evidence that spraying had any effect on the targeted pest species.

In 1958 Carson was contacted by a friend, Olga Owens Huckins, whose private bird sanctuary in Massachusetts had been sprayed—by the Commonwealth of Massachusetts—with DDT (against her wishes) to kill mosquitoes. Huckins asked Carson whether there was anybody in

Washington who could help, and this was the key event that spurred Carson into action (Waddell 2000). She began working seriously on a book, and four years later *Silent Spring* appeared, single-handedly launching the environmental movement in the United States, with devastating consequences for the agrochemical industry. In June of that year *The New Yorker* serialized ten chapters of *Silent Spring* over three successive issues, and these chapters stimulated President Kennedy to announce that his administration would examine the issues that Carson had raised. The book was hailed as revolutionary and even inspired Joni Mitchell's song "Big Yellow Taxi."

The agrochemical industry was incensed. The former secretary of agriculture Ezra Benson launched an attack on Carson's credibility by asking "why a spinster with no children was so concerned about genetics,"[44] and answering that "she was probably a communist."[45] Other opponents said that she was not a professional scientist, and that *Silent Spring* was an emotional response to male-dominated technology (Lear 1997: 430). The chemical company Monsanto circulated a spoof article—mimicking Carson's writing style—that portrayed the horrors of a world without pesticides. The scale of the industry's attack was extraordinary. As Carson's biographer Linda Lear (1997) points out, "The fury with which they [the chemical industry] attacked her reflected the accuracy of her moral charges."[46]

Carson defended herself with tough, intellectual dignity. She had support—her defenders included eminent scientists like Herman J. Muller, the Nobel Prize–winning geneticist—but they were swamped by the power and wealth of the chemical industry.

Throughout all of this, Carson rarely showed any outward sign of the cancer from which she died on 14 April 1964 (Lear 1997). Her courage at exposing the chemical companies inspired millions—yet fifty years afterward she continues to be reviled by industry.

One ornithologist who had firsthand experience of the machinations associated with the pesticide controversy was Ian Nisbet. Born in Britain and trained in maths and physics, Nisbet was a keen amateur ornithologist who, after moving to America, was mentored by Bill Drury, scientific director of the Massachusetts Audubon Society. In 1967, recognizing that Nisbet's talents complemented his own, Drury invited him to join the scientific staff of the society as an environmental scientist. This was a new and rapidly expanding field, and as Nisbet told us, "With hard work, a little background knowledge (especially in mathematics) and a dash of audacity, it was possible to become an instant expert on almost any topic du jour."[47] Drury was deeply involved in the battles over pesticide use, having been appointed to the President's Scientific Advisory Committee, as part of the review following the publication of *Silent Spring*. Nisbet soon became involved too, and in 1971 he took part in the national hearings on the proposed ban on DDT. In contrast to the situation in Britain—where decisions were made in smoke-filled rooms and usually in favor of the manufacturers—the situation in the United States was more formal. Nisbet was amazed by the interplay between science and law: "It was astounding to watch the 'experts' from the agrochemical industrial complex crumble under cross-examination and eventually admit that there had never been any scientific evidence showing that DDT actually improved crop yields or reduced costs."[48] The Nixon administration had to relent, and DDT was banned in the United States in 1972.[49]

More insidiously, the chemical manufacturers embarked upon a different strategy to garner support from the very heart of the ornithological community. In Britain, for example, Shell gave funds to organizations like the British Trust for Ornithology (BTO) and the RSPB. Often there were no strings attached, as when they funded a PhD project at the BTO to develop bird census techniques in the early 1970s.[50] As part of their campaign, Shell even named some of their North Sea oil and gas fields after birds such as Brent [Brant Goose] and Dunlin. They also subsidized the production of natural history books like *The Shell Bird Book*, written by James Fisher (1966).

When we asked Jim Flegg, one of James Fisher's contemporaries, whether Fisher had any qualms about taking cash from Shell, he said that he was so keen on communicating science that "he wouldn't have given a damn where the money came from."[51] Recognizing that industry was unlikely to be diverted from its economic goals, many ornithologists felt that they might as well take everything they could get from the industrialists. Fisher probably saw the Shell sponsorship as an excellent opportunity to publicize the widespread concern over pesticides. Previously, in a BBC Radio broadcast in March 1963, he had made his views on pesticides very clear with a powerful and positive review of *Silent Spring*, referring to the pesticide situation as a "formidable indictment of man's capacity to create uncontrollable monsters from his own inventiveness."[52] He continued: "It would seem that the most serious change in the population of birds over America stems from the uncontrolled use

of these poisons. . . . Not just insect-eating song birds—but great birds of prey like the Bald Eagle—the U.S. symbol—and the osprey."[53] It is perhaps surprising, then, that Shell invited Fisher to write "their" book. He used it as an opportunity, as is clear from correspondence between Fisher and Shell before the book went to press.

On reading Fisher's manuscript, Shell said: "The only point where we seriously differ is on the question of the organochlorine insecticides on birds. . . . We do not accept . . . that the transmission of residues via food chains was something that could reasonably have been foreseen in the early or mid-50's. . . . Nor do we accept that the use of pesticides has been taken too far with 'too few controls and tests.'"[54] Shell employee M. Le Q. Herbert referred Fisher to a *New Scientist* article, written by another Shell employee, which stated that the response to pesticides had been more emotional than scientific, that one cannot infer causality from the correlation between pesticide application and bird mortality, and that "it appears highly improbable that there is any toxic hazard to the general bird population in Britain arising from the use of insecticides of this type [DDT and dieldrin]."[55] Fisher, however, refused to compromise and wrote in *The Shell Bird Book* that "the widespread use of organic (chlorinated hydrocarbon) pesticides was proved to be a factor leading to the death and infertility of . . . Peregrines. . . . The hydrocarbon deluge . . . has provedly wrecked huge bird of prey populations already."[56]

MORE TOXIC TALES

With DDT banned, the next pesticides to come under scrutiny in the courts were aldrin and dieldrin, both manufactured by Shell at their Rocky Mountain Arsenal facility (previously a US Army base) in Colorado. When this facility closed in 1992, it was considered "one of the most contaminated places on the planet."[57] Like DDT, aldrin and dieldrin are fat-soluble, organochlorine pesticides, and both were used mainly as seed treatments to protect cereal grains against insect attack, with dieldrin also used in sheep dips up until 1966. Although less widely used than DDT, the effects of aldrin, and especially dieldrin, were more severe, often killing wildlife outright. Shell took the 1987–88 aldrin/dieldrin hearings very seriously, contesting every aspect of the case—and the whole thing lasted more than a year.

Many of the detailed wildlife studies used in that public hearing came from British birds. Ian Nisbet suggested Jim Lockie as a potential witness because Lockie had seen the devastating effects of dieldrin on breeding Golden Eagles in Scotland (Lockie et al. 1969). Shell, in turn, brought in William Brunsdon Yapp, a zoologist at the University of Birmingham in England who had written books about woodland birds. But woodland birds weren't an issue—the main casualties were seed-eating farmland birds and birds of prey that scavenged on sheep carrion. At the hearing Shell presented data on dieldrin residues in adult Red Kites—sheep carcass scavengers—found dead in Britain. The dieldrin levels were high enough to suggest that they were responsible for the kites' death. Suspicious of the amount of effort Shell put into arguing that dieldrin was *not* responsible, Nisbet wondered whether Shell expected opposing witnesses to raise this issue. That they did not was perplexing at the time, but later Nisbet suspected that only Shell had access to the relevant data. At one point during the hearing, Shell intended to use several

British government scientists as witnesses, but Derek Ratcliffe and others were refused permission to take part—possibly because their agencies were intimated by Shell. Nisbet asked his old King's College friend Tam Dalyell, a Labour MP, to ask Margaret Thatcher—who was science minister at the time—what was going on; she responded by stopping all government witnesses from testifying.[58]

Even though the negative effects of dieldrin and aldrin on wildlife were all too obvious, these pesticides were finally, in 1974, banned in the United States as dangers to human health.[59] Toxic chemical production subsequently took place elsewhere—in Venezuela, for example, for several more years—but, as Ian Nisbet told us, "All these chemicals, including DDT, are now banned under the Stockholm Convention, including continental Europe and Britain, and if they are currently being produced and used, it would be by rogue countries."[60]

Because DDT, dieldrin, and aldrin accumulate up the food chain, birds of prey—such as the Peregrine Falcon, Eurasian Sparrowhawk, Cooper's Hawk, and Bald Eagle—were among the most susceptible. When those chemicals were banned, bird numbers began to recover. Unfortunately, pesticides were not the only substances to devastate birds of prey.

The California Condor is one of the rarest birds in the world. In Pleistocene times, when giant ground sloths, mastodons, and other large mammals were abundant food sources, it had a range that stretched from Canada to Mexico, but its range contracted following the extinction of these animals approximately ten thousand years ago. By the time European settlers arrived, the condor was restricted to the Pacific coast from

California Condors at Big Sur, California.

British Columbia to Southern California. By the 1930s, Carl Koford estimated a total world population of just 60 individuals, all confined to California. The main causes of the decline were shooting, habitat degradation, and poisoning from carcasses contaminated with lead shot, and this latter factor remains the greatest threat to condor populations today. After the condor was formally listed, and then protected, with the signing of the US Endangered Species Act in 1972, the California Condor Recovery Team was formed, producing in 1975 the first recovery plan for an endangered species in the United States (Walters et al. 2010). The program was initially noninterventionist, but following recommendations by Robert Ricklefs (1978), the continuing decline in numbers led to a radical overhaul and the initiation of an intensive program involving radiotelemetry of wild birds and captive breeding.

The condor population continued to decrease; there were 22 in 1982 and just 3 wild birds by 1986.[61] As well as the effects of lead poisoning, some condors were experiencing reproductive failures associated with the ingestion of DDT (Kiff et al. 1979). In March 1986 a female laid an egg that broke immediately when she started incubation—and this was the last known breeding attempt in the wild before the Recovery Team decided to bring all of the remaining birds into captivity. Breeding programs started at the San Diego Wild Animal Park and the Los Angeles Zoo. Despite much controversy,[62] those programs were successful: the first chicks were reared in 1988, and the first releases back into the wild took place in 1992. By 2011 there were some 181 birds in the wild, and greater public awareness about the dangers of lead shot to the point where there now exists a growing trade in nonlead ammunition.

But this is not the end of the story. Sadly, the condor population can be sustained only through continued releases. There was great excitement in 2006 when a pair of condors was discovered breeding on the California coast near Big Sur. However, when researchers checked the nest, they were horrified to find eggshell fragments thinner than any they'd seen previously. Analyses confirmed that they contained DDE. How, after a forty-year ban could DDT still threaten the condors and other wildlife?

The explanation makes grim reading. The Montrose Chemical Company, based in Los Angeles, was once the world's largest manufacturer of DDT but stopped manufacture in 1982. Between the 1950s and 1970s, the company dumped around 1,700 metric tons (1,874 short tons) of untreated DDT waste into the Los Angeles sewage system. Remarkably stable, this DDT now contaminates millions of metric tons of marine sediment and is working its way up the food chain to the long-lived California sea lions on whose carcasses the coastal condors feed.[63] However, while it is inevitable that residues will continue to show up in predators for a long time, there is evidence that DDT levels are declining, since Brown Pelicans and Bald Eagles are now breeding successfully in this area.[64]

A problem similar to that induced by pesticides in the 1950s and 1960s has been the catastrophic decline in southern Asian vultures since the 1990s. Over just ten years the populations of three vulture species (all in the genus *Gyps*) crashed by a staggering 95 percent, bringing all three to the brink of extinction. Initially, this decline was thought to be the result of disease—autopsies revealed that affected birds were suffering from gout-like symptoms, with the liver and other organs covered in crystals of uric acid (Prakash et al. 2003). The reason for the illness became clear in 2004, when a team from the Peregrine Fund and BirdLife in Pakistan

established that affected vultures contained high levels of the anti-inflammatory drug diclofenac (Oaks et al. 2004). Diclofenac has been used as a human medicine for many years but was introduced as a veterinary drug in India and Pakistan only in the 1990s. Extremely cheap and widely available, diclofenac was used to treat inflammation, pain, and fever in livestock. By feeding on carcasses of animals that had been dosed with diclofenac, the birds went into rapid kidney failure and visceral gout before they died. The combination of widespread use of diclofenac and extreme sensitivity on the part of the vultures was disastrous. Apart from the loss of birds, vultures in Asia play an essential role in disposing of carcasses, both human and animal, thereby reducing the risks of disease in humans. Feral dogs, which carry rabies, have taken advantage of this food source and have increased as a result. The estimated cost to human health of the crash in Indian vultures between 1993 and 2006 was $34 billion (Markandya et al. 2008, cited in Wenny et al. 2011).

Identification of the cause of the vulture decline was followed by a ban on diclofenac in 2006, with farmers encouraged to switch to alternatives—such as meloxicam—that have no negative effects in vultures. This, coupled with captive breeding programs for the badly affected species, has improved the outlook for southern Asian vultures. In the two years following the ban the prevalence and concentration of diclofenac in livestock carcasses has dropped from 11 to 7 percent (Cuthbert et al. 2011). While this news is positive, there is still a long way to go to eliminate the threat completely: only a small proportion of livestock carcasses need contain lethal levels of diclofenac for that chemical to have an impact on vulture numbers (Green et al. 2004).

SCIENCE AND CONSERVATION

The idea of conserving or protecting a particular bird species is often emotionally motivated. This is an inevitable consequence of our enthusiasm for—and aesthetic attachment to—birds, and it explains why there's more interest in conserving birds than insects or microbes (e.g., Clark and May 2002). Some bird conservation projects are successful as a result of straightforward common sense: both the problem and its solution are obvious, as in the case of removing predators like rats or cats from seabird islands. In other cases common sense and emotional involvement are not sufficient, and a more scientific approach is necessary. The case of the Kakapo—now a model conservation story—provides an instructive example.

The Kakapo is a large, flightless, nocturnal parrot endemic to New Zealand. It is long-lived and has a lek breeding system, first described by pioneer conservationist Richard Henry in 1903. Kakapo spend much of their time on the ground, but using their beak and feet, they can climb into trees to feed on fruit, which they locate through a well-developed sense of smell.

Prior to the arrival of Polynesian settlers some two thousand years ago, the Kakapo was one of the commonest birds in New Zealand. Today it is among the rarest birds in the world. The Polynesians found the adult birds easy prey, and the dogs and *kiore* (rats) they brought with them ate the eggs and young. The rate of decline accelerated after European settlers arrived in the 1800s, through a combination of habitat loss and the introduction of yet more terrestrial predators. As the Kakapo became increasingly scarce, there was a scramble for "specimens" by European and North American museums, further accelerating the decline.

As early as 1894 it was apparent that the Kakapo was in danger of extinction, so in a government initiative led by Henry, several hundred were moved from the mainland to predator-free Resolution Island in Fiordland. Unfortunately, this first attempt was an abject failure: the island was less than 1,000 meters (0.6 miles) from the mainland, and within six years stoats had arrived and eliminated the Kakapo.

There were no further attempts to save the Kakapo until the 1950s, when the New Zealand Wildlife Service launched more than sixty expeditions to find them, mainly in Fiordland. The Wildlife Service, on noticing that the birds were absent from areas where Red Deer (another introduced species) had overgrazed the Kakapo's favorite food plants, decided that captive breeding was the best hope for saving the species. Five birds were caught in February 1961. They were transported to Mount Bruce, but all were found to be males, and within months of capture four of the birds died. Despite numerous expeditions, the next Kakapo was not captured until 1967.

With hindsight, that project was doomed to failure. The journey from the capture site to the "breeding site" at Mount Bruce took fifteen to forty hours and was extremely stressful for the birds. The aviaries at Mount Bruce, too few and too small, were located next to others occupied by different species, which further disturbed and stressed the Kakapo (Butler 1989).

In 1973 following a major review, responsibility for the Kakapo switched from the research side of the Wildlife Service to the management side, with Don Merton in charge of the field program. An ardent conservationist, Merton[65] was less keen on captive breeding than his predecessors, preferring "island transfers."

Miraculously, a new population of Kakapo was discovered in 1977 on Stewart Island, approximately 50 kilometers (31 miles) off the southern tip of South Island. Even more incredibly, Stewart Island was free of stoats, ferrets, and weasels, although it had a small number of feral cats. Stewart Island was too large for an island-wide eradication of cats, so in the 1980s all the Kakapo were moved to two other offshore islands: Little Barrier Island (where cats had been eradicated five years earlier) off the northern end of North Island and Whenua Hoa (Codfish Island), a few kilometers from Stewart Island.

Despite this monumental effort, the birds failed to thrive in their new homes—Codfish was overrun by rats that preyed on the Kakapos' eggs. By the mid-1990s the total number of Kakapo had fallen to just fifty-one individuals, leading to another urgent review (Imboden et al. 1995). The newly constituted National Kakapo Team advocated a more intrusive and intensive management regime—still in place at the time of writing (2013)—one that now appears to be working. The new regime includes (1) translocation of birds between (predator-free) islands when necessary; (2) supplementary feeding; and (3) intensive protection, observation, and management of all breeding attempts, including the artificial incubation of eggs and hand rearing of chicks. Alongside this, the Department of Conservation—which took over from the Wildlife Service in 1987—undertook a major rat eradication on Codfish Island in 1998 and a stoat eradication in 2001 on nearby Anchor Island, which is now suitable as a Kakapo sanctuary. These efforts seem to have worked, as both the frequency and success of breeding have increased (Elliot et al. 2001; Ron Moorhouse, pers. comm).

The initial phases of the Kakapo conservation effort were guided by emotion rather than scientific logic. The lack of expertise, lack of knowledge about the birds themselves, and the incredibly difficult Fiordland terrain all conspired to make this a monumentally difficult project. Yet the new management regime has been successful, and at the time of writing (January 2013) the Kakapo population stands at 131 individuals, including 11 chicks raised in 2011.

The Lord Howe Woodhen provides a rather different example of conservation in action. Lord Howe Island lies 570 kilometers (354 miles) off the east coast of Queensland, Australia. Like many oceanic islands, it once supported a range of endemic bird species, but in contrast to many islands in that region, Lord Howe was not colonized by Polynesians. On the other hand, sailors started visiting the island in the late 1700s and had settled by the 1830s, bringing with them pigs, dogs, cats, and goats. By 1924 nine of the island's thirteen endemic bird species were extinct. The woodhen survived in small numbers in the damp forests at the top of the island's two mountains, but as early as 1914 it was recognized that this was marginal habitat and provided space for only about ten territories. The population remained stable, at about 20 to 25 birds, between 1920 and 1980.

Efforts to save the woodhen started in the 1970s. The first phase involved describing the bird's biology, including differences between the sexes and territory size. Ben Miller then set about testing a set of hypotheses for why the birds were restricted to the mountaintops. A comparison of lowland and mountaintop habitats showed that neither food nor nest sites was likely to be limiting. Miller then tested several hypotheses regarding potential predators, including rats, pigs, cats, owls, feral dogs, and Pied Currawongs. As Graeme Caughley and Anne Gunn later said, "It is worth emphasizing the logic of Miller's approach. He did not succumb to building a case on a few observations alone . . . but compiled evidence for and against and then discarded a series of possible causes."[66] Pigs proved to be the main problem, and following their removal around 1980, the birds responded rapidly by moving into areas where pigs had previously been. Supplemented by captive breeding and translocation, the woodhen population had reached over 200 by 1997, occupying all available habitat. Though a decline was seen in 2001 (to 117), the population seems to have stabilized now at about 130 mature birds, and the island is thought to have a carrying capacity of about 220 birds (Bird-Life 2008).

This comparison between the woodhen and the Kakapo emphasizes that in addition to recognizing that a problem exists and that something needs to be done, conservation also requires clear, logical—scientific—thinking to best identify how problems are addressed. However, as Bill Sutherland and colleagues (2004) have pointed out, conservation's use of scientific methods has so far been inadequate. Their analogy is that conservation today is rather like medicine fifty years ago—that is, based on the "expert wisdom" of certain individuals rather than evidence based approaches that use strict criteria to assess the effectiveness of particular treatments.

POLITICS AND CONSERVATION

The greatest issue facing bird conservation is habitat loss, through destruction, fragmentation, or degradation. Vast tracts of old-growth forest have been destroyed for timber, wetlands have been drained, and grasslands converted for agriculture. Of the approximately 1.4 million square kilometers (540,543 square miles) of prairie that once stretched down from Canada through the United States and into Mexico, only 1 to 2 percent now remains in its original form. With world population growth, the pressure on environmental resources continues to grow, and this reality must be at the core of modern conservation science. The trade-offs are critical: convincing policy makers and the public that the benefits of species and habitat conservation outweigh the benefits of development is one of the greatest challenges facing conservationists. There is progress—for example, in the form of agri-environmental schemes in which farmers are provided with compensation for losses they incur by implementing wildlife-friendly measures in their farming practices. And industries, driven by the vast bodies of legislation that built up over the twentieth century, are also being forced to acknowledge that, as they continue to develop on a global scale, biodiversity preservation must be factored into their work. This recognition did not come easily, and it still forms the basis for much conflict. The following are key examples that demonstrate the conflict but also show how conservation has not only moved into the courtroom but has also widened its scope to entire ecosystems.

The Red-cockaded Woodpecker was federally listed in 1970—one of only two woodpecker species[67] to be protected under the Endangered Species Act. Its natural habitat is the mature, open, longleaf pine forests of the southeastern United States, and it is the only woodpecker species to excavate nest holes in living trees. This specialization means that the species needs trees that are sufficiently old enough and large enough to allow nesting. Unusually for woodpeckers, they are cooperative breeders, and because of this each group needs an area of about 80 hectares (200 acres). From the late 1800s to the mid-1950s, timber logging in the United States decimated the bird's native habitat; by 1946 the longleaf pine forests had decreased to about 17 percent of their original size, and by 2000 there was estimated to be just 3 percent left (Costa 2001: 310). The effects have been devastating for the woodpeckers, because, even if stands of forest remain, there is often a scarcity of suitable large nesting trees. It is estimated that there are only about six thousand groups of birds left, in small, fragmented forest patches.[68]

The Red-cockaded Woodpecker became the figurehead for forest management. In 1985 the US Forest Service was sued by conservation bodies in Texas for neglecting the woodpeckers. The court agreed, and this landmark ruling set a new precedent— essentially stating that in its timber harvesting the Forest Service was doing as much to imperil the species as if its employees had gone out with guns and started shooting the

birds. A significant overhaul of existing forestry management policy followed, with new guidelines on endangered species management and legislation and big changes in the management of public lands across much of southeastern America.

The second recovery plan, published in 2003, advocated a combination of single-species and ecosystem approaches to preserve the woodpeckers. At the species level, detailed research on the woodpecker's breeding biology led to bird translocations and the erection of artificial cavities to encourage nesting. At the ecosystem level, forests are now managed differently, with prescribed burning regimes to prevent understory encroachment and the preservation of large, old trees that have the potential to be used for nesting.

The Red-cockaded Woodpecker was a landmark case, but it was followed by an even more controversial one: the Spotted Owl. Found in the old-growth forests of northwestern America, this species was federally listed under the Endangered Species Act in 1990. More than 70 percent of its habitat has been lost to logging and land conversion, and what remains is often fragmented—a problem when an owl's natural foraging range may exceed 1,000 hectares (2,500 acres).[69] In 1989 the Interagency Scientific Committee (ISC) was convened, led by Jack Ward Thomas, and the following year the ISC published a landmark monograph on the owl and the decline of its habitat. The effects of that report were far-reaching, fueling intense debate from conservationists and those in the forestry industry: "It wasn't my ambition to become notorious—or even famous. It was obvious that this was not going to be a matter of walking in, dropping a scientific report on somebody's desk, and fading back into obscurity. This was crossing the frontier between science and management."[70] Protecting the owl meant job losses—as many as a hundred thousand—and many people were furious. The FBI told Thomas that he couldn't go home for several weeks, and his family was under guard for a week or so. Someone even left a Molotov cocktail on his doorstep.

The Bush administration stalled on implementing the ISC recommendations, fearing the loss of jobs and money from the logging industry. By 1992 there were over a dozen lawsuits and three separate court injunctions, the latter enforcing a blanket ban on logging in the old-growth forests. It was total gridlock. The owl became a political issue in the presidential election of 1992 between George H. W. Bush, Bill Clinton, and Ross Perot. Indeed, Bush may have lost his election over Spotted Owls (Thomas 2002). Bush campaigned on the issue of "owls versus jobs"; Ross Perot said, "When everybody is out of work nobody will care about spotted owls except how to cook them"[71]; and Clinton essentially said, "I feel your pain"[72] but promised to do something to develop a solution right after the election. Which he did, holding a Forest Summit in Portland in 1993 and asking Thomas to lead a team to figure out a solution. Ninety days later came the FEMAT (Forest Ecosystem Management Assessment Team) report, which developed into Clinton's Pacific Northwest Forest Plan, adopted in 1994. Logging could start again, but at a much reduced rate: one million board feet of wood could be cut from designated sites, as opposed to the five or six million board feet cut in previous years.

The Spotted Owl became the poster child for protection of an entire ecosystem: "Within my professional lifetime there was a quantum jump from a focus on individual species and individual stands of vegetation to landscape ecology and conservation biology-ecosystem management."[73] Yet even with the forests protected, the species continues to decline, with a new challenge from competition with invasive Northern Barred Owls.

These examples demonstrate how intertwined conservation science has become with government policy. The International Committee for Bird Protection (ICBP) was established in 1922. Today, the ICBP—now known as BirdLife International—is a global consortium of nongovernmental organizations, recognized as the leading authority on the state of the world's birds and their habitats. BirdLife's *World List of Threatened Birds* is accepted by the International Union for the Conservation of Nature (IUCN) as its official global Red List for birds.

The work of organizations like BirdLife has put bird conservation right at the heart of development. For example, BirdLife's *World List* is incorporated into the World Bank's operational policies; it prohibits the funding of projects that affect designated "Critical Natural Habitats," which include "areas with known high suitability for bio-diversity conservation; and sites that are critical for rare, vulnerable, migratory, or endangered species,"[74] as specified by BirdLife for birds. The World Bank policy in turn influenced the drafting of an equivalent policy[75] at the International Finance Corporation, specifying that loans will only be made to projects that meet their criteria on biodiversity and habitat preservation.

The state of the world's birds is not good, with many species in decline and 12 percent of all species at risk of imminent extinction (BirdLife 2008). It is difficult not to be pessimistic, yet concerted efforts by large numbers of organizations and individuals across the world are attempting to reduce the risk of extinction. BirdLife, for example, claims that without their efforts sixteen additional species would have gone extinct between 1994 and 2004.[76] It is also important to recognize that there have been some outstanding success stories: in North America populations of Whooping Crane, California Condor, Peregrine Falcon, and Bald Eagle have recovered to some extent. Another success story is the recovery of seabird populations following the removal of predators like cats and rats from offshore islands. The technology is now sufficiently good that a one-off investment in predator removal can provide a safe breeding island in perpetuity (Brooke 2012). However, it is worth bearing in mind that for every species "saved," ten have gone extinct.

Studies of birds have led the way for many aspects of conservation science; for example, there are almost twice as many atlases for bird species distributions than for the next most atlased taxonomic group, plants, and over eight times as many as those for mammals (Brooks et al. 2008). Birds have also been disproportionately important in the identification of sites of global conservation significance: the concept of Important Bird Areas was developed in the 1980s, and as of 2009 there were nearly 11,000[77] IBAs in over two hundred countries worldwide. This has led to similar programs being developed for butterflies, plants, and freshwater biodiversity (Brooks et al. 2008). Even though ornithology has led in some respects, massive gaps remain in our knowledge of many bird species, especially those in the Caribbean, Pacific regions, and Central America. Many threatened species are not being studied, yet without basic scientific information it is impossible to embark on a program to save them. More can also be done to disseminate and appraise conservation interventions in order to guide future decision-making, as decisions are still often made without a solid, underpinning body of evidence.

The public interest in birds has in part fostered the compilation of national breeding atlases, beginning in the 1960s, such that we now have a more detailed knowledge of the distribution and abundance of birds in Europe and North America than of any other form of wildlife (Brooks et al. 2008). The results are not encouraging, however, and demonstrate the ongoing declines of many species once considered common (Keast and Morton 1980).

Bird conservation remains a constant challenge, and our main message is that scientific knowledge of bird biology—especially from a historical perspective—is crucial if we are to slow—and ultimately halt—the decline in bird numbers.

BOX 11.1 Nigel Collar

I became a professional ornithologist and conservationist by virtue of repeated doses of knife-edge good fortune. The story owes nothing to talent or persistence. It does, however, have its roots in that involuntary love of nature that comes into a life out of nowhere—triggered by a pet or a book—and configures itself in arbitrary patterns, weak or strong, in each of us. Perhaps the best we can ever do is to give it the space it needs to make the fullest sense of our lives.

Kids of the "baby-boomer" generation—those born in the euphoric aftermath of the Second World War into the financially and emotionally rationed simplicity of the late 1940s and 1950s—were expected to have hobbies. Many of us took to collecting things like cigarette cards, comics, miniature replicas of cars, synchronized by sellers to the weekly disbursement of pocket money; but it took a tabby kitten, acquired by my household when I was four, to spark my own interest in a "hobby." When this gentle creature's tongue was bitten in half by a Common Blackbird whose capture it badly mismanaged, I sat with it for days while it recuperated on a living room chair. Soon I was imitating it, going round the house on all fours, meowing, *being* a cat. Three years later I came out of the Disney film *Vanishing Prairie* struck dumb with the love of a cat I

had never heard of before, and for months the words "cougar" and "puma" played over in my imagination as the language of poetry would come to do in my teenage years.

Soon afterward my grandfather died. I had no idea then that he loved birds, but my father, aware of my own emerging love of animals, gave me several books that had been his own dad's: *Birds in a Garden Sanctuary*, *How Birds Live*, and—infinitely the most exciting and unexpected—*A Field Guide to the Birds of Britain and Europe*, published only the year before. As I turned its pages for the first time, a door swung open on the natural world: treecreepers, nuthatches, turnstones, oystercatchers, linnets, and woodlarks all suddenly became real creatures in the sky of my imagination, with possibilities, notions, and propositions, illuminated by a burning desire to see them and by the book's tantalizingly prosaic text.

My first years of bird watching were done from this book by eyesight. When I was eleven my parents got me a cheap pair of 10x50 binoculars. Holidays in Scotland and France took on totally new dimensions. The bird-watching father of a friend at school learned of my passion, and almost every school-term Sunday afternoon in the early 1960s he would drive me out to Chew Valley Lake to look for ducks and migrant waders.

Inevitably I wanted to become a zoologist but was revolted when a girlfriend told me that she had to dissect a bull's eyeball as part of her studies. A teacher thought someone of such fastidious sentiments would do better in the study of literature, so by the time I was sixteen I had conceded that my love of nature would forever be an amateur affair of the heart. Throughout my university years, first as an undergraduate and then while doing a PhD on George Orwell, I went bird watching alone or with friends, to Fair Isle for unknown rarities and to Spain for anticipated marvels—one of them the Great Bustard, a bird that caught my imagination more strongly than any cougar—but never seriously considered that this abiding obsession could ever bear on my future career.

At the end of 1973 I needed a job—and a year to write up my thesis—and wanted for personal reasons to escape to a foreign land. A post came up at the University of Lisbon, in a country still in the grip of a deadhead dictatorship. They produced two contracts, one to start work, which we signed straight away, the other to be paid, which they withheld for six months while they checked out my political sympathies. That winter I had to keep leaving the country to renew my tourist entry visa, and each time I would drive to different points on the Spanish frontier where I could hope again to see the wondrous Great Bustard.

Arab oil sanctions after the Yom Kippur War meant that, in September 1974, the month I presented my thesis, inflation back in Britain was running at 28 percent. Academic opportunities there evaporated. By spring 1975, after a wasted winter in England, I was thinking of returning to my job at the University of Lisbon when an article in the *Guardian* reported that people trying to reintroduce Great Bustards to Salisbury Plain were puzzled why the birds would not breed, and that they were looking for someone to help them solve the problem. I wrote them a fan letter, saying I could try to organize some research (by others) on the species, since I knew where to find it. They sent me a telegram offering me the job. I thought they were taking as much of a risk as I was, so I enrolled at the Edward Grey Institute in Oxford, where my research could be watched over by real ornithologists, and my own long-term future and the Great Bustard's thereby better guaranteed.

I worked on those birds for three years. By chance at that time the great publishing enterprise *The Birds of the Western Palearctic* needed help with its bustard entries, which I supplied, using the tiny but brilliant Alexander Library in the EGI. By further chance *BWP* needed other help, too: their meticulous senior editor, Ken Simmons, required an assistant. My use of the Alexander happened to have shown what an extraordinary resource it was, and I was recruited. (Another piece of luck: the letter fell down the back of some filing cabinets where all EGI post was left, and only a colleague, looking for something urgent for himself, found it, six weeks after the offer had been sent and a day before it was due to be withdrawn.)

Meanwhile, the old International Council for Bird Preservation needed a chairman of their Bustard Specialist Group, an honorary position they offered to me. I got a few projects off the ground on some of the most threatened species—Lesser Florican, Houbara Bustard, Great Bustard—and built a small program on a shoestring. This miniature volunteer effort perhaps held sway when three years later ICBP advertised for a Red Data Book compiler. I got the job and have worked for what is now BirdLife International ever since.

It is difficult now to credit how much I have enjoyed the work with the ICBP and BLI, given the failure of conservation over the thirty years they have paid my salary. Despite massive progress with identifying and documenting threatened birds and Important Bird Areas, and growing the global partnership to stand up for these things, birds are still going extinct, their populations are declining, wilderness is in retreat, and human hegemony of the planet is virtually complete. The most depressing aspect of all this is how few people seem to care. Perhaps it is the fate of all conservationists to fall in love with the Earth, and to die brokenhearted.

AFTERWORD

You can know the name of a bird in all the languages of the world, but when you're
finished, you'll know absolutely nothing whatever about the bird. . . . So let's look
at the bird and see what it's doing—that's what counts. I learned very early the
difference between knowing the name of something and knowing something.

RICHARD FEYNMAN, US PHYSICIST, WRITER, AND EDUCATOR (1918–88)

WHY IS IT THAT WE KNOW MORE ABOUT THE
biology of birds than almost any other group
of animals? What has driven ornithology
to its current level of sophistication? In our
opinion, four things: people, education,
funding, and technology—and in that order.
It is clear from our survey of developments
since Darwin that particular individuals
have played a key role in promoting ornithology through the twentieth century: most notably Stresemann, Mayr, and Lack. Through
their personality and intellectual attributes,
these individuals changed the nature of
ornithology: Stresemann in the 1920s to
1940s for recognizing that there was more
to birds than museum studies; Ernst Mayr
in the 1940s to 1980s for being a driving
force in the modern syntheses of evolution
and systematics—incorporating natural selection and genetics—and for setting a high
scientific standard for ornithologists; and
David Lack in the 1940s to 1960s for bringing the study of bird ecology and behavior
into an individual selection framework.

Although they were not the only architects, these three individuals overturned
previous paradigms and forged a way forward. They did so mainly by their primary

research but also by promoting ornithology
as part of biology—for example, as presi-
dents of the International Ornithological
Committee—and by encouraging young scientists in various ways. In addition, each had
a broad view of ornithology, enabling them
to "think big" and address major questions.
Another attribute that we feel played a major
part in their effectiveness was that they each
produced inspirational books summarizing and synthesizing areas of interest that
encouraged others to develop their ideas:
Stresemann's *Aves* (1927–34); Mayr's *Systematics and the Origin of Species* (1942), *Animal
Species and Evolution* (1963), and *The Growth of
Biological Thought* (1982); and Lack's *Darwin's
Finches* (1947a), *Natural Regulation of Animal
Numbers* (1954), *Population Studies of Birds*
(1966), and *Ecological Adaptations for Breeding
in Birds* (1968).

At an individual level what makes someone a successful ornithologist—or indeed
any other kind of scientist? Interestingly,
the attributes that Darwin considered important in his own success—recounted in
his autobiography—are as true today as
they were then. These are (1) a "steady and
ardent"[1] love of natural science, for which

we can read sustained enthusiasm; (2) "the strongest desire to understand or explain whatever I observed,—that is, to group all facts under some general laws . . . [and] the patience to reflect or ponder for any number of years over any unexplained problem,"[2] for which we can read curiosity and tenacity; and (3) open mindedness: "I have steadily endeavoured to keep my mind free, so as to give up any hypothesis, however much beloved (and I cannot resist forming one on every subject), as soon as facts are shown to be opposed to it.[3]

Our exploration of ornithology since *Origin* suggests two other qualities that are important for success: hard work and mentors. Nearly all the ornithologists who have made significant contributions to ornithology have been incredibly hard working. Less obviously perhaps, almost all of them had mentors, one or more influential senior colleagues who proffered guidance and often eased the way through their knowledge or contacts. Ernst Mayr's mentor, for example, was Stresemann, David Lack's was Julian Huxley, and Lack himself was mentor to generations of ornithologists. To these various attributes we can add Darwin's final two: "a fair share of invention as well as of common sense."[4]

The expansion of higher education in the 1960s vastly increased the number of young people entering university in both Europe and North America. Some of these new students were interested in birds. This was a consequence of the increasing popularity of bird watching that developed through the first five decades of the 20th century, fueled by wildlife television programs and by rapidly improving field guides, binoculars, telescopes, and cameras. More people studying birds—due to significantly increased government funding for scientific research on birds, as well as the growing amateur involvement in ornithological research—meant more scientific papers, especially in Europe and North America. Ornithological journals proliferated and expanded in size, but inevitably the discipline also became more fragmented, with more specialization and more concept-oriented journals, such as those in behavioral ecology where ornithologists published their bird papers alongside those on mammals, fish, reptiles, amphibians, and invertebrates. This fragmentation and specialization is exactly what we could have predicted, because the same pattern has occurred in all expanding areas of science (de Solla Price 1963; Mayr 1982).

More ornithologists and more specialization meant that after about 1970 it became less and less likely that any one individual would have the breadth of knowledge that the pioneers—Alfred Newton, Erwin Stresemann, Ernst Mayr, or David Lack—had and therefore less likely that particular individuals would dominate the entire field. Instead, leaders have emerged in specific areas of ornithology—molecular biology and systematics, migration, physiology, behavioral ecology, or even subsets of these—some of whose work we have described and whose autobiographies we have included in these pages.

Recognizing that science and technology drive material progress, Western governments have provided substantial funds for scientific research, both applied and "pure," across a broad front throughout the twentieth century, especially since 1950 (Roberts 1999). The funding for pure, or "blue skies" research came with considerable intellectual freedom, allowing researchers to pursue their interests whether it was astrophysics, the biology of birds, or something even more esoteric. The success that this freedom engendered meant that over the next few decades some especially bright scholars were attracted into biological—including ornithological—research.

It is ironic that by the 1980s researchers had started to become victims of their own success. The increasing number of researchers led to a shortage of funds, which governments "managed" through increased bureaucratization across all branches of science, stultifying creativity in the process and generating more "safe" science than novel ideas (Smolin 2006). This trend accelerated in the early years of the twenty-first century because of the global economic downturn, resulting in both reduced research spending and less intellectual freedom, with governments directing funding toward the pressing issues of food security and energy. Superficially sensible, many have questioned the wisdom of restricting research in this way, given that—as we report in this book—the applications of scientific research can rarely

be predicted in advance. Certainly, pockets of intellectual freedom and generous funding for basic biology still exist—including, for example, Britain's Royal Society Fellowship scheme, the Max Planck Institutes, and Canada's Natural Sciences and Engineering Research Council. Even so, many researchers wonder whether the era of curiosity-led research is largely over and that in fifty years' time researchers—including ornithologists—will look back on the second half of the twentieth century with nostalgia and envy.

And what of technology? The remarkable advances brought about by the sonograph, molecular genetics (electrophoresis, PCR, fingerprinting, genomics), fMRI, radiotelemetry, geolocators, GPS units, microcomputers, and portable digital phones, recorders, mp3 players, and cameras clearly fueled some aspects of the ornithological revolution of the twentieth century. Ornithologists have been quick to exploit new technologies, possibly because, with increasing numbers of researchers, competition has intensified. Certainly, the alacrity with which

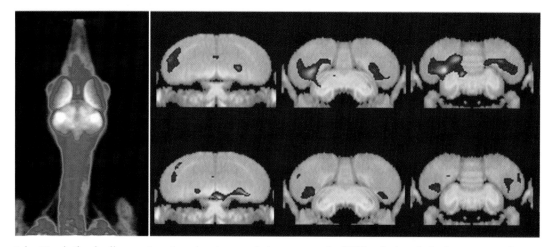

John Marzluff and colleagues (2012) used positron emission tomography (PET) to look at the brain responses of American Crows to different human faces.

ornithologists have recognized the potential of new developments has helped to keep bird studies at the forefront of many areas of zoological research (Konishi et al. 1989).

One of the most striking aspects of the development in ornithology over the past century has been the continued improvement in scientific rigor, with increasing focus on theory, hypothesis testing, statistics, and mathematical modeling. Originally this was a consequence of the shift away from purely descriptive studies of taxonomy and the move toward what in the late nineteenth century were considered to be more "philosophical" topics, such as evolution, ecology, and behavior, triggered by Darwin's work. Indeed, as Ghiselin (1969) has pointed out, Darwin was among the first to explicitly formulate and test hypotheses. Evolutionary topics have also encouraged a more theoretical approach, which in turn fostered the experimental testing of hypotheses, greater quantification, and more sophisticated statistical analyses—the latter not always to everyone's delight. Some ornithologists—Mayr is an outstanding example—also acknowledged the value of both the history and philosophy of science to inform everyday research practice (Mayr 1982).

Since Darwin's day, ornithologists have been at the cutting edge of many theoretical developments (Mendelian, population, and molecular genetics, the Modern Synthesis, the new systematics, game theory, optimal foraging theory, population dynamics, life history theory) and have employed mathematical models as a basis for generating testable hypotheses in disparate subjects (including sexual selection, island biogeography, optimal foraging, optimal clutch size, life history trade-offs, population genetics,

flight energetics and aerodynamic models, metabolic models with respect to energy budgets and energy expenditure, and phylogenetic models). It seems likely that theoretical aspects of ornithology have flourished as they have because of the wealth of empirical data and an abundance of empiricists ready to collect more data to test those models.

In writing this history, we have relied on a variety of both published and unpublished sources, including obituaries, letters, and the occasional biography. With the demise of letter-writing and the potentially transitory nature of email messages, future historians may struggle to get information. To encourage future histories of ornithology, we urge our readers to write their own biographies—long or short—and either publish them, as John Emlen (1996) and David Snow (2008) have done, or simply place them on this book's website (http://myriadbirds.com) for others to read. We also encourage ornithologists to include good-quality images of themselves with their text.

It is almost always futile to attempt to predict the future. We will resist that temptation but instead make some observations. As our voyage through the modern history of ornithology has shown, science progresses in different ways, driven by the adoption of new technologies and new ways of thinking. It should also be obvious that science is dynamic, exemplified by the idea—prevalent until the mid-1980s—that most birds were socially and sexually monogamous, which we now know is not true. Topics of interest come and go, but there is one topic whose relentless progress should be of concern to us all. As the human population continues to increase, bird populations will continue to decline. Birds provide a convenient indicator of the quality of the environment, but they also contribute immensely to the quality of that environment and of our lives. In the last hundred years or so, a lot of ornithology has been motivated by the human fascination for birds and the excitement of intellectual curiosity. But as global bird numbers continue to decline, the pursuit of knowledge for its own sake may seem like a bizarre luxury. A century from now, if someone decides to repeat our survey of the preceding hundred years, it is unlikely that they would be able to use the same title—*Ten Thousand Birds*—for at the current rate of extinction, it is more likely that there will be fewer than 9,000 bird species on the planet. The long-term health of birds and other wildlife depends on teaching our children and students to value the natural world, but it also depends on our training them to be both effective ambassadors for ornithology and first-rate scientists, so that they are able to make the right decisions.

APPENDIX I: SOME HISTORIES OF ORNITHOLOGY

The importance of bibliographical research can scarcely be over-estimated. A scientist should be wedded to the literature of his subject and as said of the ordinary marital relation, it will doubtless often prove that the partner is the better half.

—WALDO LEE MCATEE (1942), IN A PAPER ON THE MAJOR
WORKS BY SOME NORTH AMERICAN ORNITHOLOGISTS

We list here, in chronological order, a few histories of ornithology that cover some of the work done in this field since Darwin.

Fowler SP (1862). *Ornithology of the United States, Its Past and Present History.* Proceedings of the Essex Institute 2:.327–34. A useful summary of the major works to the middle of the nineteenth century.

Allen JA (1876). *Progress of Ornithology in the United States during the Last Century.* American Naturalist 10:.536-50. A summary of the major works, both books and papers, as well as an assessment of the progress that had been made since Thomas Jefferson's bird list of 1776.

Newton A (1893–96). *A Dictionary of Birds.* A and C Black: London. The Introduction (pp. 1–124) to this encyclopedia is a comprehensive, chronological, and scholarly survey of the history of ornithology to the end of the nineteenth century; includes much information on the history of systematics and regional avifaunas.

Coues E (1896). *Key to North American Birds* (4th edition). Estes and Lauriat: Boston. Pages xi–xxix are a historical overview of predominantly North American ornithology to the end of the nineteenth century.

Palmer TS (1900). *A Review of Economic Ornithology in the United States.* Yearbook of the US Department of Agriculture 1899: 259-92. An excellent overview of mainly nineteenth-century studies on the economic importance of birds with respect to agriculture, game, guano, feathers, eggs, and invasive species.

Mullens WH, Swann HK (1917). *A Bibliography of British Ornithology from the Earliest Times to the End of 1912.* Macmillan: London. As its name implies, the focus is on British ornithologists; excellent set of mini-biographies (but not a history).

Gurney JH (1921; reprinted 1972). *Early Annals of Ornithology.* Witherby: London. A selective but scholarly account, focused on Britain; chronological by century to the end of the 1800s; readily available in secondhand shops.

Chapman FM, Palmer TS (eds) (1933) *Fifty Years' Progress of American Ornithology, 1883–1933.* American Ornithologists' Union: Lancaster, PA. Covers fourteen ornithological topics by various authors; introductory and concluding information relating to the AOU.

Allen EG (1951). *The History of American Ornithology before Audubon.* Transactions of the American Philosophical Society 41: 385–591. Comprehensive, rich in detail; chronological; much more than its title suggests, including European history.

Stresemann E (1951). *Die Entwicklung der Ornithologie. Von Aristoteles bis zur Gegenwart.* F. W. Peters: Berlin. [translated into English as Stresemann, E. (1975) *Ornithology from Aristotle to the Present.* Harvard University Press:

Cambridge, MA]. A scholarly account of ornithology written in the late 1940s; in 1975 translated into English by H. J. and C. Epstein with a final chapter ("Materials for a History of American Ornithology") by Ernst Mayr; an excellent source of information on ornithologists to the mid-1900s.

Sibley CG (1955). "Ornithology." Pp. 629–59 in Kessel EL (ed.) *A Century of Progress in the Natural Sciences, 1853–1953*. California Academy of Sciences: San Francisco. An excellent, if selective, worldwide overview with sections on ornithological journals and monographs.

Farber PL (1982). *The Emergence of Ornithology as a Scientific Discipline: 1760–1850*. Reidel: Dordrecht [re-issued as Farber, P. L. (1997) *Discovering Birds: The Emergence of Ornithology as a Scientific Discipline: 1760–1850*. Johns Hopkins University Press: Baltimore]. Focuses on the towering figures of Buffon and Brisson during this period.

Davies WE, Jackson JA (1995). *Contributions to the History of North American Ornithology*. Nuttall Ornithological Club Memoir 12; Davies WE, Jackson JA (2000) *Contributions to the History of North American Ornithology*. Nuttall Ornithological Club Memoir 13. Edited volumes comprising short chapters by a wide range of authors on diverse topics, mostly rather specific, such as "ornithology in the Canadian Wildlife Service."

Voous K (1995). *In de Ban van Vogels: Ornithologisch Biografisch Woordenboek van Nederland*. Uitgeverij Scheffers: Utrecht. Ornithology in the Netherlands in the twentieth century; biographies of ornithologists born before 1950.

Barrow MVJ (1998). *A Passion for Birds: American Ornithology after Audubon*. Princeton University Press: Princeton, NJ. A detailed, readable account; includes much on the formation of North American ornithological societies and the relationship between professional and amateur ornithologists; mainly focused on mid-1800s to 1930s, and the professionalization of ornithology.

Battalio JT (1998). *The Rhetoric of Science in the Evolution of American Ornithological Discourse*.

Ablex Pub.: Stamford, CT. A unique analysis of the structure of language, arguments, images, and graphical displays in *The Auk* from 1884 to 1990.

Robin L (2001). *The Flight of the Emu: A Hundred Years of Australian Ornithology*. Melbourne University Press: Melbourne. Focuses on the Royal Australian Ornithological Society (RAOU) from 1901 to 2001; chapters are thematic and organized chronologically; spans bird watching and scientific ornithology; well illustrated.

Walters M (2003). *A Concise History of Ornithology*. Helm: London. Based largely on Stresemann (1975); formulaic, chronological structure with detailed information on various classification systems; final chapter by J. Coulson on twentieth-century ornithology added at publisher's request.

Nowak E (2005). *Wissenschaftler in turbulenten Zeiten: Erinnerungen an Ornithologen, Naturschützer und andere Naturkundler*. Stock und Stein: Schwerin. Personal memories and biographical details of around fifty mid-twentieth century European ornithologists active during the Second World War and the subsequent twenty-five years.

Bircham P (2007). *The History of Ornithology*. Collins: London. Focused mainly on British ornithology; on a wide range of disparate topics, attractively illustrated and produced.

Chansigaud V (2007). *Histoire de l'Ornithologie*. Delachaux & Niestle: Paris. [translated into English as Chansigaud V (2009) *The History of Ornithology*. New Holland: London]. A chronological account of the entire history of ornithology; richly illustrated, with some information on French ornithologists; contains a useful, illustrated timeline.

Birkhead TR (2008). *The Wisdom of Birds*. Bloomsbury: London. Structured by topic, from fertilization to development, maturation, territory acquisition, migration and longevity; from Aristotle to the twentieth century; beautifully illustrated with historically important paintings of birds by various artists.

Neumann J, et al. (2010). *Lebensbilder sächsicher Ornithologen*. Mitt. Ver. Sächs. Ornithol. [Mitteilungen des Vereins Sächsischer Ornithologen] 10, Sonderheft 3. Hohenstein-Ernstthal. Detailed biographies of all known ornithologists in Saxony, Germany.

Pittie A (2010). *Birds in Books: Three Hundred Years of South Asian Ornithology—A Bibliography*. Permanent Black: India. A comprehensive, annotated listing of about 1,700 books that contain information about the birds of Afghanistan, Bangladesh, Bhutan, India, the Maldives, Myanmar, Nepal, Pakistan, Sri Lanka, and Tibet; with a brief overview of the history of ornithology in the region since 1700 and short biographies of about 200 prominent ornithologists whose books are included in the annotated list.

APPENDIX 2: FIVE HUNDRED ORNITHOLOGISTS

WE MENTION OVER SEVEN HUNDRED ORNIthologists in this book, but space allows us to list only about five hundred here. This biographical register saves us from including birth and death dates in the main text and provides a quick source of the most basic biographical information. The information is arranged as last name, first name, initials, birth-death dates, nationality, and (if different from country of birth) the country or countries (abbreviations below) in which that person was employed as an ornithologist, and research field (categories below). We have also included a photograph of a sample of individuals whose image does not already appear in the text.

COUNTRY CODES (from http://www.immigration-usa.com/country_digraphs.html): AR = Argentina, AT = Austria, AU = Australia, BE = Belgium, CA = Canada, CN = China, CH = Switzerland, DE = Germany, DK = Denmark, EE = Estonia, FI = Finland, FR = France, HU = Hungary, IE = Ireland, IL = Israel, IT = Italy, JP = Japan, LK = Sri Lanka, LT = Lithuania, LV = Latvia, NL = the Netherlands, NO = Norway, NZ = New Zealand, PA = Panama, RU = Russia, SE = Sweden, SN = Senegal, US = United States of America, UK = United Kingdom, ZA = South Africa.

FIELDS OF STUDY: anat = anatomy, behav = behavior, coll = collector, cons = conservation, ecol = ecology, evol = evolution, feath = feathers and flight, gene = genetics, hist = history, life = life history, migr = migration, homing, and navigation, orni = ornithology in general, paleo = paleontology, phys = physiology, popn = population ecology, syst = systematics.

Abbott, Ian J 1947 AU-CA evol
Ahlquist, Jon E b? US syst
Aldrich, John W 1906–95 US popn
Alerstam, Thomas b1949 SE migr
Alexander, Horace G 1889–1989 UK orni
Alexander, Wilfred B 1885–1965 UK orni
Allen, Arthur A 1885–1964 US behav
Allen, Joel A 1838–1921 US behav
Amadon, Dean 1912–2003 US syst
Andersson, Malte 1941 SE behav
Armstrong, Edward A 1900–78 UK behav-ecol
Aschoff, Jürgen 1913–98 DE phys
Ashmole, N Philip b1934 UK ecol
Askenmo, Conny b1941 SE behav

Alerstam   Amadon   Armstrong   Aschoff   Bang   P. Bateson

Audubon, John J 1785–1851 FR-US-UK orni

Axel, Bert 1915–2001 UK orni

Bairlein, Franz b1952 DE migr

Baker, Allen J b1943 NZ-CA syst

Baker, John R 1894–1973 UK ecol

Baker, E C Stuart 1864–1944 UK behav

Baldamus, Eduard 1812–93 DE behav

Baldwin, S Prentiss 1868–1938 US orni

Bang, Betsy G 1912–2003 US phys

Barrett, John 1913–99 UK orni

Bateson, Patrick P G b1938 UK behav

Bateson, William H 1861–1926 UK gene

Beach, Frank A 1911–88 US behav

Beal, Foster E 1840–1916 US ecol

Bechstein, Johann M 1757–1822 DE orni

Beck, Rollo H 1870–1950 US coll

Bécoeur, Jean-Baptiste 1718–77 FR orni

Beebe, William 1877–1962 US orni

Beecher, William J 1914–2002 US anat

Benkman, Craig W b1956 US ecol-evol

Bennett, George 1804–93 UK anat

Bennett, Peter b1960 UK ecol-evol

Bent, Arthur C 1866–1954 US orni

Benvenuti, Silvano b1944 IT migr

Berkhoudt, Herman b1945 NL anat-phys

Bert, Paul 1833–86 FR phys

Berthold, Arnold A 1803–61 DE phys

Berthold, Peter b1939 DE phys-migr

Boag, Peter T 1953 CA evol

Bock, Walter J 1933 US anat-evol

Borelli, Giovanni A 1608–79 IT feath

Bourlière, François 1913–93 FR migr

Bowman, Robert I 1926–2009 CA-US evol-ecol

Brodkorb, W Pierce 1908–92 US paleo

Brooke, Mike de L b1950 UK cons

Broom, Robert 1866–1951 UK paleo

Brown, Jerram L b1930 US behav

Brown, Richard G B 1935–2010 UK-CA migr-cons

Bruch, Carl 1789–1857 DE syst

Brush, Alan H 1934 US feath-evol

Bumpus, Hermon C 1862–1943 US evol

Burke, Terry 1957 UK gene-evol

Burkitt, James P 1870–1959 IE orni

Burley, Nancy T b1949 US behav-evol

Butler, Pat J b1943 UK phys

Buxton, John 1912–89 UK orni

Campbell, W Bruce 1912–93 UK ecol

Carson, Rachel L 1907–64 US cons

Cayley, Neville W 1886–1950 AU orni

Chance, Edgar P 1881–1955 UK behav

Chapin, James P 1889–1964 US syst

Chapman, Frank M 1864–1945 US orni

Charmantier Anne b1977 FR-UK ecol-evol

Chiappe, Luis M b1962 US paleo

Cinat-Tomson, Hilda b?–d? LV behav

Clarke, William E 1853–1938 UK migr

Clayton, Nicky S b1962 UK-US behav

Cobb, Stanley 1887–1968 US anat

Coburn, Charles A b?–d? US behav

Cockburn, Andrew b1954 AU behav

Cody, Martin L b1941 US ecol

Collinge, Walter E 1867–1947 UK ecol

Berkhoudt

Burley

Charmantier

Clarke

Clayton

Coues

Conder, Peter 1919–93 UK orni
Cooch, F Graham 1928–2008 US-CA ecol
Cooke, Fred b1936 UK-CA gene-ecol
Cope, Edward D 1840–97 US palco
Cordeaux, John 1831–99 UK migr
Coues, Elliott 1842–99 US syst
Coulson, John C b1932 UK ecol-behav
Cracraft, Joel b1942 US syst-evol
Craig, Wallace 1876–1954 US evol
Crook, John H 1930–2011 UK ecol-evol
Croxall, John P b1946 UK cons
Cullen, Esther b1930 UK behav
Cullen, Mike 1927–2001 UK-AU behav
Cunningham, J. T. 1859–1935 UK behav
Currie, Philip J b1949 CA paleo
Cuthill, Innes C b1960 UK behav
Daan, Serge b1940 NL ecol-life
Darwin, Charles R 1809–82 UK evol
Davenport, Charles B 1866–1944 US gene
Davies, Nick B b1952 UK behav-evol
Davis, John 1916–1986 US gene
Davis, Peter b1928 UK evol
Dawson, William R b1927 US phys
Day, Lainy B b1970 US behav
De Beer, Gavin 1899–1972 UK orni
Dial, Kenneth P b1953 US feath-phys
Diamond, Jared b1937 US evol-ecol
Dill, Homer b?–d? US orni
Dong, Zhiming b1937 CN paleo
Dorst, Jean 1924–2001 FR migr
Drent, Rudi 1937–2008 NL-CA ecol-life
Drost, Rudolf 1892–1971 DE orni
Drury, William H 1921–92 US cons

Duncker, Hans 1881–1961 DE gene
Dunlop, Eric B 1887–1917 UK ecol
Dunnet, George 1928–95 UK ecol
Durham, Florence 1869 1949 UK gene
Ebbinge, Barwolt Sijbrand b1949 NL migr
Edinger, Ludwig 1855–1918 DE anat
Eibl-Eibesfeldt, Irenäus b1928 AT behav
Eifrig, Gustave 1871–1949 DE migr
Emlen, John T 1908–1997 US ecol
Emlen, Stephen T b1940 US migr-behav-evol
Endler, John A b1947 CA-US-AU behav-evol
Evans, Peter R 1937–2001 UK orni
Fabricius, Eric 1914–1994 FI behav
Farner, Donald S 1915–88 US phys-migr
Feduccia, Alan b1943 US paleo
Feinsinger, Peter b1948 US ecol
Finn, Frank 1868–1932 UK behav
Fisher, Harvey 1916–94 US anat
Fisher, James 1912–70 UK orni
Fitzpatrick, John W b1951 US behav-ecol
Flegg, Jim b1937 UK orni
Fleischer, Robert b1955 US evol-syst
Follett, Brian b1939 UK migr-phys
Fretwell, Stephen D b1942 US ecol
Friedmann, Herbert 1900–87 US behav
Fromme, Hans 1933–1980s DE? migr
Fürbringer, Max 1846–1920 DE syst
Furness, Robert W b1953 UK evol-ecol
Gadow, Hans 1855–1928 DE anat
Gagliardo, Anna b1952 IT migr
Garrod, Alfred H 1855–82 UK syst
Gätke, Heinrich 1814–97 DE migr
Gauthier, Jacques b1948 US syst-paleo

Coulson   Cracraft   Craig   Diamond   Drent   Drury

Gauthreaux, Sidney A b1940 US migr

Gegenbaur, Karl 1826–1903 DE paleo

Gibb, John A 1919–2004 UK behav-ecol

Gifford, Edward W 1887–1959 US behav

Gilbert, John 1812–45 UK behav

Gill, Frank B b1941 US syst-evol

Goethe, Friedrich 1911–2003 DE behav

Goss-Custard, John D b1942 UK behav-ecol

Gosse, Philip 1810–88 UK orni

Gould, James L b1945 US migr

Gould, John 1804–81 UK orni

Grant, Peter R b1936 UK-CA-US evol

Grant, B Rosemary b1936 UK-CA-US evol

Greenewalt, Crawford H 1902–93 US
    feath-phys

Griffin, Donald R 1915–2003 US migr-phys

Grinnell, Joseph 1877–1939 US migr-ecol

Gurney, John 1819–90 US phys

Gustafsson, Lars b1953 SE ecol-life

Gwinner, Eberhard 1938–2004 DE migr

Haffer, Jürgen 1932–2010 DE ecol-evol-hist

Hailman, Jack P b1936 US behav

Hainsworth, F Reed b? US phys

Hamilton, William D 1936–2000 UK-US
    evol

Harper, Eugene H 1867–1933? US phys

Hartert, Ernst 1859–1933 DE syst

Harvie-Brown, John 1844–1916 UK migr

Hatchwell, Ben J b1962 UK behav

Heaney, Vickie b1970 UK ecol

Heilmann, Gerhard 1859–1946 DK
    paleo-evol

Heinrich, Bernd b1940 DE-US behav

Heinroth, Kaethe 1897–1989 DE behav

Heinroth, Magdalena 1883–1932 DE behav

Heinroth, Oskar 1871–1945 DE behav

Hemenway, Harriet 1858–1960 US cons

Henry, Richard 1845–1929 IE cons

Herrick, Charles 1868–1960 US phys

Hesse, Richard 1868–1944 DE ecol

Hickey, Joseph J 1907–93 US ecol-cons

Hill, Geoffrey E b1960 US behav-evol

Hinde, Robert b1923 UK behav

Hingston, Richard 1887–1966 UK evol

Holmes, Richard T b1936 US ecol

Horn, Gabriel 1927–2012 UK phys

Horn, Henry S b1941 US behav

Howard, H Eliot 1873–1940 UK behav

Hoyle, Fred 1915–2001 UK paleo

Hudson, Peter J b1953 UK-US behav-ecol

Hudson, William 1841–1922 UK orni

Hunt, Gavin b1956 NZ behav

Hussell, David J T b1934 UK-CA ecol

Hutt, F B 1897–1991 US gene

Huxley, Julian S 1887–1975 UK
    evol-syst-behav

Huxley, Thomas H 1825–95 UK syst-evol

Irving, Laurence 1895–1979 US phys

Ising, Gustaf 1883–1960 SE migr

James, Frances b1930 US evol-ecol-paleo

Jenkins, David b1926 UK ecol

Jenner, Edward 1749–1823 UK behav

Johnsgard, Paul b1931 US behav

Johnston, Richard F b? US syst

Jones, Peter b1945 UK ecol

Jourdain, Francis 1865–1940 UK behav

Dunnet   Gagliardo   Gauthreaux   Holmes   Johnston   Kendeigh   437

Kacelnik, Alex b1946 AR-UK behav
Kaiser, Gary b1944 CA paleo
Källander, Hans b1939 SE ecol
Keast, Allen 1922–2009 AU-CA migr-evol
Keeton, William T 1933–80 US migr
Kempenaers, Bart b1967 BE behav-evol
Kendeigh, Charles 1904–86 US ecol
Kennedy, John 1893–1970 UK orni
Ketterson, Ellen b1945 US phys-ecol
Kilner, Becky b1971 UK behav
Kirkman, Frederick 1869–1945 UK behav
Klomp, Herman 1920–85 NL life-ecol
Kluijver, Huijbert 1902/3–77 NL ecol
Knudsen, Eric b1949 US phys
Koehler, Otto 1889–1974 DE behav
Koenig, Lilli 1918–94 AT behav
Koenig, Otto 1914–92 AT behav
Koenig, Walt b1950 US behav-ecol-evol
Koford, Carl 1915–79 US cons
Komdeur, Jan b1959 NL-AU behav-evol
Konishi, Masakazu b1933 JP US phys
Kooyman, Gerald b1934 US behav-phys
Kramer, Gustav 1910–59 DE migr
Krätzig, Heinrich 1912–45 DE behav
Krebs, John R b1945 UK behav-cons
Lack, David L 1910–73 UK
    behav-evol-life-ecol
Lake, Peter b1928 UK anat
Lashley, Karl 1890–1958 US migr-behav
Lawton, John b1943 UK ecol
Lee, Michael b1969 AU paleo
Lehrman, Daniel S 1919–72 US behav
Leibe, Walther b?–d? DE anat

Lessells, C M (Kate) 1956 UK ecol-evol
Lilienthal, Otto 1848–96 DE feath
Lincoln, Frederick C 1892–1960 US ecol
Linnaeus, Carl 1707 78 SE syst
Lister, Martin 1639–1712 UK ecol
Lockley, Ronald 1903–2000 UK migr
Löhrl, Hans 1911–2001 DE orni
Loisel, Gustave Antoine Armand 1864–
    1933 FR phys
Lorenz, Konrad 1903–89 AT behav
Lotem, Arnon b1962 IL behav
Lowe, Percy 1870–1948 UK evol
Lucas, Alfred M 1900 US anat-feath
Lyon, Bruce E b1956 CA-US behav-evol
MacArthur, Robert H 1930–72 US
    ecol-evol
Magrath, Robert b1957 AU behav
Manning, Aubrey b1930 UK behav
Marcgrave, George 1611–44 DE orni
Marey, Étienne Jules 1830–1904 FR feath
Marler, Peter b1928 UK-US behav
Marsh, Othniel C 1831–99 US paleo
Marshall, Jock 1911–67 AU UK migr-phys
Martin, Tom b1953 US ecol-life
Maryatt, Dorothea 1880–1928 UK gene
Marzluff, John b1958 US behav
Matthews, Geoffrey V T 1923–2013 UK migr
Mattingley, Arthur 1870–1950 AU cons
Maynard Smith, John 1920–2004 UK evol
Mayr, Ernst 1904–2005 DE evol-syst-hist
McClure, H Elliott 1910–98 US migr
McCracken, Kevin b1972 US anat-syst
McLelland, John 1800–83 US anat

Kilner   Kluijver   Koenig   Lashley   Lowe   Merton

Meek, Albert S 1871–1943 UK coll

Meinertzhagen, Richard 1878–1967 UK coll

Meise, Wilhelm 1901–2002 DE migr

Menaker, Mike b1934 US phys

Mengel, Robert 1921–90 US evol

Merkel, Friedrich 1911–2002 US migr-phys

Merton, Don 1939–2011 NZ cons

Miller, Alden H 1906–65 US orni

Miller, Robert C 1899–1984 US behav

Miller, Waldron D 1879–1929 US sys

Millikan, George b? ?? behav

Moffat, Charles 1859–1935 IE orni

Møller, Anders P 1953 DK-SE-FR
behav-evol

Monaghan, Pat b1951 UK ecol

Montagu, George 1753–1815 UK orni

Moreau, Reginald E 1897–1970 UK migr

Morel, Gérard 1925–2011 FR-SN migr

Morgan, C Lloyd 1852–1936 UK behav

Morris, Desmond b1928 UK behav

Mortensen, Hans Christian 1856–1921
DK orni

Morton, Eugene b1940 US-CA migr

Moss, Robert b1941 UK ecol

Moynihan, Martin H 1928–96 US-UK-PA
behav

Murphy, Robert C 1887–1973 US syst

Naumann, Johann A 1744–1826 DE orni

Nethersole-Thompson, Desmond 1908–89
UK behav-ecol

Nevitt, Gaby b1960 US phys

Newton, Alfred 1829–1907 UK orni

Newton, Ian b1940 UK ecol-life

Nice, Margaret M 1883–1974 US behav

Nicholson, Max 1904–2003 IE cons

Nilsson, Sven b? SE ecol

Nisbet, Ian b1934 UK cons

Noble, G Kingsley 1894–1940 US behav

Nolan, Val 1920–2008 US phys-ecol

Nopcsa, Franz 1877–1933 HU paleo-evol

Nottebohm, Fernando b1940 AR-US phys

Nur, Nadav b1953 US ecol

O'Donald, Peter b1935 UK gene-evol

Oksche, Andreas b1926 DE phys

Olson, Storrs C b1944 US syst-paleo

Orenstein, Ronald I b1946 CA behav

Orians, Gordon b1932 US behav-evol-ecol

Oring, Lewis b1938 US ecol-evol

Osborn, Henry F 1857–1935 US orni

Ostrom, John 1928–2005 US paleo

Owen, Dennis 1931–96 UK ecol

Owen, Richard 1804–92 UK paleo-anat

Owens, Ian 1967 UK-AU ecol

Padian, Kevin b1951 US paleo

Palmén, Johan A 1845–1910 FI migr

Palmer, H C 1856–1919 UK orni

Papi, Floriano b1926 migr

Parker, Patricia b? US gene-ecol-cons

Parkes, Kenneth 1922–2007 US
syst-anat-feath

Parkin, David T b1942 UK gene

Payne, Robert P b1938 US behav-evol

Peakall, David B 1931–2001 UK-US-CA cons

Pearson, Oliver 1915–2003 UK ecol

Pennycuick, Colin b1933 UK feath-phys

Pepperberg, Irene b1949 US behav

Monaghan   Nevitt   Orians   Papi   Parker   Perdeck

Perdeck, Ab C 1923–2009 NL migr
Perkins, R C L 1866–1955 UK coll
Perrins, Chris b1935 UK ecol
Peters, James L 1889–1952 US syst
Peterson, Roger T 1908–96 US cons
Pettigrew, Jack b1943 AU phys
Pettingill, Olin S 1907–2001 US orni
Piersma, Theunis b1958 NL phys-life
Piertney, Stuart B b1968 UK ecol
Pitelka, Frank 1916–2003 US ecol-behav
Pizzari, Tommaso b1972 IT-UK behav-evol
Popov, Nicolas 1888–1954 RU phys-behav
Porter, James 1873-1956 US behav
Portielje, Anton 1886–1965 NL behav
Potts, G Richard (Dick) b1939 UK ecol
Pough, Richard H 1904–2003 US cons
Preble, Edward A 1871–1957 US cons
Price, Trevor b1953 UK evol
Prum, Richard b1961 US evol-feath-anat
Pumphrey, Jerry 1906–67 UK phys
Punnett, Reginald 1875–1967 UK gene
Pycraft, William P 1868–1942 UK orni
Qiang, Ji b? CN paleo
Raspet, August 1913–60 US feath
Ratcliffe, Derek A 1929–2005 UK cons
Ray, John 1627–1705 UK orni
Rayner, Jeremy MV b1953 UK feath
Redpath, Steve M b1963 UK ecol
Regal, Philip J b1939 US paleo
Reich, Karl 1885–1970 DE gene
Reid, Jane M b1975 UK ecol
Rensch, Bernhard 1900–90 DE evol
Retzius, Gustav 1842–1919 SE anat

Rey, Eugène 1838-1909 DE orni
Richet, Charles 1850–1935 FR phys
Ricklefs, Robert E b1943 US
    hist-ecol-life-behav-phys
Riddle, Oscar 1877–1968 US phys
Ridgway, Robert 1850–1929 US orni
Robbins, Chandler S b1918 US cons
Rogers, Lesley b1942 AU phys
Rohwer, Sievert b1942 US evol-behav-anat
Romanes, George J 1848–94 CA behav
Rothschild, L Walter 1868–1937 UK syst
Rothstein, Steven I b1943 US behav-evol
Rowan, William 1891–1957 CH-CA
    migr-phys
Rowley, Ian C R 1926–2009 UK-AU
    behav-ecol
Rüppell, Werner 1908–45 DE migr
Sauer, Eleanore b? DE migr
Sauer, E G Franz 1925–79 DE migr
Schäfer, Edward 1850–1935 UK migr-phys
Schifferli, Albert 1912–2007 CH ecol
Schlegel, Hermann 1804–84 DE syst
Schleidt, Wolfgang M b1927 AT behav
Schluter, Dolph b1955 CA behav-evol
Schmidt-Koenig, Klaus 1930–2009 DE migr
Schmidt-Nielsen, Knut 1915–2007 NO phys
Schnell, Gary D b1942 US syst
Scholander, Robert 1850–1934 SE phys
Schüz, Ernst 1901–91 DE migr
Schwartzkopff, Johann 1918–95 DE
    phys-anat
Seebohm, Henry 1832–95 UK sys
Selander, Robert K b1927 US behav-evol

Perrins   Pettingill   Pitelka   Pycraft   Rothstein   Rüppell

Selous, Edmund 1857–1934 UK behav
Serventy, Dominic L 1904–88 AU phys-ecol
Seton, Ernest T 1860–1946 UK orni
Sharpe, Richard B 1847–1909 UK syst
Sheldon, Ben UK b1967 ecol-evol-behav
Shufeldt, Robert 1850–1934 US anat
Sibley, Charles G 1917–98 US gene-syst
Silverin, Bengt b1944 SE phys
Skinner, Burrhus F 1890–1990 US behav
Skutch, Alexander F 1904–2004 US-CR
    ecol-life
Smith, James N M 1944–2005 UK-CA
    behav-ecol
Smith, Susan M b1942 CA-US behav
Snow, David W 1924–2009 UK behav
Southern, Henry N (Mick) 1908–86 UK ecol
Spalding, Douglas A 1841–77 UK behav
Spottiswoode, Claire b1979 ZA behav
Stager, Kenneth E 1915–2009 US anat-behav
Starck, Dietrich 1908–2001 DE anat
Steiner, Hans b? AT evol-paleo
Stejneger, Leonhard 1851–1943 NO syst
Stettenheim, Peter R b1928 US anat-feath
Stewart, Robert E S 1913–93 US cons
Stiles, F Gary b1942 US ecol-behav
Stone, Witmer 1866–1939 US syst
Storer, Robert W 1914–2008 US syst
Stresemann, Erwin 1889–1972 DE syst-hist
Sturkie, Paul D 1909–2002 US phys
Sutter, Ernst 1914–99 CH migr
Swinton, William E 1900–94 UK-CA paleo
Swynnerton, Charles F 1877–1938 UK behav
Tebbich, Sabine b1966 AT behav

Tegetmeier, William B 1816–1912 UK orni
Temminck, Coenraad J 1778–1858 NL migr
ten Cate, Carel b1953 NL behav
Terborgh, John b1936 US cons
Théry, Marc b1961 FR ecol
Thienemann, Johannes 1863–1938 DE migr
Thomas, Jack W b1934 US cons
Thomson, Arthur Landsborough 1890–
    1977 UK migr
Thorpe, William H 1902–86 UK behav
Ticehurst, Claud B 1881–1941 UK orni
Tinbergen, Nikolaas 1907–88 NL-UK behav
Tordoff, Harrison (Bud) 1923–2008 US evol
Townsend, Charles W 1859–1934 US behav
Trivers, Robert L b1943 US evol
Tucker, Bernard W 1901–50 UK orni
Tucker, Vance 1936 US feath
Tullock, Gordon b1922 US ecol
Turek, Fred W b1947 US phys
van Balen, Hans 1930–2013 NL ecol
van Noordwijk, Arie b1949 NL gene
van Oordt, Gregorius 1892–1963 NL
    phys-migr
Van Tyne, Josselyn 1902–57 US orni
Vaurie, Charles 1906–75 FR-US syst
Verner, Jared b1934 USA behav
Verwey, Jan 1899–1981 NL behav
Videler, John J b1941 NL feath
Viguier, C b?–d? FR migr
Vinther, Jakob b1981 DK paleo-feath
Vogt, William 1902–68 US orni
von Cyon, Elias 1842–1910 LT phys
von Haartman, Lars 1919–98 FI ecol

Sheldon   Silverin   Spottiswoode   Stone   Storer   ten Cate   441

von Homeyer, Eugen 1809–89 DE orni
von Huene, Friedrich 1875–1969 DE paleo
von Middendorff, Alexander 1815–94 RU migr
von Spix, Johann B 1781–1826 DE orni
von Uexküll, Jakob 1864–1944 EE ecol
Wagner, Helmuth O 1897–1977 DE orni
Walcott, Charles D 1850–1927 US migr
Wallace, Alfred R 1858–1913 UK evol-behav
Wallraff, Hans b1930 DE migr
Ward, Peter 1934–79 UK ecol
Warner, Lucien b? ?? feath
Waterston, George 1911–80 UK orni
Watson, Adam b1930 UK ecol
Watson, John B 1878–1958 US behav
Weigold, Hugo 1886–1973 DE migr
Wenzel, Bernice b1921 US phys
Westneat, David F b1959 US behav-evol
Wetmore, Alexander 1886–1978 US syst
White, Gilbert 1720–97 UK orni
Whitman, Charles O 1842–1910 US behav
Wickramasinghe, Chandra b1939 LK paleo
Wiens, John A b1939 US ecol-cons
Wild, Martin b1941 NZ anat
Williams, Anthony b1960 UK-CA phys
Williamson, Emily 1855–1936 UK cons
Williamson, Kenneth W 1914–77 UK ecol

Willson, Mary F b1938 US behav
Willughby, Francis 1635–72 UK orni
Wilson, Alexander 1766–1813 UK-US orni
Wilson Scott B 1864–1923 UK coll
Wiltschko, Roswitha b1947 DE migr
Wiltschko, Wolfgang b1938 DE migr
Wingfield, John C b1948 UK-US phys
Winterbottom, John M 1903–84 ZA behav
Witherby, Henry (Harry) F 1873–1943 UK orni
Wittenberger, James F b1944 US behav
Wolda, Gerrit 1869–1949 NL ecol
Wolf, Larry L b1938 US behav-phys
Wood, Casey A 1856–1942 US anat
Woolfenden, Glen E 1930–2007 US behav
Worthy, Trevor b1957 NZ paleo
Wynne-Edwards, Vero C 1906–97 UK-CA ecol-evol
Xing, Xu b1969 CN paleo
Yapp, William B 1909–90 UK orni
Yeagley, Henry 1899–1996 US migr
Zahavi, Amotz b1928 IL behav
Zhou, Zhonghe b1965 CN paleo
Zimmer, John T 1889–1957 US orni
Zink, Gerhardt 1919–2003 DE migr
Zuk, Marlene b1956 US evol

Ticehurst   Van Tyne   von Haartman   Weigold   Wenzel   Willson

NOTES

PREFACE

1. This analysis of the number of publications about birds was compiled from the *Zoological Record* from 1864 to 1995—from which we tallied the number of papers each year in the Aves section—and from data at Web of Science online from 1985 to the present. In Web of Science we searched for titles containing the terms "bird *or* birds *or* avian *or* ornithology" and compared the numbers of papers from 1985 to 1995 identified by these two sources. We then calculated the number of papers that the *Zoological Record* would have listed from 1995 to the present.

2. Darwin (1887: 309).

3. Dobzhansky (1973: title).

4. We asked them to name the ten ornithologists and the ten ornithological books that they considered to be the most influential in the twentieth century. To reduce bias we tried to select correspondents from a range of countries, with representation from both genders. For the most part we sent the survey to people who had been active in the field since the 1960s, in an attempt to reduce bias toward currently active ornithologists and recent books.

5. Brooks (2011: 13).

6. Comte (1893: 19).

7. Fisher (1959: 17).

8. *The Life and Letters of Charles Darwin, Including an Autobiographical Chapter* was first published in 1887. Darwin wrote the manuscript for his family, but his son Francis, who edited it, removed several parts dealing with God and Christianity. Nora Barlow, Darwin's granddaughter, restored the omissions and published the full version in 1958, commemorating the one hundredth anniversary of *The Origin*.

9. Gill and Donsker (2012).

10. Nicholson (1959: 43).

CHAPTER 1

1. Huxley (1900: 325).

2. Quoted in Carroll (2009: 162).

3. e.g., Fastovsky and Weishampel (2005).

4. US$150,000 (US$2.1 million in today's currency—based on consumer price index at http://www.measuringworth.com/).

5. Cope and Marsh together described 136 new species of dinosaur, adding substantially to the 9 species known previously from North America.

6. All previous pterosaurs were from Europe.

7. *Ichthyornis* was actually discovered in 1870 by professor and fossil collector Benjamin Franklin Mudge, who found the fossil in Kansas. Mudge had originally worked closely with Cope, but in 1872 Marsh wrote to Mudge offering to identify any important fossils that he found and to give him full credit for the discoveries. On receiving this offer, Mudge changed the mailing label on the crate containing *Ichthyornis* and sent it to Marsh instead of Cope! Marsh did find the first *Hesperornis* fossil, in 1871, also in Kansas.

8. Darwin to Marsh, 31 August 1880; original reproduced at http://archive.peabody.yale.edu /exhibits/fossils/history/index.html; also quoted in Marsh's obituary in the *New York Times*, 19 March 1899.

9. Long before Europeans explored the American badlands, the native Lakota people of Montana noticed the abundant fossils at or near the surface, correctly deducing that these were remnants of creatures no longer with us.

10. Quoted by Robert Bakker in Ostrom obituary, 21 July 2005, in the *Los Angeles Times*.

11. From the famous Solnhofen beds near Reidenberg, Germany, where all *Archaeopteryx* fossils have been found; this specimen was originally described by Hermann von Meyer (1857) as being a pterosaur.

12. Both quotes from Carroll (2009: 173–74).

13. Darwin (1859: 280).

14. This was the jewel in a collection of about a thousand fossils owned by a local doctor, Carl Häberlein, and sold to the BMNH for £700 in total so that he could provide a dowry for his daughter. Häberlein was a keen fossil collector who often accepted fossils from the quarrymen at the rich beds at Solnhofen, in payment for his medical services.

15. The original discovery had been written up by von Meyer (1861), who named the species but provided little more than a summary of its main features.

16. Owen (1863: 46).

17. The original author of each scientific name is recorded along with the name itself to indicate its provenance; *macrura* means "long-tailed," which is appropriate here, but it was later decided that this specimen was the same species as the previously described feather, so the species name was changed back to *lithographica*.

18. Huxley (1868a: 244).

19. Huxley (1868b: title).

20. Letter to Prof. J. D. Dana, 7 Jan 1863. Darwin Correspondence Project letter 3905. http://www.darwinproject.ac.uk/entry-3905.

21. At the time he was professor and head of the Department of Applied Mathematics and Mathematical Physics at University College, Cardiff; during his career he published more than 350 papers, with more than 75 in *Nature*, and he was a leading expert on interstellar material and the origins of life from outer space.

22. Witness the recent climate change deniers.

23. e.g., Watkins et al. (1985), Hoyle et al. (1985), Spetner et al. (1988).

24. Spetner et al. (1988: 15), quoted by Nedin (1997).

25. In a bizarre twist, they claimed to have mimicked this process themselves but misrepresented the species they used for the feathers (Chambers 2002).

26. In addition to his fossil work, Heilmann also studied the local birds from his lakeside studio and built both a large aviary extending a considerable distance out over the water, where he could keep and study ducks and gulls, and a large land-based aviary for the large birds of prey (Shufeldt 1916).

27. From "An interview with Gerhard Heilmann (1940)," online at http://sumol.nl/an-interview-with-gerhard-heilmann-1940, translated into English from the original published in the Swedish newspaper *Socialdemokraten* on 24 October 1940, p. 8.

28. He designed, for example, the Danish bank note in use for more than forty years.

29. Heilmann (1926: Preface).

30. Simpson (1926), quoted in Reis (2010).

31. Heilmann (1926: 189).

32. Ibid., Preface.

33. Quoted in Reis (2010).

34. Heilmann (1926: 140).

35. Heilmann (1940: 470–71, quoted in Reis 2010).

36. Huxley (1868b: 73).

37. Cracraft (1977: 492).

38. Ibid., 488.

39. There have been at least six major hypotheses for the origins of birds proposed over the years (see figure 3 in James and Pourtless 2009).

40. Olson (2002: 1202).

41. Feduccia quoted in Prum (2002: 8).

42. Prum (2002: 5).

43. Interview with *National Geographic*, 19 June 1998 (see http://www.nationalgeographic.com/events/98/dinosaurs/interview.html).

44. M. W. Browne (19 October 1996). "Feathery Fossil Hints Dinosaur-Bird Link." *New York Times*: section 1, p. 1 of the New York edition.

45. Which became the Academy of Natural Sciences of Drexel University in 2011.

46. Interview with *National Geographic*, 19 June 1998.

47. $15,000,000 from the Ex Terra Foundation, a private nonprofit organization based in Edmonton.

48. Interview with *National Geographic*, 19 June 1998.

49. Sloan (1999: 100).

50. Quoted in Recer (1999).

51. Xu (2000: xviii).

52. Rowe et al. (2001).

53. Gentile (2001).

54. Prum (2002: 2).

55. Huxley (1868b: 74).

56. Confusingly, this is *not* the dinosaur clade that includes birds but is instead a clade of large herbivorous dinosaurs (see page 15), including the duckbills that dominated North American ecosystems in the Cretaceous. The word *Ornithischia* means "bird-hipped," with reference to their bird-like hip structure.

57. See Dyke (2011) for an excellent article about Nopcsa's life and works.

58. Nopcsa (1907: 234–35).

59. Pycraft (1910: 39).

60. His full name was Friedrich Richard von Hoinigen; von Huene named more dinosaurs in Europe than any other paleontologist of the early twentieth century.

61. Dyck (1985: 137).

62. From Shufro (2011: online at http://www.yale alumnimagazine.com/issues/2011_11 /feature_prum.html).

63. Ibid.

64. Raikow (1974).

65. Marsh (1880: 189).

66. See, for example, Bock (1965).

67. Mudge (1879) had also suggested this in his debate about the dinosaur origins of birds.

68. Xu et al. (2003) report on the discovery of *Microraptor gui*, a four-winged dromaeosaur very similar to what Beebe (1910) had drawn. The authors suggest that this specimen provides support for the idea that the precursors to birds ("proavians") passed through a tetrapteryx (literally "four-winged") stage.

69. Chukars have grown their full wing feathers by sixty days of age but do not attain full adult weight until they are four to five months old.

70. Worthington and Dial (2003).

71. Mivart (1871).

72. Quoted in Zimmer (2011: 55).

73. Caple et al. (1983: 475).

74. In those days it was in Prussia.

75. Borelli (1680) was published posthumously and contains a large section on the flight of birds, including detailed anatomical and aerodynamic analyses; most of this book was translated into English in the early 1900s (Borelli 1911).

76. About 14 meters (46 feet) high and about 61 meters (200 feet) in diameter, near Berlin.

77. Marey held *la chaire d'Histoire naturelle des corps organisés* [chair of Natural History of Organized Bodies] at the Collège de France from 1868 and published more than 150 scientific papers. For many years he was president of the Société française de photographie [French Photographic Society].

78. Anonymous (1874: 518).

79. A later version could record sixty frames per second as moving pictures, allowing him to film animals and objects in slow motion, a pioneering development in cinematography.

80. Anonymous (1882: 86).

81. Each exposure at 1/72 sec.

82. Marey (1901: 252) [original in French; translated by the authors].

83. Marey (1890: viii) [original in French; translated by the authors].

84. Raspet (1960: 199).

85. Tucker at Duke University and Pennycuick at Bristol University.

86. Pennycuick (1968a: 513).

87. This was a unique insight to apply helicopter rather than aeroplane theory to this problem.

88. A cylinder 30 centimeters (12 inches) long and 30 centimeters (12 inches) in diameter.

89. He later had them design and build equipment for the recording and analysis of birdsong. His results linking song and anatomy were published in a book (Greenewalt 1968).

90. Determined by searching Web of Science (Science Citation Index Expanded) for bird or avian or ornithology articles in *Science* or *Nature* from 2000 to 2012 (a total of 389 articles), and counting those about fossil birds and feathered dinosaurs (78). An additional 22 articles were about feathers and flight, the other topics of this chapter.

CHAPTER 2

1. Percy Lowe (1936) coined the term "Darwin's finches."

2. By then the Cal Academy already had more than 2,500 specimens of the finches in its collections, mainly from the expeditions in 1905–1906 led by Rollo Beck.

3. Bowman to Lack, 5 November 1953, CAS archives.

4. Lack to Bowman, 11 November 1953, CAS archives.

5. Bowman (1961: 1).

6. Ibid., 134–35.

7. L. C. Birch was Andrewartha's student; he was a geneticist who became a professor at the University of Sydney, but he was also a theologian who wrote widely about the relation between science and religion.

8. Andrewartha and Birch erroneously quoted Lack as saying that all differences between sympatric congeners were due to competition. Birch (1957) later agreed with Lack on the potential for competition to reinforce differences between species.

9. Bowman (1961: 275).

10. Letter from Bowman to Lack, 5 November 1953, CAS archives.

11. Keynes (1988: 352).

12. Lack (1947a: 1).

13. Bock (1963: 207).

14. Ibid.

15. Peter Grant, pers. comm., February 2012.

16. Daphne Major is about 40 hectares (about 100 acres) in surface area. Even at the peak density there were less than 1,500 Medium Ground Finches on that island during the period of the Grants' research.

17. Lowe (1936: 321).

18. Grant and Grant (2006: 226).

19. Lack (1940a: 326–27).

20. Lack (1947a: 135).

21. Mayr (1973b: 433).

22. Adults weigh about 1 kg (2.2 pounds), compared to leghorns, which weigh about 3.4 kg (7.5 pounds).

23. A fleshy flap on the side of the head, below the ear opening.

24. Hugo de Vries, Carl Correns, and Erich von Tschermak-Seysenegg are credited with rediscovering Mendel's seminal work, published originally in 1866.

25. William Keith Brooks was a prominent American embryologist who worked at Johns Hopkins University beginning in 1876.

26. Bateson (1894: 574).

27. In 1896–97 a petition to allow women to be granted university degrees was "overwhelmingly rejected" (Richmond 2001).

28. Bateson (1913: 248).

29. Comment posted on the blog *Why Evolution Is True*, 10 February 2012; available online at http://whyevolutionistrue.word press.com/2012/02/10/the-peppered-moth -story-is-solid.

30. Punnett is immortalized in the Punnett square, well known to all who have studied genetics at school; he made many contributions to the early development of genetics. He also played cricket with William Hardy and once asked him how to calculate the distribution of genotypes in a population, leading directly to the development of Hardy-Weinberg laws of genetics; in the UK a small basket is called a "punnett," named after one of Reginald's ancestors.

31. Durham was Bateson's sister-in-law.

32. Quoted in Schuster (2008: 220).

33. Dobzhansky (1937: xii).

34. This series is called the Jesup Lectures and continues to this day. In 1941 the lectures were given by Mayr and the plant geneticist Edgar Anderson (1897–1969), with the intention of producing a book on systematics from zoological and botanical perspectives. However, Anderson became ill, so Mayr wrote and published the zoological material on his own.

35. In addition to the Dobzhansky (1937) book, the other important volumes of the Modern Synthesis are Huxley (1942), Simpson (1944), Stebbins (1950), Haldane (1932), and Fisher (1930).

36. Mayr (1942: 120).

37. Mayr (1980: 415).

38. Quoted in Junker (2003: 69).

39. See Taylor and Williams (1981), Harvey (1983), and Cooke (1984).

40. Harvey (1983: 187).

41. Furness (1987: 196).

42. The two color morphs once had geographically separate breeding and wintering areas, but as their populations expanded and habitats became altered as a result of human influences, the morphs came together in their wintering areas. Because geese pair in winter quarters, and males accompany females to their nesting areas, this led to hybridization between the blue and white geese (Cooke et al. 1995).

CHAPTER 3

1. N. M. Rothschild and Sons in London.
2. About $30 million US in today's currency (see: http://www.1soft.com/todaysdollars.htm).
3. Now the Natural History Museum at Tring.
4. The collecting profession began in earnest around 1600, when boat travel made exotic locales reasonably accessible (Conniff 2010), and it was clear from some early expeditions that big discoveries were to be made. A few independent collectors plied their trade into the 1960s at least (see Heinrich 2007 for an interesting example).
5. He had (at least) two mistresses and fathered a daughter with one of them (Rothschild 1983).
6. Except the two hundred ostriches, rheas, and cassowaries (Rothschild1983).
7. About $2.9 million US in today's currency.
8. Roughly what he paid his collectors.
9. Stresemann (1975: 268).
10. Wallace on natural selection; Hudson on biogeography; Bates on mimicry.
11. Rothschild (1983: 105).
12. Ibid., 137.
13. De Réaumur and Zollman (1748: 305).
14. This practice largely ended in the 1960s for what would seem like obvious reasons, as newer, safer chemical defenses were developed, including especially paradichlorobenzene.
15. In those days this was all East Prussia.
16. Birds for Eugen von Homeyer who wrote more than 150 articles on birds and eventually donated his collection of 20,000 specimens to the Braunschweig Natural History Museum; eggs for Eugène Rey, famous for his early work on cuckoos.
17. Rothschild (1934: 352).
18. One report has it that curators and others ran to the building and saved a single cart full of type specimens and academy records when the earthquake began (http://www.terrastories.com/bearings/cal-academy-sciences).
19. Jack Dumbacher, pers. comm., December 2011.
20. Osborn to Sanford, 7 September 1922, AMNH archives.
21. Downing (2007): online at http://online.wsj.com/article/SB117106204521604170.html.
22. Gerd Heinrich (see Heinrich 2007) and John Williams in Africa are prime examples.
23. Stone (1933: 240).
24. Including John T. Zimmer to Peru and Frank M. Chapman to South America.
25. AMNH, Field Museum in Chicago, British Museum in London.
26. Examples in the 1980s and 1990s: Peru, Vietnam, and tepuis in Venezuela.
27. The standard today is to prepare the collected bird as a study skin minus one wing and one leg, a skeleton that includes that leg, a spread separate wing, and tissues preserved in alcohol or frozen at −80°C.
28. Bruce Lyon, pers. comm., 1982.
29. At the Berlin-Dahlem Waldfriedhof (Haffer 2004a).
30. Sharpe, Hans Gadow, Philip Lutley Sclater, Osbert Salvin, Ernst Hartert, Edward Hargitt, G. E. Shelley, Tommaso Salvadori, W. R. Ogilvie-Grant, Howard Saunders.
31. John Gerrard Keulemans, Joseph Smit, William Matthew Hart, Peter Smit.
32. Rothschild (1934: 353).
33. Because systematists of those days needed a staggering amount of experience to make any sense of how one species might be related to another, most systematists were quite old, having spent most of their career working on only one or a few groups that interested them.
34. Bock (2001: 805).
35. Haffer (2008: 78).
36. Mayr (1973b: 282).

37. This work continues to be published by de Gruyter and is now available online at http://www.degruyter.de/cont/fb/na/detail.cfm?id=IS-9783110216240-1.

38. "Stresemann, Erwin." *Complete Dictionary of Scientific Biography.* 2008. Encyclopedia.com. 29 August 2011 (http://www.encyclopedia.com).

39. 1922–44 and 1951–61.

40. Mayr (1973b: 283).

41. These genera were considered difficult at the time because there was no obvious way to determine which species was related to which, and taxonomists largely disagreed with each other's characterization.

42. Haffer (2007a: S146).

43. Mayr and Provine (1998: 415).

44. Orthogenesis is the idea that evolution proceeds by some internal or external driving force, what George Gaylord Simpson called "the mysterious inner force." Thus proponents reject natural selection as the engine of evolutionary change, believing, instead, that evolution is guided by some ill-defined essences that each organism possesses.

45. Depending upon the "expert" who has worked on this group.

46. Then revised and rewritten in 1884, and again in 1901.

47. Anonymous (1884: 87).

48. They met in 1884, when Coues visited Britain and delivered a lecture on the trinomial system at the British Museum.

49. American Ornithologists' Union (1886: 31).

50. Kleinschmidt was also a pastor and theologist; his "Formenkreis theory" suggested that a species could be composed of different forms (subspecies) all deriving from a single origin.

51. In Germany (Meyer, Erlanger, Schalow) and Austria (Tschusi, Hellmayr).

52. This is a very controversial topic today as we discover more about the underlying genetics of speciation.

53. At that time a junior fellow at Harvard, working with Assistant Professor Brown.

54. Wilson and Brown (1953: 100).

55. Lack (1946: 63).

56. Mayr (1951: 94).

57. Wilson and Brown (1953: 108).

58. This sort of sighting was, and still is, the goal of many birders, to see birds outside their normal ranges and to carefully document such sightings. Such sightings have proven to be useful for understanding dispersal and occasionally as harbingers of ongoing range expansion.

59. Haffer (2008: 23).

60. This made Mayr a "candidate for medicine," which meant he could return to medical school if his plans to become a zoologist fell through.

61. Mayr (1932: online at http://www.naturalhistorymag.com/picks-from-the-past/141668/a-tenderfoot-explorer-in-new-guinea).

62. Ibid.

63. Mayr (1930: 21, 22, 25).

64. Ibid.

65. Ibid.

66. Mayr (1932: online at http://www.naturalhistorymag.com/picks-from-the-past/141668/a-tenderfoot-explorer-in-new-guinea).

67. Chapman (1935: 95).

68. Chapman to Sanford, 7 May 1929, AMNH archives.

69. Mainly in *American Museum Novitates*, where editing and peer review were not exactly stringent.

70. Probably the best avifaunal work ever written, combining systematics and evolution (Mayr) with ecology and behavior (Diamond).

71. Revised and republished by Mayr alone in 1969, then in a second edition in 1991 with Peter Ashlock (Mayr and Ashlock 1991).

72. With Ronald Lockley, and the help of the Royal Navy, this film (*The Private Life of the Gannets*; see http://www.youtube.com/watch?v=lN_doZVuWEY) was made around the island of Grassholm, Pembrokeshire in Wales, and is said to be the first natural history documentary.

73. Including *Essays of a Biologist* (1923), *Essays in Popular Science* (1926), *The Stream of Life* (1926), *Animal Biology* (with J. B. S. Haldane, 1927), *Religion without Revelation* (1927), *The*

Tissue-culture King (1927) [science fiction], *Ants* (1929), *Bird-watching and Bird Behaviour* (1930), *An Introduction to Science* (with Edward Andrade, 1931–34), *What Dare I Think?: The Challenge of Modern Science to Human Action and Belief* (1931), *Africa View* (1931), *The Captive Shrew and Other Poems* (1932), *A Scientist among the Soviets* (1932), *If I Were Dictator* (1934), *Scientific Research and Social Needs* (1934), *Thomas Huxley's Diary of the Voyage of HMS Rattlesnake* (1935), *"We Europeans"* (with A. C. Haddon, 1936), *Animal Language* (1938), *The Living Thoughts of Darwin* (1939).

74. Particularly his appearances on the panel of *The Brains Trust*, a popular weekly BBC radio series begun in 1941. It was so popular that at its peak it was heard by 29 percent of the British population and generated as many as five thousand letters a week from the listening public.

75. This show ran on BBC-TV from 1952 to 1959; on his first appearance on the show, in 1955, Huxley failed to correctly identify the egg of the giant West African snail, even though this would have been familiar to generations of British zoology students. Huxley was so certain that it was the egg of a reptile that he bet the show's host, Glyn Daniel, £5 that he was right. He lost the bet but never paid up.

76. He wrote about fifty books all told and coined many terms now in common parlance among biologists at least (clade, cline, morph, ritualization).

77. Established in 1937 as an "offshoot" of the Linnaean Society of London (see *Nature* 140: 163–64, 24 July 1937), with Julian Huxley as its first chairman and fifty-three members at the start.

78. Soon after simply called the Systematics Association.

79. Huxley (1940: Foreword).

80. Mayr (1982: 276).

81. Mayr originally preferred to call this "evolutionary systematics," but in his influential book *The Growth of Biological Thought* (Mayr 1982), he reverted to the calling it the "new systematics."

82. Stresemann (1959: 542–43).

83. Nixon and Carpenter (2000: 301).

84. Mayr (1959: 293).

85. Cain (1959: 313–14).

86. Hull (1988: 119).

87. Hubbs (1953: 93–94).

88. Felsenstein (2004: 145).

89. People working in the field were mainly aware of Sibley's findings, and some had long before dismissed them as useless.

90. Corbin and Brush (1999: 810).

91. Ibid.

92. Sibley and Ahlquist (1972: 245).

93. Ahlquist (1999: 857).

94. Schodde (2000: 75).

95. And surprising, given his apparent lack of people skills.

96. John Harshman, pers. comm., 24 April 2012.

CHAPTER 4

1. Clarke (1904: 113).

2. Ibid., 120.

3. Ibid.

4. Gätke and Rosenstock (1895: 132).

5. Newton (1896: 549). Note that Newton's *Dictionary of Birds* was published in parts between 1893 and 1896 and as a single volume in 1896. For simplicity we refer to it as Newton (1896).

6. Newton (1896: 549).

7. Ibid., 561.

8. In 1879 Newton presented to the AGM of the British Association some data collected by Cordeaux and Harvie-Brown that suggested that weather was important because birds seemed to migrate only on clear nights.

9. Newton cited in Wollaston (1921:173).

10. Ibid.

11. Newton (1896: 562).

12. Ibid., 566.

13. William Tegetmeier was a Victorian authority on pigeons and poultry and one of Darwin's main informants on these topics.

14. Newton (1896: 566–67).

15. Thomson (1926: ix).

16. Wallace (1874: 459).

17. Ibid.

18. Brooks (1898: 796).

449

19. Thomson (1926: 273).
20. Thomson (1926: 275).
21. Grinnell (1931: 30–31).
22. Only in the following year, after which they disappeared.
23. Thomson (1926: 117).
24. Ibid.
25. Levey and Stiles (1992: 455).
26. Ian Newton, pers. comm., 3 September 2011.
27. Dorian Moss, pers. comm., 5 August 2011.
28. Thienemann, cited in Vaughan (2009: 79–80).
29. The Hungarian Royal Ornithological Centre (later called the Institute of Ornithology); it began ringing at several locations around the country in 1908, with the help of Johannes Thienemann, who designed the rings.
30. Conditions were basic in the 1940s; in 1954 Waterston gave the island to the National Trust for Scotland; a new observatory was built in 1969, which in turn was replaced in 2011.
31. For an overview of published atlases in European countries see http://www.euring.org /research/migration_atlases/index.html.
32. See *Canadian Atlas of Bird Banding* at http:// www.ec.gc.ca/aobc-cabb.
33. Quoted in Lewis (2004: 105–106).
34. From the European Union for Bird Ringing website: http://www.euring.org/about _euring/brochure2007/06ringers_and _centres.htm.
35. From the USGS website: http://www.pwrc .usgs.gov/bbl/homepage/howmany.cfm.
36. One study found that only 0.5 percent of banded small land birds were ever encountered again (Brewer et al. 2006).
37. The Argos satellite-based system, established in 1978, collected and processed environmental data from fixed and mobile platforms worldwide.
38. "Identifying migration routes and non-breeding areas of the globally threatened Aquatic Warbler using geolocators." From www.aquaticwarbler.net/.../Flade-Salewski _Geolocs_on_AW.pdf.
39. Buss (1946: 315).
40. Lack (1958: 287).

41. Newton (1896: 555).
42. Ibid.
43. Darwin (1871: 357).
44. Newton (1896: 570).
45. Newton (1896) also says that Gätke claims that young and old birds always migrate separately, often taking different routes, so the young birds could not be using either experience or the adults to find their way.
46. Thomson (1926: 327).
47. Pigeons are trained to home through practice flights around their loft, during which they familiarize themselves with certain landmarks.
48. Vaughan (2009: 108).
49. Mayr (1952: 397).
50. Ibid., 398.
51. Stresemann (1975: 354).
52. Watson and Lashley (1915: 60).
53. Ibid., 33.
54. Kluijver later spelled his surname as "Kluyver," but we use the original form throughout this book to minimize confusion.
55. Thomson (1936b: 474).
56. Griffin (1998: 78).
57. Jacques Loeb was a developmental biologist and physiologist who worked on the physiology of the brain. He was strictly reductionist, and he argued that biological phenomena should be explained by physical or chemical laws.
58. Griffin (1998: 79).
59. Odum (1948: 584).
60. Ibid., 596.
61. Griffin (1998: 79).
62. As Berthold (1993: 150) has said, "The decisive factor for the birds' orientation by the sun is the azimuth (the intersection between meridian and vertical circle on the horizon) and not the sun's altitude."
63. Matthews (1950).
64. Steve Emlen interview with Janis Dickinson, 3 August 2011.
65. Emlen (1981: 169).
66. Von Cyon (1900), quoted in Watson and Lashley (1915: 18).
67. Wallraff (2005: 186).
68. Wallraff (1980: 209).

69. Ibid., 221.

70. Wiltschko (1996: 115).

71. Able (1995: 598).

72. Ibid.

73. Keeton (1974).

74. Papi et al. (1978: 314–15)

75. Wolfgang and Roswitha Wiltschko interview with Jo Wimpenny, 12 August 2011.

76. Ibid.

77. Ibid.

78. DMSP—Dimethylsulfoniopropionate; DMS—Dimethyl Sulfide, a breakdown product of DMSP.

79. Gaby Nevitt, pers. comm., 11 August 2011.

80. Wolfgang and Roswitha Wiltschko interview with Jo Wimpenny, 12 August 2011.

81. As Berthold (1993: 151) says, "The earth can be viewed as a huge magnet, with the magnetic north and south poles situated close to (but not coinciding with) the geographic poles. The field lines leaving the earth in the Southern hemisphere and entering it again in the Northern hemisphere show characteristic lines of inclination changing systematically across the earth's surface. . . . Birds make use of the inclination of these field lines for orientation."

82. Wolfgang and Roswitha Wiltschko interview with Jo Wimpenny, 12 August 2011.

CHAPTER 5

1. Kaiser (2007: 5)

2. It was later considered unsafe by the Health and Safety Executive and replaced by a metal spiral staircase.

3. Lack (1963).

4. Tim Birkhead, pers. obs.

5. Lack (1973).

6. Ibid., 422.

7. Ibid., 424.

8. Ibid., 425.

9. In what appears to be an apology, Lack wrote to Howard in October 1939, saying, "I owe a very great deal to your pioneer work on territory. The paper which my father and I wrote for British Birds in 1939 was, I now realize, a superficial affair and if I were writing a similar paper again, it would have to be altered considerably." Eliot Howard archives, Oxford University.

10. Tucker (2006: 4).

11. Nelson and Haffer (2009: 120).

12. Lack (1973: 426).

13. Lack to Howard, 27 November 1935. Eliot Howard archives, Oxford University.

14. *The Birds of Cambridgeshire* (1934), *The Life of the Robin* (1943a), *The Galapagos Finches (Geospizinae): A Study in Variation* (1945).

15. The same issue of *British Birds* included a theoretical article by S. E. Brock, which Witherby felt was of little use in promoting the methods to ornithologists.

16. Alexander (1974: 27–28). By "theoretical" he probably means metaphysical, a reference to some of Howard's ideas and writing.

17. Alexander (1974: 28).

18. Kennedy (1924: 591).

19. Ticehurst (1924: 815).

20. Thorpe (1924: 815–16).

21. Nicholson to Tansley, 31 January 1927. Max Nicholson archives, Oxford University.

22. Lack (1973: 425).

23. Johnson (2004: 545).

24. Cited in ibid., 539.

25. Ticehurst also rejected one of Lack's papers on the Galápagos finches—without even seeing it! After an abstract of Lack's paper came out (in *Nature*, but stating that the complete paper would be in the Californian Academy of Science's publication), Lack was attacked by Percy Lowe (1941) for publishing the data in a non-British journal. Lack (1941: 637) replied to Lowe, saying, "Actually, I did first suggest publication in 'The Ibis,' but received so discouraging a reply from the late editor (before he saw the paper), and at the same time so pressing an invitation from the C. A. S., that I accepted the latter." Lack's plea to Mayr may have been a thinly disguised complaint about his own rejected paper.

26. Moreau (1959: 20).

27. Ibid., 29.

28. From the EGI website: http://www.zoo.ox.ac.uk/egi/.

29. Tim Birkhead, pers. obs.; the answer was the Palm-nut Vulture (vulturine fish eagle).
30. Anderson (2013) contains a list, and since 1973 a further 150 students have earned their DPhil at the EGI (EGI website).
31. Ian Newton interview with Tim Birkhead, 21 April 2010; slightly edited.
32. Ibid.
33. Lack (1968: 6).
34. Ian Newton interview with Tim Birkhead, 21 April 2010.
35. Lack (1968: 7).
36. Lack (2001: 97).
37. Lack (1968: 14).
38. She did this with the encouragement of Mayr and Stresemann, whom she visited in Germany in 1932.
39. Lack (1973: 427).
40. Officially speaking, 1915 was the starting year; however, it is said that Allen had already put a sign on his office door—which was housed in the entomology and limnology department—saying "Lab of Ornithology" when he returned to Cornell in 1912 as an instructor in zoology (John Fitzpatrick, pers. comm., 9 September 2011).
41. Chapman (1914: 284–85).
42. Pettingill (1968: 196).
43. Quoted in Barrow (2000: 200).
44. Ibid.
45. Ibid.
46. Preble was a mammalogist/ornithologist who worked at the Bureau of Biological Survey with most of the AOU old boys; at the time of his election, Preble commented, "I thought the age of miracles was past" (McAtee 1962: 735).
47. House Sparrows were first introduced to America in 1851, when eight pairs were brought over from England and released by the Brooklyn Institute; a hundred more birds were released by 1853. The sparrow quickly established itself and spread across the entire continent, achieving status as a pest species.
48. Kendeigh (1940: 6).
49. Nice (1979: 127).
50. Mrs. Nice was the first woman president of the Wilson Ornithological Club, and the second woman to receive the AOU's Brewster Medal.
51. *Bird-Banding* continued the *Bulletin of the NBBA* (Northeast Bird-Banding Association), which ran from 1925 to 1929. From 1980 it was continued under yet another new name, *Journal of Field Ornithology*.
52. Lack (1973: 430).
53. Now Tanzania.
54. Moreau (1970: 554).
55. Thorpe, quoted in Moreau (1970: 564).
56. Ian Newton, in litt., 27 Aug 2012.
57. Moreau (1972).
58. Lack (1954: 22).
59. Ibid.
60. This is the most frequently cited paper ever published in *Ibis*.
61. It was not clutch size per se, but the number of eggs laid per year (Hall 1935). Heritability was demonstrated for many other traits in poultry from the 1920s onward (see Romanoff and Romanoff 1949).
62. Peter Jones and Chris Perrins, pers. comm., January 2012.
63. Perrins and Jones also told us that much of the work of extracting and analyzing the data for their study was undertaken by biomathematician Michael Bulmer, who in the early 1970s had just acquired a large calculating machine and was keen to use it to conduct the heritability calculations. At that time it was not unusual for academics in statistics departments to undertake analyses (other than the most basic ones) for colleagues without becoming coauthors on their papers.
64. Fred Cooke, pers. comm., 22 January 2012.
65. Fisher's Fundamental Theorem: the rate of increase in fitness of any organism at any time is equal to its genetic variance in fitness at that time (Fisher 1930; Grafen 2003).
66. Somewhat surprisingly, given the amount of effort the Swiss ornithologists had put into obtaining and analyzing the data, the key paper resulting from the Swiss starling study appeared with Lack as sole author (Lack 1948). However, Albert Schifferli's son Luc confirmed that in fact Lack had coauthored several papers with Albert and others in the

Swiss journal *Ornithologische Beobachter* and that all parties were happy about this (Luc Schifferli, pers. comm., 9 February 2012).

67. Von Haartman—artist, poet, essayist, and bibliophile—was for many years Finland's leading ornithologist (Soikkeli 2000).

68. Von Haartman (1954: 453).

69. Ibid.

70. Cody (1981: 341).

71. Martin (1993: 531).

72. Williams (1966b: 689).

73. Heaney and Monaghan (1995: 364).

74. Ian Newton interview with Tim Birkhead, 21 April 2010.

75. Ibid.

76. Lack (1950: 314).

77. Chris Perrins interview with Tim Birkhead, 20 June 2011.

78. Perrins's (1970) paper "The Timing of Birds' Breeding Seasons" is the second most frequently cited paper in *Ibis*.

79. Lack (1968: 304).

80. Visser and Lessells (2001:1276).

81. Lack (1968: 306).

82. Cody (1969: 163).

83. Stearns (1977: 163).

84. Doubly labeled water is a technique allowing indirect measurement of energy expenditure. An animal's body water is labeled with isotopes of deuterium and heavy oxygen, and the washout rates of both are measured.

85. Rob Freckleton, pers. comm., 21 August 2012.

86. Van Noordwijk (2002: 589).

87. Ricklefs (2000: 7).

88. Andrew Lack, pers. comm., 20 June 2011.

89. Quoted in Lack (1973: 434).

90. Ricklefs (2000: 5).

91. Ibid.

92. Ibid.

93. Skutch (1949: 434).

94. Ricklefs (2000: 8).

95. Ibid.

96. Even though Williams's (1966a, b) ideas were published before Lack's *Ecological Adaptations*, there is very little evidence that they had much effect on Lack. Lack does cite Williams (1966a) in *Ecological Adaptations*, but it is no

more than a nod of acknowledgment, even though Chris Perrins told us that he thought Lack and Williams were on good terms.

CHAPTER 6

1. Kevin McCracken, pers. comm., 23 November 2011.

2. Conrad Eckhard (1876) discovered that it was lymph rather than blood, as in mammals, that causes erection of the duck's penis.

3. Bock and von Wahlert (1965: 269).

4. Sibley (1955: 636).

5. Fisher (1955: 57).

6. The kind of physiology we are discussing here is sometimes now referred to as "integrated systems biology" and deals with macroscopic systems, such as how hearts, lungs, guts, and gonads work, rather than with the genes and proteins that influence physiological processes.

7. Sturkie (1954: x).

8. Ibid.

9. Mike Shattock, pers. comm., 13 January 2012.

10. We sought the opinion of Len Hill, Frances Ashcroft, Mike Shattock, Kevin Fong, and David Paterson. There have been three analyses of the "trajectory" of twentieth-century studies of avian morphology and physiology: first, by Jürgen Haffer (2006a), who analyzed the content of the *Journal für Ornithologie* (later *Journal of Ornithology*) from 1920 to 2000; second, by Glen Walsberg (1993), who did the same for *The Condor* between 1900 to 1990; and third, by Bill Dawson (1995), who made a more detailed analysis of physiology studies published in *The Condor* between its inception in 1898 and 1994.

11. Mayr, in Stresemann (1975: 385).

12. Blix (1983: 917).

13. Andersen (1966: 216).

14. Millard et al. (1973: 238).

15. Jerry Kooyman, pers. comm., 29 April 2012.

16. Bennett (1834: 374–78).

17. Newton (1896: 620).

18. Marples (1932: 842).

19. Functional salt glands have been found in the roadrunners and also in some falcons. They are also often said to exist in the ostriches and the Sand Partridge, although this has been disputed.
20. Molyneux (1692: 105).
21. F. Nottebohm (20 October 2011) told us that:

    Strictly speaking, the hypoglossus nerve runs, on each side, from the medulla to the ipsilateral tracheolateralis muscle and ipsilateral syringeal muscles, including the sternotrachealis. That was established by showing that section of the nerve on one side was followed by atrophy of the same side muscles, which became thin and pale, while those on the opposite side remained bulky and red. I prefer saying that the nerves run from medulla to the syrinx because the somata of the axons are in the medulla and the nerve grows from the medulla to the muscles.

22. F. Nottebohm, pers. comm, 27 June 2012.
23. Altman (1962).
24. F. Nottebohm, pers. comm., 27 June 2012. He also told us that:

    The evidence was: (a) the labeling with birthdate markers was very similar in neurons as in other tissues known to continue to divide in adulthood; (b) we recorded from the new neurons and showed a neuronal neurophysiological and Golgi profile; (c) we showed that only a relatively small subset of neuronal classes continued to be produced in adulthood and identified the cell type; (d) we showed the new neurons responded to external stimulation and so were part of functioning circuits; (e) we identified where the new neurons were born; (f) we identified the neuronal stem cells; (g) we identified the manner of migration of the new neurons from birth site to work site (where they differentiated); (h) finally, we showed that the addition of new neurons in the adult canary HVC was part of a process of replacement.

25. Taken from http://concen.org/forum/thread-10785.html.
26. Gross (2000: 69).
27. F. Nottebohm (27 June 2012) told us that:

    The 100 year-old dogma begun by Santiago Ramon y Cajal was, by and large correct: very few, if any, new neurons are added to the adult human brain and, until we learn otherwise, those lost are not normally replaced. However, birds have taught us that recruitment of new neurons in an adult vertebrate brain can and does occur and that in some instances it is part of a process of spontaneous replacement. The avian finding that their brain has "ephemeral" neurons that live only for a period of weeks or months is, in itself, quite extraordinary. However, it is too early to tell whether the insights obtained from birds and rodents will ever be converted into a clinical application.

28. Pumphrey (1948: 194).
29. Johann Schwartzkopff was a pioneer of electrophysiology and avian sensory biology with a special interest in avian hearing. Dietrich Burkhardt (1995) in his obituary of Schwartzkopff describes him as aloof on first meeting but friendly on getting to know him. Jack Hailman (chapter 7) went to work in Schwartzkopff's lab but left after just one semester (pers. comm., 10 October 2011):

    [He] was too authoritarian for me. He thought of himself as liberal but was a prisoner of his past (and his culture). He wasn't just "formal" . . . but dictatorial. Everyone began leaving his lab: a secretary, the shop guy, at least one of the grad students, and also his other postdoc, Mark Konishi (who went to a Max Planck institute, as I recall). Even Schwartzkopff's wife eventually left him, poor bastard.

30. Their first paper, "Receptive Fields of Auditory Neurons in the Owl," was voted the best paper to appear in *Science* in 1977.
31. Collins (1884: 317).
32. J. W. Warren, "Marshall, Alan John (Jock) (1911–1967)," *Australian Dictionary of Biography*, National Centre of Biography, Australian National University, http://adb.anu.edu.au/biography/marshall-alan-john-jock-11060/text19683, accessed 25 June 2012.

33. For biographical details see obituaries in *Ibis* (110: 206) and *Emu* (69: 55).

34. See Marshall (1998).

35. The book was called *What Bird Is That?* and was first published in 1931.

36. Marshall (1998).

37. Quoted in Chisholm (1969: 56).

38. Romanes (1881: 280).

39. Marshall (1954: 2).

40. Marshall (1955), cited in Marshall (1998)).

41. Brian Lofts, quoted in Marshall (1998).

42. Mark Whittaker, pers. comm., 26 September 2011.

43. Willmer and Brunet (1985: 48).

44. Marshall (1998).

45. Sandy Middleton, pers. comm., 22 September 2011.

46. Schäfer (1907: 161).

47. Quoted in Ainley (1993: 196).

48. Ibid.

49. Ibid., 313.

50. Marshall (1961: 317).

51. Marshall (1961: 307).

52. Ibid.

53. Forbes (1949), quoted in Beach (1981: 328–29).

54. In fact the machinery driving the annual cycle resides in the central nervous system; the gonads are important only insofar as the steroids they produce alter the amplitude of the gonadotropin cycle.

55. Dewsbury (1998: 65).

56. Ibid.

57. Ibid.

58. First extracted by Corner and Allen in 1929; it was in 1934 that it was isolated as a pure crystalline form.

59. Marler (2005: 499).

60. Others included Jacques Benoit, Ivan Assenmacher, Albert Wolfson, Jürgen Aschoff, Colin Pittendrigh, Hideshi Kobayashi, and Andres Oksche.

61. King and Mewaldt (1989: 712).

62. Brian Follett interview with Tim Birkhead, 6 October 2011.

63. They later won the Nobel Prize for their work.

64. Brian Follett, pers. comm., 4 May 2012.

65. Ibid.

66. John Wingfield interview with Tim Birkhead, 6 October 2011.

67. Ibid.

68. Biographical information from John Wingfield interview with Tim Birkhead, 20 July 2010, and from Wingfield (2009: 573).

69. John Wingfield interview with Tim Birkhead, 6 October 2011.

70. Taken from an Indiana University webpage about Ketterson: http://homepages.indiana .edu/news/page/normal/3166.html.

71. Marshall and Serventy (1959: 1704).

72. Daan and Gwinner (1998: 418).

73. Ibid.

74. Drent and Daan (2006: 300).

75. Ibid., 303.

76. Barbara Helm, pers. comm., 25 October 2011.

77. Seewiesen is now directed by Bart Kempenaers and Manfred Gahr and Radolfzell by Martin Wikelski.

78. Birds may be ideal for measuring circulating hormones, but it is virtually impossible, so far, to carry out the genetic alterations that drive mouse and *Drosophila* research.

79. Williams (2012: 6).

CHAPTER 7

1. Quoted in Gray (1966); P. H. Gray was an ethologist and later wrote several papers on the history of ethology.

2. Ibid.

3. Mullens (1909: 392).

4. Selous (1901b: 193).

5. Quoted in Burkhardt (2005: 80).

6. The most complimentary review of his book, *Realities of Birdlife*, was written by American ornithologist Tom McCabe (1930), who named Selous among the British pioneers of field ornithology but wrote that these men were not progressing as a "conscious unit."

7. Kirkman to Selous, 3 January 1910. Edmund Selous archive, Oxford University.

8. Morgan to Howard, 15 November 1927. Eliot Howard archive, Oxford University.

9. Friedmann (1930: 62).

10. Nicholson (1929: 541).

11. Howard (1929: xi).

12. In William Pycraft's chapter on the grebes in Kirkman's *British Bird Book*, he in fact mentions the forthcoming work of L. Huxley on Great Crested Grebes. Leonard Huxley was Julian's father, but we have not been able to find any evidence for Leonard studying grebes.

13. This was the forerunner to the Association for the Study of Animal Behaviour (ASAB).

14. There has been some controversy about whether Pavlov used a bell (see Thomas 1997 for an overview), but it seems very likely that he did, although he probably also generated noises with tuning forks, metronomes, and whistles.

15. In 1904 Pavlov was awarded the Nobel Prize for his studies on the physiology of digestion.

16. Morgan (1932: 247–48).

17. Morgan to Howard, 15 August 1929. Eliot Howard archive, Oxford University.

18. Morgan (1894: 53).

19. Thorndike (1911: 80).

20. Allen (1909: 241).

21. Ibid.

22. Lorenz (1950: 233). In fact, Watson had hand reared birds.

23. Watson (1913: 1).

24. Quoted in Capshew (1993: 840).

25. Friedmann (1930: 62).

26. Burkhardt (2005: 137).

27. Stresemann (1975: 347).

28. Oberholser (1927).

29. Stresemann (1947: 660).

30. Ibid.

31. Lorenz (1985: 259).

32. Ibid., 260.

33. Bühler made ground-breaking advances into linguistics, as well as being Karl Popper's most influential teacher.

34. Lorenz (1985: 265).

35. Lorenz (1974).

36. See Brigandt (2005) for a detailed description of Lorenz's belief in the phylogenetic worth of instinctive behavior.

37. Lorenz (1961: 116).

38. Morris (1979: 47).

39. Quoted in Bateson (1990: 66).

40. Quoted in Haffer (2001: 67).

41. Nice (1935b: 146).

42. Ibid., 147.

43. Lorenz (1985: 268).

44. Ibid.

45. Tinbergen (1974).

46. The first was Alfred Seitz (Burkhardt 2005).

47. Tinbergen (1974).

48. Tinbergen (1984: 187).

49. Hinde (1990: 550).

50. Tinbergen (1985: 437–38).

51. Röell (2000: 100).

52. Tinbergen (1984: 22).

53. Tinbergen (1974).

54. The International Polar Year (IPY) was organized by the International Meteorological Organization to study both the poles and the newly discovered jet stream, to understand more about meteorology and geology. There were participants from forty countries. The first IPY was 1882–83, and the second was the one that the Tinbergens joined, almost fifty years later.

55. Tinbergen (1953b: 186).

56. ten Cate (2009: 786).

57. Tinbergen to Hicks, via Margaret Nice, 6 September 1945. Margaret Morse Nice archive, Cornell University.

58. The Berlin Document Center records show that Lorenz applied for membership in the Nationalsozialistische Demokratische Arbeiter Partei (NSDAP) on 1 May 1938, soon after the Anschluss. His application was accepted 28 June 1938; his membership number was 6,170,554 (Kalikow 1983).

59. Burkhardt (2005: 242).

60. During the war, he published his groundbreaking monograph on the comparative study of duck courtship (1941, 1951) and his profound assessment of "the innate forms of possible experience" (1943).

61. Lorenz to Lack, 30 June 1948. David Lack archive, Oxford University.

62. Lorenz to Thorpe, 4 March 1950. William Homan Thorpe archive, Cambridge University.

63. Burkhardt (2005: 19).

64. Whitman (1899), cited in Podos (1994: 469).

65. Burkhardt (2005: 33).

66. Thorpe (1979: 49).

67. Beach (1950: 121).

68. Burkhardt (2005: 67–68).

69. Quoted in Hirsch (1967).

70. Tinbergen to Thorpe, 8 September 1976. William Homan Thorpe archive, Cambridge University.

71. Lorenz to Thorpe, 10 August 1948. William Homan Thorpe archive, Cambridge University.

72. Thorpe (1979: 81).

73. Tinbergen (1959: 1).

74. Storer (1954: 214).

75. Hinde (1954: 161).

76. Noble is also remembered for his part in exposing Paul Kammerer's "Midwife Toad scandal" in 1926 (Koestler 1971).

77. Lehrman (1941: 87).

78. Ibid.

79. Lehrman (1953: 341).

80. Ibid., 345.

81. Ibid., 343.

82. Morris (1979); Desmond Morris, pers. comm., 11 January 2011; Aubrey Manning, pers. comm., 11 December 2010.

83. Also Karl Lashley, Hans-Lukas Teuber, and Donald O. Hebb (Rosenblatt 1995).

84. Lehrman (1953) wrote:

> He [Lorenz] states that a major effect [of unrestricted breeding] is the involution or degeneration of species-specific behavior patterns and releaser mechanisms because of degenerative mutations, which under conditions of domestication or civilization are not eliminated by natural selection. He presents this as a scientific reason for societies to erect social prohibitions to take the place of degenerated releaser mechanisms which originally kept races from interbreeding.

85. Morris (1979: 97).

86. Lorenz to Thorpe, 20 August 1954. William Homan Thorpe archive, Cambridge University.

87. Manning (1985: 291).

88. Klopfer (1999: 9).

89. Lorenz to Thorpe, 11 March 1955. William Homan Thorpe archive, Cambridge University.

90. Quoted in Hinde (1987: 630).

91. Thorpe (1979: 119).

92. Although the pair was observed for days by the most eminent ornithologists of the day, the record was thrown out by the British Birds Rarities Committee in 2005, concluding that the birds were probably an aberrant pair of Sedge Warblers.

93. Thorpe (1956: 118).

94. Potter (1945: 470).

95. Thorpe (1958a: 541).

96. Tinbergen (1951: 168).

97. Marler (1986: 16).

98. A "citation classic" is a paper that has been very highly cited—usually more than four hundred times, although this may vary according to the discipline.

99. Marler (2004: 191).

100. Lorenz to Thorpe, 30 April 1952. William Homan Thorpe archive, Cambridge University.

101. Ibid.

102. Pat Bateson interview with Jo Wimpenny, 21 April 2010.

103. Robert Hinde, pers. comm., 7 January 2010.

104. Lorenz (1985: 283).

105. Tinbergen (1985: 440).

106. Tinbergen (1985: 440).

107. Online at: http://www.nobelprize.org /nobel_prizes/medicine/laureates/1973 /press.html.

108. Online at: http://www.nobelprize.org /nobel_prizes/medicine/laureates/1973 /press.html.

109. Burkhardt (2005: 460).

110. Ibid.

111. R. Hinde, pers. comm., 12 March 2010.

CHAPTER 8

1. John Krebs, pers. comm., 24 August 2011.

2. In fact, Robert Hinde and the other ethologists at Cambridge, Pat Bateson and Bill Thorpe, were sympathetic to behavioral ecology, and in 1975 Pat Bateson initiated the

Kings College Behavioural Ecology Group (later renamed Sociobiology Group).

3. Wilbur (1979).

4. Jack Hailman interview with Bob Montgomeric and Tim Birkhead, 29 July 2011.

5. Quoted in Peck (2001: 71).

6. Skutch deplored the materialistic culture of modern society, and he chose to live in a simple way at his 178-acre farm, Los Cosingos, without electricity or a telephone line for the duration of his life there, and without running water until a gravity-operated water line was set up in 1970.

7. There were several much earlier accounts of more than a single pair of birds attending the nests of several Australian birds, including John Gould's observations at the nests of White-winged Choughs in 1846 (see Heinsohn 2009). Despite the Australians' perceptive observations, their inability to recognize their significance was due to the "theoretical blinkering" of the pioneering European naturalists in Australia (see Boland and Cockburn 2002).

8. Quoted in Peck (2001: 71).

9. Stiles (2005: 709).

10. Skutch (1935: 273).

11. Skutch (1961: 208).

12. Ibid.

13. Skutch (1987).

14. Brown (1994: 236).

15. Ibid.

16. Stearns (1976: 4).

17. Westneat and Fox (2010: 46).

18. Woolfenden and Fitzpatrick (1984: 6).

19. John Fitzpatrick interview with Jo Wimpenny, 15 September 2010.

20. John Fitzpatrick, pers. comm., 15 August 2011. Countering this, Jerry Brown (pers. comm., May 2012) told us that "he was enthusiastic about testing relevant hypotheses, not about a particular hypothesis: they [Woolfenden and Fitzpatrick] failed to realise this important difference and went around misrepresenting my position ad nauseum."

21. ". . . habitat saturation cannot by itself explain delayed breeding and dispersal." Brown (1989: 1012).

22. Before Cockburn's large-scale analysis, it was estimated that only 3 percent of birds bred cooperatively (Arnold and Owens 1998).

23. Ben Hatchwell, pers. comm., 3 March 2012.

24. Collinge (1913: iii).

25. Tullock (1971: 77).

26. Ydenberg et al. (2007: 4).

27. Schmitz (1997: 631).

28. In Krebs and Kacelnik (2007: x).

29. Ibid.

30. Davies (2000: 8)

31. Review by unknown author in *Ibis* 64 (3): 580.

32. Davis (1941: 420).

33. Edgar Chance quoted in *Emu* 26: 229.

34. This was reported in the January (1927) issue of *Ibis*.

35. Baker (1913: 385).

36. Jourdain (1925: 652).

37. Rothstein (1975: 267).

38. Swynnerton (1918: 132).

39. Lack (1968: 88).

40. Newton (1896: 123–24).

41. He also discovered the first record of avian brood parasitism—in the Asian Koel—that predated Aristotle by two thousand years (Friedmann 1964).

42. Gifford (1919: 256).

43. Irenäus Eibl-Eibesfeldt, pers. comm., 18 November 2010.

44. Darwin (1871: 105).

45. Use of the term "intelligent" will always be problematic, and a description of the philosophical and terminological issues surrounding it is far beyond the scope of this book. We do not wish to enter into a philosophical morass about the use or misuse of the term, but simply use it in its commonplace, intuitive sense.

46. Thorndike (1898: 3).

47. Thorndike (1911: 280).

48. Herrick (1924), cited in Marler and Slabbekoorn (2004: 8).

49. Porter (1904: 314).

50. Miller (1921: 122).

51. Nice (1951: 535).

52. Data from Web of Science, May 2012.

53. Thorpe (1951a: 30).

54. "Clever Hans" was a horse—owned by German maths teacher Wilhelm von Osten—that was seemingly capable of mental arithmetic. He would tap out the correct answer when asked a simple problem such as "4 + 2." Unfortunately for von Osten, who billed Hans as a "wonder" horse, psychologist Oskar Pfungst demonstrated that Hans's remarkable behavior was not the result of intelligence but instead simply the result of his responding to subtle cues unintentionally given off by von Osten or by other observers. This effect is now carefully controlled in studies of animal cognition.

55. An acronym for Avian Learning Experiment.

56. Pepperberg (2006: 81).

57. McIlvane (2001: 270).

58. Coburn and Yerkes (1915).

59. Nicky Clayton interview with Jo Wimpenny, 29 November 2010.

60. Ibid.

61. Episodic memory refers to an individual's recollection of a unique past experience. Since it is impossible to know whether animals are capable of having such conscious memory experiences, tests of this form of memory are called "episodic-like," and they focus on asking whether animals remember the what, when, and where of a past event. This is in contrast to semantic memory, which is simply "fact-learning" and unrelated to specific experience.

62. Thorpe (1951a: 36).

63. Alex Kacelnik interview with Jo Wimpenny, 17 August 2010.

64. See also the "Leaders in Animal Behaviour" autobiographical chapters in Dewsbury (1985) and Drickamer and Dewsbury (2009).

CHAPTER 9

1. In a letter to John Jenner Weir dated 27 February 1868 (letter no. 5942; accessed 11 June 2010 at Darwin Correspondence Project Database. http://www.darwin project.ac.uk/entry-5942), Darwin wrote the following:

I have always felt an intense wish to make analogous trials, but have never had an opportunity, and it is not likely that you or any one would be willing to try so troublesome an experiment. Colouring or staining the fine red breast of a bullfinch with some innocuous matter into a dingy tint would be an analogous case, and then putting him and ordinary males with a female. A friend promised, but failed, to try a converse experiment with white pigeons—viz., to stain their tails and wings with magenta or other colours, and then observe what effect such a prodigious alteration would have on their courtship.

See note 33 on page 460 for more details on Darwin's suggestions about experimentally altering tail lengths.

2. Superglues are a family of cyanoacrylate adhesives invented in 1942 by a chemist, Dr. Harry Coover at Kodak Laboratories (http://web.mit.edu/invent/iow/coover.html). Cyanoacrylates proved to be useless for lenses, but their great strength and fast-drying properties later made them popular for many applications, including mending the skin and bone of animals and even humans. Superglues are sometimes called "space age glues," but they really have nothing to do with the space program.

3. At the time of writing, 2013, Andersson's *Sexual Selection* has been cited close to eight thousand times, which is one-third as frequently as Darwin's *Origin*, published more than one hundred years earlier.

4. Darwin and Wallace (1858: 50).

5. Darwin (1871, part 2: 38).

6. Ibid., 212.

7. Audubon (1832), cited extensively in Darwin (1871).

8. Darwin (1859: 88–89).

9. Wallace to Darwin, 19 March 1868. Darwin Correspondence Project Database. http://www.darwinproject.ac.uk/entry-6024 (letter no. 6024; accessed 11 June 2010).

10. See also Kottler (1980) for an analysis of Darwin and Wallace's discussions about sexual dimorphism.

11. Wallace to Darwin, 26 April 1867. Darwin Correspondence Project Database. http://www.darwinproject.ac.uk/entry-5515 (letter no. 5515; accessed 11 June 2010).

12. See Cronin (1991: 165).

13. Mayr (1982: 596).

14. Morgan (1903: 221).

15. Selous (1906: 374–75).

16. McCabe (1930).

17. Allen (1901: 409).

18. Selous (1913: 98).

19. Howard (1907–14, part 1: 14–18).

20. Kirkman (1910–13).

21. Pycraft was an osteologist, most famous perhaps for his early and strident support for what later came to be called the Piltdown Forgery. On a more positive note, he was also a keen ornithologist who wrote several books about birds (e.g., Pycraft 1908) and was a major contributor to early ornithological and animal behavior societies.

22. According to GEL (*Ibis* 85: 109–10); apparently this was George Edward Lodge, renowned bird artist. Both Lodge and Pycraft were members of the Zoological Society of London.

23. Pycraft (1914: 140).

24. Ibid., 151.

25. Townsend spent his spare time watching birds in Massachusetts and part of one summer traveling up the coast of Labrador (Townsend 1918). He published over eighty papers on his observations and several books about birds and his travels (e.g., Townsend 1905, 1913, 1918).

26. Townsend (1920: 382).

27. Since the rejected male had been in fine condition initially, Finn (1907: 4) came to the conclusion that "grief was accountable for his end—a warning to future experimenters to remove the rejected suitor as early as possible."

28. Darwin to Weir, 27 February 1868. Darwin Correspondence Project Database. http://www.darwinproject.ac.uk/entry-5942 (letter no. 5942; accessed 11 June 2010).

29. Darwin to Tegetmeier, 5 March 1867. Darwin Correspondence Project Database. http://www.darwinproject.ac.uk/entry-5431 (letter no. 5431; accessed 11 June 2010). Tegetmeier responded (29 March 1867) by saying that a male game cock trimmed for fighting is always well received by females, so he was skeptical about manipulated males being poorly received. However, he suggested to Darwin that staining a white male pigeon magenta might be a better approach: "Do [you] not think dying a white male pigeon magenta colour which is easily done and seeing whether his wife knows him, would be of any bearing on the question."

30. See Darwin (1871, part 2: 118).

31. Darwin to Weir, 4 April 1868. Darwin Correspondence Project Database. http://www.darwinproject.ac.uk/entry-6090 (letter no. 6090; accessed 11 June 2010).

32. Darwin to Tegetmeier, 5 March 1867. It is not certain where this information about the widowbirds came from.

33. Darwin seems to have made two small errors in reporting the Long-tailed Widowbird example to Weir (letter no. 5942). First, he states that the information came from Barrow, and, second, he says that the tails were "cut." Although Barrow (1801–1804) discusses the behavior of Long-tailed Widowbirds, he does not say anything about cutting their tails. The information on the role of the male's tail in mate choice comes from C. A. Rudolphi (1812), who, citing Martin Lichtenstein, reported "that the female widow-bird (*Chera progne*) disowns the male when robbed of the long tail-feathers with which he is ornamented during the breeding-season." Again, there is no mention of their tails being cut. In *Descent* (1871, vol 2: 120) Darwin writes: "Lichtenstein, who was a good observer and had excellent opportunities of observation at the Cape of Good Hope, assured Rudolphi that the female widow-bird (*Chera progne*) disowns the male, when robbed of the long tail-feathers with which he is ornamented during the breeding-season. I presume that this observation must have been made on birds under confinement." Unfortunately, Lichtenstein provides no further details,

so it remains unclear how he acquired this information.

34. Malte Andersson, pers. comm., 31 May 2010.

35. Finn (1907: 12).

36. Cinat-Tomson (1926).

37. Fisher (1930: 137).

38. Peter O'Donald interview with Jo Wimpenny, 29 June 2010.

39. Fisher was a committed eugenicist—elected the first chairman of the Cambridge Eugenics Society in 1912—and believed that good mate choice was important for human society to advance.

40. See Huxley's introduction in Selous's *Realities of Bird Life*.

41. Letters from Selous to Huxley. Selous archive, Alexander Library, University of Oxford.

42. Letter from Thomas H. Huxley to Julian's father in 1891; see Huxley's *Memories* (1970: 22).

43. Both Selous (1901a, 1902) and Pycraft (1911, 1913) had previously studied the courtship of Great Crested Grebes, and though both were cited in Huxley's papers, he probably did not give Selous and Pycraft the credit that they were due (Burkhardt 1993).

44. Huxley (1970: 89).

45. Huxley (1914).

46. Huxley (1912b).

47. Huxley and Montague (1925: 869–70).

48. Huxley (1938: 417).

49. Ibid., 431.

50. See Bartley (1995) for a detailed analysis of Huxley's social and political interpretations of bird behavior.

51. Juliette claimed that during their honeymoon Julian was more interested in the lovemaking of grebes than in their own attempts at conjugal bliss (Huxley 1970).

52. Juliette Huxley (1987: 139).

53. Fisher (1953: 118).

54. Allen (1934: 455).

55. Lack (1943a: 60).

56. Lack (1940b: 281).

57. Lack (1968: 150).

58. This idea was later developed more fully by R. A. Fisher (1930); see also Kirkpatrick et al. (1990).

59. Lack (1968: 162).

60. Ibid.

61. Ibid., 158.

62. Perhaps unsurprisingly given that they were good friends, and that Huxley was Lack's promoter for his nomination to the Royal Society.

63. K. E. L. Simmons, in litt., to J. H. Crook; John Crook, pers. comm., May 2007.

64. John Crook, pers. comm., May 2007.

65. E. O. Wilson was professor at Harvard.

66. He also laid some of the earliest foundations for evolutionary psychology, and he followed the same track as many others, starting his research on animal behavior and moving on to consider humans.

67. John Crook, pers. comm., May 2007.

68. See Crook and Osmaston (1994) and Crook and Low (1997).

69. Lack and Emlen (1939).

70. From Orians (2009: 407). When we discussed this with Ian Newton, he commented that it was unusual for David Lack to provide such comprehensive comments, at least on PhD theses. It was possibly because Orians was still an undergraduate that he received such detailed attention.

71. Bateman was an English geneticist who worked in several laboratories over his career, mainly studying pollination and genetics of seed crops. His *Drosophila* work was done at the John Innes Horticultural Institute in Merton Park, London. Trivers embellished some of Bateman's observations (Dewsbury 2005) but nonetheless laid an enduring foundation for the modern study of sexual selection.

72. Parker (1970) also made use of Bateman's paper in developing his own ideas about sexual selection and may have also have been an inspiration to Trivers.

73. Operational Sex Ratio: "the average ratio of fertilizable females to sexually active males at any given time" (Emlen and Oring 1977).

74. Zahavi mentions that Emlen (1973) provided an earlier anticipation of this theory.

75. Zahavi (2003: 860).

76. John Maynard Smith, quoted in Gadagkar (2004: 141).

77. Born in Nørresundby, Denmark, on 26 December 1953, Møller grew up on a farm and had an interest in natural history and birds in particular: he started ringing birds while still at high school. Advised not to go into biology due to poor health, Møller trained instead as a librarian, but in 1980 he gave this up to undertake a master's in farming and birds, followed in 1982 by a PhD at the University of Aarhus, Denmark, on sexual selection in barn swallows.

78. Møller typically published around twenty papers a year when most of his successful contemporaries published only a handful. This productivity earned him the nickname "Anders Paper Miller." Since the 1980s he has published over 650 papers (usually with collaborators) and has over 28,000 citations. However, his extraordinary productivity and unusually clean results have been criticized (Abbott 2004; Borrello 2006).

79. Dawkins (1982) *The Extended Phenotype*. In this book, Dawkins argues that the effects of genes should not be considered as restricted to internal biological processes, but should be widened to anything that increases the probability of a gene being replicated. Animal artefacts such as nests, dams, and bowers fit into this category since they increase the probability of survival and/or successful reproduction.

80. Terry Burke interview with Tim Birkhead, 16 April 2010.

81. Susan Smith, pers. comm., 10 May 2010.

82. See Milam (2010) for a discussion of the historical and social development of the study of female choice.

CHAPTER 10

1. MacArthur to Lack, 7 October 1962. David Lack archives, Oxford University.

2. Sibley to Lack, 15 February 1963. David Lack archives, Oxford University.

3. Lorca-Susino (2010: 251).

4. The origin of the name Vero is unknown; Copner was a well-established family middle name.

5. The Bureau of Animal Population (BAP) was a small research institute within Oxford University, concerned largely with population studies of small mammals. It was in existence from 1932, when Elton founded it, until 1967, when Elton retired. For a full history of the BAP see Crowcroft (1991).

6. It comprised keeping a field notebook, a journal into which notes were transcribed, a detailed species account, and a catalog of collected specimens. He even suggested the type of paper and ink to use, so "that our notes will be accessible 200 years from now" (Grinnell 1958: 8).

7. Grinnell (1943: 121).

8. Ibid., 122.

9. Ibid.

10. Elton (1927: 117).

11. Ibid., 116.

12. Ibid., 117.

13. Ibid., 118.

14. Wynne-Edwards (1985: 492).

15. See Borrello (2010: 6) for details.

16. MacCoy to Wynne-Edwards, 3 May 1934. Wynne-Edwards archives, Queen's University.

17. Borrello (2010: 52).

18. Wynne-Edwards (1939: 128).

19. Ibid.

20. Ibid., 501.

21. Adam Watson, unpublished manuscript.

22. Ibid.

23. Ibid.

24. The two books led to heated debate about the importance of density-dependent, versus density-independent, factors in population biology. By 1957 the controversy was so great that population regulation became the focus of an entire Cold Spring Harbor symposium. The debate was a significant one, and had far-reaching implications for the field of biology as a whole. We direct the reader to published accounts such as that by Orians (1962) and Cooper (1993) for all the details.

25. Chitty (1957: 64).

26. Ibid.

27. Haldane (1955: 375).

28. Ibid.

29. Vogt to Lack, 17 September 1955. David Lack archives, Oxford University.

30. Adam Watson, unpublished manuscript.

31. Undated letter from Lack to Wynne-Edwards, written some time before 7 May 1955, when the follow-up letter was written.

32. Letter from Lack to Wynne-Edwards, 7 May 1955; David Lack archives, Oxford University.

33. Wynne-Edwards (1955a: 545).

34. Adam Watson, unpublished manuscript.

35. Wynne-Edwards (1955b: 434).

36. Wynne-Edwards (1980: 198).

37. Borrello (2010) incorrectly says Oxford.

38. Quoted in Borrello (2010: 67).

39. There may have been an additional reason for Southern's attack. Wynne-Edwards initially offered Southern the job of establishing the ecological field station at Culterty in 1956, an offer that Southern accepted. Then, out of the blue, Wynne-Edwards withdrew his offer. He later insisted that no firm offer had ever been made, but it seems likely that Southern found it hard to forgive him (Jenkins 2003). It looks like Wynne-Edwards initially offered the job to George Dunnet—then in Australia—who turned it down. He then offered it to Southern, but before getting a definite answer from him, he wrote again to Dunnet and asked him to reconsider. Dunnet said yes, and Wynne-Edwards was in the embarrassing position of having to withdraw the offer from Southern.

40. Dunnet to Wynne-Edwards, 26 March 1959. Wynne-Edwards archives, Queen's University.

41. Lack to Wynne-Edwards, 30 April 1959. Wynne-Edwards archives, Queen's University.

42. Dunnet to Wynne-Edwards, 26 March 1959. Wynne-Edwards archives, Queen's University.

43. Ibid.

44. Lack to Wynne-Edwards, 30 April 1959. Wynne-Edwards archives, Queen's University.

45. Dunnet to Wynne-Edwards, 26 March 1959. Wynne-Edwards archives, Queen's University.

46. David Snow, pers. comm., 16 November 2008.

47. Jenkins (2003: 97).

48. Wynne-Edwards (1985: 508).

49. Wynne-Edwards (1962: 570).

50. Paul Thompson, pers. comm., 30 January 2011.

51. Elton to Wynne-Edwards, 10 May 1963. Wynne-Edwards archives, Queen's University.

52. Dick Potts, pers. comm., 5 February 2011.

53. Ibid.

54. Ibid.

55. Ibid.; also, Ian Newton interview with Tim Birkhead, 21 April 2010.

56. Ian Newton interview with Tim Birkhead, 21 April 2010.

57. Thorpe (1974: 273).

58. Lack (1949).

59. Ibid.

60. Lack (1952: 167).

61. Ibid., 173.

62. Perrins (1978: 552).

63. Ian Newton, in litt., 27 Aug 2012.

64. Robert Hinde interview with Jo Wimpenny, 18 January 2010.

65. Gibb (1954: 541).

66. Lack (1966: 8).

67. Ibid., 280.

68. Ibid., 275.

69. Ibid.

70. Roger Bray, pers. comm., 14 November 2010.

71. Ian Newton interview with Tim Birkhead, 21 April 2010.

72. Dated 9 December 1968; courtesy Aberdeen University.

73. Adam Watson, unpublished manuscript.

74. Wynne-Edwards (1979: 6).

75. Moffat (1903: 152).

76. Dick Potts, pers. comm., 5 February 2011.

77. Peter Hudson, pers. comm., 22 February 2011.

78. Ian Newton, pers. comm., 11 February 2011.

79. The rise of behavioral ecology, with its explicit focus on individual selection in the late 1970s (chapter 8), spelled the death knell for both group selection and Wynne-Edwards, who Trivers (1985) later referred to as "the showpiece of the group selection fallacy" (see also Dawkins 1976). At the same time, however, behavioral ecology created the possibility in Wynne-Edwards's mind that kin selection might explain the selfless restraint he imagined was controlling grouse

numbers. Inspired by the kin selection life-line, Wynne-Edwards produced a second book in 1986 titled *Evolution through Group Selection*. It was a disaster, and as Pollock (1989) said, it was "universally and devastatingly panned." Matt Ridley captured the feeling of the scientific community: "What a strange man Wynne-Edwards is, to precipitate perhaps the most interesting controversy in evolutionary biology of the past quarter century, and then to write another book on the same subject which ignores the controversy completely" (Ridley 1987: 261). The red grouse sits center stage in *Evolution through Group Selection*, yet, as Watson wrote in his review, Wynne-Edwards's "account rests largely on publications from early years, omitting... most since the late 1970s, and all the grouse workers' own, published reviews of their findings.... It is unfortunate that these and other errors make the grouse chapters unreliable as a review of the bird's population ecology.... More objectivity and humility would have given the proposals a better chance of a fair hearing" (Watson 1987: 1090). Wynne-Edwards must have been devastated.

80. Coulson (2011: 11).

81. Subsequently, Cornell changed the degree to biology, as a more appropriately labeled discipline (John Fitzpatrick, pers. comm., 27 March 2012).

82. Although he did continue to be director of the lab until 1960, when Olin Pettingill took over.

83. Wilson and Hutchinson (1989: 322).

84. MacArthur (1958: 599).

85. Kendeigh (1945: 433).

86. Darwin (1859: 140).

87. Schoener (1982: 586).

88. Ibid., 594.

89. Fretwell (1975: 4).

90. Orians (2009: 408–09).

91. Ibid., 424–25.

CHAPTER 11

1. International Union for Conservation of Nature. The Red List was first started in 1963 and today is the most comprehensive assessment of the status of wild species.

2. Şekercioğlu et al. (2004: 18042). "Functionally extinct" means that their population is so small it cannot survive for long.

3. Cartwright (1792: 55).

4. The original idea of a shifting baseline is from Pauly (1995: 2); this quote is from Michael McCarthy in *The Independent*, 26 November 2010.

5. Meffe and Viederman (1995: 327).

6. Florence Merriam Bailey's *Handbook of the Birds of the Western United States* (1902) contained many elements of the modern field guide, and John Albert Leach's *An Australian Bird Book: A Pocket Book for Field Use* (1911) was also ahead of its time. But none of the earlier attempts came close to what Peterson accomplished.

7. Quoted in Price (1999: 59).

8. Hornaday (1913: 117).

9. Mattingley (1907: 72).

10. Mattingley's photographs were mentioned in *Bird Notes and News* (July 1909), with text from James Buckland; a supplement to this issue, called "The Story of the Egret," was also produced.

11. Details about Schlemmer from Unger (2004) and Rauzon (2010).

12. Unger (2004: 65).

13. Dill's exhibit (named the Laysan Island Cyclorama) opened on 15 June 1914 at Iowa's Museum of Natural History and survives to this day.

14. Quoted in Hornaday (1913: 141).

15. The expedition was run by the US Biological Survey and the Bishop Museum in Honolulu and was led by Alexander Wetmore (chapter 3).

16. Alexander (1974: 129).

17. There were other early censusing efforts. William Leon Dawson and Lynds Jones made counts of birds seen during daily outings in Ohio in the late 1890s. Edward Howe Forbush, starting in 1905, estimated population sizes of birds from responses to questionnaires by observers from all over Massachusetts. In 1907 Alfred Gross and

Harold Ray used a transect to document the number of birds per square mile in different habitats in Illinois. Christopher and Horace Alexander, in 1908, gave suggestions on how to map migratory birds in their nesting areas and suggested counting singing males. In Australia in the 1920s, J. Burton Cleland used a similar approach to census bird numbers.

18. Susan R. Drennan, in Temple and Wiens (1989: 263).

19. Jim Lowe, pers. comm., 15 April 2011, and see http://www.pwrc.usgs.gov/birds/bbc.html #contact.

20. Roaring Fork Motor Nature Trail, in the Great Smoky Mountains National Park, Tennessee.

21. The British BBS continued the previous BTO population monitoring survey, the Common Birds Census (CBC), which ran from 1962 to 2000. Data can be combined from both, allowing long-term population changes to be identified.

22. Reported by Stephen Moss (2004), after hearing Nicholson say this in a talk at a British Trust for Ornithology conference in 1993.

23. First used by Petersen (1896) to estimate the returns of tagged plaice (*Pleuronectes platessa*).

24. http://www.mabulagroundhornbill conservationproject.org.za.

25. None of these authors seems to have been aware of Howard's pioneering efforts at mapping winter distributions of warblers in Africa (chapter 4).

26. Sanderson et al. (1996: 93).

27. Although there was an accumulating wealth of information from people like Alexander Skutch, who studied the biology of resident Neotropical species (chapter 8).

28. François Vuilleumier (1982) was highly critical of the extreme heterogeneity and poor organization of this edited volume, which he nonetheless considered a landmark in migration and ecological studies. He felt that its main limitation was the lack of information on the resources required or used by migrants in their wintering areas.

29. Greenberg (1990: 640).

30. See UNEP's *Africa: Atlas of Our Changing Environment*: http://www.unep.org/dewa/africa/africaAtlas (accessed 4 June 2010).

31. BirdLife's flyway program aims to conserve migratory birds and their habitats along the migratory flyways. For more information see http://www.birdlife.org/flyways/index.html (accessed 4 June 2010).

32. See http://www.natura.org/about.html.

33. John Fanshawe, pers. comm., 1 April 2011.

34. Euan Kennedy, pers. comm., 29 January 2012. Some hybridization occurred without imprinting, possibly because there was simply a shortage of mates.

35. Dichlorodiphenyltrichloroethane. DDT is one of a group of toxic organochlorine chemicals, along with aldrin, dieldrin, and heptachlor.

36. O'Hara (1945: 18).

37. Kozlik (1946: 102).

38. "There seems to have been a big difference between the situation in Britain and the US. In Britain we used much more dieldrin per unit area than the US, and bird of prey populations crashed within 2–3 years in the late 1950s mainly from direct poisoning, whereas in the US they used much more DDT and populations generally declined more slowly, through reduced breeding success." Ian Newton, in litt., 27 August 2012.

39. Including 115 micrograms of pp-DDE (metabolite of DDT), 50 micrograms of dieldrin, 28 micrograms of heptachlor epoxide. In total there were 4 to 5 parts per million of organochloride residues in the egg.

40. Ian Newton interview with Tim Birkhead, 21 April 2010.

41. Using mass/length times breadth.

42. Ian Newton interview with Tim Birkhead, 21 April 2010.

43. Ibid.

44. Quoted in Lytle (2007: 175).

45. Ibid.

46. Lear (1997: 429). See also Oreskes and Conway (2010).

47. Ian Nisbet, pers. comm., 12 April 2010.

48. Ibid.

49. It was not the first country to do so: Hungary led the way by banning it in 1968, Norway

and Sweden did so in 1970, and Germany also banned it in 1972. In contrast, Britain did not ban it until 1984.

50. Jim Flegg, pers. comm., 29 March 2011.

51. Ibid.

52. Taken from a transcript of the interview. James Fisher archives, Natural History Museum, London.

53. Ibid.

54. Letter from Shell to Fisher, 25 July 1966. James Fisher archives, Natural History Museum, London.

55. Robinson (1966: 159).

56. Fisher (1966: 92).

57. Ian Nisbet, pers. comm., 23 February 2011.

58. Ibid., 12 April 2010.

59. Ibid., 28 February 2013.

60. Ibid.

61. There were also twenty-one in captivity at the end of 1986.

62. With a running cost of $5 million per year, this is a very expensive program, although much of the money comes from private partners (Walters et al. 2010).

63. Reported in the *New York Times*, 16 November 2010 (D3), and online: http://www.nytimes.com/2010/11/16/science/16condors.html?pagewanted=all (accessed 17 October 2011).

64. US Fish and Wildlife Service. 1999. Proposed Rule to Remove the Bald Eagle in the Lower 48 States from the List of Endangered and Threatened Wildlife. Federal Register, 6 July 1999 (64 FR 36453); US Fish and Wildlife Service. 2009. Removal of the Brown Pelican (*Pelecanus occidentalis*) from the Federal List of Endangered and Threatened Wildlife.

Federal Register, 17 November 2009 (74 FR 59444-59472).

65. For further biographical details see Ballance (2007).

66. Caughley and Gunn (1996: 79).

67. The other is the Ivory-billed Woodpecker, although it seems likely that this species has already gone extinct. More information about this species can be found at Cornell University's Laboratory of Ornithology: http://www.birds.cornell.edu/ivory (accessed 4 July 2012).

68. US Fish and Wildlife Service: http://www.fws.gov/ncsandhills/rcw.html (accessed 28 May 2012).

69. From *Northern Spotted Owls: Not Out of the Woods Yet*; published by the Seattle Audubon Society.

70. Thomas (2002: 21).

71. Ibid., 18.

72. Ibid.

73. Ibid., 16.

74. Performance Standard 6: Biodiversity conservation and sustainable management of living natural resources, published January 2012.

75. Ibid.

76. Nigel Collar, pers. comm., 29 January 2012.

77. BirdLife International: http://www.birdlife.org/action/science/sites (accessed 23 May 2012).

AFTERWORD

1. Barlow (1958: 141).

2. Ibid.

3. Ibid.

4. Ibid., 145.

REFERENCES

To save space we have not included the titles of scientific papers in this list of references. For a complete listing of titles and a downloadable file containing all of these references go to the book's website at http://myriadbirds.com.

Abbott A (2004) *Nature* 427:381. **Abbott CG** (1933) *Condor* 35:10–14. **Able KP** (1995) *Condor* 97:592–604. **Able KP, Belthoff JR** (1998) *Proceedings of the Royal Society B* 265:2063–71. **Abzhanov A et al** (2006) *Nature* 442:563–67. **Ackerman J** (1998) *National Geographic* 194 (July): 74–79. **Adler LL** (1973) *Annals of the New York Academy of Sciences* 223:41–48. **Ahlquist JE** (1999) *Auk* 116:856–60. **Ainley MG** (1993) *Restless Energy: A Biography of William Rowan, 1891–1957.* Vehicule Press: Montreal. **Alerstam T** (2003) *Nature* 421:27–28. **Alerstam T** (2011) *Journal of Ornithology* 152:S5–S23. **Alerstam T, Lindström Å** (1990) pp 331–51 in *Bird Migration: Physiology and Ecophysiology*, Gwinner E (ed), Springer-Verlag: Berlin. **Alerstam T et al** (2007) *Proceedings of the Royal Society B* 274:2523–30. **Alexander HG** (1915) *British Birds* 8:184–92. **Alexander HG** (1974) *Seventy Years of Birdwatching.* Poyser: Berkhamsted, UK. **Allee WC, Masure RH** (1936) *Journal of Comparative Psychology* 22:131–55. **Allen AA** (1914) *Abstract of Proceedings, Linnaean Society of New York* 24-25:44–128. **Allen FH** (1934) *Auk* 51:454–69. **Allen JA** (1885) *Auk* 2:129–39. **Allen JA** (1901) *Auk* 18:408–409. **Allen JA** (1909) *Auk* 26:209–14. **Allender C** (1936) *Ecology* 17:258–62. **Altman J** (1962) *Science* 135:1127–28. **Amadon D** (1950) *Bulletin of the American Museum of Natural History* 94:151–262. **Amadon D** (1966) *Systematic Biology* 15:245–49. **American Ornithologists' Union** (1886) *The Code of Nomenclature and Check-List of North American Birds Adopted by the American Ornithologists' Union; Being the Report of the Committee of the Union on Classification and Nomenclature.* American Ornithologists' Union: New York. **American Ornithologists' Union** (1957) *Check-List of North American Birds.* American Ornithologists' Union: Washington, DC. **Andersen HT** (1966) *Physiological Reviews* 46:212–43. **Anderson E** (2013) *The Life of David Lack, Father of Evolutionary Ecology.* Oxford University Press: Oxford. **Anderson RM, May RM** (1978) *Journal of Animal Ecology* 47:219–47. **Andersson M** (1982) *Nature* 299:818–20. **Andersson M** (1994) *Sexual Selection.* Princeton University Press: Princeton, NJ. **Andrewartha HG, Birch LC** (1954) *The Distribution and Abundance of Animals.* University of Chicago Press: Chicago. **Anonymous** (1874) *Nature* 10:516–19. **Anonymous** (1882) *Nature* 26:84–86. **Anonymous** (1884) *Science* 4:86–87. **Anonymous** (1917) *Ibis* 59:617. **Antonov A et al** (2006) *Behavioral Ecology and Sociobiology* 60:11–18. **Armstrong DP et al** (2002) *Journal of Biogeography* 29:609–21. **Armstrong EA** (1942) *Bird Display: An Introduction to the Study of Bird Psychology.* Lindsay Drummond: London. **Arnold KE, Owens IPF** (1998) *Proceedings of the Royal Society B* 265:739–45. **Aschoff J** (1960) *Cold Spring Harbor Symposia on Quantitative Biology* 25:11–28. **Askenmo C** (1979) *American Naturalist* 114:748–53. **Audubon JJ** (1832) *Ornithological Biography*, vol 1. Carey and Hart: Philadelphia. **Avise JC**

(2004) *Molecular Markers, Natural History, and Evolution.* Sinauer Associates Inc.: Sunderland, MA.

Bailey CEG (1950) *Ibis* 92:115–31. **Bailey FM** (1902) *Handbook of Birds of the Western United States.* Houghton Mifflin and Company: Boston. **Baird SF et al** (1874) *A History of North American Birds.* Little, Brown and Co.: Boston. **Bairlein F, Becker P** (2010) *100 Jahre Institut für Vogelforschung: Vogelwarte Helgoland.* Aula-Verlag: Wiebelsheim. **Baker ECS** (1913) *Ibis* 55:384–98. **Baker JR** (1938) *Proceedings of the Zoological Society of London* 108:557–82. **Baldamus E** (1853) *Naumannia* 3:307–25. **Baldwin FM et al** (1940) *Proceedings of the Society for Experimental Biology and Medicine* 44:373–75. **Baldwin SP** (1921) *Auk* 38:228–37. **Ball GF, Ketterson ED** (2008) *Philosophical Transactions of the Royal Society B* 363:231–46. **Ballance A** (2007) *Don Merton, the Man Who Saved the Black Robin.* Reed Publishing (NZ): Auckland. **Bang BG** (1960) *Nature* 188:547–49. **Bang BG, Cobb S** (1968) *Auk* 85:55–61. **Bang BG, Wenzel BM** (1985) pp 195–225 in *Form and Function in Birds*, vol 3, King AS, McLelland J (eds), Academic Press: London. **Bardach J et al** (1947) *A Biological Survey of Lake Opinicon.* Queen's University: Kingston, Ontario. **Barlow N** (1958) *The Autobiography of Charles Darwin 1809–1882. With the Original Omissions Restored.* Collins: London. **Barnett L** (1959) *Life* 47 (16): 96–113. **Barrow J** (1801–1804) *An Account of Travels into the Interior of Southern Africa, in the Years 1797 and 1798*, 2 vols. T. Cadell and W. Davies: London. **Barrow MV** (1998) *A Passion for Birds: American Ornithology after Audubon.* Princeton University Press: Princeton, NJ. **Bartley MM** (1995) *Journal of the History of Biology* 28:91–108. **Bateman AJ** (1948) *Heredity* 2:349–68. **Bateson PPG** (1966) *Biological Reviews* 41:177–217. **Bateson PPG** (1990) *American Psychologist* 45:65–66. **Bateson PPG** (2009) in *Leaders in Animal Behavior: The Second Generation*, Drickamer L, Dewsbury D (eds), Cambridge University Press: Cambridge. **Bateson PPG, Klopfer PH** (1989) pp v–viii in *Whither Ethology? Perspectives in Ethology*, vol 8, Bateson PPG, Klopfer PH (eds), Plenum Press: New York. **Bateson W** (1894) *Materials for the Study of Variation: Treated with Especial Regard to Discontinuity in the Origin of Species.* Macmillan: London. **Bateson W** (1913) *Problems of Genetics.* Yale University Press: New Haven, CT. **Bateson W, Punnett RC** (1911) *Journal of Genetics* 1:185–203. **Battley PF et al** (2012) *Journal of Avian Biology* 43:21–32. **Beach FA** (1948) *Hormones and Behavior.* P. B. Hoeber: New York. **Beach FA** (1950) *American Psychologist* 5:115–24. **Beach FA** (1981) *Hormones and Behavior* 15:325–76. **Beebe CW** (1915) *Zoologica* 2:39–52. **Benkman CW** (1987) *Ecological Monographs* 57:251–67. **Bennett ATD, Cuthill IC** (1994) *Vision Research* 34:1471–78. **Bennett ATD et al** (1996) *Nature* 380:433–35. **Bennett G** (1834) *Wanderings in New South Wales, Batavia, Pedir Coast, Singapore and China.* Richard Bentley: London. **Bennett PM, Owens IPF** (2002) *Evolutionary Ecology of Birds.* Oxford University Press: Oxford. **Bent AC** (1919–68) *Life Histories of North American Birds*, 23 vols. US National Museum Bulletins. **Berkhoudt H** (1985) pp 463–96 in *Form and Function in Birds*, vol 3, King AS, McLelland J (eds), Academic Press: London. **Bermingham E et al** (1992) *Proceedings of the National Academy of Sciences USA* 89:6624–28. **Bert P** (1870) pp 526–53 in *Leçons sur la physiologie comparée de la respiration*, Bert P (ed), Baillière et Fils: Paris. **Berthold P** (1993) *Bird Migration: A*

General Survey. Oxford University Press: Oxford. **Berthold P** (1996) *Control of Bird Migration*. Chapman & Hall: London. **Berthold P** (1999) *Ostrich* 70:1–11. **Berthold P** (2005) *Vogelwarte* 43:59–60. **Berthold P, Querner U** (1981) *Science* 212:77-79. **Bertram GCL et al** (1934) *Ibis* 76:816–32. **Bibby CJ et al** (2000) *Bird Census Techniques*. Academic Press: London. **Birch LC** (1957) *American Naturalist* 91:5–18. **Bird CD, Emery NJ** (2009) *Current Biology* 19:1410–14. **BirdLife International** (2008) *State of the World's Birds 2008*. BirdLife International: Cambridge. **Birkhead TR** (1998a) *Evolution* 52:1212–18. **Birkhead TR** (1998b) pp 579–622 in *Sperm Competition and Sexual Selection*, Birkhead TR, Møller AP (eds), Academic Press: London. **Birkhead TR** (2003) *The Red Canary: The Story of the First Genetically Engineered Animal*. Weidenfeld & Nicolson: London. **Birkhead TR** (2008) *The Wisdom of Birds*. Bloomsbury: London. **Birkhead TR** (2012) *Bird Sense*. Bloomsbury: London. **Birkhead TR, Biggins JD** (1998) *Behavioral Ecology* 9:253–60. **Birkhead TR, Brillard JP** (2007) *Trends in Ecology & Evolution* 22:266–72. **Birkhead TR, Charmantier I** (2013) *Archives of Natural History* 40:125–38. **Birkhead TR, Gallivan PT** (2012) *Ibis* 154:887–905. **Birkhead TR, Møller AP** (1992) *Sperm Competition in Birds: Evolutionary Causes and Consequences*. Academic Press: London. **Birkhead TR, Møller AP** (1998) *Sperm Competition and Sexual Selection*. Academic Press: London. **Birkhead TR, Monaghan P** (2010) pp 3–15 in *Evolutionary Behavioral Ecology*, Westneat DF, Fox CW (eds), Oxford University Press: New York. **Birkhead TR et al** (1995) *Proceedings of the Royal Society B* 261:285–92. **Birkhead TR et al** (2004) *Evolution* 58:416–20. **Birkhead TR et al** (2011) *Proceedings of the Royal Society B* 278:1019–24. **Bissonnette TH** (1937) *Wilson Bulletin* 49:241–70. **Blix AS, Folkow B** (1983) pp 917–45 in *Handbook of Physiology, Section 2: The Cardiovascular System*, Shepherd JT, Abboud FM (eds), American Physiological Society: Bethesda, MD. **Boag PT, Grant PR** (1981) *Science* 214:82–85. **Boag PT, Van Noordwijk AJ** (1987) pp 45–78 in *Avian Genetics*, Cooke F, Buckley PA (eds), Academic Press: London. **Bobr LW et al** (1964) *Journal of Reproduction and Fertility* 8:49–58. **Bock WJ** (1963) *Auk* 80:202–207. **Bock WJ** (1965) *Systematic Zoology* 14:272–87. **Bock WJ** (2001) *Auk* 118:805–806. **Bock WJ** (2005) *Ornithological Monographs* 58:2–16. **Bock WJ, Von Wahlert G** (1965) *Evolution* 19:269–99. **Boertmann D** (1994) *Meddelelser om Grønland, Bioscience* 38:1–63. **Boland CRJ, Cockburn A** (2002) *Emu* 102:9–17. **Bolhuis JJ** (1991) *Biological Reviews* 66:303–45. **Boomsma JJ** (2009) *Philosophical Transactions of the Royal Society B* 364:3191–3207. **Borelli GA** (1680) *De Motu Animalium* [On the Movement of Animals]. Bernabo: Rome. **Borelli GA** (1911) *The Flight of Birds*. King, Sell & Olding: London. **Borgia G et al** (1987) *Animal Behaviour* 35: 1129–39. **Borrello B** (2006) *The Scientist* 21:26–32. **Borrello ME** (2010) *Evolutionary Restraints: The Contentious History of Group Selection*. University of Chicago Press: Chicago. **Borror DJ, Reese CR** (1953) *Wilson Bulletin* 65:271–76. **Both C et al** (2006) *Nature* 441:81–83. **Botkin DB, Miller RS** (1974) *American Naturalist* 108:181–92. **Bowman RI** (1961) *University of California Publications in Zoology* 58:1–302. **Bowman RI** (1963) *Occasional Papers of the California Academy of Sciences* 44:107–40. **Boyle WA, Conway CJ** (2007) *American Naturalist* 169: 344–59. **Boyle WA et al** (2011) *Biology Letters* 7:661–63. **Branion HD** (1966) *Poultry Science*

45:211–12. **Brehm AE et al** (1876-79) *Brehm's Thierleben, Allgemeine Kunde des Thierreichs*, 2nd expanded ed, 10 vols. Verlag des Bibliographischen Instituts: Leipzig. **Brennan PLR et al** (2010) *Proceedings of the Royal Society B* 277:1309-14. **Brennan PLR et al** (2007) *PLoS One* 2:e418. **Brewer AD et al** (2006) *Canadian Atlas of Bird Banding*, vol 1: Doves, Cuckoos, and Hummingbirds through Passerines, 1921–1995, 2nd ed, online. Canadian Wildlife Service Special Publication: Ottawa. **Brigandt I** (2005) *Journal of the History of Biology* 38:571–608. **Brinkhof MWG et al** (1993) *Journal of Animal Ecology* 62:577–87. **Brittingham MC, Temple SA** (1983) *BioScience* 33:31–35. **Brodkorb P** (1963) *Bulletin of the Florida State Museum Biological Sciences* 7:179–293. **Brooke M** (1990) *The Manx Shearwater*. Poyser: London. **Brooke M** (2012) *Ibis* 154:425. **Brooks M** (2011) *Free Radicals: The Secret Anarchy of Science*. Profile Books: London. **Brooks RC, Griffith SC** (2010) pp 416–33 in *Evolutionary Behavioural Ecology*, Westneat DF, Fox CW (eds), Oxford University Press: New York. **Brooks TM et al** (2008) *Bird Conservation International* 18:S2–S12. **Brooks WK** (1883) *The Laws of Heredity: A Study of the Cause of Variation and the Origin of Living Organisms*. Murphy: Baltimore. **Brooks WK** (1898) *Popular Science Monthly* 52:784–98. **Broom R** (1913) *Proceedings of the Zoological Society of London* 1913:619–33. **Brouwer GA** (1954) *Historische Gegevens over onze vroegere Ornithologen en over de Avifauna van Nederland*. E. J. Brill: Leiden. **Brown CR, Brown MB** (1986) *Ecology* 67:1206–18. **Brown JL** (1964) *Wilson Bulletin* 76:160–69. **Brown JL** (1970) *Animal Behaviour* 18:366–78. **Brown JL** (1975) *The Evolution of Behavior*. Norton: New York. **Brown JL** (1989) *Condor* 91:1010–13. **Brown JL** (1994) *Condor* 96:232–43.

Brown JL et al (1978) *Behavioral Ecology and Sociobiology* 4:43–59. **Brown WL, Jr, Wilson EO** (1956) *Systematic Zoology* 5:49–64. **Brown-Séquard CE** (1889) *Lancet* 1889, 2:105–107. **Bruderer B, Bolt A** (2001) *Ibis* 143:178–204. **Brush AH** (1993) pp 121–62 in *Avian Biology*, vol 9, Farner DS, King JR, Parkes KC (eds), Academic Press: London. **Bumpus H** (1899) *Lectures Woods Hole Marine Biology Station* 6:20–26. **Burfield I et al** (2004) *Birds in Europe: Population Estimates, Trends and Conservation Status*. BirdLife International: Cambridge. **Burgers P, Chiappe LM** (1999) *Nature* 399:60–62. **Burke T, Bruford MW** (1987) *Nature* 327:149–52. **Burke T et al** (1989) *Nature* 338:249–51. **Burkhardt D** (1989) *Journal of Comparative Physiology A: Neuroethology, Sensory, Neural, and Behavioral Physiology* 164:787–96. **Burkhardt D** (1995) *Verhandlungen Deutsche Zoologische Gesellschaft* 88:233–34. **Burkhardt RW** (1992) pp 127–49 in *Julian Huxley: Biologist and Statesman of Science*, Waters CK, Van Helden A (eds), Rice University Press: Houston. **Burkhardt RW** (2005) *Patterns of Behaviour: Konrad Lorenz, Niko Tinbergen, and the Founding of Ethology*. University of Chicago Press: Chicago. **Burkitt JP** (1926) *British Birds* 20:91–101. **Burley N et al** (1982) *Animal Behaviour* 30:444–55. **Buss IO** (1946) *Auk* 63:315–18. **Butler D** (1989) *Quest for the Kakapo*. Heinemann Reed: Auckland, NZ. **Butler D, Merton D** (1992) *The Black Robin: Saving the World's Most Endangered Bird*. Oxford University Press: Auckland, NZ. **Butler PJ, Woakes AJ** (1979) *Journal of Experimental Biology* 79:283–300. **Butler PJ, Woakes AJ** (1980) *Journal of Experimental Biology* 85:213–26.

Cade TJ, Burnham W (2003) *The Return of the Peregrine: A North American Saga of Tenacity* and Teamwork. The Peregrine Fund: Boise, ID. **Cade TJ, Fyfe RW** (1978) pp 251–62 in *Endangered Birds: Management Techniques for Preserving Threatened Species*, Temple SA (ed), University of Wisconsin Press: Madison, WI. **Cain AJ** (1959) *Ibis* 101:302–18. **Canty N, Gould JL** (1995) *Animal Behaviour* 50:1091–95. **Caple G et al** (1983) *American Naturalist* 121:455–76. **Capshew JH** (1993) *Technology and Culture* 34:835–57. **Caro TM** (1998) *Behavioral Ecology and Conservation Biology*. Oxford University Press: New York. **Caro TM, Sherman PW** (2011) *Trends in Ecology and Evolution* 26:111–18. **Carr-Saunders AM** (1922) *The Population Problem: A Study in Human Evolution*. Clarendon Press: Oxford. **Carrick R, Dunnet GM** (1954) *Ibis* 96:356–70. **Carroll SB** (2009) *Remarkable Creatures: Epic Adventures in the Search for the Origins of Species*. Houghton Mifflin Harcourt: Boston. **Carson R** (1951) *The Sea Around Us*. Oxford University Press: New York. **Carson R** (1962) *Silent Spring*. Houghton Mifflin: Boston. **Cartwright G** (1792) *A Journal of Transactions and Events during a Residence of Nearly Sixteen Years on the Coast of Labrador*. Allin and Ridge: Newark, UK. **Caswell H** (1989) *Matrix Population Models*. Sinauer Associates: Sunderland, MA. **Catchpole CK, Slater PJB** (1995) *Bird Song: Biological Themes and Variations*, 2nd ed. Cambridge University Press: Cambridge. **Caughley G, Gunn A** (1996) *Conservation Biology in Theory and Practice*. Blackwell Science: Cambridge, MA. **Chaine AS, Lyon BE** (2008) *Science* 319:459–62. **Chambers P** (2002) *Bones of Contention: The Archaeopteryx Scandals*. John Murray: London. **Chance EP** (1922) *The Cuckoo's Secret*. Sidgwick and Jackson: London. **Chance EP** (1940) *The Truth about the Cuckoo*. Country Life: London. **Chapman FM** (1914) *Bird-Lore* 16:284–85. **Chapman FM** (1935) *Science* 81:95–97. **Charig AJ et al** (1986) *Science* 232:622–26. **Charmantier A et al** (2006) *Proceedings of the National Academy of Sciences USA* 103:6587–92. **Charmantier A et al** (2008) *Science* 320:800–803. **Charnov EL** (1976) *Theoretical Population Biology* 9:129–36. **Charnov EL, Krebs JR** (1974) *Ibis* 116:217–19. **Chatterjee S** (1997) *The Rise of Birds*. Johns Hopkins University Press: Baltimore. **Chen P et al** (1998) *Nature* 391:147–52. **Chiappe LM** (2007) *Glorified Dinosaurs: The Origin and Early Evolution of Birds*. John Wiley: Hoboken, NJ. **Chisholm AH** (1969) *Emu* 69:54–58. **Chitty D** (1957) *British Journal for the Philosophy of Science* 8:64–66. **Cinat-Tomson H** (1926) *Biologisches Zentralblatt* 46:543–52. **Clark JA, May RM** (2002) *Science* 297:191. **Clarke WE** (1904) *Ibis* 46:112–42. **Clarke WE** (1912) *Studies in Bird Migration*. Gurney and Jackson: London. **Clayton NS, Dickinson A** (1998) *Nature* 395:272–74. **Clayton NS et al** (2007) *Philosophical Transactions of the Royal Society B* 362:507–22. **Clutton-Brock TH** (1988) *Reproductive Success: Studies of Individual Variation in Contrasting Breeding Systems*. University of Chicago Press: Chicago. **Clutton-Brock TH** (2002) *Science* 296:69–72. **Clutton-Brock TH et al** (1982) *Red Deer: Behavior and Ecology of Two Sexes*. University of Chicago Press: Chicago. **Cobbold TS** (1873) *Veterinarian* 46:161–72. **Coburn CA** (1914) *Journal of Animal Behavior* 4:185–201. **Coburn CA, Yerkes RM** (1915) *Journal of Animal Behavior* 5:75–14. **Cockburn A** (1998) *Annual Review of Ecology and Systematics* 29:141–77. **Cockburn A** (2006) *Proceedings of the Royal Society B* 273:1375–83. **Cockburn A, Russell AF** (2011) *Current Biology* 21:R195–R197. **Cocker M** (2013) *Birds and People*. Random House UK: London. **Cody**

ML (1966) *Evolution* 20:174–84. **Cody ML** (1969) *Science* 163:1185–87. **Cody ML** (1974) *Competition and the Structure of Bird Communities.* Princeton University Press: Princeton, NJ. **Cody ML** (1981) *Current Contents* 23:341. **Coker CR et al** (2002) *Auk* 119:403–13. **Cole FJ** (1944) *A History of Comparative Anatomy from Aristotle to the Eighteenth Century.* Macmillan: London. **Cole LC** (1954) *Quarterly Review of Biology* 29:103–37. **Collar NJ et al** (2007) *Birds and People: Bonds in a Timeless Journey.* CEMEX: Mexico City. **Collias N, Joos M** (1953) *Behaviour* 5:175–88. **Collinge WE** (1913) *The Food of Some British Wild Birds: A Study in Economic Ornithology.* Dulau & Co.: London. **Collins JW** (1884) *Report of the Commissioner of Fish and Fisheries for 1882* 13:311–35. **Colwell MA, Oring LW** (1988) *Wilson Bulletin* 100:567–82. **Comte A** (1893) *The Positive Philosophy of Auguste Comte* [Freely Translated and Condensed by Harriet Martineau]. Kegan Paul, Trench, Trübner and Co.: London. **Conniff R** (2010) *The Species Seekers: Heroes, Fools, and the Mad Pursuit of Life on Earth.* WW Norton and Company: New York. **Cooke AS** (1973) *Environmental Pollution (1970)* 4:85–152. **Cooke F** (1984) *Science* 224:278–79. **Cooke F, McNally CM** (1975) *Behaviour* 53:151–70. **Cooke F et al** (1995) *The Snow Geese of La Pérouse Bay: Natural Selection in the Wild.* Oxford University Press: Oxford. **Cooper G** (1993) *Biology and Philosophy* 8:359–84. **Corbin KW, Brush AH** (1999) *Auk* 116:806–14. **Corfield JR et al** (2008) *Nature Protocols* 3:597–605. **Corner GW, Allen WM** (1929) *American Journal of Physiology* 88:326–46. **Cornwallis CK et al** (2010) *Nature* 466:969–72. **Cornwallis RK** (1959) *Ibis* 101:424–28. **Costa R** (2001) pp 309–21 in *Wildlife of Southern Forests: Habitat and Management*, Dickson JG (ed), Hancock House Publishers: Blaine, WA. **Coues E** (1872) *Key to North American Birds.* Naturalists' Agency: Salem, MA. **Coues E** (1896) *Key to North American Birds*, 4th ed. Estes and Lauriat: Boston. **Coulson JC** (2011) *The Kittiwake.* A&C Black: London. **Cox GW** (1985) *American Naturalist* 126:451–74. **Cracraft J** (1974) *Ibis* 116:494–521. **Cracraft J** (1977) *Wilson Bulletin* 89:488–92. **Craig W** (1911) *Journal of Morphology* 22:299–305. **Cronin H** (1991) *The Ant and the Peacock: Altruism and Sexual Selection from Darwin to Today.* Cambridge University Press: Cambridge. **Crook JH** (1964) *Behaviour Supplements* 10:1–178. **Crook JH** (1980) *The Evolution of Human Consciousness.* Oxford University Press: New York. **Crook JH, Gartlan JS** (1966) *Nature* 210:1200–1203. **Crook JH, Low J** (1997) *The Yogins of Ladakh.* Motilal Banarsidass: Delhi. **Crook JH, Osmaston H** (1994) *Himalayan Buddhist Villages: Environment, Resources, Society and Religious Life in Zagskar, Ladakh.* Motilal Banarsidass: Delhi. **Crowcroft P** (1991) *Elton's Ecologists: A History of the Bureau of Animal Population.* University of Chicago Press: Chicago. **Cullen E** (1957) *Ibis* 99:275–302. **Cunningham JT** (1900) *Sexual Dimorphism in the Animal Kingdom: A Theory of the Evolution of Secondary Sexual Characters.* Black: London. **Cunningham S et al** (2007) *Journal of Anatomy* 211:493–502. **Curio E** (1969) *Zeitschrift für Tierpsychologie* 26:394–487. **Curio E** (1996) *Trends in Ecology & Evolution* 11:260–63. **Curio E** (1998) pp 163–87 in *Behavioral Ecology and Conservation Biology*, Caro TM (ed), Oxford University Press: Oxford. **Currie PJ et al** (2004) *Feathered Dragons: Studies on the Transition from Dinosaurs to Birds.* Indiana University Press: Bloomington. **Cuthbert R et al** (2011) *PLoS One* 6:e19069. **Cuthill IC, Houston**

AI (1997) pp 97–120 in *Behavioural Ecology: An Evolutionary Approach*, 4th ed, Krebs JR, Davies N (eds), Blackwell Scientific: Oxford. **Cuthill IC et al** (2000) *Advances in the Study of Behavior* 29:159–214.

Daan S, Gwinner E (1998) *Nature* 396:418. **Danchin É et al** (2011) *Nature Reviews Genetics* 12:475–86. **Darwin CR** (1859) *On the Origin of Species by Means of Natural Selection, Or the Preservation of Favoured Races in the Struggle for Life.* John Murray: London. **Darwin CR** (1866) *On the Origin of Species by Means of Natural Selection, Or the Preservation of Favoured Races in the Struggle for Life*, 4th ed. John Murray: London. **Darwin CR** (1871) *The Descent of Man, and Selection in Relation to Sex.* John Murray: London. **Darwin CR** (1872) *The Expression of the Emotions in Man and Animals.* John Murray: London. **Darwin CR, Wallace AR** (1858) *Journal of the Proceedings of the Linnean Society of London* 3:45–62. **Darwin F** (1887) *The Life and Letters of Charles Darwin, Including an Autobiographical Chapter.* John Murray: London. **Davenport CB** (1908) *Inheritance in Canaries.* Carnegie Institution of Washington: Washington, DC. **Davies HR** (1889) *Morphologisches Jahrbuch* 15:560–645. **Davies NB** (1992) *Dunnock Behaviour and Social Evolution.* Oxford University Press: Oxford. **Davies NB** (2000) *Cuckoos, Cowbirds and Other Cheats.* Princeton University Press: Princeton, NJ. **Davies NB, Brooke M** (1989a) *Journal of Animal Ecology* 58:207–24. **Davies NB, Brooke M** (1989b) *Journal of Animal Ecology* 58:225–36. **Davis DE** (1941) *Auk* 58:420. **Davis JWF, O'Donald P** (1976) *Heredity* 36:235–44. **Davis PE, Newton I** (1981) *Journal of Animal Ecology* 50:759–72. **Dawkins R** (1976) *The Selfish Gene.* Oxford University Press: Oxford. **Dawkins R** (1982) *The Extended Phenotype.* Oxford University Press: Oxford. **Dawkins R** (1986) *The Blind Watchmaker.* Norton: New York. **Dawkins R** (2011) "Attention Governor Perry: Evolution Is a Fact." *Washington Post* (On Faith section), 23 August. **Dawson WR** (1995) *Condor* 97:838–47. **de Réaumur M, Zollman PH** (1748) *Philosophical Transactions* 45:304–20. **de Solla Price DJ** (1963) *Little Science, Big Science.* Columbia University Press: New York. **Deevey ES** (1947) *Quarterly Review of Biology* 22:283–314. **Del Hoyo J et al** (1992-2011) *Handbook of the Birds of the World.* Lynx Edicions: Barcelona. **Desjardins P, Morais R** (1990) *Journal of Molecular Biology* 212:599–634. **Desmond A, Moore J** (1991) *Darwin: The Life of a Tormented Evolutionist.* Warner Books: New York. **Dewsbury DA** (1985) *Studying Animal Behavior: Autobiographies of the Founders.* University of Chicago Press: Chicago. **Dewsbury DA** (1995) *Animal Behaviour* 49:1649–63. **Dewsbury DA** (1998) *Biographical Memoirs, National Academy of Sciences* 73:65–84. **Dewsbury DA** (2003) *American Psychologist* 58:747–52. **Dewsbury DA** (2005) *Integrative and Comparative Biology* 45:831–37. **Dial KP** (2003) *Science* 299:402–04. **Diamond J** (1978) *American Scientist* 66:322–31. **Dimond WJ, Armstrong DP** (2007) *Conservation Biology* 21:114–24. **Dobzhansky T** (1937) *Genetics and the Origin of Species.* Columbia University Press: New York. **Dobzhansky T** (1973) *American Biology Teacher* 35:125–29. **Dorshorst B et al** (2011) *PLoS Genetics* 7:e1002412. **Dorst J** (1956) *Les migrations des oiseaux; Migrations en Europe et en Asie, en Amérique du Nord, dans les régions australes, dans les régions intertropicales, migrations des oiseaux de mer, modalités des migrations, invasions d'oiseaux, hibernation,* *déterminisme physiologique, orientation des migrateurs, origine et évolution des migrations.* Payot: Paris. **Dorst J** (1962) *The Migrations of Birds* [translated by C. Sherman]. Heinemann: London. **Doucet SM et al** (2004) *Proceedings of the Royal Society B* 271:1663–70. **Doughty RW** (1975) *Feather Fashions and Bird Preservation: A Study in Nature Protection.* University of California Press: Berkeley. **Downing B** (2007) "Soldier, Birder, Writer, Faker" *Wall Street Journal* (Books section), 10 February. **Drennan SR** (1998) *Auk* 115:465–69. **Drent RH, Daan S** (1980) *Ardea* 68:225–52. **Drent RH, Daan S** (2006) *Ardea* 94:299–303. **Drickamer L, Dewsbury D** (2009) *Leaders in Animal Behavior: The Second Generation.* Cambridge University Press: Cambridge. **Dronamraju KR** (2011) *Haldane, Mayr, and Beanbag Genetics.* Oxford University Press: Oxford. **Dunlop EB** (1913) *British Birds* 7:105–14. **Durman R** (1976) *Bird Observatories in Britain and Ireland.* T & AD Poyser: Berkhamsted, UK. **Dyck J** (1985) *Zoologica Scripta* 14:137–54. **Dyke G** (2011) *Scientific American* 305:80–83.

Eberhard WG (1996) *Female Control: Sexual Selection by Cryptic Female Choice.* Princeton University Press: Princeton, NJ. **Eckhard C** (1876) *Beiträge zur Anatomie und Physiologie* 7:116–25. **Edwards SV et al** (1991) *Proceedings of the Royal Society B* 243:99–107. **Eichhorn G et al** (2009) *Journal of Animal Ecology* 78:63–72. **Eifrig G** (1924) *Auk* 41:439–44. **Ellegren H, Sheldon BC** (2008) *Nature* 452:169–75. **Elliott GP et al** (2001) *Biological Conservation* 99:121–33. **Elton CS** (1927) *Animal Ecology.* University of Chicago Press: Chicago. **Elton CS** (1963) *Nature* 197:634. **Emery NJ** (2004) pp 181–213 in *Comparative Analysis of Minds*, Watanabe S (ed), Keio University Press: Tokyo. **Emery NJ** (2006) *Philosophical Transactions of the Royal Society B* 361:23–43. **Emlen JM** (1966) *American Naturalist* 100:611–17. **Emlen JM** (1973) *Ecology, an Evolutionary Approach.* Addison-Wesley: Reading, MA. **Emlen ST** (1967) *Auk* 84:309–42. **Emlen ST** (1970) *Science* 170:1198–1201. **Emlen ST** (1981) *Auk* 98:167–72. **Emlen ST** (1982a) *American Naturalist* 119:29–39. **Emlen ST** (1982b) *American Naturalist* 119:40–53. **Emlen ST, Oring LW** (1977) *Science* 197:215–23. **Emlen ST et al** (1976) *Science* 193:505–508. **Endler JA, Day LB** (2006) *Animal Behaviour* 72:1405–16. **Endler JA, Mielke PW** (2005) *Biological Journal of the Linnean Society* 86:405–31. **Endler JA, Théry M** (1996) *American Naturalist* 148:421–52. **Erlingsson SJ** (2009) *Studies in History and Philosophy of Biology & Biomedical Science* 40:101–108. **Errington PL** (1946) *Quarterly Review of Biology* 21:144–77. **Ewen JG et al** (1999) *Animal Behaviour* 58:321–28.

Fagerström T (1987) *Oikos* 50:258–61. **Falconer DS** (1960) *Introduction to Quantitative Genetics.* Longman: London. **Farner DS** (1945) *Wilson Bulletin* 57:56–74. **Farner DS** (1950) *Condor* 52:104–22. **Farner DS** (1955) pp 198–237 in *Recent Studies in Avian Biology*, Wolfson A (ed), University Illinois Press: Urbana. **Farner DS** (1986) *American Zoologist* 26:493–501. **Farner DS, Follett BK** (1966) *Journal of Animal Science* 25:90–118. **Farner DS, King JS** (eds) (1971) *Avian Biology*, vol 1. Academic Press: New York. **Fastovsky DE, Weishampel DB** (2005) pp 265–99 in *The Evolution and Extinction of the Dinosaurs*, 2nd ed, Fastovsky DE, Weishampel DB (eds), Cambridge University Press: Cambridge. **Feduccia A** (1980) *The Age of Birds.* Harvard University Press: Cambridge, MA.

Feduccia A (1999) *The Origin and Evolution of Birds*. Yale University Press: New Haven, CT. Feduccia A (2002) *Auk* 119:1187–1201. Feduccia A (2012) *Riddle of the Feathered Dragons: Hidden Birds of China*. Yale University Press: New Haven, CT. Feinsinger P et al (1979) *American Naturalist* 113:481–97. Felsenstein J (2004) *Inferring Phylogenies*. Sinauer Associates: Sunderland, MA. Finn F (1907) *Ornithological and Other Oddities*. John Lane: London. Fisher HI (1955) pp 57–104 in *Recent Studies in Avian Biology*, Wolfson A (ed), University of Illinois Press: Urbana. Fisher J (1940) *Watching Birds*. Allen Lane: Harmondsworth, NY. Fisher J (1966) *The Shell Bird Book*. Ebury Press and Michael Joseph: London. Fisher RA (1915) *Eugenics Review* 7:184–92. Fisher RA (1918) *Philosophical Transactions of the Royal Society of Edinburgh* 52:399–433. Fisher RA (1930) *The Genetical Theory of Natural Selection*. Clarendon: Oxford. Fisher RA (1959) *Australian Journal of Science* 22:16–17. Fitzpatrick JW, Woolfenden GE (1989) pp 201–18 in *Lifetime Reproduction in Birds*, Newton I (ed), Academic Press: London. Fleischer RC et al (1998) *Molecular Ecology* 7:533–45. Follett BK et al (1974) *Proceedings of the National Academy of Sciences USA* 71:1666–69. Forbes TR (1949) *Bulletin of the History of Medicine* 23:263–67. Forsman ED et al (1984) *Wildlife Monographs* 87:3–64. Forstmeier W, Birkhead TR (2004) *Animal Behaviour* 68:1017–28. Fowler DW et al (2011) *PLoS ONE* 6:e28964. Freckleton RP (2009) *Journal of Evolutionary Biology* 22:1367–75. Fretwell SD (1975) *Annual Review of Ecology and Systematics* 6:1–13. Fretwell SD, Lucas HL (1969) *Acta biotheoretica* 19:16–36. Friedmann H (1929) *The Cowbirds: A Study in the Biology of Social Parasitism*. CC Thomas: Springfield, IL. Friedmann H (1930) *Bird-Banding* 1:61–66. Friedmann H, Kiff LF (1985) *Proceedings of the Western Foundation of Vertebrate Zoology* 2:225–304. Fürbringer M (1888) *Untersuchungen zur Morphologie und Systematik der Vögel, zugleich ein Beiträge zur Anatomie der Stütz-& Bewegungsorgane*. T. von Holkema: Amsterdam. Fürbringer M (1902) *Zur vergleichenden Anatomie des Brustschulterapparates und der Schultermuskeln (Band V): Vögel*. Gustav Fischer: Jena. Furness RW (1987) *The Skuas*. T & AD Poyser: Calton, UK.

Gadagkar R (2004) *Journal of Biosciences* 29:139–41. Gadow HF (1891–93) *Vogel*, parts 1 and 2. Bronn's Classen und Ordnungen des Thierreichs: Leipzig. Gadow HF, Selenka E (1891) pp 1–1008 in *Klassen und Ordnungen des Tier-Reichs, wissenschaftlich dargestellt in Wort und Bild*, Bd. 6, Bronn HG (ed), C. F Winter: Leipzig. Gagliardo A et al (2009) *Journal of Experimental Biology* 212:3119–24. Gallagher TF, Koch FC (1929) *Journal of Biological Chemistry* 842:495–500. Garfield B (2007) *The Meinertzhagen Mystery: The Life and Legend of a Colossal Fraud*. Potomac Books Inc.: Washington, DC. Garner WW, Allard HA (1920) *Journal of Agricultural Research* 18:553–606. Garrod AH (1876) *Proceedings of the Zoological Society of London* 44:506–26. Gaston S, Menaker M (1968) *Science* 160:1125–27. Gätke H (1891) *Die Vogelwarte Helgoland*. Joh. Heinr. Meyer: Braunschweig. Gätke H, Rosenstock R (1895) *Heligoland as an Ornithological Observatory: The Result of Fifty Years' Experience*. D. Douglas: Edinburgh. Gause GF (1934) *The Struggle for Existence*. Williams & Wilkins Co.: Baltimore. Gauthier J (1986) pp 1–55 in *The Origin of Birds and the Evolution of Flight*, Padian K (ed), California Academy of Sciences: San Francisco. **Gauthreaux SA** (1992) pp 96–100 in *Ecology and Conservation of Neotropical Migrant Landbirds*, Hagan JM, Johnston DW (eds), Smithsonian Institution Press: Washington. **Gauthreaux SA** (1996) *Condor* 98:442–53. **Gauthreaux SA et al** (2003) pp 335–46 in *Avian Migration*, Berthold P, Gwinner E, Sonnenschein E (eds), Springer-Verlag: Berlin. **Gehring D et al** (2012) *Proceedings of the Royal Society B* 279:4230–35. **Geist NR, Feduccia A** (2000) *American Zoologist* 40:664–75. **Gentile OF** (2001) "At last, his theory flies: but resistance till now to bird-dinosaur link still irks ex-Yale scientist." *Hartford Courant*, 5 May. **Ghiselin M** (1969) *The Triumph of the Darwinian Method*. University of California Press: Berkeley. **Gibb J** (1954) *Ibis* 96:513–53. **Gibbs HL et al** (2000) *Nature* 407:183–86. **Gifford EW** (1919) *Proceedings of the California Academy of Sciences, 4th series* 2:189–258. **Gilbert AB** (1979) pp 237–360 in *Form and Function in Birds*, vol 1, King AS, McLelland J (eds), Academic Press: London. **Gill Jr RE et al** (2009) *Proceedings of the Royal Society B* 276:447–57. **Gill F, Donsker D** (2012) *IOC World Bird Names (V 3.1)*. Retrieved online 8 September 2012 from http://www.worldbirdnames. org. **Glue DE** (1972) *Proceedings XVth International Ornithological Congress*:647–648. **Goethe F** (1940) *Zeitschrift für Tierpsychologie* 4:165–67. **Goldschmidt R** (1940) *The Material Basis of Evolution*. Yale University Press: New Haven, CT. **Goss-Custard JD** (1977) *Animal Behaviour* 25:10–29. **Gould J** (1840–48) *The Birds of Australia*. R & JE Taylor: London. **Gould SJ, Lewontin RC** (1979) *Proceedings of the Royal Society B* 205:581–98. **Grafen A** (1990) *Journal of Theoretical Biology* 144:517–46. **Grafen A** (2003) *Journal of the Royal Statistical Society D* 52:319–29. **Grant**

PR (1965a) *Postilla* 90:1–106. **Grant PR** (1965b) *Systematic Zoology* 14:47–52. **Grant PR** (1965c) *Ibis* 107:350–56. **Grant PR** (1966) *Wilson Bulletin* 78:266–78. **Grant PR** (1968) *Systematic Zoology* 17:319–33. **Grant PR** (1972) *Biological Journal of the Linnean Society* 4:39–68. **Grant PR** (1975) *Evolutionary Biology* 8:237–37. **Grant PR** (1986) *Ecology and Evolution of Darwin's Finches*. Princeton University Press: Princeton, NJ. **Grant PR, Grant BR** (2006) *Science* 313:224–26. **Grant PR, Grant BR** (2010) *Philosophical Transactions of the Royal Society B* 365:1065–76. **Gray PH** (1966) *Journal of the History of the Behavioral Sciences* 2:330–34. **Gray V** (2006) "Something in the Genes: Walter Rothschild, Zoological Collector Extraordinaire." Lecture Delivered at Royal College of Surgeons, October 25.Retrieved online 28 August 2011 from http://www.rcseng.ac.uk/museums /events/audio-video-and-transcripts/Docs /victor_gray_transcript_october06.pdf. **Green JA et al** (2003) *Journal of Experimental Biology* 206:43–57. **Green R et al** (2004) *Journal of Applied Ecology* 41:793–800. **Greenberg R** (1990) *Auk* 107:640–42. **Greenewalt CH** (1955) *Auk* 72:1–5. **Greenewalt CH** (1960) *Hummingbirds*. Doubleday: New York. **Greenewalt CH** (1962) *Smithsonian Miscellaneous Collections* 144:1–46. **Greenewalt CH** (1968) *Bird Song: Acoustics and Physiology*. Smithsonian Institution Press: Washington, DC. **Greenewalt CH** (1975) *Transactions of the American Philosophical Society* 65:3–67. **Griffin DR** (1944) *Quarterly Review of Biology* 19:15–31. **Griffin DR** (1952) *Biological Reviews of the Cambridge Philosophical Society* 27:359–93. **Griffin DR** (1998) pp 68–93 in *The History of Neuroscience in Autobiography*, vol 2, Squire LS (ed). Academic Press: San Diego. **Griffiths PE** (2004) *Biology &*

Philosophy 19:609–31. **Grim T** (2007) *Proceedings of the Royal Society B* 274:373-81. **Grinnell J** (1912) *Pacific Coast Avifauna* 8:1–23. **Grinnell J** (1922) *Science* 56:671–76. **Grinnell J** (1931) *Auk* 48:22–32. **Grinnell J** (1943) *Joseph Grinnell's Philosophy of Nature: Selected Writings of a Western Naturalist.* University of California Press: Berkeley. **Groothuis TG, Schwabl H** (2008) *Philosophical Transactions of the Royal Society B* 363:1647–61. **Gross CG** (2000) *Nature Reviews Neuroscience* 1:67–72. **Guilford T et al** (2009) *Proceedings of the Royal Society B* 276:1215–23. **Gustafsson L** (1986) *American Naturalist* 128:761–64. **Gustafsson L, Sutherland WJ** (1988) *Nature* 335:813–15. **Gwinner E, Benzinger I** (1978) *Journal of Comparative Physiology A: Neuroethology, Sensory, Neural, and Behavioral Physiology* 127:209–13. **Gwinner E, Helm B** (2003) pp 81–95 in *Avian Migration,* Berthold P, Gwinner E, Sonnenschein E (eds), Springer-Verlag: Berlin.

Hackett SJ et al (2008) *Science* 320:1763–68. **Haddrath O, Baker AJ** (2001) *Proceedings of the Royal Society B* 268:939–45. **Haeckel E** (1866) *Generelle Morphologie der Organismen.* G. Reimer: Berlin. **Haeckel E** (1895) *Systematische Phylogenie der Wirbelthiere (Vertebrata).* G. Reimer: Berlin. **Haffer J** (1974) *Avian Speciation in Tropical South America with a Systematic Survey of the Toucans (Ramphastidae) and Jacamars (Galbulidae).* Nuttall Ornithological Club: Cambridge, MA. **Haffer J** (1997) *"We Must Lead the Way on New Paths": The Work and Correspondence of Hartert, Stresemann and Ernst Mayr—International Ornithologists.* Hölzinger: Ludwigsburg, DE. **Haffer J** (2001) *Journal für Ornithologie* 142:27–93. **Haffer J** (2004a) *Acta Historica Leopoldina* 34:399–427. **Haffer J** (2004b) *Journal of Ornithology*
145:163–76. **Haffer J** (2006a) *Blätter aus dem Naumann-Museum* 25:1–55. **Haffer J** (2006b) *Acta Zoologica Sinica* 52:415–20. **Haffer J** (2007a) *Journal of Ornithology* 148:S125–S153. **Haffer J** (2007b) *Ornithology, Evolution, and Philosophy: The Life and Science of Ernst Mayr, 1904–2005.* Springer-Verlag: Berlin. **Haffer J** (2008) *Archives of Natural History* 35:76–87. **Haffer J, Bairlein F** (2004) *Journal of Ornithology* 145:161–62. **Hailman JP** (1965) *Wilson Bulletin* 77:346–62. **Haldane JBS** (1932) *The Causes of Evolution.* Longmans, Green, & Co.: London. **Haldane JBS** (1954) *British Journal for Animal Behaviour* 2:1–1. **Haldane JBS** (1955) *Ibis* 97:375–77. **Haldane JBS** (1956) *British Journal for Animal Behaviour* 4:162–64. **Hall GO** (1935) *Poultry Science* 14:323–29. **Hall KRL, Crook JH** (eds) (1970) *Social Behaviour in Birds and Mammals: Essays on the Social Ethology of Animals and Man.* Academic Press: London. **Hamilton WD** (1963) *American Naturalist* 97:354–56. **Hamilton WD** (1964a) *Journal of Theoretical Biology* 7:1–16. **Hamilton WD** (1964b) *Journal of Theoretical Biology* 7:17–52. **Hamilton WD, Zuk M** (1982) *Science* 218:384–87. **Hamilton WJ, Orians GH** (1965) *Condor* 67:361–82. **Hammond N** (2004) "Lemon, Margaretta Louisa (1860-1953)." *Oxford Dictionary of National Biography.* Retrieved online 6 January 2013 from http://www.oxforddnb.com/view/article/53037. **Hardy AC** (1973) *Ibis* 115:434–36. **Harper DGC** (1982) *Animal Behaviour* 30:575–84. **Harshman J et al** (2008) *Proceedings of the National Academy of Sciences USA* 105:13462–67. **Hartert E** (1903–32) *Die Vögel der paläarktischen Fauna: Systematische übersicht der in Europa, Nord-Asien und der Mittelmeerregion vorkommenden Vögel.* Friedländer: Berlin. **Hartley PHT** (1953) *Journal of Animal Ecology* 22:261–88. **Harvey PH** (1983) *New*

Scientist 100: 187–88. **Harvey PH, Pagel M** (1991) *The Comparative Method in Evolutionary Biology.* Oxford University Press: Oxford. **Hatchwell BJ** (2009) *Philosophical Transactions of the Royal Society B* 364:3217–27. **Hatchwell BJ, Komdeur J** (2000) *Animal Behaviour* 59:1079–86. **Heaney V, Monaghan P** (1995) *Proceedings of the Royal Society B* 261:361–65. **Hebert PDN, Gregory TR** (2005) *Systematic Biology* 54:852–59. **Hegner RE, Wingfield JC** (1986) *Hormones and Behavior* 20:294–312. **Hegner RE, Wingfield JC** (1987) *Auk* 104:462–69. **Heilmann G** (1913a) *Dansk Ornitologisk Forenings Tidsskrift* 7:1–71. **Heilmann G** (1913b) *Dansk Ornitologisk Forenings Tidsskrift* 8:1–92. **Heilmann G** (1914a) *Dansk Ornitologisk Forenings Tidsskrift* 9:1–91. **Heilmann G** (1914b) *Dansk Ornitologisk Forenings Tidsskrift* 9:92–160. **Heilmann G** (1916a) *Vor Nuværende Viden Om Fuglenes Afstamning.* Dansk Ornitologisk Tidsskrift: Copenhagen. **Heilmann G** (1916b) *Fuglenes Afstamning* [Special Bound Copy Edition]. Dansk Ornitologisk Tidsskrift: Copenhagen. **Heilmann G** (1916c) *Dansk Ornitologisk Forenings Tidsskrift* 10:1–91. **Heilmann G** (1926) *The Origin of Birds.* HF & G Witherby: London. **Heilmann G** (1940) *Universet og Traditionen.* Ejnar Munksgaards Forlag: Copenhagen. **Heilmann G, Manniche ALV** (1928) *Danmarks Fugleliv,* 3 vols. Gyldendal: Copenhagen. **Heinrich B** (1995) *Auk* 112:994–1001. **Heinrich B** (2007) *The Snoring Bird: My Family's Journey through a Century of Biology.* Ecco: New York. **Heinroth O** (1911) *Proceedings of the 5th International Ornithological Congress, Berlin* 589–709. **Heinroth O** (1917) *Journal für Ornithologie* 65:116–20. **Heinroth O** (1922) *Journal für Ornithologie* 70: 172–275. **Heinroth O** (1930) *Sitzungsberichte der Gesellschaft naturforschendender Freunde zu Berlin* 1929:333–42. **Heinroth O, Heinroth M** (1924–34) *Die Vögel Mitteleuropas,* 4 vols. Bermühler: Berlin. **Heinsohn R** (2009) pp 223–39 in *Boom and Bust: Bird Stories for a Dry Country,* Robin L, Heinsohn R, Joseph L (eds), CSIRO Publishing: Canberra. **Hellmayr CE** (1938) *Fieldiana Zoology* 13:1–430. **Hennig W** (1950) *Grundzüge einer Theorie der phylogenetischen Systematik.* Deutscher Zentralverlag: Berlin. **Hennig W** (1965) *Annual Review of Entomology* 10:97–116. **Hennig W** (1966) *Phylogenetic Systematics.* University of Illinois Press: Urbana. **Herrick CJ** (1924) *Neurological Foundations of Animal Behavior.* Metropolitan Books/Henry Holt and Company: New York. **Hesse R** (1922) *Sitzungsberichte des Naturhistorischen Vereins Bonn für 1922–23:* A13–17. **Hesse R et al** (1937) *Ecological Animal Geography.* John Wiley: New York. **Hickey JJ** (1943) *A Guide to Bird Watching.* Oxford University Press: Oxford. **Hill GE** (1990) *Animal Behaviour* 40:563–72. **Hill GE** (1991) *Nature* 350: 337–39. **Hill GE, McGraw KJ** (2006) *Bird Coloration: Function and Evolution.* Harvard University Press: Cambridge, MA. **Hill GE, Montgomerie R** (1994) *Proceedings of the Royal Society B* 258:47–52. **Hinde RA** (1954) *Ibis* 96:160–61. **Hinde RA** (1966) *Animal Behaviour: A Synthesis of Ethology and Comparative Psychology.* Cambridge University Press: Cambridge. **Hinde RA** (1987) *Biographical Memoirs of Fellows of the Royal Society* 33:621–39. **Hinde RA** (1990) *Biographical Memoirs of Fellows of the Royal Society* 36:549–65. **Hingston RWG** (1933) *The Meaning of Animal Color and Adornment.* Edward Arnold: London. **Hirsch J** (1967) *American Psychologist* 22:118–30. **Hoffmann K** (1954) *Zeitschrift für Tierpsychologie* 11:453–75. **Holmes RT** (2007) *Ibis* 149:2–13. **Holmes RT, Sherry TW** (2001) *Auk* 118:589–609.

Hölzer C et al (1995) pp 198–206 in *Research and Captive Propagation*, Ganslosser U, Hodges JK, Kaumanns W (eds), Filander Verlag: Fürth, DE. **Holzhaider JC et al** (2011) *Animal Behaviour* 81:83–92. **Horgan J** (1996) *The End of Science.* Broadway Books: New York. **Hornaday WT** (1913) *Our Vanishing Wild Life: Its Extermination and Preservation.* C Scribner's Sons: New York. **Horwich RH** (1989) *Zoo Biology* 8:379–90. **Houston AI et al** (1980) *Behavioral Ecology and Sociobiology* 6:169–75. **Howard E** (1907–14) *The British Warblers.* RH Porter: London. **Howard E** (1920) *Territory in Bird Life.* Murray: London. **Howard E** (1929) *An Introduction to the Study of Bird Behaviour.* Cambridge University Press: Cambridge. **Hoyle F et al** (1985) *British Journal of Photography* 132:693-94. **Hubbs CL** (1953) *Systematic Zoology* 2:93–94. **Hudson PJ** (1986) *Red Grouse, the Biology and Management of a Wild Gamebird.* The Game Conservancy Trust: Fordingbridge, UK. **Hudson PJ et al** (1998) *Science* 282:2256–58. **Hudson PJ et al** (1992) *Journal of Animal Ecology* 61:477–86. **Hull D** (1988) *Science as a Process.* University of Chicago Press: Chicago. **Hunt GR** (1996) *Nature* 379:249–51. **Hurst CH** (1895) *Natural Science* 6:112–22, 180–86, 244–48. **Huth HH, Burkhardt D** (1972) *Naturwissenschaften* 59:650. **Huxley JS** (1912a) *Biologisches Zentralblatt* 32:621–23. **Huxley JS** (1912b) *Proceedings of the Zoological Society of London* 1912:647–55. **Huxley JS** (1914) *Proceedings of the Zoological Society of London* 1914:491–562. **Huxley JS** (1938) *American Naturalist* 72:416–33. **Huxley JS** (1940) *The New Systematics.* Clarendon Press: Oxford. **Huxley JS** (1942) *Evolution: The Modern Synthesis.* Harper: New York. **Huxley JS** (1970) *Memories.* Allen and Unwin: London. **Huxley JS, Montague FA** (1925) *Ibis* 12:868–97. **Huxley J** (1987) *Leaves of the Tulip Tree.* Oxford University Press: Oxford. **Huxley L** (1900) *Life and Letters of Thomas Henry Huxley*, vol 1. D. Appleton and Company: New York. **Huxley TH** (1867) *Proceedings of the Zoological Society of London* 1867:415–72. **Huxley TH** (1868a) *Proceedings of the Royal Society of London* 16:243–48. **Huxley TH** (1868b) *Annals and Magazine of Natural History*, 4th ser 2:66–75. **Huxley TH** (1870) *Quarterly Journal of the Geological Society* 26:12–31.

Imboden C et al (1995) "Review of the Kakapo Recovery Programme: Report for the New Zealand Department of Conservation." Unpublished report (January 1995), Department of Conservation: Wellington, NZ. **Ising G** (1945) *Ark Matematik Astronomi Och Fysik* 32A:(N. 18). **Iwasa Y et al** (1991) *Evolution* 45:1431–42.

James FC, Pourtless IV JA (2009) *Ornithological Monographs* 66:1–78. **Jenkins D** (2003) *Of Partridges & Peacocks—and of Things about Which I Know Nothing: A Life in Wildlife Ecology & Wildlife Management.* TLA Publications: Austin, TX. **Jenner E** (1788) *Philosophical Transactions of the Royal Society of London* 78:219–37. **Jetz W, Rubenstein DR** (2011) *Current Biology* 21:72–78. **Ji Q et al** (1998) *Nature* 393:753–61. **Ji Q, Ji S** (1996) *Chinese Geology* 10:30–33. **Johnson K** (2004) *Journal of the History of Biology* 37:515–55. **Jones IL, Hunter FM** (1993) *Nature* 362:238–39. **Jonker RM et al** (2010) *PLoS ONE* 5:e11369. **Jonker RM et al** (2011) *Behavioral Ecology* 22:326–31. **Jourdain FCR** (1925) *Proceedings of the Zoological Society of London* 95:639–67. **Junker T** (2003) *Avian Science* 3:65–73.

Kacelnik A (1984) *Journal of Animal Ecology* 53:283–99. **Kaiser GW** (2007) *The Inner Bird:*

Anatomy and Evolution. University of British Columbia Press: Vancouver. **Kalikow TJ** (1983) *Journal of the History of Biology* 16:39–73. **Källander H** (1974) *Ibis* 116:365–67. **Kattan GH** (1995) *Auk* 112:335–42. **Kamil A, Sargent T** (eds) (1981) *Foraging Behaviour: Ecological, Ethological and Psychological Approaches*. Garland STPM Press: New York. **Keast A, Morton ES** (1980) *Migrant Birds in the Neotropics: Ecology, Behavior, Distribution, and Conservation*. Smithsonian Institution Press: Washington, DC. **Keeton WT** (1969) *Science* 165:922–28. **Keeton WT** (1971) *Proceedings of the National Academy of Sciences USA* 68:102–106. **Keeton WT** (1974) *Monitore Zoologico Italiano (NS)* 8:227–34. **Keeton WT, Brown AI** (1976) *Journal of Comparative Physiology A: Neuroethology, Sensory, Neural, and Behavioral Physiology* 105:259–66. **Kempenaers B, Schlicht E** (2010) pp 359–411 in *Animal Behaviour: Evolution and Mechanisms*, Kappeler P (ed), Springer: Berlin. **Kempenaers B et al** (1992) *Nature* 357:494–96. **Kendeigh SC** (1940) *Auk* 57:1–12. **Kendeigh SC** (1945) *Auk* 62:418–36. **Kendeigh SC, Baldwin SP** (1937) *Ecological Monographs* 7:91–123. **Kennedy ES** (2009) *Extinction Vulnerability in Two Small, Chronically Inbred Populations of Chatham Island Black Robin* Petroica traversi. PhD thesis, Lincoln University, Christchurch, New Zealand. **Kennedy J** (1924) *Ibis* 66:591–92. **Kennedy JS** (1992) *The New Anthropomorphism*. Cambridge University Press: Cambridge. **Kenward B et al** (2005) *Nature* 433:121–21. **Ketterson ED, Nolan Jr V** (1992) *American Naturalist* 140:S33–S62. **Ketterson ED et al** (1996) *Ibis* 138:70–86. **Keynes RD** (1988) *Charles Darwin's Beagle Diary*. Cambridge University Press: Cambridge. **Kiff LF et al** (1979) *Condor* 81:166–72. **King AS** (1981) pp 107–48 in *Form and Function in Birds*, vol 2, King AS, McLelland J (eds), Academic Press: London. **King AS, McLelland J** (1979-89) *Form and Function in Birds*, 4 vols. Academic Press: London. **King JR, Mewaldt LR** (1989) *Auk* 106:710–13. **Kirkman FB** (1910-13) *The British Bird Book*. TC and EC Jack: London. **Kirkpatrick M et al** (1990) *Evolution* 44:180–93. **Klicka J, Zink RM** (1997) *Science* 277:1666–69. **Klinghammer E** (1967) pp 5–42 in *Early Behavior: Comparative and Developmental Approaches*, Stevenson HW, Hess EH, Rheingold HL (eds), John Wiley: New York. **Klomp H** (1970) *Ardea* 58:1–124. **Klopfer PH** (1999) *Politics and People in Ethology: Personal Reflections on the Study of Animal Behavior*. Bucknell University Press: Lewisburg, PA. **Kluijver HN** (1935) *Ardea* 24:227–39. **Kluijver HN** (1951) *Ardea* 39:1–135. **Knudsen EI, Konishi M** (1978) *Science* 200:795–97. **Knudsen EI et al** (1977) *Science* 198:1278–80. **Koehler O** (1949) *Verhandlungen der Deutschen Zoologischen Gesellschaft* 1949:219–38. **Koehler O** (1950) *Bulletin of Animal Behavior* 9:41–45. **Koenig WD, Mumme RL** (1987) *Population Ecology of the Cooperatively Breeding Acorn Woodpecker*. Princeton University Press: Princeton, NJ. **Koestler A** (1971) *The Case of the Midwife Toad*. Random House: New York. **Kokko H** (1999) *Journal of Animal Ecology* 68:940–50. **Komdeur J** (1992) *Nature* 358:493–95. **Komdeur J** (1994) *Behavioral Ecology and Sociobiology* 34:175–86. **Komdeur J, Deerenberg C** (1997) pp 262–76 in *Behavioral Approaches to Conservation in the Wild*, Clemmons JR, Buchholz R (eds), Cambridge University Press: Cambridge. **Konishi M** (1965) *Zeitschrift für Tierpsychologie* 22:770–83. **Konishi M et al** (1989) *Science* 246:465–72. **Kottler MJ** (1980) *Proceedings of the American Philosophical Society* 124:203–26. **Kozlik FM** (1946) *Passenger*

Pigeon 8:99–103. **Kramer G** (1951) *Proceedings of the 10th International Ornithological Congress, Uppsala*:269–80. **Kramer G** (1953) *Journal für Ornithologie* 94:201–19. **Krätzig H** (1940) *Journal für Ornithologie* 88:139–65. **Krebs JR, Davies NB** (1978) *Behavioural Ecology: An Evolutionary Approach*. Blackwell Scientific Publications: Oxford. **Krebs JR, Kacelnik A** (2007) pp ix–xii in *Foraging: Behavior and Ecology*, Stephens DW, Brown JS, Ydenberg RC (eds), University of Chicago Press: Chicago. **Krebs JR, May RM** (1990) *Nature* 343:310–11. **Krebs JR, Sjölander S** (1992) *Biographical Memoirs of Fellows of the Royal Society* 38:211–28. **Krebs JR et al** (1974) *Animal Behaviour* 22:953–64. **Krebs JR et al** (1977) *Animal Behaviour* 25:30–38. **Kritsky G** (1992) *Archives of Natural History* 19:407–10. **Krüger O, Davies NB** (2002) *Proceedings of the Royal Society B* 269:375–81. **Krüger O, Davies NB** (2004) *Behavioral Ecology* 15:210–18. **Kruuk H** (2003) *Niko's Nature: The Life of Niko Tinbergen and His Science of Animal Behaviour*. Oxford University Press: Oxford. **Kruuk LEB** (2004) *Philosophical Transactions of the Royal Society B* 359:873–90. **Kuhn TS** (1962) *The Structure of Scientific Revolutions*. University of Chicago Press: Chicago. **Kükenthal WG** (1923) *Handbuch der Zoologie*. W de Gruyter: Berlin.

Lack DL (1934) *The Birds of Cambridgeshire*. Cambridge Bird Club: Cambridge. **Lack DL** (1940a) *Nature* 146:324–27. **Lack DL** (1940b) *Condor* 42:269–86. **Lack DL** (1941) *Ibis* 83:637–38. **Lack DL** (1943a) *The Life of the Robin*. HF and G Witherby: London. **Lack DL** (1943b) *British Birds* 37:122–30, 143–50. **Lack DL** (1945) *Occasional Papers of the California Academy of Sciences* 21:1–159. **Lack DL** (1946) *Bulletin of the British Ornithologists' Club* 66:55–65. **Lack DL** (1947a) *Darwin's Finches*. Cambridge University Press: Cambridge. **Lack DL** (1947b) *Ibis* 89:302–52. **Lack DL** (1948) *Evolution* 2:95–110. **Lack DL** (1949) *Bird Notes* 23:231–36. **Lack DL** (1952) *Ibis* 94:167–73. **Lack DL** (1954) *The Natural Regulation of Animal Numbers*. Clarendon Press: Oxford. **Lack DL** (1956) *Swifts in a Tower*. Methuen: London. **Lack DL** (1958) *Ibis* 100:286–87. **Lack DL** (1959) *Ibis* 101:71–81. **Lack DL** (1963) *Bird Notes* 30:258–90. **Lack DL** (1966) *Population Studies of Birds*. Oxford University Press: Oxford. **Lack DL** (1968) *Ecological Adaptations for Breeding in Birds*. Methuen: London. **Lack DL** (1971) *Ecological Isolation in Birds*. Blackwell Scientific Publications: Oxford. **Lack DL** (1973) *Ibis* 115:421–41. **Lack DL, Emlen JT** (1939) *Condor* 41:225–30. **Lack DL, Lack L** (1933) *British Birds* 27:179–99. **Lack DL, Lockley RM** (1938) *British Birds* 31:242–48. **Lack DL, Varley GC** (1945) *Nature* 156:446. **Lack P** (2001) *British Wildlife* 13:95–100. **Lake PE** (1975) *Symposium of the Zoological Society of London* 35:225–44. **Langmore NE et al** (2003) *Nature* 422:157–60. **Leach JA, Tate F** (1911) *An Australian Bird Book: A Pocket Book for Field Use*. Whitcombe and Tombs: Melbourne. **Lear L** (1997) *Rachel Carson: Witness for Nature*. Henry Holt and Company: New York. **Lebreton JD et al** (1992) *Ecological Monographs* 62:67–118. **LeCroy M** (2005) *Ornithological Monographs* 58:30–49. **Lee MSY, Worthy TH** (2012) *Biology Letters* 8:299–303. **Lehrman DS** (1953) *Quarterly Review of Biology* 28:337–63. **Lehrman DS** (1964) *Scientific American* 211:48–54. **Leitch GF** (1953) *North Queensland Naturalist* 21:4. **Leonard ML et al** (1989) *Behavioral Ecology and Sociobiology* 25:357–61. **Lerner HRL et al** (2011) *Current Biology* 21:1838–44. **Levey**

DJ, Stiles FG (1992) *American Naturalist* 140:447–76. Levins R (1966) *American Scientist* 54:421–31. Lewis JC (1995) "Whooping Crane: *Grus americana.*" in *The Birds of North America*, Poole A, Gill FB (eds), American Ornithologists' Union: Philadelphia. Lewis ML (2004) *Inventing Global Ecology: Tracking the Biodiversity Ideal in India, 1947–1997.* Ohio University Press: Athens, OH. Li Q-G et al (2010) *Science* 327:1369–72. Lilienthal O (1889) *Der Vogelflug als Grundlage der Fliegekunst: Ein Beitrag zur Systematik der Flugtechnik.* R. Gaertners Verlagsbuchhandlung: Berlin. Lincoln FC (1930) *United States Department of Agriculture Circular* 118:1–4. Linnaeus C (1735) *Systema Naturae Sive Regna Tria Naturae Systematice Proposita per Classes, Ordines, Genera, & Species.* Lugduni Batavorum, Apud Theodorum Haak: Leiden. Lockie JD et al (1969) *Journal of Applied Ecology* 6:381–89. Long JA, Schouten P (2008) *Feathered Dinosaurs: The Origin of Birds.* Oxford University Press: New York. Lorca-Susino M (2010) *The Euro in the 21st Century: Economic Crisis and Financial Uproar.* Ashgate Publishing Ltd.: Farham, Surrey, UK. Lorenz K (1927) *Journal für Ornithologie* 75:511–19. Lorenz K (1935) *Journal für Ornithologie* 83:137–13. Lorenz K (1937a) *Naturwissenschaften* 25:289–300. Lorenz K (1937b) *Auk* 54:245–73. Lorenz K (1939) *Verhandlungen der Deutschen Zoologischen Gesellschaft, Zoologischer Anzeiger Supplementband* 12:69–102. Lorenz K (1941) *Journal für Ornithologie* 89:194–294. Lorenz K (1943) *Zeitschrift für Tierpsychologie* 5:235–409. Lorenz K (1950) *Symposia of the Society for Experimental Biology* 4:221–68. Lorenz K (1951–53) *Avicultural Magazine* 57:157–82; 58:8–17, 61–72, 86–94, 172–84; 59:24–34, 80–91. Lorenz K (1961) *King Solomon's Ring: New Light on Animal Ways.* Methuen: London. Lorenz K (1974) pp 176–84 in *Les Prix Nobel en 1973*, Odelberg W (ed), The Nobel Foundation: Stockholm. Lorenz K (1982) *The Foundations of Ethology: The Principal Ideas and Discoveries in Animal Behavior.* Simon and Schuster: New York. Lorenz K (1985) pp 258–87 in *Studying Animal Behavior: Autobiographies of the Founders*, Dewsbury DA (ed), University of Chicago Press: Chicago. Lorenz K, Tinbergen N (1939) *Zeitschrift für Tierpsychologie* 2:1–29. Lotem A (1993) *Nature* 362:743–45. Lovat L (1911) *The Grouse in Health and in Disease.* Smith & Elder: London. Lovette IJ et al (2010) *Molecular Phylogenetics and Evolution* 57:753–70. Lowe PR (1936) *Ibis* 1936:310–21. Lowe PR (1941) *Ibis* 5:315–17. Lucas AM, Stettenheim PR (1972) *Avian Anatomy: Integument*, parts I and II. *Agriculture Handbook 362*. US Department of Agriculture: Washington, DC. Lucas FA (1922) *Animals of the Past: An Account of Some of the Creatures of the Ancient World.* American Museum of Natural History: New York. Lütken CF (1893) *Dyeryget: En Haand- og Laerebog* [The Animal Kingdom: A Manual and Learning Book]. Salomonson: Copenhagen. Lytle MH (2007) *The Gentle Subversive.* Oxford University Press: New York.

MacArthur RH (1957) *Proceedings of the National Academy of Sciences USA* 43:293–95. MacArthur RH (1958) *Ecology* 39:599–619. MacArthur RH, Pianka ER (1966) *American Naturalist* 100:603–609. MacGillivray W (1837) *A History of British Birds, Indigenous and Migratory: Including Their Organization, Habits, and Relations; Remarks on Classification and Nomenclature; an Account of the Principal Organs of Birds and Observations Relative to Practical Ornithology.* Vol 1, *Raptores, Scrapers,*

Or Gallinaceous Birds; Gemitores, Cooers, Or Pigeons; Deglubitores, Huskers, Or Conirostral Birds; Vagatores, Wanderers, Or Crows and Allied Genera. Scott, Webster, and Geary: London. **Magat M, Brown C** (2009) *Proceedings of the Royal Society B* 276:4155–62. **Magrath RD** (1991) *Nature* 353:611. **Maier EJ, Bowmaker JK** (1993) *Journal of Comparative Physiology A: Neuroethology, Sensory, Neural, and Behavioral Physiology* 172:295–301. **Mallison WT** (1973) *The Balfour Declaration: An Appraisal in International Law.* Association of Arab-American University Graduates: Massachusetts. **Manning A** (1985) pp 289–314 in *Studying Animal Behaviour: Autobiographies of the Founders,* Dewsbury DA (ed), University of Chicago Press: Chicago. **Marey É-J** (1873) *La machine animale: Locomotion terrestre et aérienne.* Germer Baillière: Paris. **Marey É-J** (1882) *La nature: Revue des sciences et de leurs applications aux arts et à l'industrie* 22 April:326–30. **Marey É-J** (1890) *Physiologie du movement: Le vol des oiseaux.* G. Masson: Paris. **Marey É-J** (1901) *La nature: Revue des sciences et de leurs applications aux arts et à l'industrie* 7 September:232–34. **Markandya A et al** (2008) *Ecological Economics* 67:194–204. **Marler P** (1955) *Nature* 176:6–8. **Marler P** (1957) *Behaviour* 11:13–39. **Marler P** (1986) *Current Contents/Agriculture Biology & Environmental Sciences* 11:16. **Marler P** (2004) *Anais da Academia Brasileira de Ciências* 76:189–200. **Marler P** (2005) *Hormones and Behavior* 47:493–502. **Marler P, Slabbekoorn HW** (2004) *Nature's Music: The Science of Birdsong.* Academic Press: San Diego. **Marples BJ** (1932) *Proceedings of the Zoological Society of London* 102:829–44. **Marsh OC** (1880) *Report of the Geological Exploration of the Fortieth Parallel,* vol 7. Professional papers of the Engineer Department, US Army, no. 18.

Marshall AJ (1954) *Bower-Birds: Their Displays and Breeding Cycles, a Preliminary Statement.* Clarendon Press: Oxford. **Marshall AJ** (1960) *Biology and Comparative Physiology of Birds,* vol 1. Academic Press: New York, London. **Marshall AJ** (1961) *Biology and Comparative Physiology of Birds,* vol 2. Academic Press: New York. **Marshall JC** (1998) "Jock Marshall: One Armed Warrior." Retrieved online 6 January 2013 from http://www.asap .unimelb.edu.au/bsparcs/exhib/marshall /marshall.htm. **Marshall AJ, Serventy DL** (1958) *Journal of Experimental Biology* 35:666–70. **Marshall AJ, Serventy DL** (1959) *Nature* 184:1704–1705. **Martin TE** (1993) *BioScience* 43:523–32. **Martin TE** (1995) *Ecological Monographs* 65:101–27. **Marzluff JM et al** (2012) *Proceedings of the National Academy of Sciences USA* 109:15912–17. **Matthews GVT** (1950) "The Navigational Basis of Homing in Birds." PhD thesis, Cambridge University, Cambridge. **Matthews GVT** (1953) *Journal of Experimental Biology* 30:243–67. **Mattingley AHE** (1907) *Emu* 7:71–73. **Maxwell JM, Jamieson IG** (1997) *Conservation Biology* 11:683–91. **Mayfield H** (1961) *Wilson Bulletin* 73:255–61. **Maynard Smith J** (1964) *Nature* 201:1145–47. **Mayr E** (1923a) *Ornithologische Monatsberichte* 31:135–36. **Mayr E** (1923b) *Ornithologische Monatsberichte* 31:136. **Mayr E** (1926) *Journal für Ornithologie* 74:571–671. **Mayr E** (1927) *Journal für Ornithologie* 75:596–619. **Mayr E** (1931) *American Museum Novitates* 469:1–10. **Mayr E** (1932) *Natural History* 32:83–97. **Mayr E** (1942) *Systematics and the Origin of Species, from the Viewpoint of a Zoologist.* Columbia University Press: New York. **Mayr E** (1951) *Proceedings of the Xth International Ornithological Congress:* 91–131. **Mayr E** (1952) *Biological Reviews* 27:394–400. **Mayr E** (1959) *Ibis* 101:293–302. **Mayr E**

REFERENCES

(1963) *Animal Species and Evolution*. Belknap Press: Cambridge, MA. **Mayr E** (1969) *Principles of Systematic Zoology*. McGraw-Hill: New York. **Mayr E** (1973a) *Ibis* 115:282–84. **Mayr E** (1973b) *Ibis* 115:432–34. **Mayr E** (1975) pp 365–96 in *Ornithology from Aristotle to the Present*, Stresemann E (ed), Harvard University Press: Cambridge, MA. **Mayr E** (1980) pp 413–23 in *The Evolutionary Synthesis. Perspectives on the Unification of Biology*, Mayr E, Provine WB (eds), Harvard University Press: Cambridge, MA. **Mayr E** (1982) *The Growth of Biological Thought*. Belknap Press: Cambridge, MA. **Mayr E** (1983) pp 1–21 in *Perspectives in Ornithology*, Brush AH, Clarke Jr. GA (eds), Cambridge University Press: Cambridge. **Mayr E, Ashlock PD** (1991) *Principles of Systematic Zoology*, 2nd ed. McGraw-Hill: New York. **Mayr E, Diamond JM** (2001) *The Birds of Northern Melanesia: Speciation, Ecology and Biogeography*. Oxford University Press: New York. **Mayr E, Meise W** (1930) *Vogelzug* 1:149–72. **Mayr E, Provine WB** (1998) *The Evolutionary Synthesis: Perspectives on the Unification of Biology*. Harvard University Press: Cambridge, MA. **Mayr E et al** (1953) *Methods and Principles of Systematic Zoology*. McGraw-Hill: New York. **McAtee WL** (1942) *Wilson Bulletin* 54:140. **McAtee WL** (1962) *Auk* 79:730–42. **McCabe RA, Deutsch HF** (1952) *Auk* 69:1–18. **McCabe TT** (1930) *Condor* 32:218–19. **McCafferty G** (2002) *They Had No Choice: Racing Pigeons at War*. Tempus Publishing: Charleston, SC. **McCarter J et al** (2001) *Science* 294:2099–101. **McCarthy M** (2010) "The End of Abundance." *The Independent* 26 November, London. **McClure HE** (1974) *Migration and Survival of the Birds of Asia*. US Army Medical Component, SEATO Medical Research Laboratory: Bangkok.

McClure HE, Ratanaworabhan N (1973) *Some Ectoparasites of the Birds of Asia*. Jintana Printing Ltd.: Bangkok. **McCracken KG** (2000) *Auk* 117:820–25. **McCracken KG et al** (2001) *Nature* 413:128. **McDonald DB et al** (1996) *Ecology* 77:2373–81. **McIlvane WJ** (2001) *Quarterly Review of Biology* 76:269–70. **McKellar RC et al** (2011) *Science* 333:1619–22. **McKinney F et al** (1983) *Behaviour* 86:250–94. **McLean IG et al** (1999) *Biological Conservation* 87:123–30. **Meffe GK, Viederman S** (1995) *Wildlife Society Bulletin* 23:327–32. **Meijer T, Drent R** (1999) *Ibis* 141:399–414. **Menaker M et al** (1970) *Proceedings of the National Academy of Sciences USA* 67:320-25. **Mengel RM** (1964) *Living Bird* 3:9–44. **Milam EL** (2010) *Looking for a Few Good Males: Female Choice in Evolutionary Biology*. Johns Hopkins University Press: Baltimore, MD. **Milgrom M** (2010) *Still Life: Adventures in Taxidermy*. Houghton Mifflin Harcourt: Boston, MA. **Millard RW et al** (1973) *Comparative Biochemistry and Physiology Part A: Physiology* 46:227–40. **Miller AH** (1956) *Evolution* 10:262–77. **Miller RC** (1921) *Condor* 23:121–27. **Millikan GC, Bowman RI** (1967) *Living Bird* 6:23–41. **Mitchell PC** (1901) *Transactions of the Linnean Society of London*, 2nd series, *Zoology* 8:173–275. **Mivart SGJ** (1871) *On the Genesis of Species*. D Appleton and Company: New York. **Moffat CB** (1903) *Irish Naturalist* 12:152–66. **Moksnes A, Røskaft E** (1995) *Journal of Zoology* 236: 625–48. **Møller AP** (1988) *Nature* 332:640–42. **Møller AP** (1989) *Nature* 339:132–35. **Møller AP** (1989) *Oikos* 56:421–23. **Møller AP** (1990) *Journal of Evolutionary Biology* 3:319–28. **Møller AP** (1992) *Nature* 357:238–40. **Molyneux W** (1692) *Dioptric Nova*. Tooke: London. **Monaghan P et al** (1998) *Proceedings of the Royal Society B* 265:

1731–35. **Montagu G** (1802) *Ornithological Dictionary; Or, Alphabetical Synopsis of British Birds.* J White: London. **Montgomerie R** (2009) *Auk* 126:477–84. **Montgomerie R, Birkhead T** (2005) *ISBE Newsletter* 17:16–24. **Montgomerie R, Briskie JV** (2007) pp 115–48 in *Reproductive Biology and Phylogeny of Birds, Part A: Phylogeny, Morphology, Hormones and Fertilization*, Jamieson BGM (ed), Science Publishers: Enfield, NH. **Mora CV et al** (2004) *Nature* 432:508–11. **More AG** (1865) *Ibis*, New Series 7:1–27, 119–42, 425–58. **Moreau RE** (1937) *Proceedings of the Zoological Society of London A, General and Experimental* 107:331–46. **Moreau RE** (1944) *Ibis* 86:286–347. **Moreau RE** (1959) *Ibis* 101:19–38. **Moreau RE** (1970) *Ibis* 112:549–64. **Moreau RE** (1972) *The Palaearctic-African Bird Migration Systems.* Academic Press: New York. **Morel G, Bourlière F** (1962) *Terre et Vie* 102:371–93. **Morgan CL** (1894) *An Introduction to Comparative Psychology.* W. Scott: London. **Morgan CL** (1932) pp 237–64 in *A History of Psychology in Autobiography*, vol 2, Murchison C (ed), Clark University Press: Worcester, MA. **Morgan TH** (1903) *Evolution and Adaptation.* Macmillan: New York. **Morris D** (1979) *Animal Days.* William Morrow: New York. **Morton ES** (1975) *American Naturalist* 109:17–34. **Moss R et al** (1994) *Proceedings of the Royal Society B* 258:175–80. **Moss R, Watson A** (1991) *Ibis* 133:113–20. **Moss S** (2004) *A Bird in the Bush: A Social History of Birdwatching.* Aurum Press: London. **Mouritsen H, Hore PJ** (2012) *Current Opinion in Neurobiology* 22:343–52. **Mudge BF** (1879) *Kansas City Review of Science and Industry* 3:224–26. **Mueller HC, Parker PG** (1980) *Behaviour* 74:101–13. **Mullens WH** (1909) *British Birds* 2:389–99. **Mumme RL** (1992) *Behavioral Ecology and Sociobiology* 31:319–28. **Mundy NI** (2005) *Proceedings of the Royal Society B* 272:1633–40. **Munro U et al** (1997) *Naturwissenschaften* 84:26–28. **Murphy RC** (1936) *Oceanic Birds of South America: A Study of Species of the Related Coasts and Seas, Including the American Quadrant of Antarctica, Based Upon the Brewster-Sanford Collection in the American Museum of Natural History*, 2 vols. Macmillan: New York.

Nedin C (1997) "On *Archaeopteryx*, Astronomers, and Forgery." *The Talk Origins Archive.* Retrieved online 11 November 2012 from http://www.talkorigins.org/faqs/archaeopteryx/forgery.html. **Nelson EC, Haffer J** (2009) *Archives of Natural History* 36:107–28. **Nevitt GA** (2000) *Biological Bulletin* 198:245–53. **Newton A** (1861) *Ibis* 3:190–96. **Newton A** (1892) *Nature* 45:465–49. **Newton A** (1896) *A Dictionary of Birds.* Adam and Charles Black: London. **Newton I** (1979) *Population Ecology of Raptors.* T and AD Poyser: Berkhamsted, UK. **Newton I** (1986) *The Sparrowhawk.* T and AD Poyser: Berkhamsted, UK. **Newton I** (1988) pp 201–19 in *Reproductive Success: Studies of Individual Variation in Contrasting Breeding Systems*, Clutton-Brock TH (ed), University of Chicago Press: Chicago. **Newton I** (1989) *Lifetime Reproduction in Birds.* Academic Press: San Diego. **Newton I** (1995) *Journal of Animal Ecology* 64:675–95. **Newton I** (1998a) *Population Limitation in Birds.* Academic Press: London. **Newton I** (1998b) *Biographical Memoirs of Fellows of the Royal Society* 44:473–84. **Newton I** (2003) *The Speciation and Biogeography of Birds.* Academic Press: London. **Newton I** (2008) *The Migration Ecology of Birds.* Academic Press: London. **Newton I, Rothery P** (1997) *Ecology* 78:1000–1008. **Nguembock B et al** (2008) *Molecular Phylogenetics and Evolution*

48:396–407. **Nice MM** (1933) *Journal für Ornithologie* 81:552–95. **Nice MM** (1934) *Journal für Ornithologie* 82:1–95. **Nice MM** (1935a) *Bird-Banding* 6:90–96. **Nice MM** (1935b) *Bird-Banding* 6:146–47. **Nice MM** (1937) *Transactions of the Linnaean Society of New York* 4:1–247. **Nice MM** (1939) *Auk* 56:255–62. **Nice MM** (1943) *Transactions of the Linnaean Society of New York* 6:1–328. **Nice MM** (1951) *Auk* 68:533–35. **Nice MM** (1979) *Research Is a Passion with Me.* Consolidated Amethyst Communications Inc.: Toronto. **Nicholson AJ** (1933) *Journal of Animal Ecology* 2:132–78. **Nicholson EM** (1926) *Birds in England: An Account of the State of Our Bird-Life, and a Criticism of Bird Protection.* Chapman and Hall: London. **Nicholson EM** (1927) *How Birds Live.* Williams and Norgate: London. **Nicholson EM** (1929) *Ibis* 71:540–41. **Nicholson EM** (1931) *The Art of Birdwatching.* HF and G. Witherby: London. **Nicholson EM** (1959) *Ibis* 101:39–45. **Nilsson J-Å et al** (1999) *Proceedings of the 22nd International Ornithological Congress, Durban,* 234–47. **Nilsson SG** (1984) *Ornis Scandinavica* 15:167–75. **Nixon KC, Carpenter JM** (2000) *Cladistics* 16:298–318. **Noble GK, Lehrman D** (1940) *Auk* 57:22–43. **Nolan V** (1978) *Ornithological Monographs* 26:1–595. **Nopcsa F** (1907) *Proceedings of the Zoological Society of London* 77:223–36. **Nopcsa F** (1923) *Proceedings of the Zoological Society of London* 93:463–77. **Nottebohm F** (1969) *Ibis* 111:386–87. **Nottebohm F** (1971) *Journal of Experimental Zoology* 177:229–61. **Nottebohm F** (1981) *Science* 214:1368–70. **Nottebohm F, Arnold AP** (1976) *Science* 194:211–13. **Nottebohm F, Nottebohm ME** (1976) *Journal of Comparative Physiology* 108:171–92. **Nottebohm F et al** (1976) *Journal of Comparative Neurology* 165:457–86. **Nur N** (1984) *Journal of Animal Ecology* 53:479–96.

O'Donald P (1959) *Nature* 183:1210–11. **O'Donald P** (1960) *Heredity* 15:79–85. **O'Donald P** (1972) *Nature* 238:403–04. **O'Donald P** (1973) *Heredity* 31:145–56. **O'Donald P** (1980) *Genetic Models of Sexual Selection.* Cambridge University Press: Cambridge. **O'Donald P** (1983) *The Arctic Skua, a Study of the Ecology and Evolution of a Seabird.* Cambridge University Press: Cambridge, UK. **O'Hara J** (1945) *The New Yorker* 26 May:18. **Oaks JL et al** (2004) *Nature* 427:630–33. **Oberholser HC** (1927) *Science* 65:164. **Ockenden N et al** (2012) *Bird Study* 59:111–25. **Odum HT** (1948) *Auk* 65:584–97. **Olson SC** (2002) *Auk* 119:1202–1205. **Orenstein RI** (1972) *Auk* 89:674–76. **Oreskes N, Conway EM** (2010) *Merchants of Doubt: How a Handful of Scientists Obscured the Truth on Issues from Tobacco Smoke to Global Warming.* Bloomsbury Press: New York. **Orians GH** (1962) *American Naturalist* 96:257–63. **Orians GH** (1969) *American Naturalist* 103:589–603. **Orians GH** (1980) *Some Adaptations of Marsh-Nesting Blackbirds.* Princeton University Press: Princeton, NJ. **Orians GH** (2009) in *Leaders in Animal Behavior: The Second Generation,* Drickamer L, Dewsbury D (eds), Cambridge University Press: Cambridge. **Orians GH, Kuhlman F** (1956) *Condor* 58:371–85. **Orians GH, Pearson NE** (1979) pp 154–77 in *Analysis of Ecological Systems,* Horn DJ, Mitchell RD, Stairs GR (eds), Ohio University Press: Athens, OH. **Ostrom JH** (1969) *Peabody Museum of Natural History Bulletin* 30:1–165. **Ostrom JH** (1974) *Quarterly Review of Biology* 49:27–47. **Ostrom JH** (1975) *Annual Review of Earth and Planetary Sciences* 3:55–77. **Owen R** (1863) *Philosophical Transactions of the Royal Society of London* 153:33–47.

Paley W (1802) *Natural Theology.* Wilks and Taylor: London. **Palmén JA** (1876) *Ueber* *die Zugstrassen der Vögel.* W. Engelmann: Leipzig. **Papi F et al** (1971) *Monitore zoologico italiano (NS)* 5:265–67. **Papi F et al** (1972) *Monitore zoologico italiano (NS)* 6:85–95. **Papi F et al** (1978) *Journal of Comparative Physiology A: Neuroethology, Sensory, Neural, and Behavioral Physiology* 128:303–17. **Parker E** (1935) *The Ethics of Egg Collecting.* The Field: London. **Parker GA** (1970) *Biological Reviews of the Cambridge Philosophical Society* 45:525–67. **Parker GA** (1984) pp 1–60 in *Sperm Competition and the Evolution of Animal Mating Strategies,* Smith RL (ed), Academic Press: Orlando, FL. **Parker GA** (2006) pp 23–56 in *Essays in Animal Behaviour,* Lucas JR, Simmons LW (eds), Elsevier: Amsterdam. **Parkes KC** (1966) *Living Bird* 5:77–86. **Paul GS** (2002) *Dinosaurs of the Air: The Evolution and Loss of Flight in Dinosaurs and Birds.* Johns Hopkins University Press: Baltimore, MD. **Payne RB** (1974) *Evolution* 28:169–81. **Peakall DB et al** (1973) *Comparative and General Pharmacology* 4:305–13. **Peaker M, Linzell JL** (1975) *Salt Glands in Birds and Reptiles.* Cambridge University Press: Cambridge. **Pearson OP** (1950) *Condor* 52:145–52. **Peck RM** (2001) *Natural History* 110:70–71. **Pennycuick CJ** (1968a) *Journal of Experimental Biology* 49:509–26. **Pennycuick CJ** (1968b) *Journal of Experimental Biology* 49:527–55. **Pepperberg IM** (1999) *The Alex Studies: Cognitive and Communicative Abilities of Grey Parrots.* Harvard University Press: Cambridge, MA. **Pepperberg IM** (2006) *Applied Animal Behaviour Science* 100:77–86. **Pepperberg IM** (2008) *Alex and Me: How a Scientist and a Parrot Discovered a Hidden World of Animal Intelligence—and Formed a Deep Bond in the Process.* HarperCollins: New York. **Perdeck AC** (1958) *Ardea* 46:1–37. **Perkin H** (1972) *Higher Education Quarterly* 1:111–20. **Perrins CM** (1965) *Journal of Animal Ecology* 34:601–47. **Perrins CM** (1966) *Ibis* 108:132–35. **Perrins CM** (1970) *Ibis* 112:242–55. **Perrins CM** (1978) *Ibis* 120:552. **Perrins CM, Jones PJ** (1974) *Condor* 76:225–29. **Perry GC** (2004) *Welfare of the Laying Hen.* CABI: Cambridge, MA. **Peters JL** (1931–87) *Check-List of Birds of the World,* 16 vols. Harvard University Press: Cambridge, MA. **Petersen CGJ** (1896) *Report of the Danish Biological Station* 6:1–48. **Peterson RT** (1934) *A Field Guide to the Birds, Giving Field Marks of All Species Found in Eastern North America.* Houghton Mifflin Company: Boston. **Peterson RT** (1941) *A Field Guide to Western Birds.* Houghton Mifflin Company: Boston. **Peterson RT et al** (1954) *A Field Guide to the Birds of Britain and Europe.* Houghton Mifflin Company: Boston. **Pettingill OS** (1968) *Auk* 85:192–202. **Pianka ER** (1970) *American Naturalist* 104:592–97. **Pierce GJ, Ollason JG** (1987) *Oikos* 49:111–18. **Piersma T, Van Gils JA** (2011) *The Flexible Phenotype: A Body-Centred Integration of Ecology, Physiology, and Behaviour.* Oxford University Press: Oxford. **Piersma T et al** (1998) *Proceedings of the Royal Society B* 265:1377–83. **Piersma T et al** (2005) *Annals of the New York Academy of Sciences* 1046:282–93. **Piertney SB et al** (2008) *Molecular Ecology* 17:2544–51. **Pitelka FA** (1986) *American Birds* 40:385–87. **Pizzari T, Birkhead TR** (2000) *Nature* 405:787–89. **Podos J** (1994) *Ethology, Ecology & Evolution* 6:467–80. **Pollock GB** (1989) *Journal of Evolutionary Biology* 2:205–21. **Porter JP** (1904) *American Journal of Psychology* 15:313–46. **Porter JP** (1906) *American Journal of Psychology* 17:248–71. **Potter RK** (1945) *Science* 102:463–70. **Prakash V et al** (2003) *Biological Conservation* 109:381–90. **Pratt D** (2005) *The Hawaiian Honeycreepers: Drepanidinae.* Oxford University Press: Oxford. **Price J** (1999) *Flight Maps: Adventures with Nature in Modern America.* Basic Books: New York.

Price T (2008) *Speciation in Birds.* Roberts and Co.: Colorado. Provine WB (1971) *The Origins of Theoretical Population Genetics.* University of Chicago Press: Chicago. Pruett-Jones SG, Lewis MJ (1990) *Nature* 348:541–42. Prum RO (1999) *Journal of Experimental Zoology* 285:291–306. Prum RO (2002) *Auk* 119:1–17. Prum RO, Brush AH (2002) *Quarterly Review of Biology* 77:261–95. Pumphrey RJ (1948) *Ibis* 90:171–99. Pycraft WP (1894) *Natural Science* 4:350–60, 437–48. Pycraft WP (1908) *A Book of Birds.* Sidney Appleton: London. Pycraft WP (1910) *A History of Birds.* Methuen Co.: London. Pycraft WP (1911) *Field* 118:823–24. Pycraft WP (1914) *The Courtship of Animals.* Hutchinson & Co. Ltd: London.

Qiang J et al (1998) *Nature* 393:753–61. Quinn TW et al (1987) *Nature* 326:392–94.

Raby CR et al (2007) *Nature* 445:919–21. Raikow R (1974) *Bird-Banding* 45:76–79. Ralph CJ, Scott JM (1981) *Estimating Numbers of Terrestrial Birds.* Proceedings of an international symposium held at Asilomar, California. Ralph CL et al (1967) *Comparative Biochemistry and Physiology* 22:591–99. Rappole JH (1995) *The Ecology of Migrant Birds: A Neotropical Perspective.* Smithsonian Institution Press: Washington, DC. Rasmussen PC, Collar NJ (1999) *Ibis* 141:11–21. Raspet A (1950) *Aeronautical Engineering Review* 9:14–17. Raspet A (1960) *Science* 132:191–200. Rauzon MJ (2010) *Elepaio* 70:25–28. Ray J (1676) *Ornithologiae libri tres: In quibus aves omnes hactenus cognitae in methodum naturis suis convenientem redactae accurate describuntur.* John Martyn: London. Ray J (1678) *The Ornithology of Francis Willughby.* John Martyn: London. Ray J

(1691) *The Wisdom of God Manifested in the Works of Creation.* S. Smith: London. Rayner JMV (1979) *Journal of Experimental Biology* 80:17–54. Razran GHS (1933) *Psychological Bulletin* 30:261–324. Read AF (1987) *Nature* 328:68–70. Recer P (1999) "Evidence Found of Flying Dinosaur." *Chicago Sun-Times* 15 October. Redpath SM et al (2006) *Ecology Letters* 9:410–18. Regal PJ (1975) *Quarterly Review of Biology* 50:35–66. Reiner A et al (2004) *Journal of Comparative Neurology* 473:377–414. Reis CJ (2010) *Interdisciplinary Science Reviews* 35:69–91. Reisz RR et al (2012) *Proceedings of the National Academy of Sciences USA* 109:2428–33. Rensch B (1929) *Das Prinzip geographischer Rassenkreise und das Prinzip der Artbildung.* Borntraeger: Berlin. Rensch B (1933) *Verhandlungen der deutschen Zoologischen Gesellschaft (Zoologischer Anzeiger Suppl 6)* 35:19–83. Rensch B (1938) *Proceedings of the Eighth International Ornithological Congress,* 305–11. Retzius G (1909) *Biologische Untersuchungen, Neue Folge* 14:89–122. Reyer HU (1980) *Behavioral Ecology and Sociobiology* 6:219–27. Reyer HU et al (1997) *Behavioral Ecology* 8:534–43. Reynolds JD, Cooke F (1988) *Animal Behaviour* 36:1788–95. Richardson DS et al (2002) *Evolution* 56:2313–21. Richet C (1894) *Compte rendu et mémoires de la Société de Biologie* 1:244–45. Richmond ML (2001) *Isis* 92:55–90. Ricklefs RE (1978) *National Audubon Society Conservation Report* 6:1–27. Ricklefs RE (2000) *Condor* 102:3–8. Ricklefs RE (2004) *Wilson Bulletin* 116:119–33. Riddle O et al (1932) *Experimental Biology and Medicine* 29:1211–12. Riddle O et al (1933) *American Journal of Physiology—Legacy Content* 105:191–216. Ridgely RS (1976) *A Guide to the Birds of Panama.* Princeton University Press: Princeton, NJ. Ridgely RS, Gwynne JA (1989) *A Guide to the Birds of*

Panama, 2nd ed. Princeton University Press: Princeton, NJ. **Ridgway R** (1901–47) *The Birds of North and Middle America.* US National Museum: Washington, DC. **Ridley M** (1987) *Ethology* 74:260–61. **Riley GM, Witschi E** (1938) *Endocrinology* 23:618–24. **Ritchie J** (1931) *Beasts and Birds as Farm Pests.* Oliver and Boyd: Edinburgh. **Ritz T et al** (2000) *Biophysical Journal* 78:707–18. **Robbins CS et al** (1989) *Proceedings of the National Academy of Sciences USA* 86:7658–62. **Robbins CS, Stewart RE** (1949) *Journal of Wildlife Management* 13:11–16. **Roberts JM** (1999) *Twentieth Century.* Allen Lane: London. **Robin L** (2001) *The Flight of the Emu: A Hundred Years of Australian Ornithology, 1901–2001.* Melbourne University Publishing: Melbourne. **Robins JD, Schnell GD** (1971) *Auk* 88:567–90. **Robinson J** (1966) *New Scientist* 29:159–60. **Rochon-Duvigneaud A** (1943) *Les yeux et la vision des vertébrés.* Masson: Paris. **Rogers LJ** (1982) *Nature* 297:223–25. **Rogers LJ, Anson JM** (1979) *Pharmacology Biochemistry and Behavior* 10:679–86. **Rohwer S** (1977) *Behaviour* 61:107–29. **Rolshausen G et al** (2009) *Current Biology* 19:2097–101. **Romanes GJ** (1882) *Animal Intelligence.* D Appleton and Company: New York. **Romanoff AL** (1960) *The Avian Embryo.* Macmillan: New York. **Romanoff AL, Romanoff AJ** (1949) *The Avian Egg.* Cornell University Press: Ithaca, NY. **Rookmaaker LC et al** (2006) *Archives of Natural History* 33:146–58. **Rosenblatt JS** (1995) *Biographical Memoirs, National Academy of Sciences (US)* 66:227–45. **Rosenfeld CS** (2010) *Biology of Reproduction* 82:473–88. **Rosenthal EJ** (2008) *Birdwatcher: The Life of Roger Tory Peterson.* Lyons Press: Guilford, CT. **Rothschild M** (1983) *Dear Lord Rothschild: Birds, Butterflies, and History.* Balaban: Glenside, PA. **Rothschild W** (1893–1900) *The Avifauna of Laysan and Neighbouring Islands; With a Complete History to Date of the Birds of the Hawaiian Possessions.* R. H. Porter: London. **Rothschild W** (1900) *Transactions of the Zoological Society of London* 15:109–48. **Rothschild W** (1907) *Extinct Birds: An Attempt to Unite in One Volume a Short Account of Those Birds Which Have Become Extinct in Historical Times: That Is, Within the Last Six Or Seven Hundred Years: To Which Are Added a Few Which Still Exist, But Are on the Verge of Extinction.* Hutchinson & Co.: London. **Rothschild W** (1934) *Ibis* 1934:350–77. **Rothstein SI** (1975) *Condor* 77:250–71. **Rowan W** (1929) *Proceedings of the Boston Society of Natural History* 19:151–208. **Rowan W** (1930) *Proceedings of the National Academy of Sciences USA* 16:520–25. **Rowan W** (1931) *The Riddle of Migration.* Williams and Wilkins Company: Baltimore, MD. **Rowan W** (1932) *Proceedings of the National Academy of Sciences USA* 18:639–54. **Rowe T et al** (2001) *Nature* 410:539–40. **Röell DR** (2000) *The World of Instinct: Niko Tinbergen and the Rise of Ethology in the Netherlands (1920–1950).* Uitgeverij van Gorcum: Assen, NL. **Rudolphi KA** (1812) *Beyträge zur Anthropologie und allgemeinen Naturgeschichte.* Haude and Speuer: Berlin. **Rüppell W** (1934) *Journal für Ornithologie* 83:462–24. **Rüppell W** (1937) *Journal für Ornithologie* 85:120–35.

Sadovinkova MP (1923) *Journal of Comparative Psychology* 3:123–29. **Salomonsen F** (1951) *Proceedings of the Xth International Ornithological Congress, Uppsala,*515–26. **Salomonsen F** (1979) *Meddelelser om Grønland* 204: 1–234. **Salvi RJ et al** (2008) *Hair Cell Regeneration, Repair, and Protection.* Springer Verlag: New York. **Sanderson FJ et al** (2006) *Biological*

Conservation 131:93–105. **Sardell RJ et al** (2011) *Proceedings of the Royal Society B* 278:3251–59. **Sato K et al** (2011) *Journal of Experimental Biology* 214:2854–63. **Sato NJ et al** (2010) *Biology Letters* 6:67–69. **Sauer EGF** (1958) *Scientific American* 199:42–47. **Savill PS et al** (2010) *Wytham Woods: Oxford's Ecological Laboratory.* Oxford University Press: Oxford. **Schäfer EA** (1907) *Nature* 77:159–63. **Schlegel H** (1844) *Kritische Übersicht der Europäischen Vögel.* Arnz und Comp: Leiden. **Schleidt WM** (1961) *Zeitschrift für Tierpsychologie* 18:534–60. **Schleidt W et al** (2011) *Journal of Comparative Psychology* 125:121–33. **Schluter D** (2000) *The Ecology of Adaptive Radiation.* Oxford University Press: Oxford. **Schluter D et al** (1985) *Science* 227:1056–58. **Schmidt-Koenig K** (1990) Experientia (*Cellular and Molecular Life Sciences*) 46:336–42. **Schmidt-Nielsen K** (1960) *Circulation* 21:955–67. **Schmidt-Nielsen K et al** (1957) *Federation Proceedings* 16:113–14. **Schmitz OJ** (1997) *Evolutionary Ecology* 11:631–32. **Schodde R** (2000) *Emu* 100:75–76. **Schoener TW** (1982) *American Scientist* 70:586–95. **Schoener TW** (1989) pp 79–113 in *Ecological Concepts: The Contribution of Ecology to an Understanding of the Natural World*, Cherrett JM, Bradshaw AD (eds), vol 29 of *Symposium of the British Ecological Society.* Blackwell Scientific Publications: London. **Scholander PF** (1940) *Hvalrådets Skrifter; Norske Videnskaps-Akad Oslo* 22:1–131. **Schulten K et al** (1978) *Zeitschrift für Physikalische Chemie* 111:1–5. **Schulze-Hagen K et al** (2009) *Journal of Ornithology* 150:1–16. **Schuster P** (2008) pp 217–41 in *Boltzmann's Legacy*, Gallavotti G, Reiter WL, Yngvason J (eds), European Mathematical Society: Zurich. **Schüz E** (1949) *Vogelwarte* 15:63–78. **Schüz E** (1933) *Verhandlungen der Ornithologische Gesellschaft in Bayern* 20:191–225. **Schüz E, Weigold H** (1931) *Atlas des Vogelzugs nach den Beringungsergebnissen bei paläarktischen Vögeln.* Friedländer and Sohn: Berlin. **Schwabl H** (1983) *Journal für Ornithologie* 124:101–16. **Schwabl H** (1993) *Proceedings of the National Academy of Sciences USA* 90:11446–50. **Segerstråle U** (2000) *Defenders of the Truth: The Battle for Science in the Sociobiology Debate and Beyond.* Oxford University Press: Oxford. **Seibt U, Wickler W** (2006) *Ethology* 112:493–502. **Şekercioğlu CH et al** (2004) *Proceedings of the National Academy of Sciences USA* 101:18042–47. **Selander RK** (1964) *University of California Publications in Zoology* 74:1–305. **Selous E** (1901a) *Zoologist* 5:161–83. **Selous E** (1901b) *Bird Watching.* JM Dent and Co.: London. **Selous E** (1902) *Zoologist* 6:133–44. **Selous E** (1906) *Zoologist* 10:201–19. **Selous E** (1913) *Naturalist* 1913:96–98. **Selous E** (1927) *Realities of Bird Life. Being Extracts from the Diaries of a Life-Loving Naturalist.* Constable: London. **Selous E** (1933) *Evolution of Habit in Birds.* Constable: London. **Sharpe RB** (1874–98) *Catalogue of the Birds in the British Museum.* Trustees of the British Museum: London. **Sharpe RB** (1891–98) *Monograph of the Paradiseidae, Or Birds of Paradise, and Ptilonorhynchidae, Or Bower-Birds*, vol 1. Taylor and Francis (text) and Mintern Brothers (plates) for Henry Sotheran & Co.: London. **Sheldon BC et al** (1997) *Proceedings of the Royal Society B* 264:297–302. **Sheldon BC et al** (2003) *Evolution* 57:406–20. **Shipman P** (1998) *Taking Wing: Archaeopteryx and the Evolution of Bird Flight.* Simon and Schuster: New York. **Shufeldt RW** (1914) *Auk* 31:287–89. **Shufeldt RW** (1916) *Auk* 33:457–58. **Shufro C** (2011) "The Bird-Filled World of Richard Prum: How an Ornithologist Discovered New Kinds of Color, Proved T. Rex

Had Feathers, and Answered the Question 'What Is Art?'" *Yale Alumni Magazine* November/December. **Sibley CG** (1955) pp 629–59 in *A Century of Progress in the Natural Sciences*, Miller RC, Kessel EL, Papenfuss GF (eds), California Academy of Sciences: San Francisco. **Sibley CG** (1970) *Peabody Museum of Natural History Bulletin* 32:1–131. **Sibley CG, Ahlquist JE** (1972) *Peabody Museum of Natural History Bulletin* 39:1–276. **Sibley CG, Ahlquist JE** (1990) *Phylogeny and Classification of Birds: A Study in Molecular Evolution*. Yale University Press: New Haven, CT. **Siegel PB et al** (2006) *Poultry Science* 85:2050–60. **Silverin B** (1980) *Animal Behaviour* 28:906–12. **Simpson GG** (1926) *American Journal of Science* 5:453–54. **Simpson GG** (1944) *Tempo and Mode in Evolution*. Columbia University Press: New York. **Simpson GG** (1946) *Bulletin of the American Museum of Natural History* 87:1–99. **Simpson GG** (1949) *The Meaning of Evolution*. Yale University Press: New Haven, CT. **Skinner BF** (1962) *Journal of the Experimental Analysis of Behavior* 5:531–33. **Skutch AF** (1935) *Auk* 52:257–73. **Skutch AF** (1949) *Ibis* 91:430–55. **Skutch AF** (1961) *Condor* 63:198–226. **Skutch AF** (1987) *Helpers at Birds' Nests: A Worldwide Survey of Cooperative Breeding and Related Behavior*. University of Iowa Press: Iowa. **Sloan CP** (1999) *National Geographic* 196 (November): 99–107. **Smith EFG et al** (1991) *Ibis* 133:227–35. **Smith HG, Nilsson J-Å** (1987) *Auk* 104:109–15. **Smith HG et al** (1991) *Behavioral Ecology* 2:90–98. **Smith JNM, Keller LF, Marr AB, Arcese P** (eds) (2006) *Conservation and Biology of Small Populations; the Song Sparrows of Mandarte Island*. Oxford University Press: Oxford. **Smith MS** (2004) *American Journal of Physiology—Endocrinology and Metabolism* 287:E813–E814. **Smith SM**

(1988) *Behaviour* 107:15–23. **Smolin L** (2006) *The Trouble with Physics*. Houghton-Mifflin: Boston. **Sneath PHA, Sokal RR** (1973) *Numerical Taxonomy. The Principles and Practice of Numerical Classification*. WH Freeman and Co.: San Francisco. **Snow DW** (1990) *Ibis* 132:621–22. **Snow DW** (2008) *Birds in Our Life*. William Sessions Limited: York, UK. **Snyder LL** (1958) *Collecting Birds and Conservation*. Department of Ornithology, Royal Ontario Museum. **Snyder NFR, Snyder H** (2000) *The California Condor: A Saga of Natural History and Conservation*. Academic Press: London. **Soikkeli M** (2000) *Auk* 117:1029–30. **Southern HN** (1954) pp 219–32 in *Evolution as a Process*, Huxley JS, Hardy AC, Ford EB (eds), George Allen and Unwin: London. **Spetner LM et al** (1988) *British Journal of Photography* 135:14–17. **Spottiswoode CN, Stevens M** (2010) *Proceedings of the National Academy of Sciences USA* 107:8672–76. **Spottiswoode CN, Stevens M** (2012) *American Naturalist* 179:633–48. **Stapput K et al** (2010) *Current Biology* 20:1259–62. **Steadman R, Levy C** (2012) *Extinct Boids*. Bloomsbury: New York. **Stearns SC** (1976) *Quarterly Review of Biology* 51:3–47. **Stearns SC** (1977) *Annual Review of Ecology and Systematics* 8:145–71. **Stebbins GL** (1982) pp 12–14 in *Perspectives on Evolution*, Milkman R (ed), Sinauer Associates: Sunderland, MA. **Stebbins GL** (1950) *Variation and Evolution in Plants*. Columbia University Press: New York. **Steiner H** (1917) *Jenaische Zeitschrift für Naturwissenschaft* 55 (NF 48): 221–496. **Stejneger LH** (1885) pp 1–195 and 369–547 in *The Standard Natural History*, vol 4, Kingsley JS (ed), S. E. Cassino: Boston. **Stephens DW et al** (2007) *Foraging: Behavior and Ecology*. University of Chicago Press: Chicago. **Stephens DW, Krebs JR** (1986) *Foraging Theory*.

Princeton University Press: Princeton, NJ. **Stewart RE, Aldrich JW** (1951) *Auk* 68:471–82. **Stewart RE et al.** (1946) *Journal of Wildlife Management* 10:195–201. **Stiles FG** (2005) *Auk* 122:708–10. **Stone W** (1933) *Auk* 50:237–40. **Storer R** (1954) *Auk* 71:213–14. **Streets TH** (1870) *Proceedings of the Academy of Natural Sciences of Philadelphia* 22:84–88. **Stresemann E** (1923) *Journal für Ornithologie* 71:517–25. **Stresemann E** (1927–34) *Aves* [4 vols.], *Handbuch der Zoologie*, Kükenthal W, Krumbach T (eds), W de Gruyter and Co.: Berlin. **Stresemann E** (1935) *Ardea* 24:213–26. **Stresemann E** (1947) *Auk* 64:660–61. **Stresemann E** (1951) *Die Entwicklung der Ornithologie, von Aristoteles bis zur Gegenwart.* FW Peters: Berlin. **Stresemann E** (1959) *Auk* 76:542–44. **Stresemann E** (1967) *Proceedings of the 14th International Ornithological Congress, Oxford*, 75–80. **Stresemann E** (1975) *Ornithology from Aristotle to the Present.* Harvard University Press: Cambridge, MA. **Stroud CP** (1953) *Systematic Zoology* 2:76–92. **Sturkie PD** (1954) *Avian Physiology.* Comstock Publishing Company: Ithaca, NY. **Stutchbury BJ** (2007) *Silence of the Songbirds.* HarperCollins: Toronto. **Suddendorf T et al** (2009) *Philosophical Transactions of the Royal Society B* 364:1317–24. **Sulloway FJ** (1982) *Journal of the History of Biology* 15:1–53. **Sutherland WJ et al** (2004) *Bird Ecology and Conservation: A Handbook of Techniques.* Oxford University Press: New York. **Sutter E** (1957) *Revue Suisse de Zoologie* 64:294–303. **Swinton WE** (1958) *Fossil Birds.* British Museum (Natural History): London. **Switek B** (2010) pp 251–63 in *Dinosaurs and Other Extinct Saurians: A Historical Perspective*, Moody RTJ, Buffetaut E, Naish D, Martill DM (eds), Geological Society Special Publications 343: London. **Swynnerton CFM** (1918) *Ibis* 60:127–54.

Takahashi A et al (2004) *Proceedings of the Royal Society B* 271:S281–S282. **Tarr CL, Fleischer RC** (1995) pp 147–59 in *Hawaiian Biogeography: Evolution on a Hotspot Archipelago*, Wagner WL, Funk VA (eds), Smithsonian Institution Press: Washington, DC. **Taylor AH et al** (2007) *Current Biology* 17:1504–07. **Taylor HJ** (1931) *Wilson Bulletin* 43:177–89. **Taylor PD, Williams GC** (1981) *Quarterly Review of Biology* 56:305–13. **Tebbich S et al** (2001) *Proceedings of the Royal Society B* 268:2189–93. **Tebbich S et al** (2002) *Ecology Letters* 5:656–64. **Temple SA** (1974) *General and Comparative Endocrinology* 22:470–79. **Temple SA, Wiens JA** (1989) *American Birds* 43:260–70. **ten Cate C** (2009) *Animal Behaviour* 77:785–94, 795–802. **Terborgh J** (1989) *Where Have All the Birds Gone?* Princeton University Press: Princeton, NJ. **Thomas JW** (2002) *An Interview with Jack Ward Thomas.* Forest History Society: Durham, NC. **Thomas RK** (1997) *American Journal of Psychology* 110:115–25. **Thomson AL** (1926) *Problems of Bird-Migration.* HF and G Witherby: London. **Thomson AL** (1936a) *Bird Migration: A Short Account.* HF and G Witherby: London. **Thomson AL** (1936b) *Ibis* 78:472–530. **Thomson AL** (1950) *Ibis* 92:173–84. **Thomson AL** (1964) *A New Dictionary of Birds.* Nelson: London. **Thomson KS** (2010) "Natural History Museum Collections in the 21st Century." An ActionBioscience.org original article, retrieved online 6 January 2013 from http://www.mnhn.ul.pt/pls/portal/docs/1/336039.PDF. **Thorndike EL** (1898) *Psychological Monographs: General and Applied* 2:1–109. **Thorndike EL** (1911) *Animal Intelligence.* Macmillan: New York. **Thornhill R** (1983) *American Naturalist* 122:765–88. **Thorpe WH** (1924) *Ibis* 66:815–16. **Thorpe**

WH (1951a,b) *Ibis* 93:1–52, 252–96. Thorpe WH (1954) *Nature* 173:465–69. Thorpe WH (1956) *Learning and Instinct in Animals.* Taylor and Francis: London. Thorpe WH (1958a) *Ibis* 100:535–70. Thorpe WH (1958b) *Nature* 182:554–57. Thorpe WH (1974) *Biographical Memoirs of Fellows of the Royal Society* 20:271–93. Thorpe WH (1979) *The Origins and Rise of Ethology.* Heinemann: London. Ticehurst C (1924) *Ibis* 66:813–15. Tinbergen N (1939) *Bird Lore* 41:23–30. Tinbergen N (1948) *Wilson Bulletin* 60:6–51. Tinbergen N (1951) *The Study of Instinct.* Clarendon Press: Oxford. Tinbergen N (1953a) *Social Behaviour in Animals.* Methuen: London. Tinbergen N (1953b) *The Herring Gull's World: A Study of the Social Behaviour of Birds.* Collins: London. Tinbergen N (1959) *Behaviour* 15:1–70. Tinbergen N (1963) *Zeitschrift für Tierpsychologie* 20:410–33. Tinbergen N (1974) pp 196–218 in *Les Prix Nobel en 1973,* Odelberg W (ed), The Nobel Foundation: Stockholm. Tinbergen N (1984) *Curious Naturalists.* University of Massachusetts Press: Boston. Tinbergen N (1985) pp 431–64 in *Studying Animal Behavior: Autobiographies of the Founders,* Dewsbury DA (ed), University of Chicago Press: Chicago. Tinbergen N, Perdeck AC (1950) *Behaviour* 3:1–39. Tobalske BW et al (2003) *Nature* 421:363–66. Todd JT, Morris EK (1986) *Behavior Analyst* 9:71–88. Townsend CW (1905) *The Birds of Essex County, Massachusetts.* The Club: Cambridge, MA. Townsend CW (1913) *Sand Dunes and Salt Marshes.* D Estes and Company: Boston. Townsend CW (1918) *In Audubon's Labrador.* Houghton Mifflin Company: Boston. Townsend CW (1920) *Auk* 37:380–93. Trivers RL (1972) pp 136–79 in *Sexual Selection and the Descent of Man, 1871–1971,* Campbell BG (ed), Aldline-Atherton: Chicago. Trivers RL (1985) *Social Evolution.* Benjamin Cummings Publishing: Menlo Park, CA. Trivers RL (2002) *Natural Selection and Social Theory: Selected Papers of Robert Trivers.* Oxford University Press: New York. Tucker N (2006) "My Best Teacher: Interview—Eva Ibbotson." *Times Education Supplement 2006 3 March* 4675:4. Tucker VA (1966) *Science* 154:150–51. Tucker VA (1968) *Journal of Experimental Biology* 48:67–87. Tullock G (1971) *American Naturalist* 105:77–80. Turek FW et al (1976) *Science* 194:1441–43.

Unger TE (2004) *Max Schlemmer, Hawaii's King of Laysan Island.* iUniverse: New York.

van Noordwijk AJ (2002) *Nature* 418:588–89. van Oordt GJ (1928) *Tijdschrift der Nederlandsche Dierkundige Vereeniging* 3:25–30. Vardanis Y et al (2011) *Biology Letters* 7:502–05. Vaughan R (2009) *Wings and Rings: A History of Bird Migration Studies in Europe.* Isabelline Books: Falmouth, UK. Vaurie C (1959) *The Birds of the Palearctic Fauna: Passeriformes.* HF & G Witherby: London. Vaurie C (1965) *The Birds of the Palearctic Fauna: Non-Passeriformes.* HF & G. Witherby: London. Venette N (1697) *Traité du rossignol.* Chez Charles de Sercy: Paris. Verhulst S, Nilsson J-Å (2008) *Philosophical Transactions of the Royal Society B* 363:399–410. Versluys J (1910) *Zoologische Järbucher Abteilung für Allgemeine Zoologie und Physiologie der Tiere* 30:175–260. Verwey J (1930) *Zoologische Järbucher Abteilung für Allgemeine Zoologie und Physiologie der Tiere* 48:1–120. Videler JJ (2006) *Avian Flight.* Oxford University Press: Oxford. Viguier C (1882) *Revue philosophique de la France et de l'étranger* 14:1–36. Vinther J et al (2008) *Biology Letters* 4:522–25. Vinther J et al (2009) *Geological Society of America, Abstracts with Programs* 41:32. Vinther J et al

(2010) *Biology Letters* 6:128–31. **Visser ME, Lessells CM** (2001) *Proceedings of the Royal Society B* 268:1271–77. **von Cyon E** (1900) *Archiv für die gesamte Physiologie des Menschen und der Tiere* 79:211–302. **von Haartman L** (1954) *Proceedings of the 11th International Ornithological Congress, Basel*:450–53. **von Huene F** (1914) *Geologische und paläontologische Abhandlungen (NF)* 13:1–53. **von Meyer H** (1857) *Neues Jahrbuch für Geologie und Paläontologie* 1857:532–43. **von Meyer H** (1861) *Neues Jahrbuch für Geologie und Paläontologie* 1861:678–79. **von Middendorff AF** (1859) *Mémoires de l'Académie Impériale des Sciences de St Pétersbourg, avec l'histoire de l'Académie*, series 6, 8:1–143. **Vuilleumier F** (1982) *Auk* 99:801–04.

Waddell C (2000) pp 1–16 in *And No Birds Sing: Rhetorical Analyses of Rachel Carson's Silent Spring*, Waddell C (ed), Southern Illinois University Press: Carbondale. **Wagner HO** (1930) *Journal of Comparative Physiology A: Neuroethology, Sensory, Neural, and Behavioral Physiology* 12:703–24. **Wagner JA** (1862) *Annals and Magazine of Natural History (London)*, series 3, 9:261–67. **Wagner PD, Paterson DJ** (2011) *Journal of Physiology* 589:3899–900. **Walcott C et al** (1979) *Science* 205:1027–29. **Wallace AR** (1874) *Nature* 10:459. **Wallace AR** (1889) *Darwinism: An Exposition of the Theory of Natural Selection with Some of Its Applications*. Macmillan: London. **Wallace AR** (1892) *Natural Science* 1:749–50. **Wallander J, Andersson M** (2002) *Oecologia* 130:391–95. **Wallis HM** (1926) *Ibis* 68:628–29. **Wallraff HG** (1980) *Journal of Comparative Physiology A: Neuroethology, Sensory, Neural, and Behavioral Physiology* 139:209–24. **Walsberg GE** (1993) *Condor* 95:748–57. **Walters JR et al** (2010) *Auk* 127:969–1001. **Warner LH** (1931) *Quarterly Review of Biology* 6:84–98. **Watkins**

RS et al (1985) *British Journal of Photography* 132:264–66. **Watson A** (1987) *Journal of Animal Ecology* 56:1089–90. **Watson A, Parr R** (1981) *Ornis Scandinavica* 12:55–61. **Watson JB** (1913) *Psychological Review* 20:158–77. **Watson JB, Lashley KS** (1915) *Papers from the Department of Marine Biology of the Carnegie Institution of Washington* 211:5–104. **Weidensaul S** (2007) *Of a Feather: A Brief History of American Birding*. Houghton Mifflin Harcourt: Orlando, FL. **Weiner J** (1994) *The Beak of the Finch*. Alfred Knopf: New York. **Weir AAS et al** (2002) *Science* 297:981. **Weir AAS et al** (2004) *Proceedings of the Royal Society B* 271:S344–S346. **Weir JT, Schluter D** (2004) *Proceedings of the Royal Society B* 271:1881–87. **Weir JT, Schluter D** (2007) *Science* 315:1574–76. **Wells HG et al** (1929–30) *The Science of Life: A Summary of Contemporary Knowledge about Life and Its Possibilities*. Amalgamated Press: London. **Wenny DG et al** (2011) *Auk* 128:1–14. **Werdenich D, Huber L** (2006) *Animal Behaviour* 71:855–63. **Wernham C et al** (2002) *The Migration Atlas: Movements of the Birds of Britain and Ireland*. T & AD Poyser: London. **Westneat DF** (1987) *Animal Behaviour* 35:877–86. **Westneat DF, Fox C** (2010) *Evolutionary Behavioral Ecology*. Oxford University Press: New York. **Westneat DF, Stewart IRK** (2003) *Annual Review of Ecology Evolution and Systematics* 34:365–96. **Wetmore A** (1930) *Proceedings of the United States National Museum* 76:1–8. **Wetmore A** (1940) *Smithsonian Miscellaneous Collections* 99 (4): 1–81. **Wetmore A, Miller WD** (1926) *Auk* 43:337–46. **Wetton JH et al** (1987) *Nature* 327:147–49. **Whitman CO** (1899) *Biological Lectures from the Marine Biological Laboratory Wood's Hole, Massachusetts* 1898:285–38. **Whitteridge G** (1971) *William Harvey and the Circulation of* the Blood. Macdonald: London. **Wiens JA** (1977) *American Scientist* 65:590–97. **Wilbur HM** (1979) *Science* 205:781. **Wilkins JS** (2009) *Species: A History of the Idea.* University of California Press: Berkeley. **Williams GC** (1966a) *Adaptation and Natural Selection: A Critique of Some Current Evolutionary Thought.* Princeton University Press: Princeton, NJ. **Williams GC** (1966b) *American Naturalist* 100:687–90. **Williams T** (2012) *Physiological Adaptations for Breeding in Birds.* Princeton University Press: Princeton, NJ. **Williston SW** (1879) *Kansas City Review of Science and Industry* 3:457–60. **Willmer EN, Brunet PCJ** (1985) *Biographical Memoirs of Fellows of the Royal Society* 31:32–63. **Wilson EO** (1975) *Sociobiology: The New Synthesis.* Belknap Press: Cambridge, MA. **Wilson EO, Brown Jr WL** (1953) *Systematic Zoology* 2:97–111. **Wilson EO, Hutchinson GE** (1989) *Biographical Memoirs of the National Academy of Sciences* 58:319–27. **Wilson SB, Evans AH** (1890–99) *Aves Hawaiienses: The Birds of the Sandwich Islands* [with Supplements by Hans Gadow]. RH Porter: London. **Wiltschko R** (1996) *Journal of Experimental Biology* 199:113–19. **Wiltschko R, Wiltschko W** (1998) *Naturwissenschaften* 85:164–67. **Wiltschko W, Wiltschko R** (1972) *Science* 176:62–64. **Wiltschko W et al** (1987) *Naturwissenschaften* 74:196–98. **Wiltschko W et al** (2002) *Nature* 419:467–70. **Wimpenny JH et al** (2009) *PLoS One* 4:e6471. **Wimpenny JH et al** (2011) *Animal Cognition* 14:459–64. **Wingfield JC** (2009) pp 561–92 in *Leaders in Animal Behavior: The Second Generation,* Drickamer L, Dewsbury D (eds), Cambridge University Press: Cambridge. **Wingfield JC, Farner DS** (1976) *Condor* 78:570–73. **Wingfield JC et al** (1990) *American Naturalist* 136:829–46.

Winterbottom JM (1929) *Journal of Genetics* 21:367–87. **Witherby HF et al** (1938-41) *The Handbook of British Birds.* HF and G Witherby: London. **Witschi E** (1935) *Wilson Bulletin* 47:177–88. **Woakes AJ, Butler PJ** (1983) *Journal of Experimental Biology* 107:311–29. **Wollaston AFR** (1921) *Life of Alfred Newton: Professor of Comparative Anatomy, Cambridge University, 1866–1907.* J. Murray: London. **Wood C** (1917) *The Fundus Oculi of Birds.* Lakeside Press: Chicago. **Woolfenden GE, Fitzpatrick JW** (1984) *The Florida Scrub Jay: Demography of a Cooperative-Breeding Bird.* Princeton University Press: Princeton, NJ. **Worthington J, Dial K** (2003) "These Wings Were Made for Running: An Interview with Kenneth Dial," *Cabinet* 11. **Wright AA** (1972) *Journal of the Experimental Analysis of Behavior* 17:325–37. **Wright S** (1931) *Genetics* 16:97–159. **Wu L-Q, Dickman D** (2012) *Science* 336:1054–57. **Wynne-Edwards VC** (1929) *British Birds* 23:138–53, 170–80. **Wynne-Edwards VC** (1931) *British Birds* 24:346–53. **Wynne-Edwards VC** (1935) *Proceedings of the Boston Society of Natural History* 40:233–346. **Wynne-Edwards VC** (1939) *Proceedings of the Zoological Society of London* A109:127–32. **Wynne-Edwards VC** (1955a) *Acta of the Eleventh International Congress of Ornithology,* 540–47. **Wynne-Edwards VC** (1955b) *Discovery: A Monthly Popular Journal of Knowledge* 16:433–35. **Wynne-Edwards VC** (1962) *Animal Dispersion in Relation to Social Behaviour.* Hafner Publishing Co.: New York. **Wynne-Edwards VC** (1979) *Devon Birds* 32:3–9. **Wynne-Edwards VC** (1980) *Current Contents/Agriculture Biology & Environmental Sciences* 25:198. **Wynne-Edwards VC** (1985) pp 487–512 in *Studying Animal Behavior: Autobiographies of the Founders,* Dewsbury DA (ed), University of Chicago Press:

Chicago. **Wynne-Edwards VC** (1986) *Evolution Through Group Selection*. Blackwell Scientific Publications: Palo Alto, CA.

Xing L et al (2012) *PLoS ONE* 7:e44012. **Xu X** (2000) *National Geographic* 197[3]:xviii. **Xu X et al** (1999) *Nature* 399:350–54. **Xu X et al** (2000) *Nature* 408:705–708. **Xu X et al** (2003) *Nature* 421:335–40. **Xu X et al** (2011) *Nature* 475:465–70.

Ydenberg RC et al (2007) pp 1–28 in *Foraging: Behavior and Ecology*, Stephens DW, Brown JS, Ydenberg RC (eds), University of Chicago Press: Chicago. **Yeagley HL** (1947) *Journal of Applied Physics* 18:1035–63. **Young RL et al** (2011) *Developmental Dynamics* 240:1042–53.

Zahavi A (1975) *Journal of Theoretical Biology* 53:205–14. **Zahavi A** (2003) *Animal Behaviour* 65:859–63. **Zelenitsky DK et al** (2012) *Science* 338:510–14. **Zimmer C** (2011) *National Geographic* 219 (February): 32–57. **Zink G, Bairlein F** (1995) *Der Zug europäischer Singvögel: Ein Atlas der Wiederfunde beringter Vögel*. Aula-Verlag: Wiesbaden. **Zwarts L et al** (2009) *Living on the Edge: Wetlands and Birds in a Changing Sahel*. KNNV Publishing: Zeist, NL.

INDEX

Pages with photographs. drawings, or graphics are shown in *italics*.

IMAGE CREDITS

53. Illustration from Bowman (1961: 313), courtesy California Academy of Sciences Archives, San Francisco.

58. Photo by Eric Hosking, courtesy Proceedings of the International Ornithological Congress and Ernst Mayr estate.

62. Photo by Eric Mills, courtesy Ian Newton and Fred Cooke.

69. Photo by Denise Applewhite, courtesy Peter Grant.

72. Photo by Els Atema, courtesy Arie van Noordwijk.

CHAPTER 3

74. Paintings by John Gerrard Keulemans compiled from Rothschild (1900: plates 22, 38, 29, 33), reproduced with permission from Zoological Society of London.

76. Photo by J. T. Newman from *The Picture Magazine* 6: 56 (1895).

79. LEFT: Ridgway photo by Ulke Brothers (April 1873); woodcock illustration from Coues (1896: 619). RIGHT: Sibley photo by Peter Stettenheim; Beck photo courtesy California Academy of Sciences Archives, San Francisco; Hartert photo from Royal Ontario Museum, Toronto.

81. Photo courtesy California Academy of Sciences Archives, San Francisco.

82. Photo from Bodleian Libraries, University of Oxford.

83. Photo courtesy Natural History Museum, London.

88. Photo from Historische Bild- u. Schriftgutsammlungen (Sigel: MfN, HBSB), Bestand: Zool. Mus. Signatur: Orn. 100,1 / Orn. 173,4 / B I/2115, courtesy Museum für Naturkunde, Berlin.

95. Photo courtesy Smithsonian Institution Archives, Washington, DC, image MNH-17021.

97. Photo courtesy American Museum of Natural History, New York.

100. Photo courtesy American Museum of Natural History, New York.

102. TOP: illustration from Mitchell (1901: plate 21). BOTTOM: illustration from Mac-Gillivray (1837: 99).

109. Illustration by Darren Naish.

113. Photo courtesy American Museum of Natural History, New York.

CHAPTER 4

116. Painting by Guy Tudor, with permission from the artist.

120. Painting by Mary Eagle Clarke from Clarke (1912: frontispiece).

121. Painting by Sir Peter Scott, with permission from Dafila Scott and Scottish Ornithologists' Club.

124. LEFT: brains illustration from Rowan (1931: 21); Rowan photo from *The Auk* 75: plate 20 facing page 387 (1958), with permission from American Ornithologists' Union; blackcap photo courtesy Peter Berthold; Gätke photo from Gätke and Rosenstock (1895: 588); Newton photo courtesy Balfour & Newton Libraries, University of Cambridge, Department of Zoology; RIGHT: Wiltschkos photo by Jo Wimpenny; Fair Isle photo by Deryk Shaw; radar image courtesy US National Weather Service; Watson photo from *Johns Hopkins Gazette* 30 (2001); Courish Spit photo by Vadim Kantor.

126. Photo courtesy Max Planck Institute of Ornithology, Vogelwarte Radolfzell.

128. Photos courtesy Zoological Museum, University of Copenhagen.

129. Photo courtesy Max Planck Institute of Ornithology, Vogelwarte Radolfzell.

136. Photo courtesy Max Planck Institute of Ornithology, Vogelwarte Radolfzell.

139. Photo by Kevin Lane.

140. Redrawn from Guilford et al. (2009: fig. 3), courtesy Tim Guilford and Jessica Meade.

144. Photo courtesy Max Planck Institute of Ornithology, Vogelwarte Radolfzell.

145. LEFT: photo courtesy Max Planck Institute of Ornithology, Vogelwarte Radolfzell. RIGHT: illustration by Tim Birkhead.

147. Photo courtesy William T. Keeton family archives.

153. TOP: photo by Phil Battley. BOTTOM: maps modified slightly from Battley et al. (2012: figs. 1, 2), courtesy Phil Battley with permission from *Journal of Avian Biology*.

155. Photo by Uli Querner, courtesy Peter Berthold.

157. Photo by Dee Dee Hatch, courtesy Steve Emlen.

CHAPTER 5

160. Paintings by Eric Ennion, courtesy Susan Ennion.

162. LEFT: photo by Bernard Genton. RIGHT: photo by Juan Gonzales.

164. Photo courtesy Alexander Library of Ornithology, Bodleian Libraries, University of Oxford.

166. LEFT: sparrow drawing by Jamie N. M. Smith from Smith et al. (2006: 89); Nice photo from Nice (1979: frontispiece). RIGHT: swallow photo by P-G Bentz; nightjar clutch photo by Tim Birkhead; coot clutch photo by Bruce Lyon; Allen photo courtesy *Living Bird*, Cornell Lab of Ornithology.

171. Photo courtesy Edward Grey Institute, Oxford University.

172. Drawing by Robert Gillmor from Lack (1968: 52).

177. Photo from *Ibis* 112: plate 11 (1970), with permission from British Ornithologists' Union.

178. Photo by Warwick Tarboton.

181. TOP: map courtesy Edward Grey Institute, Oxford University. BOTTOM: photo by David Tipling.

183. Photo courtesy Martin Cody.

188. Photo by Natalino Fenech.

194. Photo courtesy Robert Ricklefs and University of Missouri-St. Louis.

197. Photo by Roger Walsey, courtesy Kate Lessells.

CHAPTER 6

200. Painting by David Miller, with permission from the artist.

202. Photo by Kevin McCracken from McCracken et al. (2001: 128).

205. LEFT: fundus illustration from Wood (1917); Wood illustration from *Library Notes* (2005), Galter Health Library, Northwestern University; musculature illustration from Fürbringer (1902: figs. 198–202); Max Fürbringer photo from http://www.ub.uni-heidelberg.de/helios/digi/anatomie/fuerbringer.html. RIGHT: Wingfield photo courtesy John Wingfield; salt glands illustration by Tim Birkhead; Riddle photo from US National Library of Medicine; bird sperm illustration from Retzius (1909: fig. 36); Retzius photo from Dictionary of Swedish National Biography.

217. Photo courtesy National Library of Australia, Canberra, image MS 1465.

219. Photo by David Tipling.

224. Drawing from Witschi (1935: 179), with permission from *Wilson Journal of Ornithology*.

229. Photo by Russell G. Forster, courtesy Brian Follett.

234. Photo from *The Auk* 122: 1014 (2005), with permission from American Ornithologists' Union.

237. Photo courtesy Fernando Nottebohm.

238. Photo by Ann C. Nolan, courtesy Ellen Ketterson.

241. Photo by Jan van de Kam, courtesy Theunis Piersma.

243. Photo by Jan van de Kam.

CHAPTER 7

244. Painting by Robert Bateman, with permission from the artist.

247. LEFT: Selous photo courtesy Andrew Selous; guillemots drawing from Selous (1901b: 30). RIGHT: photo of Tinbergen and Lorenz by Irenäus Eibl-Eibesfeldt; sonogram by Bob Montgomerie; goose drawings based on photos in Lorenz and Tinbergen (1939); Morgan photo from private collection of Bob Montgomerie.

248. Photo by W. H. Thorpe, courtesy Margaret Schuelein.

250. Letter courtesy Alexander Library of Ornithology, Bodleian Libraries, University of Oxford.

251. TOP LEFT: photo courtesy Michael Howard. RIGHT: illustration by George Lodge from Howard (1920: 110). BOTTOM LEFT: illustration from Howard (1929: plate 5 facing page 78).

255. Photo from www.all-about-psychology. com.
257. Photos from Heinroth and Heinroth (1924–34, vol. 4: plates 12, 10).
258. Photo courtesy Konrad Lorenz Archive, Konrad Lorenz Institute, Altenberg.
263. Photo by Bas Teunis.
264. Illustration by Tim Birkhead, based on illustrations in (TOP) Tinbergen (1951: 32), and (BOTTOM) *Popular Mechanics* (1969).
267. Photo from Wikipedia.
271. Photo by Alex Badyaev, tenbestphotos.com.
272. Photo courtesy Desmond Morris.
274. Photo courtesy Paul Heavens.
275. Photo courtesy Margaret Schuelein.
277. Photo courtesy Paul Heavens.
278. Photo by Otto Koehler, courtesy Judith Marler.
283. Photo by Jo Wimpenny.

CHAPTER 8

286. Painting by Raymond Ching, with permission from Caroline Ching.
288. Photo courtesy Nick Davies.
291. LEFT: cuckoo illustration from Pycraft (1919: facing page 106); wagtails illustration from Howard (1920: 87). RIGHT: crow photo by Jolyon Troscianko; Fitzpatrick photo courtesy John Fitzpatrick; Woolfenden photo from *Ibis* 150: 444 (2008), with permission from British Ornithologists' Union; jay photo by David Tipling; cuckoo photo courtesy Karl Schulze-Hagen.
292. Photo courtesy Tropical Science Center, Costa Rica.
294. Photo by John Fitzpatrick.
299. Photo by John Krebs.
302. Photo by Oldo Mikulica.
304. Photo courtesy Alexander Library of Ornithology, Bodleian Libraries, University of Oxford.
306. Photo courtesy Smithsonian Institution Archives, Washington, DC, image SIA2008-2368.
307. Photos by Hugh Chittenden (LEFT, birds), Claire Spottiswoode (eggs), and Alan Manson (RIGHT, bird).

314. Photos courtesy California Academy of Sciences Archives, San Francisco.
317. Photo by Tim Birkhead, courtesy Andrew Cockburn.
319. Photo by Oliver Krüger, courtesy Nick Davies.

CHAPTER 9

322. Painting by William Matthew Hart, from Sharpe (1891–98, vol. 1: plate 293), reproduced with permission from Zoological Society of London.
324. Photo by Uno Unger, courtesy Malte Andersson.
326. LEFT: Parker photo courtesy Geoff Parker; Trivers photo courtesy Robert Trivers; skylark illustration from Finn (1907: 30); Wallace photo from private collection of Bob Montgomerie; Darwin photo from private collection of Bob Montgomerie. RIGHT: Smith photo by Doris Atkinson, courtesy Susan Smith; budgerigar photo by Jiří Kirk; grebe photos by Pauline Leggat.
329. Photo by David Tipling.
333. Photo by Paul Jones.
335. Photo courtesy Alexander Library of Ornithology, Bodleian Libraries, University of Oxford.
338. Photo by M. Crook, courtesy Stamati Crook.
340. Photo by David Tipling.
343. Photo by Alex Badyaev, tenbestphotos.com.
346. Drawing by David Quinn from Davies (1992: 164), with permission from David Quinn and Oxford University Press.
350. Photo by Avishag Zahavi, courtesy Amotz Zahavi.
351. Photo by Doris O'Donald, courtesy Peter O'Donald.

CHAPTER 10

354. Painting by Rodger McPhail, *Male Grouse Challenging*, oil on canvas, 24 x 30 inches, with permission from Tryon Gallery, London.
356. Drawing by V. C. Wynne-Edwards, from Vero C. Wynne-Edwards *fonds*, locator

5137.1, box 8, courtesy Queen's University Archives.

358. LEFT: Grinnell photo by Michelle Nijhus from The Bancroft Library, University of California, Berkeley; grouse illustration from Lovatt (1911: plate 16); warbler illustration from Howard (1908, part 2: frontispiece). RIGHT: grouse photo by Tim Birkhead; graph by Bob Montgomerie from data supplied by Peter Hudson; warbler photo by David Wendelken; heron photo by David Tipling; tit photo by David Tipling.

365. Photo by Paul Thompson.

367. Photo courtesy the Lack family and Alexander Library of Ornithology, Bodleian Libraries, University of Oxford.

373. Photo by Allan Cameron, courtesy Iain James Cameron.

379. Photo courtesy Edward Grey Institute, Oxford University.

383. Photo courtesy Ian Newton.

385. Photo courtesy Peter Hudson and Penn State University Archives.

CHAPTER II

388. Painting by Ralph Steadman, *Guadalupe Caracara*, from Steadman and Levy (2012), with permission from the artist and Ceri Levy.

392. LEFT: Kakapo photo by Tim Birkhead; Ratcliffe photo courtesy Des Thompson; Hickey photo from *The Auk* III: 44 (1994), with permission from American Ornithologists' Union; RSPB logo courtesy Royal Society for the Protection of Birds; Audubon logo courtesy Audubon Society. RIGHT: BirdLife logo courtesy BirdLife International; Partners in Flight logo from http://www.partnersinflight.org; Mallard photo by Tim Birkhead; CITES logo courtesy CITES Secretariat; pigeon illustration from Rothschild (1907: plate 22); Chapman drawing courtesy American Museum of Natural History, New York; duck illustration from Rothschild (1907: plate 36).

393. Photo by T. Winfield Eastwood, courtesy Muffy Aldrich.

395. Photo from Rothschild (1893–1900: plate 36).

397. Illustration from Eaton catalog (1913: 36).

406. Photo by Klaus Nigge.

407. Photo by Tim Birkhead.

410. Photo by Rex Gary Schmidt, courtesy US Fish and Wildlife Service.

414. Photo by Melissa Foley.

422. Photo by Dolors Buxó, courtesy Nigel Collar.

AFTERWORD

427. Images courtesy John Marzluff.

APPENDIX 2

Alerstam. Photo by Jo Wimpenny.

Amadon. Photo courtesy American Museum of Natural History, New York.

Armstrong. Photo courtesy Alexander Library of Ornithology, Bodleian Libraries, University of Oxford.

Aschoff. Photo from *Ibis* 142: 181 (2000), with permission from British Ornithologists' Union.

Bang. Photo courtesy Molly Bang.

P. Bateson. Photo courtesy Patrick Bateson.

Berkhoudt. Photo by Tim Birkhead, courtesy Herman Berkhoudt.

Burley. Photo by Richard Symanski, courtesy Nancy Burley.

Charmantier. Photo by Sandra Bouwhuis, courtesy Anne Charmantier.

Clarke. Photo courtesy Alexander Library of Ornithology, Bodleian Libraries, University of Oxford.

Clayton. Photo courtesy Nicky Clayton.

Coues. Photo courtesy Balfour & Newton Libraries, University of Cambridge, Department of Zoology.

Coulson. Photo courtesy John Coulson.

Cracraft. Photo by Bob Montgomerie.

Craig. Photo courtesy Special Collections, Fogler Library, University of Maine.

Diamond. Photo by Jochen Braun, courtesy Jared Diamond.

Drent. Photo by Adam Watson.

Drury. Photo from *The Auk* 113: 931 (1996), with permission from American Ornithologists' Union.

523

Dunnet. Photo courtesy Paul M. Thompson and University of Aberdeen.

Gagliardo. Photo by Grigori Tertitski, courtesy Anna Gagliardo.

Gauthreaux. Photo by Peter Stettenheim, courtesy Sidney Gauthreaux.

Holmes. Photo courtesy Richard Holmes and Dartmouth College archives.

Johnston. Photo by Peter Stettenheim.

Kendeigh. Photo from *The Auk* 104: 58 (1987), with permission from American Ornithologists' Union.

Kilner. Photo by Claire Spottiswoode, courtesy Rebecca Kilner.

Kluijver. Photo from *het Vogeljaar* 25: 224 (1977), courtesy *het Vogeljaar*.

Koenig. Photo by Dianne Tessaglia-Hymes, courtesy Walt Koenig.

Lashley. Photo from http://www.browsebiography .com/bio-karl_lashley.html.

Lowe. Photo from *Ibis* 91: plate 1 facing page 147 (1949), with permission from British Ornithologists' Union.

Merton. Photo courtesy Ron Moorhouse.

Monaghan. Photo by Neil Metcalfe.

Nevitt. Photo by Brian Hoover, courtesy Gabrielle Nevitt.

Orians. Photo by Elizabeth Orians, courtesy Gordon Orians.

Papi. Photo by Roberto Guidi, courtesy Anna Gagliardo.

Parker. Photo by Greg Linton, courtesy Patty Parker and Saint Louis Zoo.

Perdeck. Photo from Voous (1995: 352).

Perrins. Photo by Mary Perrins, courtesy Christopher Perrins.

Pettingill. Photo from *The Auk* 119: 1105 (2002), with permission from American Ornithologists' Union.

Pitelka. Photo from *The Auk* 121: 964 (2004), with permission from American Ornithologists' Union.

Pycraft. Photo courtesy Balfour & Newton Libraries, University of Cambridge, Department of Zoology.

Rothstein. Photo by Bruce Lyon.

Rüppell. Photo courtesy Karl Schulze-Hagen.

Sheldon. Photo by Tim Birkhead, courtesy Ben Sheldon.

Silverin. Photo courtesy Bengt Silverin.

Spottiswoode. Photo by Tim Birkhead, courtesy Claire Spottiswoode.

Stone. Photo from *The Auk* 58: plate 10 following page 299 (1941), with permission from American Ornithologists' Union.

Storer. Photo courtesy Philip Myers.

ten Cate. Photo by Tim Birkhead, courtesy Carel ten Cate.

Ticehurst. Photo from *Ibis* 83: plate 4 (1941), with permission from British Ornithologists' Union.

Van Tyne. Photo courtesy University of Michigan Museum of Zoology.

von Haartman. Photo by Peter Stettenheim.

Weigold. Photo from Bairlein and Becker (2010), courtesy Franz Bairlein.

Wenzel. Photo courtesy Bernice Wenzel.

Willson. Photo by Skip Gray, courtesy Mary Willson.